Laboratory Applications in Microbiology

A CASE STUDY APPROACH

Second Edition

Barry Chess
Pasadena City College

The McGraw-Hill Companies

Connect
Learn
Succeed™

LABORATORY APPLICATIONS IN MICROBIOLOGY: A CASE STUDY APPROACH, SECOND EDITION

Published by McGraw-Hill, a business unit of The McGraw-Hill Companies, Inc., 1221 Avenue of the Americas, New York, NY 10020. Copyright © 2012 by The McGraw-Hill Companies, Inc. All rights reserved. Previous editions © 2009. No part of this publication may be reproduced or distributed in any form or by any means, or stored in a database or retrieval system, without the prior written consent of The McGraw-Hill Companies, Inc., including, but not limited to, in any network or other electronic storage or transmission, or broadcast for distance learning.

Some ancillaries, including electronic and print components, may not be available to customers outside the United States.

This book is printed on acid-free paper.

Printed in the United States of America.

2 3 4 5 6 7 8 9 0 QDB/QDB 1 0 9 8 7 6 5 4 3 2

ISBN 978-0-07-340237-6
MHID 0-07-340237-0

Vice President & Editor-in-Chief: *Marty Lange*
Vice President EDP/Central Publishing Services: *Kimberly Meriwether David*
Sponsoring Editor: *Lynn Breithaupt*
Marketing Manager: *Amy Reed*
Developmental Editor: *Darlene M. Schueller*
Senior Project Manager: *Lisa A. Bruflodt*
Design Coordinator: *Brenda A. Rolwes*
Cover Designer: *Studio Montage, St. Louis, Missouri*
Photo Research Coordinator: *Lori Hancock*
Cover Images: *(Scientist Peering into Microscope):* © Getty Images RF; *(Arrangements of Rod-Shaped Bacteria, Bacillus Anthracis Bacterial Colonies, Posteroanterior (PA) Chest X-Ray):* © Center for Disease Control
Buyer: *Susan K. Culbertson*
Media Project Manager: *Linda Avenarius*
Compositor: *Laserwords Private Limited*
Typeface: *10/12 Palatino*
Printer: *Quad/Graphics*

All credits appearing on page or at the end of the book are considered to be an extension of the copyright page.

Some of the laboratory experiments included in this text may be hazardous if materials are handled improperly or if procedures are conducted incorrectly. Safety precautions are necessary when you are working with microorganisms, chemicals, glass test tubes, hot water baths, sharp instruments, and the like, or for any procedures that generally require caution. Your school may have set regulations regarding safety procedures that your instructor will explain to you. Should you have any problems with materials or procedures, please ask your instructor for help.

Contents

Preface

"Did you see the interesting article in the newspaper about . . . ?" Every instructor of microbiology has started a lecture with these, or similar, words. And always for the same reason: to impress upon our students that microbiology is not only important but is, in fact, relevant to all our lives. For microbiologists, and those who already intend to enter the field, this battle of relevance has already been won, but for the vast majority of students who are taking microbiology as a prelude to nursing, dental hygiene, or another allied health profession, the science only seems to come alive when linked to everyday experiences. Usually we, as microbiologists, recognize the story in the paper or the blurb on the radio as a living example of the impact of microbiology on our lives and try to share that with our students. Unfortunately, news stories do not always adhere to our syllabi, and many of these wonderful teaching moments go unexploited.

For Whom Is This Lab Manual Written?

Written for students entering the allied health fields, *Laboratory Applications in Microbiology: A Case Study Approach*, is designed to use real-life examples, or case studies, as the basis for exercises in the laboratory. Over the past few years, the number of lecture texts utilizing case studies has grown rapidly, and for good reason—case studies work! This book is the only lab manual focusing on this means of instruction, an approach particularly applicable to the microbiology laboratory. All the microbiological theory in the world means little if students cannot understand the importance of a Gram stain, antibiogram, or other laboratory procedure.

What Sets This Lab Manual Apart?

This book was created to make the microbiology lab a more valuable experience by reconnecting the **what** and **how** of microbiology with the sometimes forgotten **why**. Although Latin names, complex media, and complicated assays will always be a part of the curriculum, the context of the exercises has been expanded so that the reason for completing a specific task will be clear from the outset. Several features of the book are used to accomplish this goal and serve to distinguish it from other microbiology lab manuals.

Case Studies

The first 39 exercises begin with actual cases taken from the scientific literature. After reading each case, several questions help to define the most important issues and how they should be addressed. As the exercise is completed, new techniques, media, and observational tools are introduced, all with the goal of solving the issues presented in the case. Evidence has shown that the use of case studies boosts learning, develops critical thinking skills, increases retention of students in the classroom, and even reduces the incidence of academic dishonesty. Simply put, students learn more, learn faster, and retain more with case studies than with traditional instruction methods. Although this seems obvious to those of us who cannot wait to share the day's news story with our class, the results are backed up by empirical evidence. In one study focused on instructors who use cases, 97 percent reported that students who were taught with cases learned new ways to think about an issue; 95 percent reported that students took a more active part in the learning process; and 92 percent reported that students were more engaged in classes, truly remarkable numbers.

Changes to the Second Edition

The second edition of *Laboratory Applications in Microbiology* has incorporated a number of new features designed to keep the manual up-to-date, make it easier to use, and improve the experience of both students and instructors. These changes include:

- Five new exercises have been added to the manual, bringing to 39 the number of case study-supported exercises. The new exercises cover **DNA profiling, ABO blood typing, differential white blood cell count, slide agglutination**, and the use of **ELISA**.
- A photo atlas containing more than 300 images has been added to the manual, ensuring that students have a clear visual reference for every exercise, test result, and organism they will encounter.
- Student learning outcomes (SLOs) have been included for the first 39 exercises in the manual, outlining the skills and theory a student should master as they complete each exercise.

- Seventeen completely new case studies have been included in the manual, while eight others have been revised to improve clarity or include updated information.
- The importance of laboratory safety has been emphasized through the inclusion of an additional case study to Exercise 1.
- Exercise 5—Identification and Classification of Fungi—has been moved so that it falls immediately after exercises devoted to the identification of algae and protozoa, providing a more unified look at eukaryotic microorganisms.
- The procedures in Exercise 17—Lethal Effects of Ultraviolet Light—and Exercise 18—Evaluation of Disinfectants—have been modified to produce more consistent results.
- Exercise 23—Morbidity and Mortality Weekly Reports—contains a particularly interesting case study on the use of Google to track H1N1 influenza. Additionally, Table 23.1 has been updated to reflect changes in nationally notifiable infectious diseases.
- Recipes for media have been removed from the exercise themselves but may still be found in Appendix D for those who desire them.
- All microscopic images now include a reference to the magnification used to obtain the image, allowing students to compare the images they obtain with those in the book.

Progression of Exercises Promotes Active Learning

Material in each of the first 39 exercises has been carefully organized so that students develop a solid intellectual base beginning with a particular technique, moving through the case study, and finally applying new knowledge to unique situations beyond the case study. Immediately following the case and introductory material, pre-lab questions help students to focus on the important aspects of the case, developing a framework for what they will need to do **prior to the lab,** most of which require two or three periods. Between the multiday labs, questions are posed to ensure that the students understand what they have just done, the results they should expect, and the significance of those results. Post-lab questions require applying the knowledge gained from the case study to broader questions, facilitating the connection between the exercise and the real-life application.

While the first 39 exercises focus on case studies, the **why** of microbiology, the **how** of the subject has not been forgotten. The final 56 exercises serve as a thorough compendium of common microbiological methods. These exercises are presented in such a way that students will develop critical thinking skills simply by deciding on a particular course of action. All similar techniques, such as selective and differential media or biochemical tests, are grouped together, and each exercise begins with a brief overview. By reviewing these, a

student may select an appropriate test, media, or staining technique from the many available, ensuring that they have decided on not only what information they need, but how to go about getting it. Written to clearly guide students while also pointing out the importance of a particular technique, this portion of the manual provides detailed, well-illustrated procedures that stand by themselves or can be used in conjunction with the case studies in the front of the book. This is particularly helpful when undertaking unknowns, as each student's unknown culture will require a unique set of procedures for complete identification. A data sheet in the appendix provides a single location for students to record their test results, reinforcing the importance of record keeping in the laboratory.

Extensive Flowcharts for Bacterial Identification

Exercise 39 introduces the concept of bacterial identification, using a case study recounting the recognition of *Legionella pneumophila* as the causative agent of Legionnaires' disease. Within this exercise, 31 flowcharts are used to help identify bacterial unknowns commonly seen in the microbiology laboratory, a far more extensive collection than the one or two found in most manuals. This exercise also serves as an introduction to the techniques section of the manual, allowing students to quickly decide which diagnostic techniques are applicable to their particular unknown culture.

A Self-Contained Resource for the Microbiology Laboratory

In the workplace, allied health professionals are expected to evaluate a situation and find a solution using whatever resources are available to them. This book serves as a self-contained resource, with everything a student needs to solve a problem in the microbiology laboratory. A **glossary** provides definitions of all microbiological terms used in the book, a rarity in the field. **Appendices** contain the formula of every media and reagent used, in addition to **tutorials** covering universal techniques such as the use of pipettes and spectrophotometers as well as the preparation of media. Each exercise also includes a **link to applicable websites,** such as the CDC homepage for each pathogenic microorganism encountered. In short, this book will help students develop the ability to solve problems.

Teaching and Learning Supplements

Digital Tools for Your Lab Course

New! McGraw-Hill Connect Microbiology for *Chess' Laboratory Applications in Microbiology* allows instructors and students to use art and animations for assignments and lectures. Instructors now have access to a variety of new resources including assignable and gradable lab questions from the lab manual, additional pre- and post-lab activities, case study activities, interactive questions based on atlas images, lab skill videos, and more. In addition, digital images, PowerPoint slides, and instructor resources are available through Connect. Visit www.mcgrawhillconnect.com.

McGraw-Hill Higher Education and Blackboard® have teamed up.

Blackboard, the web-based course management system, has partnered with McGraw-Hill to better allow students and faculty to use online materials and activities to complement face-to-face teaching. Blackboard features exciting social learning and teaching tools that foster more logical, visually impactful and active learning opportunities for students. You'll transform your closed-door classrooms into communities where students remain connected to their educational experience 24 hours a day.

This partnership allows you and your students access to McGraw-Hill's Connect™ and Create™ right from within your Blackboard course—all with one single sign-on.

Not only do you get single sign-on with Connect and Create, you also get deep integration of McGraw-Hill content and content engines right in Blackboard. Whether you're choosing a book for your course or building Connect assignments, all the tools you need are right where you want them—inside of Blackboard.

Gradebooks are now seamless. When a student completes an integrated Connect assignment, the grade for that assignment automatically (and instantly) feeds your Blackboard grade center.

McGraw-Hill and Blackboard can now offer you easy access to industry leading technology and content, whether your campus hosts it, or we do. Be sure to ask your local McGraw-Hill representative for details.

Digital Lecture Capture. Tegrity Campus™ is a service that allows class time to be any time by automatically capturing every lecture in a searchable video format for students to review at their convenience. Educators know that the more students can see, hear, and experience class resources, the better they learn. Help turn all your students' study time into learning moments by supplying them with your lecture videos.

Electronic Book—GO GREEN!

Green. . . . it's on everybody's mind these days. It's not only about saving trees, it's also about saving money. If you or your students are ready for an alternative, McGraw-Hill eBooks offer a less expensive and eco-friendly alternative to traditional printed textbooks and laboratory manuals. This laboratory manual is available as an eBook at www. CourseSmart.com. At CourseSmart, your students can take advantage of significant savings off the cost of a printed textbook or laboratory manual, reduce their impact on the environment, and gain access to powerful web tools for learning. CourseSmart eBooks can be viewed online or downloaded to a computer. The eBooks allow students to do full text searches, add highlighting and notes, and share notes with classmates. Contact your McGraw-Hill sales representative or visit www.CourseSmart.com to learn more.

Personalize Your Lab

Craft your teaching resources to match the way you teach! With McGraw-Hill Create™, www.mcgrawhillcreate.com, you can easily rearrange chapters, combine material from other content sources, and quickly upload content you have written like your course syllabus or teaching notes. Find the content you need in Create by searching through thousands of leading McGraw-Hill textbooks. Arrange your book to fit your teaching style. Create even allows you to personalize your book's appearance by selecting the cover and adding your name, school, and course information. Order a Create book and you'll receive a complimentary print review copy in 3–5 business days or a complimentary electronic review copy (eComp) via email in minutes. Go to www.mcgrawhillcreate.com today and register to experience how McGraw-Hill Create empowers you to teach *your* students *your* way.

Student Resource

Annual Editions: Microbiology 10/11 (0-07-738608-6) is a series of over 65 volumes, each designed to provide convenient, inexpensive access to a wide range of current articles from some of the most respected magazines, newspapers, and journals published today. *Annual Editions* are updated on a regular basis through a continuous monitoring of over 300 periodical sources. The articles selected are authored by prominent scholars, researchers, and commentators writing for a general audience. The *Annual Editions* volumes have a number of common organizational features designed to make them particularly useful in the classroom: a general introduction; an annotated table of contents; a topic guide; an annotated listing of selected World Wide Web sites; and a brief overview for each section. Visit www.mhhe.com/cls for more details.

Acknowledgments

If writing the first edition of a book is a little like giving birth (with apologies to all the moms out there), then I suppose producing the second edition is a bit like raising a child. You want to be sure you've done everything right so that when you send him on his way he returns with compliments, and not complaints. Well, so far there have been mostly compliments, the complaints have been generally minor, and to take the parenting metaphor one step further, it has, for sure, taken a village to raise this child. Thanks are due.

As always, the first thanks go out to my students, who are the ultimate arbiters of whether anything I say or do is good or bad. I know you didn't sign up to be test subjects for every idea that pops into my head, but there is no way I could do this without your good-natured feedback. Please know that you have helped create a better book. In the lab at PCC, where we seem to have grown in the last few years from a small mom-and-pop outlet to a microbiology superstore, a great number of people have supplied ideas, critiques, and an unending supply of bad jokes (that's you Ray) that have helped shape this book. Special thanks to Jessica Igoe, Sonya Valentine, John Stantzos, and the aforementioned Ray Burke. Of course, nothing happens in the lab without the support of Mary Timmer, laboratory technician of the gods, who has the ability to keep us all on track, supplied with what we need before we even know we need it. Finally, it would seem karmically unwise not to thank Dr. Dave Douglass, who has made sure the microbiology program has remained well supported over the last several years. If I haven't mentioned it, thanks for the new microscopes!

Of course saying that one person "wrote" a book glosses over the contributions of the many people who had a hand in bringing it to life. Believe me, I could write all I want and without the people at McGraw-Hill, this book would still be a file in my computer. My deepest appreciation to sponsoring editor Lynn Breithaupt, developmental editor Darlene Schueller, project manager Lisa Bruflodt, and marketing manager Amy Reed. Enjoyed the photos? Thank Lori Hancock and Danny Meldung for their hard work, I certainly have. While I may have written the words, all of these people had a hand in creating the book.

Lastly, there are three people who deserve more thanks than I could ever deliver. On a practical note, my sons were pressed into service as a hand model and lighting director for several of the photos you'll see in the pages that follow while my wife took on single parenting duties whenever an "oh!" box loaded with manuscript landed on the front porch. Everyone got used to me being at the center of a pile of papers at all hours of the day and night, and did so with good cheer, knowing not to rearrange things, when to offer encouragement, and when to offer cheesesteak. Safura, Noah, and Josh, you three mean the world to me, and without your support, I couldn't have done this. I love you all very much.

Reviewers

Throughout the revision process, I was lucky enough to have on hand the opinions of a team of wonderful microbiologists, who provided their take on the content, procedures, depth, and even order of the exercises in the manual. They have helped to make this a better book than I could ever have written on my own, and I thank them wholeheartedly.

Debra M. Adair
Paradise Valley Community College

Anthony Arment
Central State University

Michelle Badon
University of Texas at Arlington

Ranjit Banerjee
New York Medical College

Tesfaye Belay
Bluefield State College

Eric L. Buckles
Dillard University

Erin A. Christensen
Middlesex County College

Iris M. Cook
Westchester Community College

Lauren Cross
Wor-Wic Community College

Natasha Dean
La Sierra University

Kristiann Dougherty
Valencia Community College

Denise Ferguson
Carolinas College of Health Sciences

Robert Gessner
Valencia Community College

Carl Hamby
New York Medical College

Daniece Harris-Williams
Hinds Community College Rankin Campus

Julie A. Huggins
Arkansas State University

Dena Johnson
Tarrant County College

James Masuoka
Midwestern State University

Karin Melkonian
Long Island University – C.W. Post Campus

Murad M. Odeh
South Texas College

Connie Pitman
University of Colorado at Colorado Springs

Narayanan Rajendran
Kentucky State University

Lois Sealy
Valencia Community College

Peter Sheridan
Idaho State University

Jane Slone
Cedar Valley College/ DCCCD

Janice Yoder Smith
Tarrant County College Northwest Campus

Gabriel Swenson
Paine College

Stephen Wagner
Stephen F. Austin State University

Van Wheat
South Texas College

Dean E. Willis
University of North Florida

Robin Woodard
University of Virginia

Linda Young
Ohio Northern University

Jianmin Zhong
Humboldt State University

Brenda Zink
Northeastern Junior College

To the Student

As an introductory student in microbiology, you may find that the reasons behind a particular exercise appear overly complex. Such is the nature of science, but the reasons should, at the very least, be apparent. The first step in closing the chasm between the scientific and the everyday is to understand, always, how each step relates to the overall objective. It is just as important to understand **why** you are doing something as it is to understand **what** it is you are doing. If you can master both the **why** and the **what,** then your success in microbiology will be assured.

This book was written with you in mind. In other words, everything in this book has been written to support something. That means that the introductory material helps to explain the case study, the photos and diagrams are used to clarify procedures, the glossary contains definitions of microbiological terms, and websites are provided if you would like further information on a topic. When you are using this book, please, **use** this book. If the meaning of a sentence is unclear, look to the accompanying figure; if a word is a mystery, use the glossary; if space is provided for a detailed drawing, give it your best shot—it will all be important soon. A well-used book becomes weathered as knowledge moves from the book to the reader, and a lab book is no different in this regard. Dog-eared pages, drawings, notes, and circled definitions are all part of learning, and the physical process of making the book yours parallels the intellectual process of making the information yours. This is as true with microbiology as it is with any other interest, job, or hobby. Take the steps to own the book, and you'll own the information within.

About the Author

Barry Chess has been a microbiologist at Pasadena City College since 1996. He received his Bachelors and Masters degrees from California State University, Los Angeles, and completed several years of post-graduate work at the University of California, Irvine, where his research centered on the expression of genes involved in the development of muscle and bone.

At Pasadena City College, Barry developed a new course in human genetics and helped to found a biotechnology program at the campus. He regularly teaches courses in microbiology, biology, and genetics and has supervised students completing independent research projects in biology and microbiology. Over the past several years Barry's interests have begun to focus on innovative methods of teaching that lead to greater student understanding. He has written and reviewed cases for the National Center for Case Study Teaching in Science and presented papers and talks on the use of case studies in the classroom. He is a coauthor of the microbiology textbook *Foundations in Microbiology*, having recently joined Kathy Talaro on the project.

Barry is a member of the American Society for Microbiology and regularly attends meetings in his fields of interest, both to keep current of changes in the discipline and to exchange teaching and learning strategies with others in the field.

Safety Considerations in the Microbiology Laboratory

CASE SYNOPSES

Laboratory-Acquired Infection with *Escherichia coli* O157:H7—New York, 2004

In March 2004, a laboratory worker experienced diarrhea 3 days after handling an *E. coli* O157:H7 isolate. This strain of the bacteria can cause hemolytic uremic syndrome, a type of hemolytic anemia that produces kidney damage and may lead to renal failure. Laboratory tests revealed the presence of *E. coli* O157:H7 in her feces and genetic analysis showed that the isolated strain was identical to the strain she had worked with in the laboratory.

Ocular Vaccinia Infection of a Laboratory Worker—Philadelphia, 2004

A graduate student was examined by an ophthalmologist after complaining of itching, swelling, tearing, and redness of the left eye. The symptoms began five days prior to the visit and were originally attributed to viral conjunctivitis by the campus health service. As the condition worsened, the student was referred to a specialty eye hospital, where she stated that she worked with vaccinia virus as part of her graduate research in immunology. Scraping of vesicles near the corner of the eye showed the presence of vaccinia virus. The treatment included antibacterial and antiviral medications, pain medications, and intravenous vaccinia immune globulin. The patient was discharged after nine days in the hospital although full recovery took several weeks.

Laboratory Researcher Dies after Suffering Burns—Los Angeles, California, 2009

Sheri Sangji, a 23-year-old research assistant at UCLA, was transferring a small quantity of t-butyl lithium from one container to another when a plastic syringe came apart in her hands, splashing her with a chemical compound that ignites instantly when exposed to air. She received second and third degree burns over 43% of her body and died 18 days later.

Resolution of the Cases appears on page 6

Spina, N., Zansky, S., Dumas, N., and Kondracki, S. 2005. Four laboratory-associated cases of infection with *Escherichia coli* O157:H7. *J. Clini. Microbiol.* 43: 2938–2939.

CDC. 2006. Ocular vaccinia infection in laboratory worker, Philadelphia, 2004. *Emerging Infect. Dis.* 12: 134–137.

Christensen, Kim. May 5, 2009. State Fines UCLA in Fatal Lab Fire. Los Angeles Times.

STUDENT LEARNING OUTCOMES

After completing this exercise, you should be able to:

1. Demonstrate proper primary and secondary containment procedures.
2. Explain the procedures involved in dealing with a laboratory emergency.

INTRODUCTION

The microbiology laboratory presents a number of unique challenges. Not only are the normal hazards of a laboratory environment (flames, caustic chemicals, and glassware) present, but so too are infectious organisms. In fact, the microbiology lab is devoted to growing and studying the very organisms that may cause us such harm! As the cases above illustrate, laboratory workers are by no means immune to infection in the laboratory, and it is not an exaggeration to say that proper safety procedures can be a matter of life and death.

Safe laboratory procedures revolve around containment of microorganisms. **Primary containment** concerns the protection of personnel and the laboratory environment from exposure to infectious microbes. Proper microbiological techniques, such as the safe transport and disposal of cultures, along with the correct use of personal safety equipment (e.g., gloves and safety glasses) go a long way toward accomplishing the goal of personal containment. **Secondary containment** deals with protecting the outside environment from exposure to infectious organisms and depends principally on the design of the laboratory and the availability of equipment. There is generally little that can be done to influence the physical aspects of the laboratory other than not disabling safety features, such as keeping open a door that should remain closed, turning off an exhaust fan, or removing a fire extinguisher.

The type of organisms dealt with in the laboratory will dictate the safety precautions that are mandated. Working with deadly viruses obviously requires a greater degree of vigilance than working with bacteria that are nonpathogenic. To clarify exactly what techniques and equipment should be used, microorganisms are classified into one of four biosafety levels (BSL-1 through BSL-4) based on their ease of transmission and pathogenicity. Each level has a set of minimum standards with regard to laboratory practices, equipment, and facilities. At one end, BSL-1 organisms generally do not cause disease in a healthy person and require very few specialized techniques. BSL-4 organisms, in contrast, are easily transmitted and cause life-threatening diseases. BSL-4 laboratories are the stuff of science fiction, with full-body spacesuits, respirators, and showers upon exiting the laboratory. Table 1.1 summarizes the recommended biosafety levels for infectious agents.

The vast majority of introductory microbiology laboratories are designed to handle BSL-1 and -2 rated organisms, and the rules that apply in these laboratories have a very common sense and feel about them. Your instructor may modify these rules based on college, municipal, or state regulations as well as the organisms you are likely to work with as part of your course. Adhering to these guidelines will help to ensure your safety in the microbiology lab.

Prior to the Lab

- Dress appropriately for the lab. No open toed shoes or sandals. Clothing with baggy sleeves that could catch fire or hinder your movements should be avoided.

- Know where the safety equipment is in the lab. Note the location of the eye wash, safety shower, fire extinguisher, and first aid kit. Take a moment to learn their operation; remember, if you need to use the eye wash, you very well may not be able to see at the time!

During the Lab

- Always wear a lab coat while in the lab. Although you may not be working yourself, another person in the lab could have an accident. This garment should only be used during lab and should remain in the lab. Even discounting potential biohazards, a lab coat will protect your clothing. There is a reason many of the chemicals you will be working with are called *stains*.

- Wash your hands prior to beginning the lab and just before leaving as well. Also wash when removing gloves and if you feel you may have contaminated yourself. If your laboratory sink has a hands-free method of activating the flow of water (such as foot pedals), use it.

- Tie back any long hair, it is both a source of contamination and a fire hazard.

- Disinfect your benchtop with amphyl, Lysol, or 10% bleach prior to beginning work and just before leaving the laboratory. If time permits, allow the disinfectant to evaporate rather than wiping the surface of your bench.

TABLE 1.1	Summary of biosafety levels for selected infectious agents				
BSL	Health risk	Practices	Primary barriers	Secondary barriers	Organisms (Selected examples)
1	Not known to cause disease in healthy individuals.	Open bench microbiology.	None required.	Open benchtops and sinks.	*Micrococcus luteus* and *Bacillus megaterium*.
2	Can cause disease in healthy people, but organisms are easily contained.	Limited lab access, and biohazard warning signs.	Gloves, lab coat, eye protection, and/or face shield as needed.	BSL-1 plus: • Access to autoclave.	*Escherichia coli* and *Staphylococcus aureus* (most human pathogens).
3	Can cause severe disease, especially when inhaled.	BSL-2 plus: • Controlled access to lab. • No unsterilized material can leave the lab. • Decontamination of clothes prior to laundering.	BSL-2 plus: • Biosafety cabinets used for all manipulations.	BSL-2 plus: • Access to self-closing double doors. • Negative pressure (air flows into lab from outside). • Exhausted air not recirculated.	*Mycobacterium tuberculosis*, HIV, and *Yersinia pestis*.
4	Highly virulent microbes posing extreme risk to humans, especially when inhaled.	BSL-3 plus: • Clothing must be changed before entering the lab, and personnel must shower upon exiting the lab. • All material is decontaminated prior to leaving the facility.	BSL-3 plus: • All procedures are conducted in complete isolation BSCs or in class I or II BSCs along with full body, positive-pressure suits with supplied air.	BSL-3 plus: • Isolated building or lab. • Isolated laboratory systems (air supply and exhaust, vacuum, and decontamination).	Lassa fever virus, Ebola virus, and Marburg virus.

- Keep clutter on your bench to a minimum. Store book bags, purses, and other unneeded items where they will not consume precious bench space and where they will be less likely to be contaminated by an inadvertent spill.
- Nothing should go into your mouth while you are in the laboratory. Do not smoke, eat, or drink in the lab, even if no work is being done at the time.
- Skin and eyes represent a common portal of entry for pathogens. Do not apply makeup, and never handle contact lenses in the lab.
- Organize your workplace before beginning (Figure 1.1). Store culture tubes upright in a rack, never on their side. Caps on tubes are generally not liquid tight, and liquid will leak out (even from a solid culture), leading to contamination.
- The open flame produced by a Bunsen burner presents an obvious danger in the laboratory. Burners should be set up away from overhead equipment or shelving and the immediate area should be free of combustible materials such as notes or books. Prior to lighting the burner, quickly inspect the hose for holes, cracks, or leaks, and be sure it fits securely on both the gas valve and the burner.

Figure 1.2 Autoclaves use steam under pressure to sterilize biohazardous materials prior to disposal.

Disposal of Contaminated Materials

Most material in the microbiology laboratory must be decontaminated prior to being disposed of or reused, and this is most often accomplished using an autoclave (Figure 1.2), which uses steam under high pressure to kill even the most resistant organisms. After decontamination, culture tubes, glass pipettes, and the like are washed and reused. Plastic Petri dishes melt during the decontamination process and are discarded after autoclaving along with single-use items such as tongue depressors, needles, and swabs. In general,

disposal of lab materials depends on whether or not it will be reused. In any event, the contents of plates or tubes should never be touched by hand.

- Dispose of plastic Petri dishes, swabs, disposable gloves, and similar nonreusable items in the biohazard bag (Figure 1.3a). Petri dishes should be taped closed, but there is no need to remove labels or tape from items.
- Reusable supplies such as culture tubes and glass pipettes should have all labels removed and placed in a rack or container designated for autoclaving.
- Used microscopes slides should be placed in a container for autoclaving or soaked in a disinfectant solution for a minimum of 30 min before being discarded.

Safety Considerations

- Be realistic if you feel you shouldn't be in lab because of health concerns. Conditions that may leave you vulnerable to infection such as a short-term illness, being immunocompromised, taking immunosuppressant drugs, or being pregnant should be candidly discussed with your instructor.
- Always wear gloves when handling blood or blood products. Blood-borne pathogens have special procedures associated with them, and work of this type should only be done with the explicit knowledge of your instructor.
- Wash with an antiseptic if your skin is exposed to microorganisms as a result of a spill.

Figure 1.1 Personal protection in the laboratory includes the use of a lab coat, gloves, and eye protection. Also note that long hair is tied back and the work area is free of clutter.

(a)

(b)

Figure 1.3 All disposable, potentially infectious waste should be placed in a biohazard container, with needles, slides, tongue depressors, and anything else that could penetrate a plastic bag restricted to disposal in a hard-sided receptacle. The international biohazard symbol on these containers not only marks the contents for autoclaving prior to disposal but also cautions anyone in the room as to the possibly hazardous nature of the items inside the container.

- Dispose of broken glass, needles, lancets, wooden applicators, and any other object that could penetrate the skin, in a hard-sided sharps container (see Figure 1.3b). Do not overfill the container, and never, ever force objects into the container.
- Make use of fume hoods when undertaking procedures where noxious chemicals may be released during heating (Figure 1.4).
- In the event of a spill, notify your instructor immediately. Broken glass and bacterial cultures are a hazardous combination. With the instructor's approval, cover the spill with paper towels and saturate the towels with disinfectant. After 20 min, carefully wipe up the spill and discard the paper towels in the biohazard container for autoclaving. Discard the broken glass in the sharps container.

Figure 1.4 The use of a fume hood does not provide an isolated environment for work with microorganisms but rather reduces exposure to irritating or dangerous chemicals in the laboratory.

PRE-LAB QUESTIONS

1. *Classify each of the following agents as BSL class 1, 2, 3, or 4 based on its description.*

Description	Severity of disease	Mode of transmission	BSL level
Marburg virus	Lethal in 25%–100% of cases. No effective treatment.	Direct contact or body fluids.	4
Mycobacterium tuberculosis	Severe, but treatable respiratory disease.	Respiratory droplets.	3
Bacillus subtilis	Does not cause disease in immunocompetent persons.	Not easily transmitted.	1
Clostridium tetani	Can be lethal in nonprotected individuals. Vaccine provides protection.	Anaerobic sites (deep puncture wounds).	2

2. *Explain, as specifically as possible, how each of the following helps to enhance safety in the microbiology lab.*

Negative air flow (i.e., air flows into the laboratory rather than out)

gets rid of bad air

Gloves, safety glasses, and lab coat

Protect our body from any chemical harm that could be caused.

Vaccination of laboratory workers

Just in case you get pocked with a rusty metal or something goes wrong.

Foot pedal activation of sinks

So chemicals on hands before washing dont get on the knobs and contaminate other people

Prohibitions on eating and drinking in the lab

chemicals in air and on table could get on the food & cause harm to your body.

3. *How would you properly dispose of each of the following items?*

A Petri dish containing a fungal culture

Should be taped closed & be put in the biohazard bag.

A glass culture tube containing a bacterial culture

Labels removed & placed on a rack or container ready for autoclaving

A hypodermic needle used to draw blood

be placed in a biohazard Sharps container

A spill containing broken glass and a bacterial culture

Cover spill with paper towels that are disinfected for 20 minutes and then discard towels in biohazard container for autoclaving. Discard the glass in the sharps container.

4. *Identify the location of each of the following pieces of safety equipment in your laboratory.*

Lab coat storage

Fire extinguisher

by back door exit.

Disinfectant

First aid kit

by eye wash station

Biohazardous waste disposal

Sharps container

RESOLUTION OF THE CASES

Laboratory-Acquired Infection with *Escherichia coli* O157:H7—New York, 2004

In the interviews, the laboratory worker stated that she always wore latex gloves and a buttoned lab coat and that she always washed her hands immediately after removing her gloves. She also declared that she wore gloves while working on the computer and answering the telephone, a practice that most probably led to cross contamination of these surfaces. Items such as these should only be handled after gloves have been removed and hands thoroughly washed. The investigators in this case concluded that the low infectious dose of *E. coli* O157:H7 and its ability to survive on inanimate objects contributed to laboratory transmission in this case.

Ocular Vaccinia Infection of a Laboratory Worker—Philadelphia, 2004

A review of laboratory practices and experimental protocols showed that while the laboratory generally followed established safety guidelines, several opportunities for exposure to active virus still existed; protective eyewear was rarely worn, used pipettes were not always disinfected before being removed from the biosafety cabinet, and samples containing small amounts of live virus were occasionally manipulated outside of a biosafety cabinet. Most importantly, no member of the lab had been vaccinated against vaccinia virus in the past 10 years, as recommended by the Centers for Disease Control.

Genetic analysis of the vaccinia virus recovered from the student showed it to be a strain found only in the student's laboratory that had been used by the student 5 days prior to the onset of symptoms. At one point during this experiment, the student removed plates from the biosafety cabinet and carried them to another room where the covers were briefly removed. During this portion of the experiment, the student did not wear eye protection, and she was unable to remember if she wore gloves.

Laboratory Researcher Dies after Suffering Burns—Los Angeles, California, 2009

Sangji was a recent college graduate who had only been working in the lab at UCLA for a few months when the incident occurred. An investigation by the California Division of Occupational Safety and Health concluded that she had not been properly trained for the procedure she was undertaking or what to do in the event she caught fire. Sangji was not wearing a protective lab coat and was dressed in a nylon sweater described as "solid gasoline" by a lab safety expert. A previous inspection of the lab by UCLA safety personnel turned up several safety violations, which had not been corrected at the time of the accident. Investigation is ongoing, but some people feel the evidence is compelling. Said Neal Langerman, the lab safety expert mentioned above, "Poor training, poor technique, lack of supervision, and improper method. . . . She died, didn't she? It speaks for itself."

EXERCISE 1 REVIEW QUESTIONS

1. For each of the cases seen here, postulate where the breakdown in laboratory safety occurred and suggest how it could be corrected.

   E. coli infection - the lady should have taken her gloves off which were contaminated before using objects like the computer keyboard and the phone.

Laboratory fire   was not wearing a lab coat, and had
Lack of proper training, techniquet
Supervision.

Vaccinia infection   Student should of been wearing Cloves and
glasses the whole time while being in contact with
the experiment.

REFERENCE

U.S. Department of Health and Human Services. 1999. *Biosafety in Microbiological and Biomedical Laboratories (BMBL), 4th ed.*
http://www.cdc.gov/OD/ohs/biosfty/bmbl4/bmbl4toc.htm.

NOTES

Microscopy and Measurement of Microscopic Specimens

CASE SYNOPSES

Excerpt of letters from Anton van Leeuwenhoek to the Royal Society of London for the Improvement of Natural Knowledge

September 7, 1674: "Passing just lately over this lake . . . and examining this water next day, I found floating therein divers earthy particles, and some green streaks, spirally

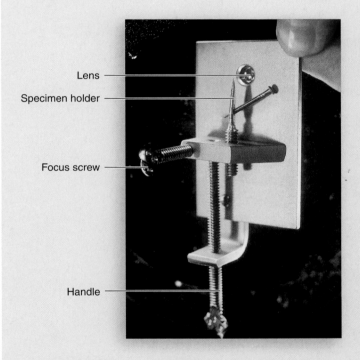

Lens

Specimen holder

Focus screw

Handle

A brass replica of one of Leeuwenhoek's original microscopes.

wound serpentwise, and orderly arranged, after the manner of the copper or tin worms. . . . The whole circumference of each of these streaks was about the thickness of a hair of one's head."

September 17, 1683: [On observing plaque between the teeth] ". . . a little white matter, which is as thick as if 'twere batter. . . . I then most always saw, with great wonder, that in the said matter there were many very little living animalcules, very prettily a-moving. The biggest sort . . . had a very strong and swift motion, and shot through the water like a pike does through the water. The second sort . . . oft-times spun round like a top . . . and these were far more in number . . . an unbelievably great company of living animalcules, a-swimming more nimbly than any I had ever seen up to this time . . . in such enormous numbers, that all the water . . . seemed to be alive."

December 25, 1702: "In structure these little animals were fashioned like a bell, and at the round opening they made such a stir, that the particles in the water thereabout were set in motion thereby. . . . And though I must have seen quite 20 of these little animals on their long tails alongside one another very gently moving, with outstretched bodies and straightened-out tails; yet in an instant, as it were, they pulled their bodies and their tails together, and no sooner had they contracted their bodies and tails, than they began to stick their tails out again very leisurely, and stayed thus some time continuing their gentle motion: which sight I found mightily diverting."

Resolution of the Cases appears on page 18

Leeuwenhoek, Antoni. 1999. *The Collected Letters of Antoni van Leeuwenhoek*. Palm, L.C. (ed.). Boca Raton: CRC Press.

STUDENT LEARNING OUTCOMES

After completing this exercise, you should be able to:

1. Properly use and care for a brightfield microscope.
2. Explain the theories underlying optical microscopy.

INTRODUCTION

As the excerpts illustrate, many of the commonly encountered things of everyday life, be they pondwater, plaque, pus, or milk, are teeming with microorganisms, and being able to adequately view these creatures is essential to a thorough understanding of microbiology. It should come as no surprise that the proper use of a microscope is a central skill in microbiology, and the placement of this laboratory exercise near the very beginning of this book is no accident.

LIGHT MICROSCOPE

Components of the Light Microscope

The instrument most commonly seen in microbiology labs is the brightfield microscope, so named because when objects are examined, they appear as dark objects in a bright visual field. Although your microscope may differ slightly in appearance from the one seen in this exercise, in theory and practice, the same rules will apply. The components and functions of each of the parts found in a typical brightfield microscope (Figure 2.1) are outlined here.

Framework The frame of a microscope consists of the arm and base. Keep in mind when holding the microscope that these are the only two parts built to support the weight of the microscope.

Lamp A light source is located in the base of a microscope. A rotating wheel or knob can be used to adjust the voltage received by the lamp, which in turn adjusts the intensity of the light. Many microscopes will also have a blue filter that can be placed over the light source to reduce the intensity of the light and increase the resolution of the microscope.

Diaphragm The diaphragm is an adjustable disc with a hole in the center. The size of the hole can be varied to allow more or less light to pass to the slide by use of a dial or lever.

Condenser The first of three lens systems found on all microscopes, the condenser is located beneath the stage and is usually contained in the same housing as the diaphragm. The condenser collects and focuses light on the specimen being studied. Although the condenser can be raised or lowered, best results in the microbiology lab will be obtained when the condenser is kept at its highest point, just below the level of the stage.

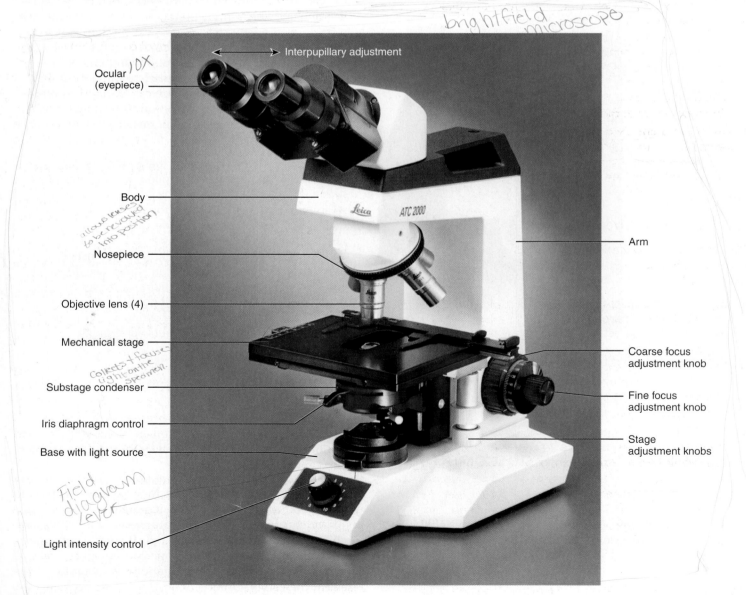

Figure 2.1 Components of a typical compound microscope.

Stage The platform that supports the slide is known as the stage. Most microscopes have a clamping device, the mechanical stage, that allows the slide to be held and moved with more precision.

Objective Lens Three or four objective lenses, more commonly referred to simply as objectives, are found just above the stage. The objectives are attached to a revolving **nosepiece** that allows the lenses to be rotated into position. Most microscopes will have three objectives with magnifications of 10x, 45x, and 100x, designated as **low power, high dry,** and **oil immersion,** respectively. Occasionally a 4x **scanning** objective will also be present, but it tends to be of little use in most microbiology labs.

Ocular Lenses The third set of lenses, those closest to your eyes, are the ocular lenses. In most instances, these lenses have a magnification of 10x. **Binocular** microscopes have two sets of lenses while **monocular** microscopes have only a single ocular. Binocular microscopes will also have a means of adjusting the distance between the oculars. On some microscopes, the oculars will simply pivot in and out, while on others a knurled wheel adjusts the distance between the lenses. One ocular may also have a small ring that allows the focus of that ocular to be adjusted independently of the rest of the microscope.

Focus Adjustment Two concentric focusing knobs are located on each side of the microscope. The large outer knob is the **coarse focus adjustment,** while the smaller, inner knob is the **fine focus adjustment.**

Care of the Light Microscope

Microscopes are delicate instruments, and care must be taken in their use. Some general rules related to microscope care include:

Transport The microscope should always be held with two hands. One hand should grasp the microscope around the arm while the second supports the instrument from the bottom. The biggest danger in carrying a microscope with one hand is not that it will be dropped but rather that the scope will collide with the corner of a lab bench or other piece of furniture. Once at your bench, place the scope gently on the table.

Electric Cord Dangling electric cords are never a good idea, and more than one microscope has been pulled off a table when a hand, foot, or backpack has become entangled in the cord. Keep excess cord secured or wrapped loosely around the base of the microscope. Likewise, water and electrical devices are a poor mix. Keeping any excess cord secured or wrapped loosely around the base of your microscope decreases the chances of it becoming entangled or coming in contact with water.

Protection Against Dust and Chemicals If a dustcover is provided for your microscope, be sure to cover the instrument while staining or undertaking any other procedure that could splatter your scope with dyes or chemicals. Also be sure to cover the scope at the end of the period.

Lens Cleaning The number one cause of unacceptable images is dirty microscope lenses. Besides impeding your ability to see, dust, oil, and other contaminants will eventually damage the lenses, which are often the most expensive part of the microscope to replace.

Lens Care

The importance of lens care cannot be overstated. Dirty or scratched lenses will limit the degree of resolution achievable with your microscope, and the delicate nature of optical glass means that scrupulous attention to detail is required when attempting any cleaning. When cleaning any of the three lens systems on your microscope, follow the guidelines below as well as any lab-specific instructions you may receive.

Use only lint-free optical tissues or cotton swabs to clean lenses. Other tissues or cloths may contain microscopic grit or lint that can damage lenses. If the lenses are quite dirty, a small amount of alcohol or xylene can be used to remove oily residue. Other solvents such as acetone may be acceptable but can damage the mounting cement used on objective lenses. Check with your instructor as to which solvents are acceptable for cleaning lenses.

Ocular lenses are often the recipients of thumb prints, dust, and mascara, and so are often quite dirty. The easiest method of determining if the ocular is clean is to rotate it and see if the dirt rotates as well. If the dirt rotates, clean the ocular; if it remains stationary, check the objective lenses. Occasionally dust and other debris will make their way to the inside of the ocular. If this is the case, the ocular can be removed and compressed air can be used to dislodge the particles. Whenever an ocular is removed from the microscope, a piece of lens tissue should be used to cover the ocular opening. Although removing an ocular is a straightforward procedure, consult your instructor before disassembling any part of your microscope.

Objective lenses are commonly soiled with material from slides or fingers. If a properly focused image ever appears unclear or cloudy, it is safe to assume that the objective lenses are soiled. Clean the lenses using lens paper or a cotton swab moistened with a solvent acceptable in your laboratory. Use a circular motion beginning at the center of the lens and working outward.

Condensers generally accumulate less dirt than oculars or objectives, but their upward facing surfaces do tend to collect dust. An occasional wiping with a piece of lens tissue is enough to keep them clean in most cases. Prior to returning your microscope at the end of the day, it is essential to clean the immersion oil from all the objective lenses. When left on the lenses for an extended period, the oil can soften the cement holding the lenses in place, rendering the lens unusable.

THEORY

Microscopy has two related goals. The first goal is *magnification,* or the creation of a larger-than-life image. This image is created as light first passes through the condenser lens focusing the light on the specimen. The light refracts, or bends, as it passes through the objective lens, creating the real image. The virtual image is then created as light from the real image is magnified once more as it passes through the ocular lens. The virtual image then passes into the eye where it is eventually interpreted by the brain (Figure 2.2). The total magnification of any specimen is easily calculated by multiplying the magnifications of the objective lens and the ocular lens.

$$\begin{array}{c} \text{Total} \\ \text{magnification} \end{array} = \begin{array}{c} \text{Magnification} \\ \text{of objective} \\ \text{lens} \end{array} \times \begin{array}{c} \text{Magnification} \\ \text{of ocular lens} \end{array}$$

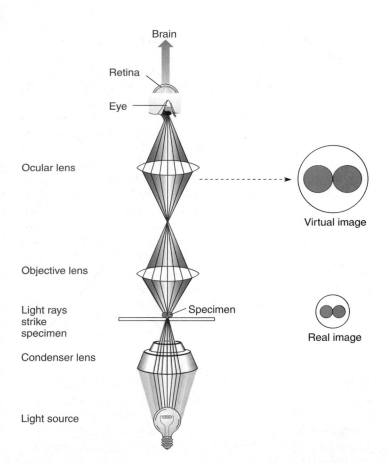

Figure 2.2 Formation of a microscopic image. As light passes through the condenser, it is focused on the specimen. Light leaving the specimen is refracted by the objective lens, forming the real image. The real image is refracted again, by the ocular lens, to form the virtual image that strikes the retina and is interpreted by the brain. Notice that the virtual image is reversed both left-to-right and top-to-bottom with respect to the specimen.

Figure 2.3 The characteristics of each objective are inscribed on the barrel of the lens. In this case, the centermost objective has a magnification of 40x and a numerical aperture of 0.65.

For example, a 10x ocular used in combination with a 45x objective lens would provide a total magnification of 450x. Conveniently, the magnification of both ocular and objective lenses is marked on the lens (Figure 2.3).

The second goal of microscopy is **resolution,** which can be most easily understood as the clarity of an image produced by a set of lenses. The **resolving power** offers a quantifiable means of measuring the ability of a lens system to resolve detail and is defined as the smallest distance between two points that can still be distinguished as two separate entities. Although the calculation of resolution is quite complex and depends on a number of factors, the real world (as opposed to theoretical) limits of resolution can be calculated using the following equation:

$$R = 0.61\lambda/NA$$

where R refers to the resolving power of the optical system, λ is the wavelength of the light used, and NA is the numerical aperture of the objective being used. The numerical aperture is a measure of a lenses ability to gather light and resolve fine specimen detail at a fixed object distance. Numerical aperture is dependent on several aspects of the lens and is quite complex to calculate; it usually ranges from 0.4 (low power) to 1.25 (oil immersion) on most microscopes. Fortunately numerical aperture, like magnification, is inscribed on the barrel of each objective lens (Figure 2.3).

Resolution is also dependent on the wavelength of light used for examining a specimen, and as the equation presented earlier reveals, when the wavelength of light decreases, the resolution increases (i.e., the resolving power gets smaller). Most microscopes use filters to produce blue or green light because these wavelengths are among the shortest seen clearly by humans. Plugging numbers into the equation shows that, using a wavelength of 500 nm and a lens with a numerical aperture of 1.25, the maximum resolution obtainable will be about 244 nm (0.244 μm). This means that two points on a specimen that are less than 0.244 μm apart are not resolvable and will be seen as a single entity.

Ensuring that enough light passes through the sample and enters the objective lens is both necessary to produce an adequate image and is more complicated than one would initially think. Four factors go a long way toward guaranteeing that your images have maximum resolution and adequate contrast.

- **Blue Light** The shorter wavelengths produced when a blue filter is placed over the light source will increase the resolution of the lens.

- **Condenser Position** The condenser should be raised to its uppermost position. This maximizes the amount of light entering the objective lens and minimizes the amount lost to refraction.

- **Diaphragm Position** The diaphragm should be stopped down just enough to provide an acceptable image. Although closing the diaphragm increases the contrast, it also decreases the numerical aperture. The best results are usually obtained by beginning an examination on low power with the diaphragm almost completely closed. As higher power objectives are inserted into the light path, open the diaphragm to provide more light. By the time the oil immersion lens is in place, the diaphragm should be almost completely open to maximize resolution. Light intensity should be adjusted using the voltage control of your microscope, not the diaphragm. If you continually feel the need to close the diaphragm when viewing a specimen through the oil immersion lens, a neutral density filter (one that blocks all colors of light equally) should be inserted into the light path.

- **Immersion Oil** This clear mineral oil has the same refractive index as glass and, when used between the slide and oil immersion lens, prevents the loss of light rays due to diffraction (Figure 2.4).

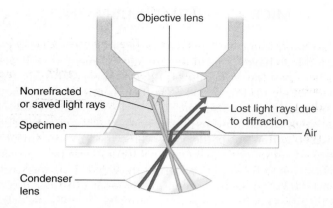

Figure 2.4 Immersion oil has the same refractive index as glass, which prevents the loss of light rays due to diffraction.

OTHER LIGHT MICROSCOPES

Brightfield microscopes are good general-purpose instruments and quite often the only microscope found in many laboratories. One disadvantage of this type of microscope is that images are generally poor if the specimen is lacking contrast (Figure 2.5a). Although stains are often used to increase the contrast of microbial specimens, this almost always results in the death of the cell. Modification of the basic structure of a brightfield microscope has resulted in other types of light microscopes with unique properties, including the ability to produce good images from specimens with little contrast. A **darkfield** microscope blocks most of the light passing through the condenser so that only those light rays reflecting off of an object in the microscopic field are used to form the image. The result is a brightly illuminated object being viewed against a darkened field (Figure 2.5b). **Phase-contrast** microscopes employ a complex optical system in which slight differences in light refraction are converted to much larger differences in contrast. Because cellular structures refract light to various degrees, these structures appear as differences in light intensity, resulting in a high contrast, detailed image that is especially useful when examining living specimens (see Figure 2.5c).

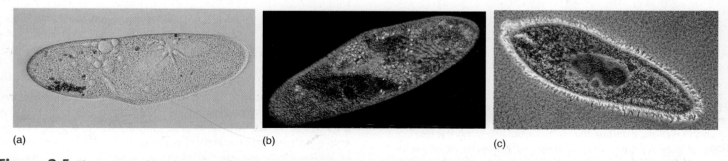

(a)   (b)   (c)

Figure 2.5 Three views of a eukaryotic cell. This paramecium was examined while still alive using a (a) brightfield (400x), (b) darkfield (400x), and (c) phase-contrast (400x) microscope. Notice the differences in the microscopic field and degree of detail afforded by each type of microscope.

MICROSCOPIC MEASUREMENTS

In addition to a clear image of a specimen, it is sometimes desirable to have an accurate size measurement as well, and microscopes can be easily outfitted with ocular and stage micrometers for this purpose. An **ocular micrometer** is simply a circular glass disc that has a series of regularly spaced markings etched onto its upper surface. The micrometer is installed within one of the oculars anytime a measurement is required, although in practice most laboratories find it easier to have an extra ocular with the micrometer permanently attached. When measurements are necessary, the ocular containing the micrometer is simply switched out with one of the oculars in the microscope. The distance between the markings on an ocular micrometer have no meaning until the ocular is paired with an objective lens and calibrated using a **stage micrometer.** The stage micrometer resembles a slide except that it has markings etched upon it that are exactly 0.1 mm (100 μm) and 0.01 mm (10 μm) apart (Figure 2.6).

To calibrate the ocular micrometer for use with a specific objective, the scales on the two micrometers must be superimposed on one another and the number of ocular gradations per stage gradation (10 or 100 μm) is then determined. If, when using the high power lens, for example, seven ocular divisions align with one stage division, then each ocular division equals 10 μm/7 ocular divisions or 1.43 μm/ocular division. This process is summarized in Figure 2.7. Once the ocular micrometer is calibrated for use with a specific objective lens, the stage micrometer is removed and replaced with a slide containing the organism to be measured. By counting the number of ocular gradations and multiplying by the distance between the gradations, the size of an unknown specimen is easily calculated.

Because of the thickness of the stage micrometer, it is generally impossible to directly calibrate an ocular micrometer for use with the oil immersion lens. In this case, calibration can be accomplished mathematically. Using the example from the previous paragraph, one ocular division equals 1.43 μm when viewed using the high power (40x) objective. When using the oil immersion (100x) objective, an object should appear 2.5 times larger than when viewed under high power (100/40 = 2.5), but the inscribed marks on the ocular haven't changed at all. Therefore, the distance between two marks on the ocular micrometer now span a distance 2.5 times as great as they did under the high power lens. To calibrate the ocular micrometer for use with the oil immersion lens, all that is needed is to divide the size of each ocular division (1.43 μm in the previous example) by the difference in the magnification of each lens (100/40 or 2.5):

1.43 μm per ocular unit (high power)/2.5 = 0.57 μm (oil)

PRE-LAB QUESTIONS

1. *Define the following terms:*

Magnification   *the creating of a larger than life image.*

Resolution   *the clarity of an image produced by a set of lenses.*

Contrast

Ocular micrometer   *a circular glass disc that has a series of regularly spaced markings etched onto its upper surface.*

Objective

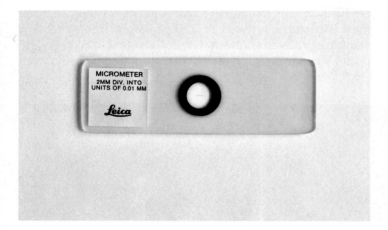

Figure 2.6 A stage micrometer has markings etched onto its surface that are a precise distance from one another. The label on the micrometer, in this case 0.01 mm, indicates that the distance between two adjacent lines is 0.01 mm (10 μm).

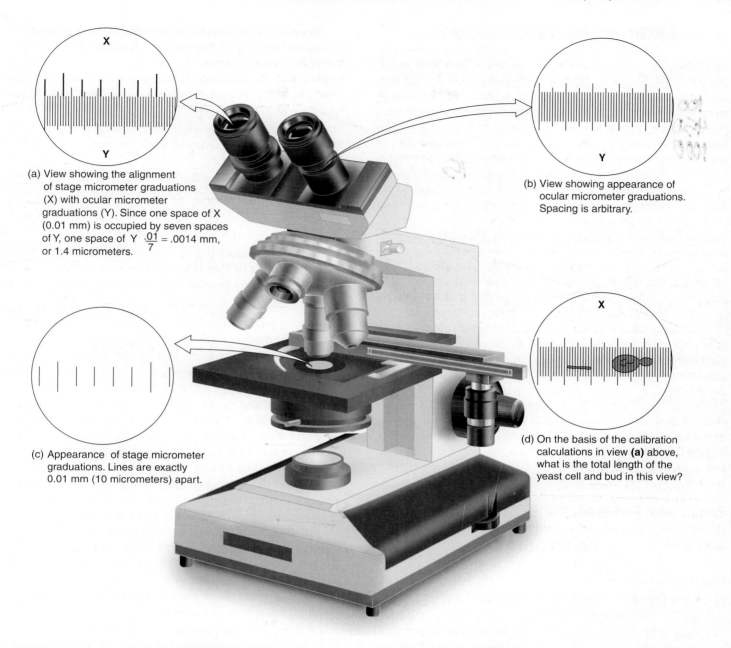

(a) View showing the alignment of stage micrometer graduations (X) with ocular micrometer graduations (Y). Since one space of X (0.01 mm) is occupied by seven spaces of Y, one space of Y $\frac{.01}{7}$ = .0014 mm, or 1.4 micrometers.

(b) View showing appearance of ocular micrometer graduations. Spacing is arbitrary.

(c) Appearance of stage micrometer graduations. Lines are exactly 0.01 mm (10 micrometers) apart.

(d) On the basis of the calibration calculations in view **(a)** above, what is the total length of the yeast cell and bud in this view?

Figure 2.7 Procedure for the use of stage and ocular micrometers to measure microscopic specimens.

2. *How should the condenser and diaphragm be adjusted for optimum viewing?*

3. *Explain how to properly clean the lenses on a microscope.*

4. *Complete each of the tables below.*

Total magnification	Ocular magnification	Objective magnification
100	10x	10x
450	10x	45x
1000	10x	100x
750x	10x	75
225x		45x

Resolution	Wavelength	Numerical aperture
	500 μm	0.4
	700 μm	0.4
	500 μm	1.25

MATERIALS

Each student should have:

A microscope

Prepared slides of:

Bacteria

Vorticella or *Spirogyra*

Paramecium

Access to stage and ocular micrometers

PROCEDURE

Part I: Observations

1. Carefully carry your microscope to your work area and place it gently on the benchtop.

2. Plug in the microscope and turn on the light source, keeping the voltage control (brightness) at a minimum. Be sure the condenser is raised to its maximum and the diaphragm is almost completely closed. Increase the voltage to the lamp until the illumination is at a comfortable level.

3. Place a slide on the stage, holding it in place with the stage clips. Be sure that the specimen is on the upper side of the slide. If the specimen is on the lower surface of the slide, you will be able to focus when using low power and high dry, but not when using the oil immersion lens.

4. Center the slide over the light beam emanating from the condenser. Move the lowest power objective (usually 10x) into position, listening for an audible click as it slips into place.

5. Bring the image into focus using first the coarse focusing knob and then the fine focusing knob to obtain the sharpest possible image. If you are using a

binocular microscope, use the diopter adjustment to compensate for differences in visual acuity between your eyes as follows:

- Close the eye with the adjustable ocular, and focus for your open eye using the coarse and fine adjustment knobs.
- Using only the eye with the adjustable ocular, turn the diopter adjustment until the image is in sharp focus.

6. Adjust the interpupillary distance to match the distance between your eyes by looking though both oculars and slowly adjusting the distance between them until a single image is seen.

If you will be measuring specimens in today's lab, follow the instructions in Part II to calibrate the ocular micrometer before proceeding. Otherwise, continue with the steps below.

7. Place a prepared slide on the stage of your microscope. Use the mechanical stage to hold it in place.

8. Scan the slide using the low power objective, selecting a potentially interesting area for further examination in greater detail.

9. Move the slide so that the area you wish to examine is centered in the microscopic field. Your microscope is **parcentric**, meaning that once a specimen is located in the center of the field, it will remain centered when changing objectives.

10. Swing the next highest objective into place, making sure that it clicks into position. Most microscopes are **parfocal**, meaning that if a specimen is focused with one objective, the image should remain sharp as the objectives are changed. You will have to make small adjustments to both the fine focus and diaphragm for each new objective.

11. When viewing bacteria, you will need to use the oil immersion lens. Do so as follows:

- After viewing the specimen with the high dry lens and obtaining a clear image, swing the lens out of the way and add a single drop of immersion oil to the slide directly above the condenser. Rotate the oil immersion (100x) lens into place, making sure that the end of the lens is submerged in the oil. Open the diaphragm almost all the way to get the highest quality image.

12. For each of your slides, record drawings of the specimens you observed in the space provided. Label each drawing with the name of the sample, total magnification, and size if measurements are being made.

13. When you are done for the period, clean any oil off of the lenses of you microscope as well as off the stage, focusing knobs, etc.

14. Rotate the low power objective into position, center the mechanical stage, lower the lamp's voltage to a minimum, and switch it off.

15. Wrap or gather the electrical cord according to the rules of your lab and return your microscope to its storage area.

Part II: Measurements

1. Following the instructions for your particular laboratory, insert an ocular containing a micrometer into your microscope.

2. Move the lowest power objective into place, making sure it clicks into position.

3. Place a stage micrometer onto the stage and hold it in place with the stage clips. Move the stage micrometer, and rotate the ocular micrometer until the two are superimposed on one another with the left sides of each aligned.

4. Record the number of ocular gradations per stage gradations in the accompanying table.

5. Repeat the process with the high power lens in place.

6. Repeat the process with the oil immersion lens, or calibrate the ocular micrometer for use with the oil immersion lens mathematically.

Lens	Ocular units (OU)	Distance (from stage micrometer)	Distance/OU
Scanning			
Low power			
High dry			
Oil immersion	100X		

7. After completing calibration of the ocular micrometer, return to the procedure above.

RESULTS

Record your results. For each specimen, provide as accurate a drawing as possible. Include the total magnification used to observe the specimen (i.e., 100x, 450x, or 1000x) and indicate the size of each specimen if measurements were taken.

Specimen #1

0   .5   1.0   1.5   2.0

Scale

40X

Specimen #2

Letter e

40X

Specimen #3

red   yellow   blue

40X

Specimen #4

40X

RESOLUTION OF THE CASES

From Leeuwenhoek's descriptions as well as detailed examination of his samples made with modern instruments, it is thought that the organisms he described as "green streaks, spirally wound . . ." were the algae *Spirogyra*, those "fashioned like a bell . . ." were the algae *Vorticella,* and the inhabitants of plaque were various species of bacteria (Figure 2.8). Leeuwenhoek's microscopic examination of everyday items and his descriptions of the microorganisms he found occurred over 150 years before the so-called Golden Age of Microbiology. Algae similar to those he saw are still tracked today because of their potential to cause a variety of neuro-

logical illnesses and alter the biology of lakes and oceans as well as their ability to cause a variety of neurological illnesses. His description of the myriad bacteria found between his own teeth is probably the first ever description of a biofilm, a type of bacterial community essential to the formation of dental caries (cavities) and a number of other diseases. Leeuwenhoek's correspondence with the Royal Society of London spanned more than 50 years. Prior to his death in 1723, he was credited with being the first to observe protozoa, bacteria, spermatozoa, and the banded pattern of muscle fibers.

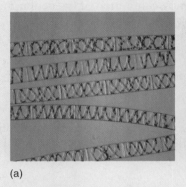
(a)

(b)

(c)

Figure 2.8   (a) *Spirogyra,* an algae; (b) *Vorticella,* a protozoa; and (c) various species of bacteria.

EXERCISE 2 REVIEW QUESTIONS

1. What is meant by the terms *parcentric* and *parfocal?*

parfocal
↳ Specimen focused c̄ with one objective, the image
   should remain sharp as objectives are changed

parcentric
↳ Specimen stays in middle of field
   when objective changes

2. In order to make specimens easier to see, why don't microscope makers use 100x ocular lenses?

3. What is the total magnification of a microscope with a 45x objective lens and a 15x ocular?

4. For each combination of numerical aperture and wavelength, calculate the resolution and determine if two points found at specified distances apart would be resolvable as discrete points?

Wavelength of light	Numerical aperature	Resolution (μm)	Distance between points	Resolvable (yes or no)
500 nm	0.8		400	
500 nm	0.9		325	
500 nm	1.2		297	
600 nm	0.8		517	
600 nm	1.2		253	

5. For a microscope on which 17 ocular units align with 100 μm on the stage micrometer when the 4x objective is used, how many micrometers are there per ocular unit when using the (a) 4x objective, (b) 10x objective, (c) 45x objective, and (d) 97x objective?

REFERENCE

University of California Museum of Paleontology. *Antony Leeuwenhoek (1632–1723).*
    http://www.ucmp.berkeley.edu/history/leeuwenhoek.html.

Identification and Classification of Algae

CASE SYNOPSIS

Oregon Harmful Algal Bloom Monitoring Project

In June 2005, a ban on clamming was instituted along much of the Oregon coast (Figure 3.1) after razor clam in that area were found to contain high levels of domoic acid, a naturally occurring toxin produced by algae in the genus *Pseudonitzschia*. Filter-feeding mollusks, such as clams and mussels, accumulate high levels of domoic acid during periods of robust algal growth known as *blooms*. Ingestion of domoic acid by humans causes amnesiac shellfish poisoning, which is marked by headache, dizziness, nausea, confusion, and potentially permanent loss of short-term memory. In severe cases, respiratory paralysis and death may occur within a day.

A different kind of shellfish illness, paralytic shellfish poisoning, results from ingesting saxitoxins, which are, like domoic acid, produced by certain species of algae. In this case, algae in the genus *Alexandrium* produce the toxin, which then accumulates in mussels, clams, scallops, oysters, crabs, and lobsters during

WARNING TOXIC SHELLFISH

Shellfish from this area are unsafe to eat due to shellfish toxins.
DO NOT EAT CLAMS, MUSSELS, OR SCALLOPS

¡CUIDADO! CRUSTÁCEOS TÓXICOS CRUSTÁCEOS SACADO DE ESTA ZONA SON INSEGUROS DE COMER Y PUEDEN RESULTAR EN PARÁLISIS. NO COME ALMEJAS, OSTRAS, MEJILLONES O MOLUSCOS.

OREGON STATE DEPARTMENT OF AGRICULTURE FOR FURTHER INFORMATION: (503) 986-4720

The Oregon Dept. of Fish & Wildlife requires a clam fishery be closed when health advisories are issued.

Figure 3.1 As a result of ingesting some species of algae, shellfish accumulate high levels of toxins, making them hazardous to humans who consume them.

periods of greater than usual algal growth. Ingestion of saxitoxin by humans can lead to numbness, paralysis, disorientation, and death due to respiratory failure. Neither domoic acid nor saxitoxin is affected by temperature, so cooking or freezing has no effect on the toxin.

Shortly after the 2005 shellfish harvesting closure, the Oregon Harmful Algal Bloom Monitoring Project was initiated. The project monitors water at five locations along the Oregon coast, retrieving samples every week or two (depending on the site) and examining each sample for the presence of algal species that produce domoic acid or saxitoxin. When sudden blooms lead to high levels of harmful algae, specific harvesting controls can be instituted. In Oregon, beaches are closed to clamming when domoic acid levels reach 20 parts per million (ppm) in randomly selected clams. Projects like this operate throughout the United States to ensure the safety of harvested seafood.

Resolution of the Case appears on page 30

STUDENT LEARNING OUTCOMES

After completing this exercise, you should be able to:

1. Prepare a wet mount of a liquid sample.
2. Differentiate algal samples based on physical appearance and physiological traits.

INTRODUCTION

Rather than being a well-defined taxonomic group with a clear evolutionary history, the algae represent a diverse collection of eukaryotic organisms. Historically, the **Subkingdom Algae** encompasses all of the photosynthetic members of the **Kingdom Protista,** with further subdivisions

based on characteristics that are easily observed in the laboratory. Organisms classified as algae may be unicellular, colonial, or filamentous in nature (Figure 3.2), and lack complete tissue differentiation, so that stems, roots, and leaves are not seen. Although the vast majority of species grow submerged in water, algae can be found anywhere sufficient moisture and sunlight, along with favorable temperatures, exist.

Modern genetic analysis, however, has shown that classification schemes based on morphological characteristics are not the best representation of the evolutionary relatedness between organisms. Put another way, the classification system historically used to categorize algae was fatally flawed. Currently, many different schemes are vying for supremacy, with no single one being universally accepted. In this lab, we will be using a hybrid system that categorizes algae into groups based on (1) the type of photosynthetic pigments present, (2) the chemical makeup of the cell wall, (3) cellular

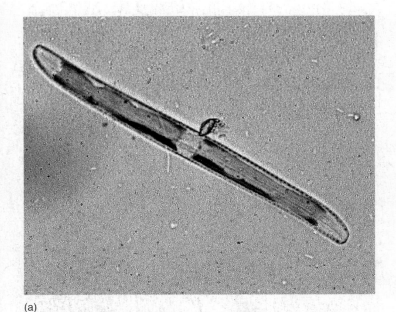

(a)

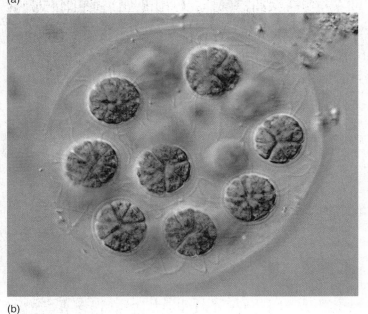

(b)

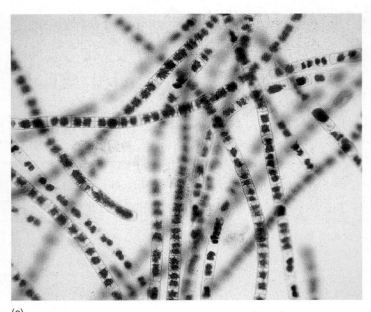

(c)

Figure 3.2 Algae may exhibit a (a) unicellular (diatoms 400x), (b) colonial (*Pandorina*, 400x), or (c) filamentous (*Zygnema*, 400x) structure.

morphology, and (4) the form in which food is stored. It is important to note that this system separates algae into informal groups, rather than formal hierarchal levels (Figure 3.3).

ORGANIZATION OF ALGAE

Supergroup Excavata

Group 1: Euglenozoa

This group is populated by flagellated algae that seem to straddle the boundary between algae and protozoans, and in fact many microbiologists group them with the animal-like protozoa. The **euglenoids** lack a cell wall, instead possessing a pellicle that provides them with a definite form while at the same time allowing a great deal of flexibility. Not all members of the division are photosynthetic, but those that are contain

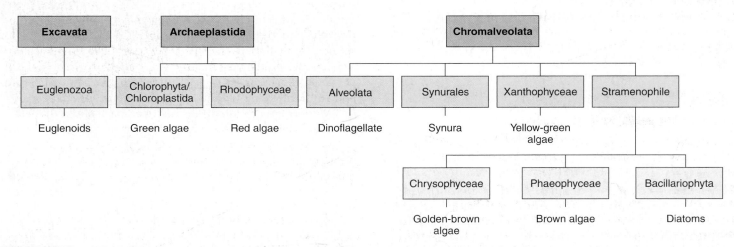

Figure 3.3 Classification of algae into supergroups (dark blue), groups (medium blue), and subgroups (pale blue). Common names appear below each group or subgroup.

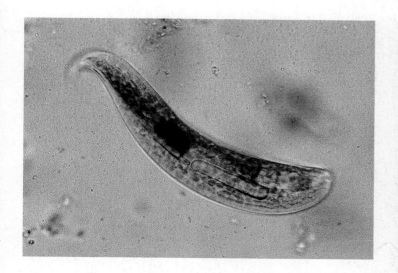

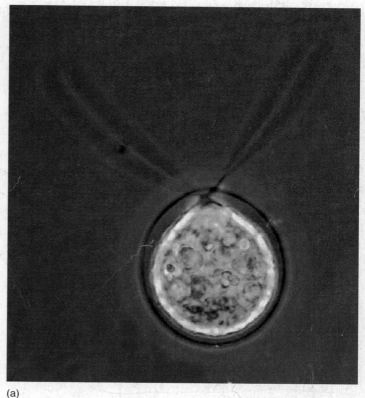

(a)

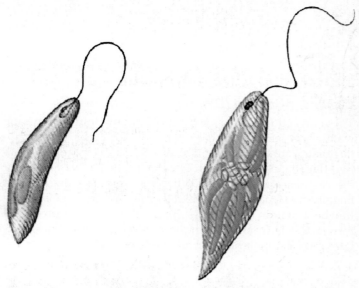

Figure 3.4 Euglena, a flagellated alga in the supergroup Excavata (400x). The red, light-sensitive stigma allows the cell to detect light, ensuring that it gains as much energy from photosynthesis as possible.

chlorophylls a and **b** and have a photosensitive red spot, or **stigma,** that permits them to detect the presence of light (Figure 3.4). The ability to sense light and move to where it is most intense allows them to gain the maximum amount of energy possible through photosynthesis. Excess food is stored in the form of **paramylon,** a lipopolysaccharide.

Supergroup Archaeplastida

Group 2: Chlorophyta/Chloroplastida - Green algae

This large division containing over 7000 species is commonly seen in ponds, neglected swimming pools, and fishtanks. They are a very diverse group with members found in fresh and salt water and exist both on their own as well as associated with other organisms. Molecular studies have shown the members of this group to be closely related to land plants. Energy from photosynthesis is stored in the form of **starch,** and the deep green color they exhibit is the result of the **chlorophylls a** and **b** they use for photosynthesis. The most

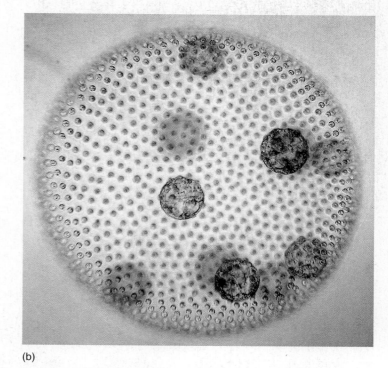

(b)

Figure 3.5 Commonly seen members of the Chlorophyta/Chloroplastida. (a) The unicellular Chlamydomonas (400x) and (b) the colonial Volvox (200x). Note the daughter colonies within the (Volvex).

common species include the single-celled Chlamydomonas along with various colonial forms such as Volvox and Pandorina (Figure 3.5) also commonly observed. The majority of species have cell walls made of **cellulose.**

(a)

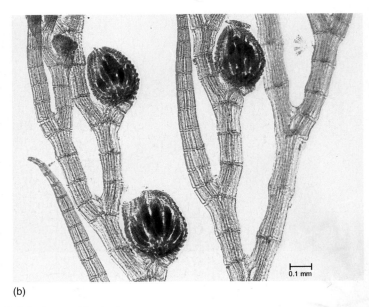

(b)

Figure 3.6 (a) *Callophyllis* and (b) *Polysiphonia* (100x). Red algae typical of the group Rhodophyceae.

Group 3: Rhodophyceae – red algea

Members of this group are marine seaweeds (Figure 3.6). In addition to **chlorophyll a** and orange **caratenoid** pigments, members of the Rhodophyceae contain two photosynthetic pigments known as **phycobilins**, the red **phycoerythrin**, and the blue **phycocyanin.** These two pigments allow the red algae to photosynthesize at depths of over 100 m, where the wavelengths of light that penetrate to these depths are unavailable to **chlorophyll a.** After light energy is absorbed by one of the two phycobilins, it is transferred to **chlorophyll a;** the resulting chemical energy is eventually stored in the form of starch. Many species are economically useful, serving as the source of carrageenan, a food additive, and **agar,** a solidifying agent for microbiological media.

Supergroup Chromalveolata

Group 4: Alveolata (Dinoflagellates)

These unicellular algae are most commonly found in marine waters and serve as the base of many food webs. Most dinoflagellates are covered by two protective plates and possess two flagella, each of which rests within a groove encircling the organism (Figure 3.7). When moving, the longitudinal flagellum extends behind the cell while the transverse flagellum beats within its groove, spinning the cell as it is propelled forward. It is from this unique motion that the dinoflagellates received their name (from the Greek *dinein*, "to whirl").

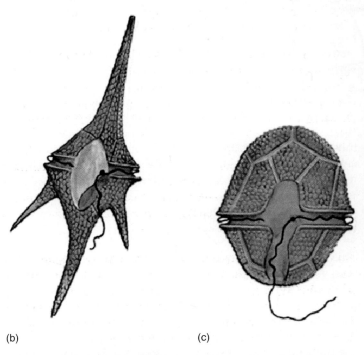

(a)

(b)

(c)

Figure 3.7 (a) and (b) *Ceratium* (400x), a common member of the algal group Alveolata (dinoflagellates). Notice the position of the flagella within a groove encircling the organism. (c) *Peridinium,* another common dinoflagellate.

Figure 3.8 *Synura* (400x), the most common member of the group Synurales. The cells are covered with silica scales, and the flagella (inset) may not always be visible in the microscope.

Most dinoflagellates possess **chlorophylls a** and **c** as well as **beta-carotenes** and **xanthophylls.** The yellow and orange pigments lend an orange color to the water during population blooms known as red tides, sometimes poisoning fish and other marine organisms that feed on the algae. Depending on the species, foods are stored as starches or oils.

Group 5: Synurales

This group of mostly flagellated algae contains **chlorophylls a** and **c** as well as several types of xanthin pigments. The most commonly seen genus is *Synura* (Figure 3.8).

Group 6: Xanthophyceae (Yellow-Green Algae)

Chlorophylls a and **c** as well as several different **xanthin** pigments give the members of this group a yellow-green color. Of the 600 species in the group, only three genera (*Botrydium, Tribonema,* and *Vaucheria*) are commonly seen (Figure 3.9).

Group 7: Stramenophile

Members of this group all contain varying amounts of the brown pigment **fucoxanthin** and consequently display a distinctly brown color. The type of cell wall, other photosynthetic pigments, and form of food storage are used to divide the group into three subgroups.

Subgroup Chrysophyceae: The Golden-Brown Algae   This

large subgroup contains over 6000 species. In addition to **chlorophylls a** and **c,** members of this division commonly possess the brown pigment **fucoxanthin** as well as orange **beta-carotene** and various yellow **xanthophylls.** Together, this combination of photosynthetic pigments gives most species in the division a golden-brown appearance. Energy

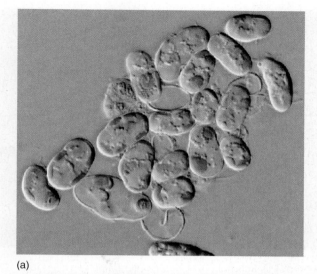

(a)

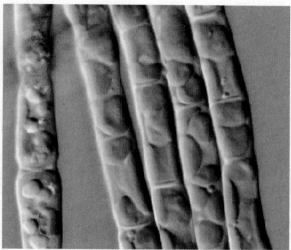

(b)

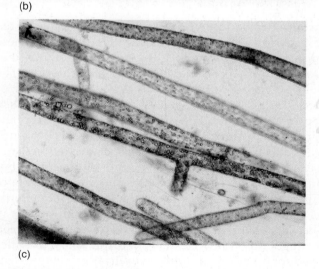

(c)

Figure 3.9 The three common members of the Xanthophyceae: (a) *Botrydium* (200x), (b) *Tribonema* (400x), and (c) *Vaucheria* (100x).

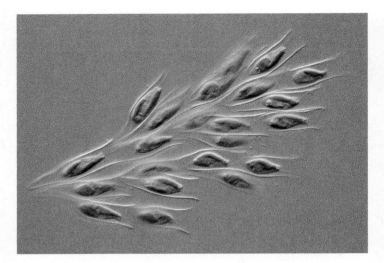

Figure 3.10 Subgroup Chrysophyceae. *Dinobryon* owes its brownish color to the pigment fucoxanthin (400x).

produced from photosynthesis is stored as either oil or laminarian, a polysaccharide. Common members include *Dinobryon* (Figure 3.10).

Subgroup Phaeophyceae: The Brown Algae   This group consists almost entirely of multicellular marine organisms, and most brown to green seaweeds are members of this group. Like the golden-brown algae, members of the Phaeophyceae contain **chlorophylls a** and **c, beta-carotene, xanthophylls,** and **fucoxanthin,** but the greater amount of fucoxanthin lends these organisms a decidedly brownish hue. Food storage is in the form of laminarian, a polysaccharide, and mannitol, a sugar alcohol. *Macrocystis*, the largest of all seaweeds, is a common example (Figure 3.11).

Subgroup Bacillariophyta: The Diatoms   Some of the most beautiful algae are diatoms (Figure 3.12). These organisms have outer shells called **frustules,** composed of cellulose and of silicon dioxide or calcium chloride. The frustule is composed of two halves, with the larger half, or epitheca, fitting over the smaller half, or hypotheca, like the top and bottom of a Petri dish. Diatoms are some of the most numerous organisms in the sea and serve as a primary food source for filter feeders like clams, oysters, and mussels as well as many other marine organisms. After death, the frustules of dead diatoms sink to the bottom of the lakes and oceans where they form deposits of material called diatomaceous earth, which is used industrially in polishes, abrasives, and filters such as those used to keep swimming pool water clean. While the silica frustules are used for a variety of industrial purposes, many scientists believe that the oil generated by diatoms via photosynthesis accounts for much of our current petroleum reserves.

Figure 3.11 Giant members of the subgroup Phaeophyceae can grow to be over 200 m in length. This kelp forest is made up of the algae *Macrocystis pyrifera*.

PRE-LAB QUESTIONS

1. *Define the following terms:*

Photosynthesis

Photosynthetic pigment

Flagella

Frustule

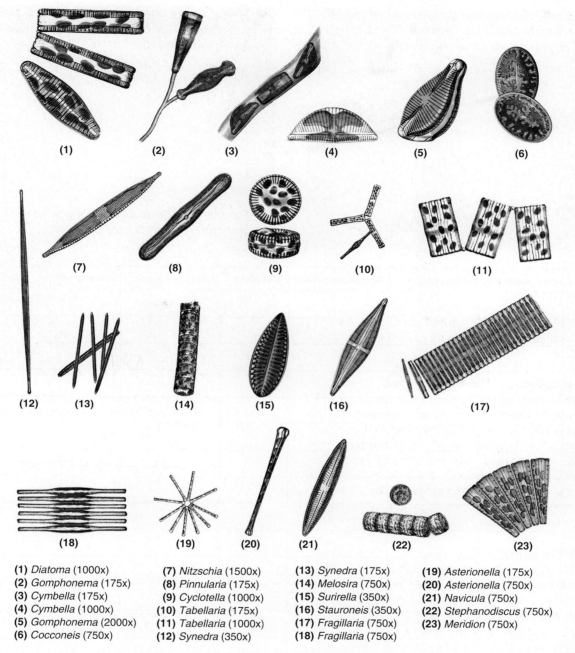

Figure 3.12 Diatoms are some of the most immediately recognizable algal organisms and can be classified by the specific shape of their rigid frustules.

(1) *Diatoma* (1000x)
(2) *Gomphonema* (175x)
(3) *Cymbella* (175x)
(4) *Cymbella* (1000x)
(5) *Gomphonema* (2000x)
(6) *Cocconeis* (750x)

(7) *Nitzschia* (1500x)
(8) *Pinnularia* (175x)
(9) *Cyclotella* (1000x)
(10) *Tabellaria* (175x)
(11) *Tabellaria* (1000x)
(12) *Synedra* (350x)

(13) *Synedra* (175x)
(14) *Melosira* (750x)
(15) *Surirella* (350x)
(16) *Stauroneis* (350x)
(17) *Fragillaria* (750x)
(18) *Fragillaria* (750x)

(19) *Asterionella* (175x)
(20) *Asterionella* (750x)
(21) *Navicula* (750x)
(22) *Stephanodiscus* (750x)
(23) *Meridion* (750x)

2. Why are "red tides" red? Why are these tides seen at some times and not others? What algal division would be most likely to produce a "green tide"?

3. The number of cases of seafood poisoning is greatest in the summer months. Besides the fact that people are more likely to harvest seafood when the weather is warm, why else would illnesses due to ingestion of harmful algae be more prevalent in the summer?

4. The number and size of harmful algal blooms seem to be correlated to an increased use of fertilizers. Speculate on a possible connection between these two events.

5. Several months after beaches are closed to clamming, the same beaches can be declared safe and reopened. Why are unsafe clams later deemed safe?

6. For each algal group seen below, place an "x" in the box if the characteristic indicated is present.

	Euglenozoa	Chlorophyta	Rhodophyceae	Alveolata	Chrysophyceae	Phaeophyceae	Bacillariophyta
Chlorophyll a							
Chlorophyll b							
Chlorophyll c							
Fucoxanthin							
Xanthophylls							
Carotenes							
Phycoerythrin							
Phycocyanin							

Pellicle							
Flagella							
Cell wall							

Starch							
Paramylon							
Leucosin							
Oils							
Laminarian							
Mannitol							

MATERIALS

Each student should obtain:
Several bottles containing pond water samples
Microscope slides and cover slips
Pasteur pipettes and forceps

PROCEDURE

Make wet mount slides of each of the water samples available. A wet mount is a simple preparation method that allows the observation of live specimens. Prepare each wet mount as follows:

1. If not already done, thoroughly clean a slide and cover slip with soap and water.
2. Use the pipette to extract a small amount of sample from the bottom of the sample jar. Very few organisms will be found swimming in the middle of the sample.
3. If needed, use forceps to transfer a small amount of filamentous algae to a slide.
4. Place a cover slip gently on the sample.
5. Examine the slide under low power, looking for areas of interest. Adjust the diaphragm to reduce the light reaching your slide.
6. When you encounter a specimen you would like to study in more detail, swing the high-dry objective into position.
7. Use the pictures and descriptions found here, along with any supplementary books in your lab, to identify the organisms you encounter.

RESULTS

Record your results. As you examine each sample, draw as accurately as possible those aspects (including colors) that would be most important in identifying a particular algae. Include the total magnification used to observe each specimen (100x or 450x).

Sample #1

Sample #2

Sample #3

Sample #4

Sample #5

Sample #6

RESOLUTION OF THE CASE

Widespread shellfish harvesting restriction are a serious disruption to coastal communities, negatively affecting harvesters and related businesses, along with tourism in the area. Because of this, officials try to keep harvest control measures as geographically limited and short-lived as possible. On June 21, 2006, due in part to ongoing water sampling by the Oregon Harmful Bloom Monitoring Project, the entire Oregon coast was opened to razor clamming for the first time in 4 years (although short stretches of beach were temporarily closed later in the summer).

A primary reason for the increased number of cases of shellfish illness in the summer months is that algal growth is always greater when supported by longer periods of daylight and warmer water temperatures. In addition, algal blooms often occur when phosphorus and nitrogen, which are common ingredients in fertilizers, accumulate in the water. Fertilizers used on land leach into the groundwater and eventually find their way to open bodies of water, where they induce abnormally robust growth of algal populations, like the one seen here. When algal levels decrease, the toxins eventually leach out of the shellfish, but it can take weeks to months before a beach may be safely reopened.

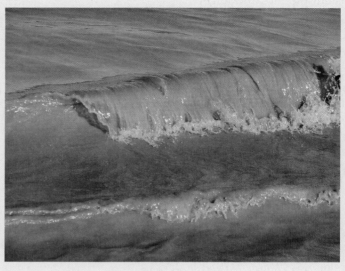

A harmful algae bloom in Lake Erie.

EXERCISE 3 REVIEW QUESTIONS

1. Water for algae analysis may be collected in one of two ways. In the first method, an 8-oz. jar is directly filled with water while in the second, a net with a large opening at one end and a collection jar at the other is towed through the water, collecting algae in the jar while water passes through the fine mesh of the net. How would the algal populations collected with each method differ from one another? Why would one method be chosen over another?

2. A sample of seawater contained organisms with the following descriptions. What group is represented by each description?

  a. A colonial algae, bright green in color with cell walls made of cellulose.

  b. A portion of a large organism (over 3 m), brown in color, which stores food as mannitol.

  c. Several small organisms with nearly symmetrical shells. Further analysis indicates that the shells are made of silicon dioxide and that the organisms are rich in oil.

  d. A sample taken from a jar in which the water has an orange tint to it. Organisms in the jar have a pair of flagella and seem to rotate as they move through the water.

  e. A sample in which the flagellated organisms are marked by a single red spot at one end.

  f. An alga containing both red and blue pigments.

Evolutionary theory says that organisms which are closely related will share more characteristics than those that are more distantly related, and scientists often construct evolutionary trees to represent these relationships. Use Figure 3.3 and the information in the exercise to answer the following questions.

3. Why are the green algae and red algae classified in separate groups but found within the same supergroup?

4. Assuming that a single ancestral species gave rise to the Alveolata, Synurales, Xanthophyceae, and Stramenophiles, what type of photosynthetic pigments did it possess? How do you know this?

5. Why are the brown algae, golden-brown algae, and diatoms assumed to be members of the Stramenophile group and not one of the other six algal groups.

REFERENCE

Oregon Department of Fish and Wildlife: Shellfish Safety Closures. http://oregon.gov/ODA/FSD/shellfish_status.shtml.

CASE STUDY EXERCISE

Survey of Medically Important Protozoa

CASE SYNOPSIS

Acanthamoeba keratitis—Multiple States

In the first half of 2006, the Illinois Department of Public Health (IDPH) was investigating an increase in the number of cases of *Acanthamoeba* keratitis (AK), a rare but potentially blinding infection of the cornea. The initial investigation focused on a single ophthalmology center in Illinois and found that cases of AK had indeed increased over the preceding 3 years. Beginning with this information, the Centers for Disease Control (CDC) initiated a multistate survey of ophthalmology centers nationwide to determine whether cases of AK were increasing throughout the United States.

Acanthamoeba keratitis is an infection of the cornea caused by a free-living protozoon (*Acanthamoeba*) that is found throughout the environment. The protozoon is commonly found in water (including tap water), sewage, cooling towers and heating/ventilation/air conditioning systems. An estimated 85% of AK cases occur in otherwise healthy people who wear contact lenses, including those who follow recommended lens-care practices. Activities known to increase the risk of infection include not adequately disinfecting lenses, wearing lenses while swimming, bathing, or using hot tubs, or coming into contact with contaminated water. Persons with corneal trauma also have an increased risk of infection. Previous estimates of AK infection in the United States have historically ranged between one and two cases per million contact lens wearers.

In March 2007, the CDC revealed that data from a wide geographic area indicated that the number of cases of culture-confirmed *Acanthamoeba* keratitis had been rising since 2004 throughout the United States. On March 16, 2007, an investigation was launched to determine what risk factors were associated with this increase in AK cases.

Resolution of the Case appears on Page 39

CDC. 2007. *Acanthomoeba* keratitis—multiple states, 2005–2007. *Morbidity and Mortality Weekly Report,* 56 (Dispatch): 1–3.

STUDENT LEARNING OUTCOMES

After completing this exercise, you should be able to:

1. Prepare a wet mount of a liquid sample.
2. Differentiate protozoan samples using a light microscope.

INTRODUCTION

Like their evolutionary cousins, the algae, the protozoans represent a diverse group of eukaryotic organisms. Until recently, the protozoans were considered to be the "animal-like" subkingdom of the Kingdom Protista while the algae represented the "plant-like" subkingdom. This view has changed dramatically as molecular biology has allowed us to look at an organism's underlying genetics and not just its outward appearance. Currently, major taxonomic shifts are underway that will change the manner in which these organisms are classified, but for now, we will use a classical system of division that is still utilized by most micro-biologists. In this system, the Kingdom Protista is populated by single-celled eukaryotic organisms that lack tissue specialization. Further partitioning of the Kingdom is accomplished by assigning photosynthetic protists (i.e., plant-like protists) to the subkingdom Algae while animal-like protists (i.e., those unable to photosynthesize) are assigned to the subkingdom Protozoa. The subkingdom Protozoa is itself divided into four groups (three phyla) based on the type of motility exhibited by its members. Most protozoans exist in two distinct forms in different parts of their life cycle: the **trophozoite** stage represents the active, feeding form of the organism while the **cyst** stage represents the inactive form. Classification of protozoans is based primarily on the type of motility observed in the trophozoite stage, but identification can often be made based on an examination of the cyst (much like identifying a bird from its egg). Other characteristics used in identification include the shape and size of the organism, the physical appearance and location of cellular structures such as flagella and cilia, the presence of special organelles used for feeding, and the number of nuclei. The organizational hierarchy of the subkingdom Protozoa can be seen in Figure 4.1. Each group within the Kingdom is discussed below.

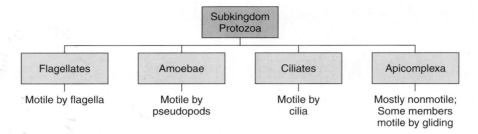

Figure 4.1 The classical organization of the subkingdom Protozoa relies on characteristics such as motility that are easily seen in the laboratory.

ORGANIZATION OF PROTOZOA

Flagellates

This group contains organisms that are motile by means of a flagella alone or a combination of flagellar movement and amoeboid motion. *Giardia lamblia* is one of the most important human parasites of this group. Giardia is spread through ingestion of water containing *G. lamblia* cysts, generally as a result of fecal contamination. The hardiness of the cyst allows the organism to pass through the stomach and initiate an infection in the small intestine, where its four pairs of flagella help it to bind tightly to epithelial cells. Although most infections are asymptomatic, Giardia is best known for causing diarrhea, especially in backpackers who drink from what seem to be pristine streams. Diagnosis of infection is made by identifying trophozoites or cysts in stool specimens.

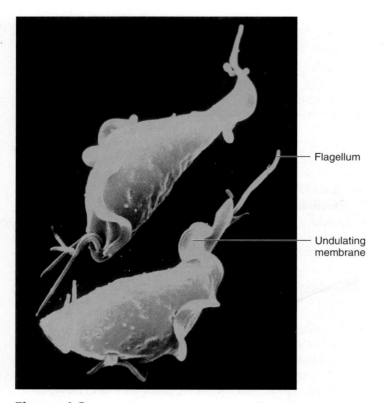

Figure 4.2 Scanning electron micrograph of *Trichomonas vaginalis* (2000x). One flagellum trails behind the organism while the other remains attached to the cell, forming the undulating membrane.

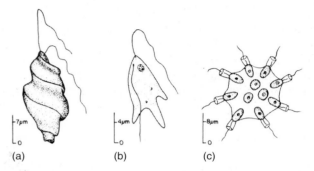

Figure 4.3 Common freshwater flagellated protozoans: (a) *Heteronema,* (b) *Cercomonas,* and (c) *Protospongia.*

A second commonly seen flagellate is *Trichomonas vaginalis* (Figure 4.2), the cause of trichomoniasis in humans, a disease marked by inflammation of the genitourinary tract. An important difference between these two parasites is that *T. vaginalis* does not form cysts, leaving the more delicate trophozoite as the infective form. Because the trophozoite is far less hardy than a cyst, *T. vaginalis* must be passed directly from person to person, and because of this, trichomoniasis is classified as a sexually transmitted disease. Figure 4.3 illustrates a variety of flagellates commonly found in fresh water.

Amoebas

The amoebas use flowing projections of cytoplasm called **pseudopods** as their major locomotor organelle, although some species have flagellated reproductive states (Figure 4.4). Even though most amoebas are free living and not infectious, some human parasites do exist. *Entamoeba histolytica* is the cause of amoebic dysentery, a gastrointestinal disease associated with poor sanitation and one of the most prevalent protozoan infections in the world. Infection begins with the ingestion of the organism, most often via contaminated food and water. While trophozoites are destroyed by the low pH of the stomach, cysts can continue and enter the small intestine where excystation and mitosis result in eight trophozoites being produced from each ingested cyst. The disease itself may range from asymptomatic to quite severe, with signs and symptoms including diarrhea, nausea, vomiting, blood and mucous in the feces, and hepatitis. Cysts are shed in the feces, helping to complete the oral-fecal route. Diagnosis is made by identifying cysts or trophozoites in stool specimens (Figure 4.5).

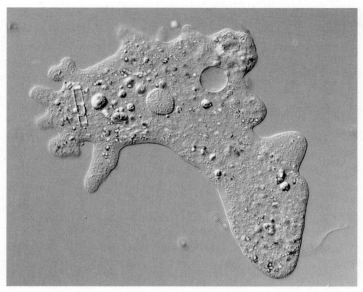

(a)

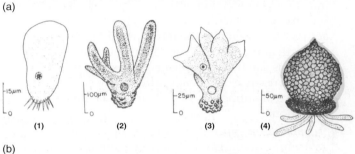

(b)

Figure 4.4 (a) Amoeba move using flowing projections of cytoplasm called pseudopods (1000x). (b) Common freshwater amoeba include (1) *Trichamoeba,* (2) *Amoeba,* (3) *Mayorella,* and (4) *Difflugia.*

Acanthamoeba is another free-living amoeba that can cause disease in humans. Its common portals of entry include the conjunctiva, broken skin, and, less often, the lungs and urogenital tract. If the infection is restricted to the eye, *Acanthamoeba* keratitis can cause damage to the cornea, leading to blindness. Granulomatous amebic encephalitis, a serious infection of the brain and spinal cord, can result if the organism spreads throughout the body from any portal of entry. This type of infection is most often seen in persons with compromised immune systems and is almost always fatal. Diagnosis is made by identifying trophozoites or cysts in cerebrospinal fluid or corneal tissue.

Ciliates

Cilia are small hairlike projections that are used by some protists for motility as well as to direct food into a specialized feeding structure. While cilia and flagella share a common organizational structure, a cell typically possesses only a few long flagella while the far shorter cilia usually number in the thousands, completely covering the cell. The vast majority of ciliates are free living and harmless and generally exhibit the most complex structures and behaviors of all protists (Figure 4.6). The most commonly encountered parasite is *Balantidium coli,* which can cause bloody- and

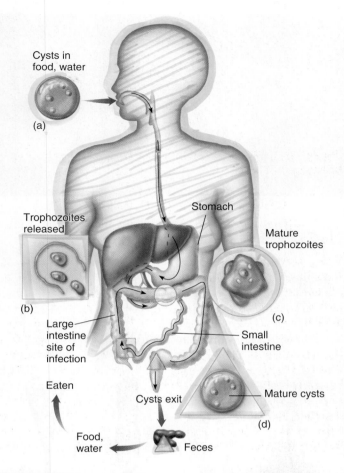

Figure 4.5 Stages in the infection and transmission of amebic dysentery. (a) Ingested cysts of *Entamoeba histolytica* are tough enough to withstand the low pH of the stomach, (b) allowing trophozoites to emerge in the small intestine, and (c) eventually move to the wall of the large intestine. Mature cysts are released in the feces, (d) allowing the organism to spread when fecal matter contaminates food or water. Identification of cysts in the feces allows definitive diagnosis of the disease.

mucus-filled diarrhea, although most infections are thought to be asymptomatic.

Apicomplexans

The apicomplexans consists of organisms that share a unique structure called an apical complex. All members of the phylum are parasitic and, with the exception of male gametes, nonmotile. Apicomplexans have complex life cycles that include both sexual and asexual stages. Several human pathogens exist, including:

Plasmodium spp. Five species of Plasmodium cause malaria in humans, *P. falciparum, P. malariae, P. ovale, P. vivax,* and *P. knowlesi.* The life cycle is split between human tissues (red blood cells and liver) and an insect vector, the female *Anopheles* mosquito. Diagnosis, as well as determination of the infectious species can be accomplished by examining red blood cells for the presence of the parasite (Figure 4.7).

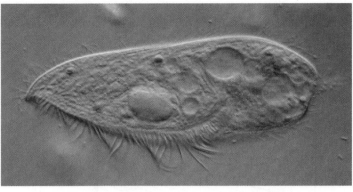

(a)

(b)

Figure 4.6 (a) Interference contrast image of *Blepharisma,* a common freshwater ciliate (450x). (b) Commonly observed ciliates: (1) *Paramecium,* (2) *Lacrymaria,* (3) *Litonotus,* (4) *Loxodes,* (5) *Blepharisma,* (6) *Coleps,* (7) *Condylostoma,* (8) *Stentor,* (9) *Vorticella,* (10) *Carchesium,* (11) *Zoothamnium,* (12) *Stylonychia,* (13) *Onychodromus,* (14) *Hypotrichidium,* (15) *Euplotes,* and (16) *Didinium.*

Trypanosome spp.   Two species within this genus deserve mention. *T. brucei* is the causative agent of African sleeping sickness, which causes 50,000 deaths yearly. The parasite is carried by the tsetse fly, which is geographically limited to the African continent. *T. cruzi* is the etiologic agent of Chagas disease, which causes several thousand deaths yearly. The vector in this case is the reduviid bug, which spreads the trypanosome through its feces. Once again, geographic limitation of the vector results in Chagas disease being relatively common in Central and South America, Mexico, and even the southern United States, but rare elsewhere. In both cases, diagnosis can be made by examining the blood for the infectious agent or through serological tests that detect antibodies to the organism.

Toxoplasma gondii   The life cycle of this apicomplexan is split between the intestines of cats and humans. While infection via ingestion of the organism is generally not serious, an infected mother can pass the infection across the placenta to the fetus, potentially causing stillbirth and brain or liver damage. Laboratory diagnosis can be accomplished by examining the blood for the presence of *T. gondii,* although testing for the presence of antibodies to the parasite is more often done.

Cryptosporidium spp.   Cryptosporidiosis is a gastrointestinal disease resulting from infection with any of several species of this protozoan, which are commonly found in water contaminated with feces. Recreational water, such as swimming pools and fountains, are especially susceptible because the water is recycled and common disinfection techniques, such as chlorination, are ineffective against the organism. Visualization of the trophozoite or cyst in a stool sample is the most common diagnostic test.

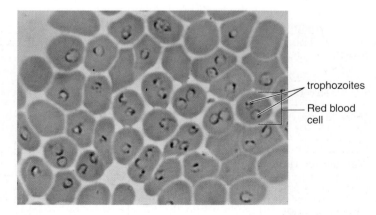

trophozoites

Red blood cell

Figure 4.7 Infection with the apicomplexan *Plasmodium falciparum* causes malaria in humans. Diagnosis can be accomplished by examining red blood cells for the presence of the trophozoite form of the organism (1000x).

PRE-LAB QUESTIONS

1. *Define the following terms:*

Tissue specialization

Flagellum

Cilia

Pseudopod

Vector

Parasite

2. *Differentiate between pseudopod, amoeba, and amoeboid motion.*

3. *Do amoebas have cell walls? Justify your answer.*

4. *What is the difference between a trophozoite and a cyst, and why is one generally more infective than the other?*

5. *For each group found in the table, place an "x" in the box if the indicated structure or behavior is present.*

	Flagellate	Amoeba	Ciliate	Apicomplexan
Nucleus				
Cell membrane				
Motility seen in most members				
Are generally free living				
Are generally parasitic				
Move using flagella				
Move using cilia				
Move using pseudopods				
Are generally nonmotile				

MATERIALS

Each group should have access to the following:
Several different samples of pond water
Cultures of *Amoeba*
Prepared slides of *Trypanosoma sp.*
Prepared slides of *Plasmodium sp.*
Microscope slides and cover slips
Pasteur pipettes

PROCEDURE

1. Remove a small volume of water from one of the pond samples or the *Amoeba* culture. Take your sample from the bottom of the container because this is where most organisms will be found.

2. Place one or two drops onto your slide, and cover with a cover slip. Return the pipette to the original sample.

3. Scan the slide using the low power lens. Use your microscopes voltage regulator and diaphragm to keep the light as low as possible.

4. When you see an organism that you would like to examine more closely, swing your high power lens into place, and adjust the diaphragm to get as clear an image as possible.

5. Identify the organisms you see to the level of the phylum, and record this information in the space provided.

6. Examine the prepared slides on low and high power. Record your results.

RESULTS

Record your results. For each sample, provide a drawing that indicates the major features seen in the organism. Also indicate the total magnification used to observe the specimen (i.e., 100x and 450x). Based on the features you observe, identify each organism to at least the level of the phylum.

Sample #1

Sample #2

Sample #3

Sample #4

Sample #5

Sample #6

RESOLUTION OF THE CASE

In May 2007, the Centers for Disease Control released a preliminary report. Beginning January 1, 2005, a total of 138 persons had been diagnosed with *Acanthamoeba* keratitis, and this diagnosis had been supported by obtaining corneal specimens that yielded *Acanthomoeba*. Of the first 46 culture-confirmed patients, 42 wore contact lenses, and of these 16 patients (35%) reported swimming and 35 (83%) reported showering while wearing their lenses during the month prior to the onset of symptoms. Further investigation revealed that the use of a particular brand of lens cleaning solution (Advanced Medical Optics Complete® Moisture Plus™ or AMOCMP) was statistically linked to AK. Solution production information (lot number) was available for 10 patients who reported using AMOCMP; no single lot number was repeated, suggesting that contamination of a specific production run was not to blame.

While the investigation of *Acanthamoeba* keratitis cases continues, the CDC has recommended that persons who wear contact lenses and use AMOCMP should (1) stop using the solution immediately and discard any remaining solution, (2) discard their current pair of contact lenses and current lens storage containers, and (3) become informed as to the symptoms of AK and see their ophthalmologist if they experience any signs of eye infection. In response to the initial results of the investigation, the makers of AMOCMP initiated a voluntary recall of the product.

EXERCISE 4 REVIEW QUESTIONS

1. Postulate a reason that this particular brand of contact lens solution was linked to cases of *Acanthamoeba* keratitis. Recall that no contamination was found in the lens solution itself and that *Acanthamoeba* is widely distributed in the environment.

2. *Acanthamoeba* is found as both a trophozoite and cyst, but both forms seem to be equally effective at causing *Acanthamoeba* keratitis infections. In contrast, amebic dysentery is due almost exclusively to infection with the cyst form of *Entamoeba histolytica* even though the organism can be found in the environment as both a trophozoite and a cyst. Why is this so?

3. Of 46 culture-confirmed cases investigated by the CDC, 42 wore contact lenses while 4 reported no contact lens use at all. How do people who do not wear contact lenses contract AK, and why does wearing lenses increase the rate of infection so dramatically?

Identification and Classification of Fungi

CASE SYNOPSIS

Outbreak of Histoplasmosis among Travelers Returning from El Salvador, 2008

Have you ever been on a mission trip with your church or youth group? Each year, thousands of Americans travel every year to other countries—or to disaster areas within their own country, (such as New Orleans after Hurricane Katrina)—where they pitch in to perform all kinds of ordinary, but important, manual tasks, such as simple construction, renovation, or flood clean up. For one such trip a group of church volunteers from Pennsylvania and Virginia traveled to a church in Nueva San Salvador. A total of 35 volunteers went to El Salvador, traveling in three groups between January 3 and February 20, 2008. The trip to El Salvador did not turn out well. Twenty of the volunteers came down with a serious respiratory disease resembling acute influenza within 3 to 25 days of arriving in El Salvador.

To diagnose the disease and figure out how the patients had acquired it, public health officials began investigating the activities of all the volunteers, those affected by the illness as well as those unaffected. The volunteers had helped clean indoor and outdoor renovation sites, install electrical and plumbing components, build additional rooms onto the church, replace the roof, and excavate the septic tank. In addition, each of the mission groups had taken one day off during their stay to visit a local beach or lake. The church volunteers in El Salvador were doing heavy cleaning both indoors and outdoors, as well as working with soil (cleaning renovation sites, excavating the septic tank). Those activities point to the possible presence of *Histoplasma,* a fungus that often grows with the aid of bat and bird excrement in places such as El Salvador, where it is endemic. Fungal spores can become airborne when sweeping, digging, or vacuuming stirs up dust or dirt.

Resolution of the Case appears on page 50

CDC. 2008. Outbreak of histoplasmosis among travelers returning from El Salvador, Pennsylvania, and Virginia, 2008. *Morbidity and Mortality Weekly Report,* 57(50): 1349–1353.

STUDENT LEARNING OUTCOMES

After completing this exercise, you should be able to:

1. Differentiate fungal samples based on macroscopic and microscopic characteristics.

INTRODUCTION

The Kingdom Myceteae (Fungi) is a large group of nonmotile, eukaryotic organisms that can be found in a variety of forms, including mushrooms, puffballs, molds, and yeasts, all of which appear very different but share many biological features. All members of the Kingdom are **heterotrophic,** meaning they require an organic source of carbon, as opposed to **autotrophic** (photosynthetic) organisms that can utilize CO_2 from the atmosphere. Most often, dead organisms serve as the primary source of carbon for most fungi; in fact, along with the bacteria, fungi are the primary decomposers of organic compounds on earth. This decomposition allows for the reuse of carbon and nitrogen by other organisms and prevents the planet from being swamped with organic waste. The Greek prefix *myco* means fungus and helps to explain the fact that the study of fungus is mycology and a person who studies fungus is a mycologist.

The common characteristics of fungi provide ample means to differentiate them from plants and animals. Unlike plants, fungi are unable to photosynthesize and contain the polysaccharide chitin in their cell walls, rather than cellulose that is commonly seen in plants. Unlike animals, fungi must digest their food before ingesting it. By secreting digestive enzymes, they can digest organic material in the surrounding environment. The simple nutrients that result are then imported into the cell.

Fungi can be found in a variety of forms, many of which should already be familiar. The larger mushrooms, puffballs and, truffles (Figure 5.1) are sometimes referred to as **macrofungi** while microscopic fungi are referred to as yeasts or molds, depending on their structure. Yeasts are single-celled fungi, with a spherical to oval shape and a typical diameter of 3–5 μm. The most common means of reproduction in yeasts is budding, in which a new, small cell forms from an older larger one. When a string of these new cells forms, the result is a **pseudohypha** (Figure 5.2a). Two

Figure 5.1 Mushrooms represent the fruiting body of a fungus.

common yeasts with human importance are *Saccharomyces cerevisiae* (see Figure 5.2b), which is used in the manufacturing of bread and beer, and *Candida albicans,* part of the normal flora of the respiratory, gastrointestinal, and female reproductive tract. Overgrowth of *Candida* can lead to conditions

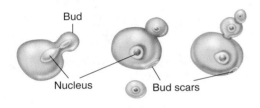

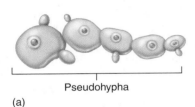

(a)

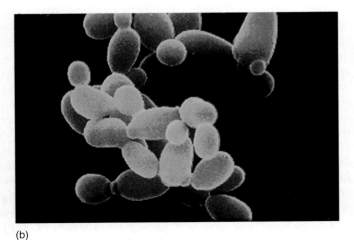

(b)

Figure 5.2 (a) As a yeast cell reproduces, it creates buds, eventually resulting in the formation of a pseudohypha. (b) Scanning electron micrograph of *Saccharomyces cerevisiae,* the yeast normally used for baking bread and brewing beer (x2500).

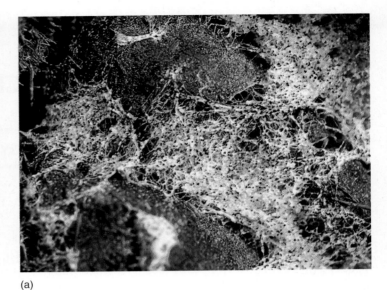

(a)

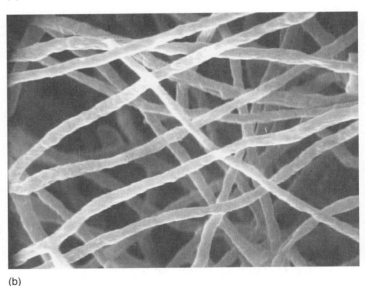

(b)

Figure 5.3 (a) Mycelium of a basidiomycota growing on soil. (b) Scanning electron micrograph of a mycelium reveals the structure of the hyphae (x5000).

such as thrush and vulva vaginitis (inflammation of the vulva or vagina) in otherwise healthy persons while more serious systemic infections may be seen in patients who are immunocompromised, diabetic, or have indwelling medical devices such as catheters or shunts.

When fungi grow as strings of long, filamentous cells, they are referred to as *molds.* A single filament is known as a *hypha* (plural, hyphae) while a large intermeshed mat of hypha is called a *mycelium* (Figure 5.3). Hypha can be categorized as *septate* if individual cells in the hypha are separated by walls or as *nonseptate* if the hypha is continuous (Figure 5.4). Each hypha in a mycelium is capable of digesting and absorbing nutrients, making a fungus almost perfectly adapted for eating and ensuring that fungi are found in nearly every habitat on earth. Some fungi are referred to as *dimorphic,* capable of growing as yeasts or molds,

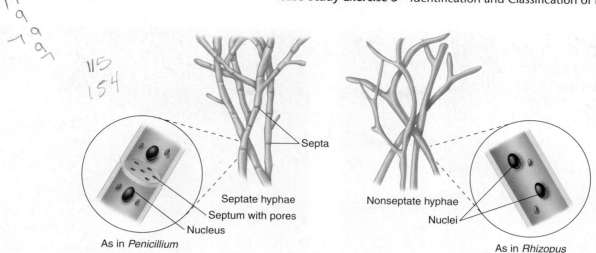

Figure 5.4 Fungal hyphae may be septate or nonseptate, and this characteristic can be used in the laboratory to help identify an unknown fungal species.

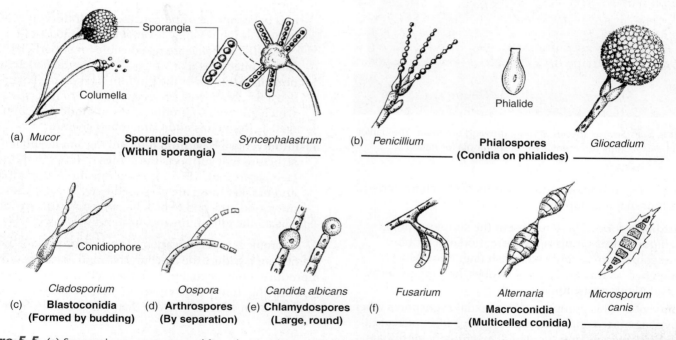

Figure 5.5 (a) Sporangiospores are asexual fungal spores borne within a dedicated structure known as a sporangium, while spores that form on specialized hyphae are called conidia. (b) Phialospores sit atop a pear-shaped cell called a phialide. (c) Blastoconidia are formed by budding from preexisting conidia. (d) Arthrospores form from preexisting hyphal cells. (e) Chlamydospores are large, thick walled structures commonly found on older cultures of many fungi. (f) Any multicellular conidia is known as a macroconidia.

depending on the environmental conditions. Many fungal pathogens grow in the environment as molds and then convert to a yeast form within the body.

Fungal reproduction is complex, with both sexual and asexual methods of reproduction being used by most species. Asexual reproduction occurs when haploid spores are released from a single parent. If these spores are contained within a saclike container called a **sporangium,** the spores are referred to as **sporangiospores. Conidia** are spores that are not enclosed within a spore-bearing sac but rather form from a special fertile hyphae. Both types of spores are seen in Figure 5.5. Asexual spores give rise to new fungi that are exactly like the parent that produced them.

Sexual spore formation is more complex but offers the promise of more variability in the offspring. By combining genes from two parents, the offspring produced will vary in form and function both from each other and from both parents. This variability provides at least some of the offspring with advantageous traits and increases the overall survival of the species. Three types of fungal spores are seen in fungi. These are illustrated in Figure 5.6.

Morphological characteristics such as the formation of spores is commonly used in the laboratory to identify and classify fungi, even as molecular techniques play a greater role in the process. Currently, the fungi are divided into

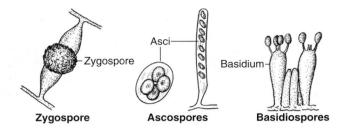

Figure 5.6 Type of sexual spores seen in fungi.

Figure 5.7 Several species of fungi growing on bread. The darker areas of each colony represent areas of sporangia formation.

eight groups, four major and four minor. The characteristics of the four major groups are:

Chytridiomycota   These represent the simplest fungi. Unique to this group is the formation of a motile **zoospore** with a single whiplash flagellum. Most members of this group are saprobes, helping to decay dead plants and animals.

Zygomycota   This group produces sexual **zygospores** when hyphae of two opposite mating strains fuse to create a diploid zygote. This zygote will eventually germinate, giving rise to a sporangium filled with sexual spores. The common bread mold *Rhizopus* belongs to the Zygomycota. A cursory look at a *Rhizopus* colony can give a clue to its age; the whitish mycelium will grow darker as dark sporangia develop (Figure 5.7).

Ascomycota   Spores formed within a fungal sac, or **ascus,** are the hallmark of this group. The ascus develops from the union of two hypha of opposite mating type. Meiosis, sometimes followed by mitosis, results in the formation of four to eight haploid ascospores (Figure 5.8). Many species of great importance to humans are members of the Ascomycota. Species of *Aspergillus* can be found growing on fruit and bread, and industrial uses of the fungus include the production of citric acid and soy sauce. Some species of *Aspergillus* are opportunistic pathogens and are responsible for aspergillosis, a collection of mostly respiratory diseases of varying

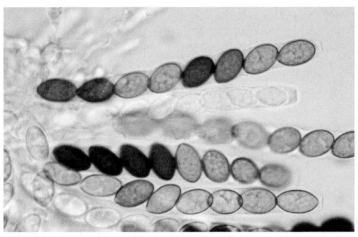

Figure 5.8 *Sordaria fimicola*. Four or eight ascospores are contained within a saclike sporangia.

severity. *Histoplasma capsulatum* is associated with bird and bat droppings and causes a respiratory infection known as Ohio Valley fever, while members of the genus *Trichophyton* are responsible for superficial fungal infections like athletes foot (Figure 5.9). Other notable members of the phylum include the baker's yeast *Saccharomyces cerevisiae* and *Penicillium,* species of which are used to produce the antibiotic penicillin as well as Roquefort and Camembert cheeses.

Basidiomycota   **Basidium** refers to the small clublike structure that supports the spores of fungi in this group (see Figure 5.6). Most common mushrooms, puffballs, and bracket fungi are classified here. The basidiospores are found attached to basidia, which in turn are located within the gills of the mushroom.

A problem in classification has arisen because of the ability of fungi within the Ascomycota and Basidiomycota to reproduce both sexually and asexually. Because sexually reproducing fungi (teleomorphs) may appear at different times and on different media than the asexual (anamorph)

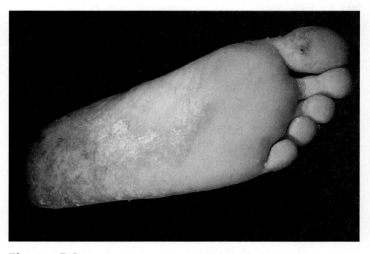

Figure 5.9 The foot of a patient with tinea pedis ('ringworm') due to *Trichophyton rubrum*.

form, in many cases the same fungal species has been "discovered," named, and studied twice. Rather than trying to correct every past reference in the scientific literature, an artificial group, the Deuteromycota, was created to house the 15,000 species in the anamorph stage that have been extensively studied. The problem with this process, however, is that a fungus can be known by two completely different names, both of which are correct. A simple way of thinking of this is to consider a person who takes the name of their spouse after marriage; both names refer to the same person, but until this fact is known, some confusion will no doubt ensue. *Coccidioides immitis*, the cause of San Joaquin Valley fever, and

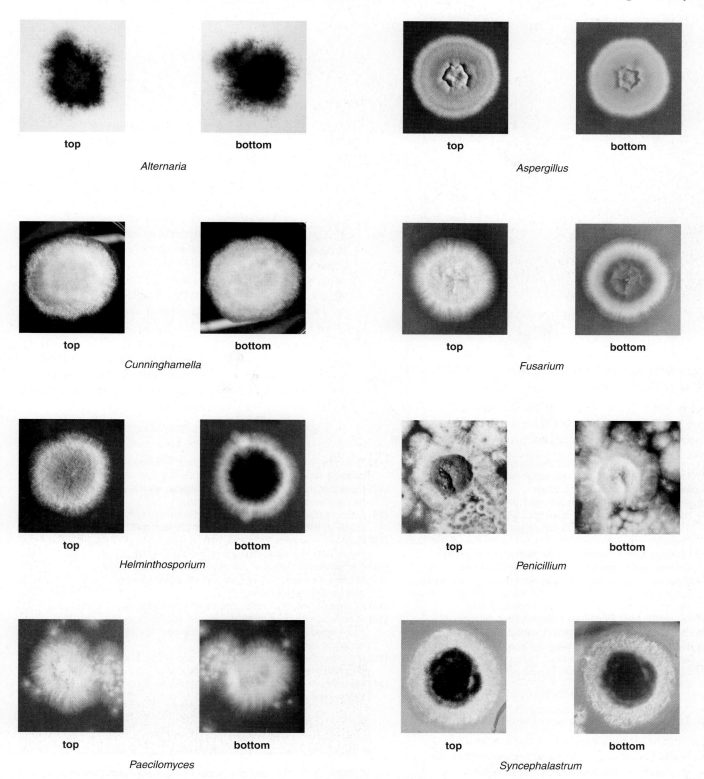

Figure 5.10 Colony characteristics of some common molds.

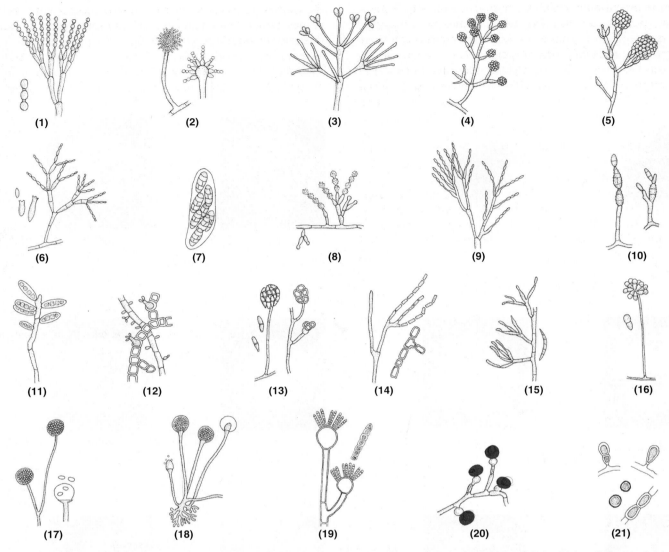

Figure 5.11 Microscopic appearance of some common molds.

(1) *Penicillium*– bluish-green; brush arrangement of phialospores.

(2) *Aspergillus*– bluish-green with sulfur-yellow areas on the surface. *Aspergillus niger* is black.

(3) *Verticillium*– pinkish-brown, elliptical microconidia.

(4) *Trichoderma*– green, resemble *Penicillium* macroscopically.

(5) *Gliocadium*– dark-green; conidia (phialospores) borne on phialides, similar to *Penicillium;* grows faster than *Penicillium.*

(6) *Cladosporium (Hormodendrum)*– light-green to grayish surface; gray to black back surface; blastoconidia.

(7) *Pleospora*– tan to green surface with brown to black back; ascospores shown are produced in sacs borne within brown, flask-shaped fruiting bodies called pseudothecia.

(8) *Scopulariopsis*– light-brown; rough-walled microconidia.

(9) *Paecilomyces*– yellowish-brown; elliptical microconidia.

(10) *Alternaria*– dark greenish-black surface with gray periphery; black on reverse side; chains of macroconidia.

(11) *Bipolaris*– black surface with grayish periphery; macroconidia shown.

(12) *Pullularia*– black, shiny, leathery surface; thick-walled; budding spores.

(13) *Diplosporium*– buff-colored wooly surface; reverse side has red center surrounded by brown.

(14) *Oospora (Geotrichum)*– buff-colored surface; hyphae break up into thin-walled rectangular arthrospores.

(15) *Fusarium*– variants, of yellow, orange, red, and purple colonies; sickle-shaped macroconidia.

(16) *Trichothecium*– white to pink surface; two-celled conidia.

(17) *Mucor*– a zygomycete; sporangia with a slimy texture; spores with dark pigment.

(18) *Rhizopus*– a zygomycete; spores with dark pigment.

(19) *Syncephalastrum*– a zygomycete; sporangiophores bear rod-shaped sporangioles, each containing a row of spherical spores.

(20) *Nigrospora*– conidia black, globose, one-celled, borne on a flattened, colorless vesicle at the end of a conidiophore.

(21) *Montospora*– dark-gray center with light-gray periphery; yellow-brown conidia.

Blastomyces dermatitidis, the cause of North American blastomycosis (Chicago disease), are two examples of pathogenic fungi that exist as both anamorphs and teleomorphs.

While biochemical, physiological, and genetic techniques are often used for conclusive identification, initial classification of a fungus is often accomplished through visual examination of a sample. First, the color, texture, and morphology of a mold colony, both top and bottom, is examined (Figure 5.10). The texture of a fungal colony can be described as leathery, velvety, cottony, or powdery while common morphologies include flat, verrucose (rough, hilly), and cerebriform (brainlike). Color is self-explanatory with the caveat that the color of a colony generally will change as it ages and forms sporangia. After the initial macroscopic assessment, microscopic examination of hyphae (septate or not), sporangia, and spores can be used to complete the initial identification (Figure 5.11).

PRE-LAB QUESTIONS

1. *Define the following terms:*

Saprobe

Dimorphic

Parasite

2. *Differentiate between a hypha, a pseudohypha, and a mycelium.*

3. *Explain the difference between a sporangiospore and a conidiospore. Sketch an example of molds producing each type of structure.*

4. *Sketch an example of septate hypha and nonseptate hypha.*

MATERIALS

Each group should obtain:
Sealed Sabouraud dextrose agar plate cultures of:
    Rhizopus sp.
    Penicillium sp.
    Aspergillus sp.

These plates should be sealed with parafilm or tape. Do not remove the lids from the plates because the number of spores released could constitute a hazard.

Sabouraud dextrose agar plate culture of:
    Saccharomyces cerevisiae
Prepared slides of:
    Aspergillus sp.
    Rhizopus sp.
    Penicillium sp.
Methylene blue, iodine, or lactophenol cotton blue

PROCEDURE

Saccharomyces cerevisiae

1. Prepare a wet mount of *Saccharomyces cerevisiae.* The procedure for preparing a wet mount is found in Exercise 4.
   - Begin by adding a single drop of dye to the center of the slide.
   - Use an inoculating needle to aseptically add a small amount of yeast to the dye.
   - Gently place a cover slip atop the dye.

2. Examine the slide using low power. When you find an area of great interest, swing your high power objective into position. Sketch a portion of the microscopic field, looking especially for budding cells and pseudohyphae.

Rhizopus sp.

1. Examine a colony of *Rhizopus* macroscopically. Do not remove the cover from the plate. Sketch the colony, and provide a description of the color (both top and bottom of the colony), texture (leathery, velvety, cottony, or powdery), and surface morphology (flat, verrucose, or cerebriform) as described in the Introduction.

2. Using a dissecting microscope, examine the colony and sketch the details of the mycelium, indicating hyphae and sporangia.

3. Using a compound microscope, examine prepared slides of *Rhizopus.* Sketch a portion of the microscopic field, being certain to identify sporangiophores, sporangia, and spores. Label each of these structures.

Penicillium sp.

1. Examine a colony of *Penicillium* macroscopically. Do not remove the cover from the plate. Sketch the colony, and provide a description of the color (both top and bottom of the colony), texture (leathery, velvety, cottony, or powdery) and surface morphology (flat, verrucose, or cerebriform) as described in the Introduction.

2. Using a dissecting microscope, examine the colony and sketch the details of the mycelium, indicating hyphae and conidia.

3. Using a compound microscope, examine prepared slides of *Penicillium.* Sketch a portion of the microscopic field, being certain to identify hyphae, conidiophores, and conidia. Label each of these structures.

Aspergillis

1. Examine a colony of *Aspergillis* macroscopically. Do not remove the cover from the plate. Sketch the colony, and provide a description of the color (both top and bottom of the colony), texture (leathery, velvety, cottony, or powdery) and surface morphology (flat, verrucose, or cerebriform) as described in the Introduction.

2. Using a dissecting microscope, examine the colony and sketch the details of the mycelium, indicating hyphae and conidia.

3. Using a compound microscope, examine prepared slides of *Aspergillis.* Sketch a portion of the microscopic field, being certain to identify hyphae, conidiophores, and conidia. Label each of these structures.

RESULTS

Record your results. For each sample, provide a drawing that indicates the major features seen in both macroscopic and microscopic views of the organism. Also indicate the total magnification used to observe the specimen, whether it is 1x (naked eye), 20x or 40x (dissecting microscope), or 100x or 450x (compound microscope), as well as any colony morphology.

S. cerevisiae (_____x)

Rhizopus sp.
(_____x)

Rhizopus sp.
(_____x)

Rhizopus sp.
(_____x)

Penicillium sp.
(_____x)

Penicillium sp.
(_____x)

Penicillium sp.
(_____x)

Aspergillus sp.
(_____x)

Aspergillus sp.
(_____x)

Aspergillus sp.
(_____x)

RESOLUTION OF THE CASE

Case File Wrap-Up

The first time one of the church volunteers in El Salvador reported respiratory problems, a physician performed a chest x-ray. Although there are no specific radiographic signs that point definitively to histoplasmosis, this patient exhibited clear signs of inflammation, and the physician suspected *Histoplasma* because it is endemic to Central and South America as well as to eastern Asia, Australia, and the midwestern United States. The diagnosis was confirmed by conducting ELISA tests, which use antibodies to detect specific organisms in bodily fluids. In this case, *Histoplasma* was detected in either the blood or urine of all 20 patients.

Interestingly, histoplasmosis is highly prevalent in the Ohio River Valley of the United States. The majority of people living in this area are thought to have antibodies to the fungus as a result of prior exposure, even though they may never have shown symptoms of the disease. Such persons may have been protected from the infection if they had taken a similar mission trip!

EXERCISE 5 REVIEW QUESTIONS

1. Describe and/or draw the principle differences between a yeast and a mold.

2. Briefly, what are the advantages and disadvantages of sexual reproduction in fungi.

3. Briefly, what are the major characteristics used to differentiate:

    a. Fungi from plants?

    b. Fungi from animals?

    c. Fungi from protozoa?

    d. Deuteromycota from each of the other fungal groups?

    e. The four major groups (Chytridiomycota, Zygomycota, Ascomycota, and Basidiomycota) from one another?

4. What is a mycosis?

5. What specific steps do you think could be taken to reduce the likelihood of a similar outbreak on a future trip?

REFERENCE

CDC. *Histoplasmosis.* http://www.cdc.gov/nczved/divisions/dfbmd/diseases/histoplasmosis.

Ubiquity of Microorganisms

CASE SYNOPSES

Sorcerer II Global Ocean-Sampling Expedition— 2007–2009

In 2000, genomic researcher J. Craig Venter stood with physician and geneticist Francis Collins and U.S. President Bill Clinton to announce that the Human Genome Project, a worldwide effort to identify all the genes in a human being, was essentially complete. Two years later, Venter was aboard his 95-foot sailboat, the *Sorcerer* II, "fishing" for new genomes to map—those of microorganisms living in the ocean.

As the *Sorcerer* II sailed the Sargasso Sea, Venter and his assistants collected 200-liter samples of seawater and filtered them so that only organisms 1–3 μm in size were retained. They then froze these life forms onto filter paper and sent them to Venter's facility in Rockville, MD, for analysis. Using molecular biology techniques first developed for the Human Genome Project, Venter hoped to classify the new life forms by identifying novel genes without having to coax organisms to grow in the lab. Venter's efforts were so successful that many people compared his voyage to that of the British naturalist Charles Darwin, which had occurred over 170 years earlier and led to Darwin's theory of evolution, a premise that underlies nearly every aspect of biology today. In 1831, Charles Darwin embarked on a 5-year voyage around the globe on a ship called the HMS *Beagle*. While on this journey, Darwin identified many never-before-seen plant and animal species. Eventually his studies of these organisms led to the development of his theory of evolution by natural selection, which states, in part, that as the genetic material of living beings changes over time, new life forms with unique structures and functions are produced. Traits that favor the survival of an organism, such as the ability to metabolize a new food source, are retained and passed on to the organism's descendents.

The Sorcerer II traveled thousands of miles collecting sea-dwelling microorganisms. Genetic analysis of the samples recovered during the first voyage revealed over 1800 new species.

Resolution of the Cases appears on page 57

S. Yooseph. 2007. The Sorcerer II global ocean sampling expedition. *PLoS_Biology*, Mar. 13; 5(3): e16.

STUDENT LEARNING OUTCOMES

After completing this exercise, you should be able to:

1. Recognize microbial growth in microbiological media.
2. Differentiate between bacterial and fungal growth on solid media.

INTRODUCTION

Microbiology involves, obviously, the study of microorganisms, but the history of exactly what microbes were, what they do, and where they might be found has been unexpectedly complex. The century between 1850 and 1950 is often referred to as the "Golden Age of Microbiology" because this is when the importance of bacteria, protists, and fungi in our daily lives was first appreciated. For example, prior to the early 1900s, doctors, even surgeons, neglected to wash their hands between patients, their primary complaint being that it was simply too time consuming.

Saying that the "Golden Age" ended in 1950 however was shortsighted, to say the least. Over the next half century, new bacteria continued to be discovered, viruses such as HIV became more prevalent, and entirely new infectious agents—such as the prions responsible for mad cow disease—emerged. The understanding that microbes are part of our daily life, living in our food and drink, inhabiting our bodies, and floating in the air we breathe is not nearly as surprising now as it was a century ago. Furthermore, we realize that the type and number of microbes is significant, going out of our way to buy yogurt with active (microbial) cultures, keeping milk refrigerated to prevent the growth of any more bacteria than is absolutely necessary, and paying the really big bucks for mushrooms and truffles (both fungi).

Along with specifically culturing microbes to take advantage of their abilities, we are also interested in preventing the growth of unwanted microbes, with the most obvious examples being taken from the area of health care. One important difference is that in a hospital we are primarily concerned with the spread of **pathogenic** microbes while in the laboratory, any unwanted microbe is capable of ruining an experiment, meaning that laboratory requirements are often far more stringent than those seen in the hospital setting. The first step in determining what techniques and practices will be effective in preventing the spread of microbes is to assess their presence in the environment.

Although microorganisms may be present in a variety of locations, their detection generally requires that they be removed and transferred to growth media for culturing and (potential) identification. Today you will choose two different environmental sites to sample and will use one sample to inoculate a plate of nutrient agar and the other to inoculate a tube of nutrient broth.

PRE-LAB QUESTIONS

1. *Define the following terms:*

Sterile

Ubiquitous

Pathogen

Inoculation

Turbid

PERIOD ONE

MATERIALS

Each student should obtain:

One tube of trypticase soy broth
One plate of trypticase soy agar (TSA)
Sterile swabs
Marking pen

PROCEDURE

1. For the plate of solid media, use the exposure method assigned to you in Table 6.1. For the broth tube, sample any item you like.
2. Choose two objects or areas for sampling.
3. Label a tube of trypticase soy broth with your name, lab time, and the identity of the item or area to be sampled.
4. To inoculate liquid media, draw a sterile swab across the surface to be tested, and then transfer the swab to the media. If the surface you've chosen to sample is

TABLE 6.1

Student number	Exposure method for TSA plate
1, 10, 19	Press fingertips gently against the media
2, 11, 20	Press lips gently against media
3, 12, 21	Comb hair or scratch scalp over media for 20 sec
4, 13, 22	Blow dust onto exposed media
5, 14, 23	Expose plate to air within the laboratory for 30 min
6, 15, 24	Expose plate to air outside of the laboratory for 30 min
7, 16, 25	Expose plate to air outside the building for 30 min
8, 17, 26	Press a coin or paper money gently against the media, then remove
9, 18, 27	Any method not previously mentioned

dry, dip the swab into the broth, and then use the moistened swab to sample the surface.

5. Label the bottom of a plate of trypticase soy agar, around the edge, with your name, lab time, and the identity of your second item to be sampled.

6. In the results section, record the appearance of both the broth and the agar prior to inoculation (color, clarity, consistency, etc.). A quick note on handling media. Be gentle when inoculating the nutrient agar so as not to crack it; agar has the consistency of firm gelatin and will break if it is handled too roughly.

7. Incubate the inoculated media at 30°C for at least 48 h.

QUESTIONS—PERIOD ONE

1. How would you recognize microbial growth in a liquid media? What about a solid media?

2. What is a colony? What is the relationship between a cell and a colony?

3. What additional information does a solid culture provide compared to a liquid culture?

4. What are some characteristics of bacterial and fungal colonies that can be used to differentiate between the two?

PERIOD TWO

PROCEDURE

Retrieve your cultures from the incubator. Examine both the liquid media (which may require gentle shaking of the tube) and solid media. In what way have the cultures changed from their preincubation state? Exercises 40 and 41 may help you interpret any growth seen in your cultures.

RESULTS

Record your results in the table as well as on a similar table or computer spreadsheet your instructor has provided.

	Appearance of media prior to incubation	Appearance of media after incubation
Trypticase soy broth		
Trypticase soy agar		

Plates	Colony counts		Broth	
Plate exposure method	**Bacteria**	**Fungi**	**Source**	**Result (+/−)**
╫			╌	
╫	✗		╫	

RESOLUTION OF THE CASE

J. Craig Venter's initial efforts led to the discovery of 1.2 million new genes and 1,800 new species. He heads an organization called the Institute for Biological Energy Alternatives. One of the institute's goals is to create synthetic organisms tailor-made for a specific purpose, such as synthesizing chemicals, degrading waste products, or producing energy. It stands to reason that Venter's discovery of new species will increase the potential for even more useful products, both naturally occurring and manmade.

Based on the extraordinary success of the first *Sorcerer* II voyage, a more extensive second voyage visited many different locations around the world, including the Galápagos Islands, where Darwin made many of his observations.

J. Craig Venter's second voyage led to the discovery of 20 million new genes and thousands of new protein families. Of particular interest to Venter were a group of genes called *rhodopsins,* which help bacteria capture energy from the sun. Venter hopes these bacteria may one day be used as an alternative energy source. He articulated this hope in a 2007 interview when he said, "We really need to find an alternative to taking carbon out of the ground, burning it, and putting it into the atmosphere. That is the single biggest contribution I could make."

On March 19, 2009 , the *Sorcerer* II left her home port of San Diego for a third voyage. Further exciting discoveries seem likely.

EXERCISE 6 REVIEW QUESTIONS

1. Did any of the sites sampled by the class appear to be sterile? If so, do you think this result would hold true if the same sites were to be sampled again?

2. Can any conclusions be drawn as to the types of habitats likely to contain the most microbes?

3. Did all the colonies on your plates look the same? Can any conclusions be drawn from this observation?

7

Aseptic Techniques

CASE SYNOPSIS

Multiple Misdiagnoses of Tuberculosis Resulting from Laboratory Error—Wisconsin 1996

The pulmonary disease tuberculosis (TB) is caused by infection with the bacterium *Mycobacterium tuberculosis.* Although skin tests and chest x-rays indicate probable infection, a diagnosis of tuberculosis is confirmed only when *Mycobacterium tuberculosis* is isolated from patient specimens. During 1996, the Wisconsin State Division of Health (DOH) became aware of 5 incidents that potentially led to a misdiagnosis of tuberculosis in 11 persons. Because of the expense, strain on healthcare resources, and patient stigmatization associated with an inaccurate diagnosis of TB, the DOH launched an investigation into the cause of these events. The first of these cases is summarized here.

In March 1996, a medical laboratory forwarded two *M. tuberculosis* positive specimens, one from patient 1 and the other from patient 2, to the Wisconsin State Laboratory of Hygiene (SLH) for further testing. When the treating physician of patient 1 was informed that *M. tuberculosis* had been isolated from a sputum sample of his patient, he informed the SLH that TB had not been suspected, that another specimen obtained from the patient had yielded *Streptococcus pneumoniae,* and that treatment for pneumococcal pneumonia had resolved all the symptoms, indicating that the *M. tuberculosis* results was a false positive.

Resolution of the Case appears on page 64

CDC. 1997. Multiple misdiagnoses of tuberculosis resulting from laboratory error—Wisconsin, 1996. *Morbidity and Mortality Weekly Report,* 46(34): 797–801.

STUDENT LEARNING OUTCOMES

After completing this exercise, you should be able to:

1. Correctly handle and inoculate microbiological media so as to minimize the possibility of contamination.

INTRODUCTION

Studying microorganisms is fundamentally different than studying any other type of living organism. Microorganisms are found virtually everywhere, including soil, water, air, dust, food, and most body surfaces, even when the body is healthy. Furthermore, these same microbes are generally invisible to the naked eye, meaning their presence is rarely apparent, even when a sample is heavily populated with microbes. The combination of the ubiquitous nature of microbes and their "invisibility" means that the microbiologist must use specific aseptic techniques to exclude unwanted microbes. The proper use of these techniques cannot be overemphasized; without them the study of microbiology is a hopeless exercise. Additionally, these techniques must be mastered quickly: the organisms you work with will not care that you've only recently begun to study microbiology! Although there are no physical manipulations associated with this case, it should serve as a reference to potentially be used with every other exercise in this

manual. It is recommended that this exercise be combined with Exercise 8 (Pure Culture Techniques) to provide you with an opportunity to demonstrate your newfound skills.

PRE-LAB QUESTIONS

1. Define the following terms:

   Ubiquitous

   Aseptic

   Sterile

   Sputum

2. Explain what is meant by a false positive result. How does this differ from a false negative result?

MATERIALS

Each student should obtain:

A test tube rack

A Bunsen burner

An inoculating loop and needle

PROCEDURE

General Considerations

The following practices should be followed under all circumstances. They are intended to promote safety in the laboratory and reduce contamination.

- Keep your work area uncluttered and organized. Your benchtop will most likely be crowded with inoculating tools, stains, test tube rack, Bunsen burner, lab book, notebook, etc. Keep other items to a minimum. Label tubes as instructed; this may include using tape, paper held on with rubber bands, or laboratory marker. Petri dishes may be labeled directly on the base of the plate.

- Test tubes should always be placed in a test tube rack when not in use. Use the rack to transport the tubes as well. Plates should be carried in a sleeve if there are more than can be comfortably carried by hand, generally two or three. (Figure 7.1)

- Regulate the flame of your Bunsen burner by altering the amount of air entering the bottom of the burner. Adjust the amount of air entering the burner until the flame has both an inner and outer cone. The hottest part of the flame is the tip of the inner cone, and this is the area that is used for sterilization of inoculating tools. The outer, less intense flame is adequate for flaming of tubes and heat fixing of bacterial smears (Figure 7.2).

- Inoculating loops and needles should be held lightly with your dominant hand while culture tubes should be held with the nondominant hand.

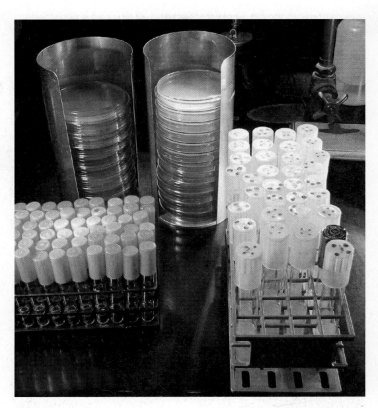

Figure 7.1 Metal sleeves should be used whenever you have more than a few Petri dishes. Tubes of media should always be carried in racks to ensure that they remain upright.

- Sterilize inoculating tools by passing them through the flame until each portion of the wire has achieved an orange color. Begin flaming about 2–3 cm up the handle and draw the tool backward through the flame until each part of the wire has been heated to a uniform orange. Flaming from handle to tip allows the end of the wire to heat up slowly and helps prevent the formation of media- and bacteria-containing aerosols (see Figure 7.2).

- Remove caps from tubes by grasping them with the smallest two fingers of your dominant hand and using these fingers to hold the cap while inoculating. Tight fitting or screw-on caps may have to be loosened before beginning this procedure (Figure 7.3).

- The neck of open tubes should be sterilized immediately after removing the cap and immediately prior to replacing it. This can be accomplished by passing the tube through the flame two or three times. Open tubes should always be held at an angle to minimize the possibility of airborne contamination (see Figure 7.3).

- Prior to transfer, inoculating loops and needles must be allowed to cool for 10–15 sec. If a hot inoculating tool is plunged into media containing bacteria, an aerosol may be created, releasing bacteria into the air.

Figure 7.2 The hottest portion of a Bunsen burner's flame is found at the edge of the inner blue cone. Inoculating loops and needles should be sterilized in this area. The area outside the cone is slightly cooler but is acceptable for flaming of tubes during inoculations.

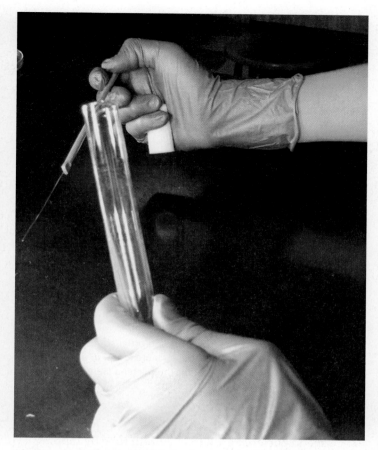

Figure 7.3 The last two fingers of your dominant hand should be used to hold the cap of any tube you are working with. The remaining three fingers are used to manipulate the inoculating tool.

- When transferring organisms to or from a tube, hold the inoculating tool steady and move the tube. Reducing the movement of the loop or needle lessens the possibility of bacteria ending up where they are not wanted.
- When transferring organisms to or from a Petri dish, use the lid to protect against airborne contamination (Figure 7.4).

Specific Transfer Methods

Because the same aseptic techniques are applicable in virtually all transfers, specific procedures will vary little from situation to situation. The most important aspect is the choice of inoculating tool, with a loop being used when bacterial growth is being **removed from a liquid culture** and a needle being used when growth is **removed from a solid culture,** such as a slant or plate. Notice that the media to which growth will be transferred is not usually a concern. Although abridged step by step instructions appear below, be sure to use proper aseptic technique as outlined above.

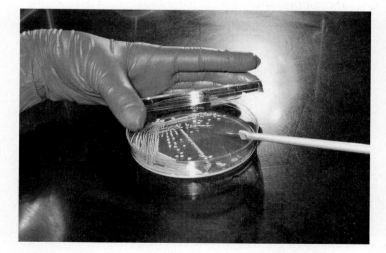

Figure 7.4 When working with a Petri dish, the lid should be used as a shield to protect against the incorporation of unwanted organisms.

Figure 7.5 Gently mix broth cultures in order to suspend the organisms within. On the left, the bacteria are found in a mass on the bottom of the tube while on the right the bacteria are well distributed throughout the media.

Obtaining Growth from a Broth

1. Suspend bacterial growth in a broth tube by tapping the tube with your fingers until a vortex of cells is seen swirling in the tube (Figure 7.5).
2. Flame an inoculating loop, and allow it to cool.
3. Remove the cap from the broth tube, and flame the neck of the tube.
4. Holding the loop still, move the tube up and around the loop until the loop is submerged in the broth. Carefully withdraw the tube from the loop.
5. Flame the neck of the tube, and replace the cap.
6. Transfer the bacteria to the intended media.

Obtaining Growth from a Slant

1. Flame an inoculating needle, and allow it to cool.
2. Remove the cap from the slant, and flame the neck of the tube.
3. Holding the needle still, move the tube up and around the loop until it is positioned above the growth. Carefully touch the needle to the bacteria obtaining a tiny (often invisible) quantity of growth. Withdraw the tube from the needle (Figure 7.6).
4. Flame the neck of the tube, and replace the cap.
5. Transfer the bacteria to the intended media.

Obtaining Growth from an Agar Plate

1. Flame an inoculating needle, and allow it to cool. Touching the needle briefly to an uninoculated portion of the plate will ensure that it is cool.

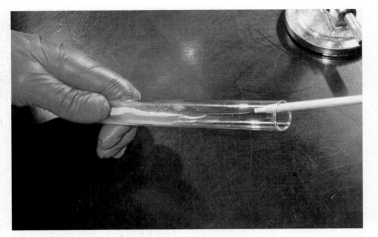

Figure 7.6 A needle is used to remove a small quantity of bacterial growth from a slant.

2. Remove the cover from an agar plate, but hold it over the plate to act as a shield against airborne contamination (see Figure 7.4).
3. Carefully touch the needle to an isolated bacterial colony, obtaining a tiny (often invisible) quantity of growth. Withdraw the needle from the plate, and replace the cover.
4. Transfer the bacteria to the intended media.

Inoculating an Agar Slant with a Fishtail Streak

1. Remove the cap from the slant, holding it with the smallest finger of your dominant hand.
2. Flame the neck of the tube.
3. Keeping the tube tilted (to guard against airborne contamination), slide the tube up and around the needle or loop. The surface of the agar inside the tube should be facing upward.
4. Touch the needle or loop to the surface of the agar, and while moving the tool in a zigzag pattern, slowly withdraw the tube (Figure 7.7).

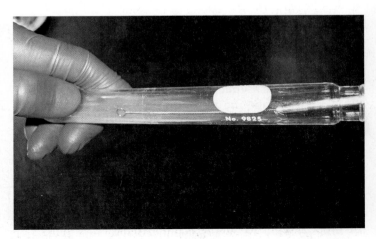

Figure 7.7 A loop is used to inoculate an agar slant with a broth culture.

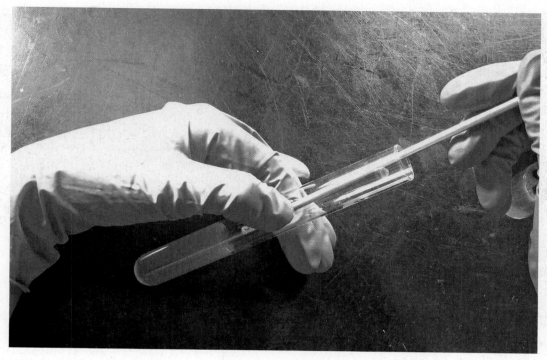

Figure 7.8 A needle is inserted nearly to the bottom of the media when inoculating a deep. Note that the tube is held nearly parallel to the work surface, which reduces the chance of contamination.

5. Flame the neck of the tube, and replace the cap.
6. Sterilize the inoculating tool in the flame of the Bunsen burner.

Inoculating an Agar Deep

1. Remove the cap from the deep, holding it with the smallest finger of your dominant hand.
2. Flame the neck of the tube.
3. Keeping the tube tilted (to guard against contamination), slide the tube up and around the needle.
4. Insert the needle into the agar until it is about 1 cm from the bottom of the tube (Figure 7.8). Carefully remove the needle from the agar so that it follows the same pathway exiting the agar as it did entering it.
5. Flame the neck of the tube, and replace the cap.
6. Sterilize the inoculating needle in the flame of the Bunsen burner.

Inoculating a Broth

1. Remove the cap from the broth tube, holding it with the smallest finger of your dominant hand.
2. Flame the neck of the tube.
3. Keeping the tube tilted (to guard against airborne contamination), slide the tube up and around the needle or loop until it is submerged in the broth. Swirl the inoculating tool several times to ensure the transfer of the cells.
4. Slide the tube away from the inoculating tool. If using a loop, touch it to the side of the tube to remove as much

broth as possible (Figure 7.9). This reduces the chance of aerosol formation when the loop is flamed and the chances of a dripping culture contaminating your work surface.
5. Flame the neck of the tube, and replace the cap.
6. Sterilize the inoculating tool in the flame of the Bunsen burner.

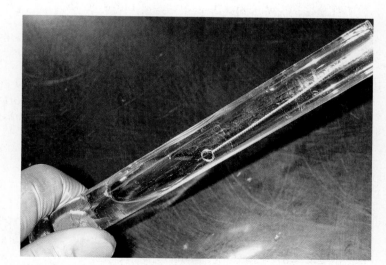

Figure 7.9 When using a loop to sample a liquid culture, touching the loop to the inner surface of the tube momentarily will remove excess broth, reducing the chances of a dripping culture contaminating your work surface and the formation of aerosols when the loop is flamed.

RESOLUTION OF THE CASE

DNA fingerprinting of the isolates from patients 1 and 2 revealed that they were identical, indicating one of three possibilities: (1) both patients were infected through contact with a common source, (2) one patient infected another, or (3) laboratory error resulted in one specimen being contaminated with another. A detailed patient history revealed that patients 1 and 2 had no previous contact with one another and had no potential sources of exposure in common. Records from the medical laboratory showed that the sample from patient 1 was processed immediately after the sample from patient 2, who had clinically obvious TB. Based on this information, the Division of Health concluded that laboratory error led to cross contamination between the two specimens.

The other four incidents followed a similar script, with easily preventable laboratory error being to blame in each case. In one instance, poor handling of a test culture containing *M. tuberculosis,* was allowed to cross contaminate several other samples, resulting in seven false positive results. Ironically, the test sample was regularly included to assess laboratory proficiency and quality control of automatic testing machinery.

These five incidents were responsible for a total of eleven false positive results. By the time the results were recognized as erroneous, eight patients had received unnecessary medical treatment including hospitalization in respiratory isolation, bronchoscopy, or anti-TB medication. A total of 108 of the patients, family and social contacts were tested for tuberculosis, with all 108 testing negative. In addition, 328 hospital employees and patients were tested for tuberculosis, using either skin tests or chest x-rays, again with no positive results. It was estimated that local and state health department staff expended 240 person-hours and hospital staff 330 person-hours as a result of these episodes.

EXERCISE 7 REVIEW QUESTIONS

1. In the case seen here, how did contamination most likely occur?

2. As part of their quality control procedures, most laboratories test multiple samples from a single patient. How would this strategy help prevent false positives from being reported?

3. *Mycobacterium* is able to survive long periods on inanimate objects. How would this property increase the tendency of the bacterium to be a troublesome contaminant in the laboratory?

4. Describe how a bacterial sample would be obtained from and inoculated into each of the following types of media.

Agar slant:

Agar plate:

Broth:

REFERENCE

CDC. 1997. Multiple misdiagnoses of Tuberculosis resulting from laboratory error—Wisconsin 1996. *Morbidity and Mortality Weekly Report,* 46(34) 797–801.

NOTES

Pure Culture Techniques

CASE SYNOPSIS

Neonatal Tetanus—Montana, 1998

On March 21, 1998, a 9-day-old newborn with no previous medical history was evaluated in an emergency room. Her parents indicated the baby had not nursed in 10 h and had difficulty opening her mouth. They also noticed a foul-smelling discharge emanating from the umbilical stump over the previous 1–2 days. Doctors noted that the infant exhibited trismus as well as generally increased muscle tone. Based on these clinical characteristics, the infant was diagnosed with neonatal tetanus.

Resolution of the Case appears on page 76

CDC. 1998. Neonatal tetanus—Montana, 1998. *Morbidity and Mortality Weekly Report,* 47: 928–930.

STUDENT LEARNING OUTCOMES

After completing this exercise, you should be able to:
1. Correctly isolate a pure bacterial culture from a mixed culture using appropriate techniques.

INTRODUCTION

It is an unfortunate fact of life that bacteria must be studied as a single species but are rarely found this way in the environment. In the real world, whether it is the normal flora of the body, a wound, sewage, soil, or water, bacteria are found in mixed populations made up of many different species. Properly studying a single species requires, however, that it be isolated from other species in the population to obtain a **pure culture.** Robert Koch, one of the earliest medical microbiologists, knew that if he wanted to prove that a specific disease was due to infection with a specific microbe, he would have to isolate the bacteria in question prior to any meaningful characterization. Techniques developed in his laboratory to produce pure cultures are still routinely used today, and mastery of these techniques is essential to success in the laboratory.

Although many methods can be used to obtain a pure culture, all have the same theoretical foundation. In a successful separation, one cell in a mixed population of bacteria will be separated from all others and immobilized atop or within a solid growth media. As this separated cell continues to reproduce over many generations, it will give rise to a single **colony** containing hundreds of thousands of cells, all of which are derived from a single progenitor (Figure 8.1). Each colony can now be considered a pure culture and can be used for further study of the bacterium. In this exercise, three different techniques are provided to accomplish the isolation; your instructor will tell you which procedures you will perform.

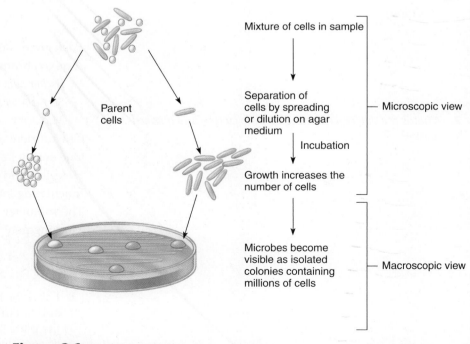

Figure 8.1 Regardless of the method used, all pure culture techniques depend on separating individual cells of a culture from one another. Once separated, each cell will produce a colony containing many hundreds of thousands of cells, all of which are descendents of the single isolated cell.

PRE-LAB QUESTIONS

1. *Define the following terms:*

Trismus

Aerobic

Anaerobic

Sequelae

Colony

2. *What is the difference between a pure culture and a mixed culture?*

3. *How can a pure or mixed culture become contaminated? How can this be prevented?*

PERIOD ONE

Three methods are described here: the streak-plate method, the loop dilution method, and the spread-plate method.

I. STREAK-PLATE METHOD

This method is the most economical in terms of time and materials, requiring just a few minutes and only a single plate of media. Its main drawback is that a certain degree of skill is required, which takes time to fully develop. The streak-plate method of isolation is outlined in Figure 8.2.

MATERIALS

Each group should obtain:
A mixed broth culture containing:
   Escherichia coli
   Serratia marcescens
   Micrococcus luteus
One nutrient agar pour
One empty Petri dish
Hot plate and beaker
Inoculating loop
Thermometer
Marking pen

PROCEDURE

1. Label the bottom of your Petri dish with your name and lab time. Place the label around the periphery of the plate so your view of the bacterial colonies will be as complete as possible after the incubation.

2. Place the agar tube into a beaker containing enough water to cover the agar in the tube. Bring the water in

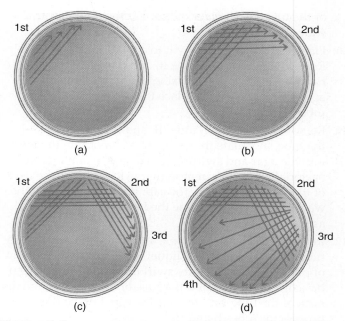

Figure 8.2 (a) In a streak plate, a single loopful of a culture is spread four or five times in the first quadrant. (b) The loop is flamed and allowed to cool for about 10 sec. Touching the loop to an uninoculated area of the plate will ensure that it is cool. Once cool, five or six streaks are made from quadrant 1 into quadrant 2. (c) Following flaming and cooling of the loop, six to seven streaks are made from quadrant 2 into quadrant 3. (d) After flaming and cooling once more, several streaks are made from quadrant 3, using up all of the uninoculated space on the plate. Finally, the loop is flamed before being placed aside.

the beaker to a boil and allow 5 min for the agar to liquefy. Remove the beaker from the hot plate and allow the agar to cool to 50°C. Adding fresh water to the beaker will allow it to cool faster.

3. Remove the cap and flame the neck of the tube. Carefully pour the liquefied agar into the bottom of the Petri plate, swirling gently if needed to ensure that the media covers the entire bottom of the plate. Allow 10 min for the plate to fully solidify.

4. Carefully agitate the tube containing the mixed culture until the bacteria is suspended in the media.

5. Flame the loop to red hot, and allow it to cool. Remove the cap from the culture tube, and while holding the cap with the pinky finger of your dominant hand, flame the neck of the tube.

6. Remove a single loopful of broth from the tube.

7. Flame the neck of the tube and replace the cap.

8. Streak the organism onto the plate as shown in Figure 8.2. Hold the Petri dish cover over the plate to guard against airborne contamination as you work. Use as little pressure as possible to avoid gouging the medium. Begin by applying four or five parallel streaks in the first quadrant.

9. Flame the loop, and allow it to cool. Touching the loop briefly to an uninoculated area of the plate will ensure that it has cooled sufficiently. Make five to six streaks extending from the first quadrant to the second. Flame the loop, and allow it to cool.

10. Make six to seven streaks extending from the second quadrant to the third. Flame the loop, and allow it to cool.

11. Make as many streaks as is practical beginning in the third quadrant and extending outward. Use as much of the uninoculated area of the plate as possible but do not allow the loop to enter either of the first two quadrants.

12. Flame the loop before placing it aside.

13. Incubate the plate in an inverted position at 30°C for 48 h. Inverting the plates prevents condensation (which may have accumulated on the lid of the plate) from dropping onto the agar and causing the organisms to spread over the agar surface, ruining the entire isolation procedure.

II. LOOP DILUTION (POUR PLATE) METHOD

This method consumes more time and materials than does a streak plate, but the results produced, even by the beginning student, are generally quite good. The loop dilution method of isolation is outlined in Figure 8.3.

MATERIALS

Each group should obtain:
A mixed broth culture containing:
   Escherichia coli
   Serratia marcescens
   Micrococcus luteus
Three nutrient agar pours
Three empty Petri dishes
Hot plate
Inoculating loop
Thermometer

PROCEDURE

1. Label the bottom of your Petri dishes with your name and lab time. Also label each dish I, II, or III. Place the label around the periphery so your view of the plate will be as complete as possible after the incubation.

2. Label the agar tubes I, II, and III, and place them into a beaker containing enough water to cover the agar in the tubes. Bring the water in the beaker to a boil, and allow 5 min for the agar to liquefy. Remove the beaker from the hot plate, and allow the agar to cool to 50°C. Adding fresh water to the beaker will allow it to cool

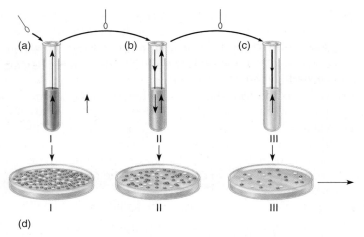

Figure 8.3 (a) In a loop dilution, a single loopful of a culture is used to inoculate a tube of melted agar that has been cooled to 50°C. The loop is used to stir the media so that bacteria cells are spread evenly throughout the tube. (b) After flaming the loop and allowing it to cool for about 10 sec, a loopful of media from tube I is removed and used to inoculate tube II. Again, the loop should be used to distribute the bacteria throughout the tube. (c) The loop is flamed and allowed to cool once again, and then used to transfer a loopful of media from tube II to tube III. Flame the loop before setting it down. (d) Once all three tubes have been inoculated, flame the neck of tube I and pour the agar into the bottom of the first Petri dish, carefully swirling the dish to ensure that the media covers the entire bottom of the plate. Do the same for tubes II and III.

faster. Leave the melted agar tubes in the warm water as this will maintain them at 50°C.

3. Carefully agitate the tube containing the mixed culture until the bacteria is suspended in the media.

4. Flame the loop to red hot, and allow it to cool. Remove the cap from the culture tube, and while holding the cap with the pinky finger of your dominant hand, flame the neck of the tube.

5. Remove a single loopful of broth from the tube.

6. Flame the neck of the tube, and replace the cap.

7. Remove the cap from tube I, and flame the neck of the tube. Inoculate the media by submerging the loop into the agar and gently swirling it to dislodge the bacterial cells. After removing the loop, flame the neck of the tube and replace the cap. Place the tube back in the water-filled beaker.

8. Flame the loop and allow it to cool for 5–10 sec. Remove the cap from tube I, and flame the neck of the tube. Carefully remove a single loopful of inoculated agar. Flame the tube, replace the cap, and return the tube to the beaker.

9. Remove the cap from tube II, and flame the neck of the tube. Inoculate the media by submerging the loop into the agar and swirling it to dislodge and mix the bacterial cells. After removing the loop, flame the neck

of the tube and replace the cap. Return the tube to the water-filled beaker.

10. Flame the loop and allow it to cool for 5–10 sec. Remove the cap from tube II, and flame the neck of the tube. Carefully remove a single loopful of inoculated agar. Flame the tube, replace the cap, and return the tube to the beaker.

11. Remove the cap from tube III, and flame the neck of the tube. Inoculate the media by submerging the loop into the agar and swirling it to dislodge and mix the bacterial cells. After removing the loop, flame the neck of the tube and replace the cap. Return the tube to the water-filled beaker. Flame the loop before placing it aside.

12. Remove the cap from tube I, and flame the neck of the tube. Carefully pour the liquefied agar into the bottom of plate number I. Gently swirl the plate until the agar has completely covered the base of the plate. Flame the neck of the tube, and replace the cap. Repeat this process for tubes II and III.

13. After the media has completely solidified, incubate the plates in an inverted position at 30°C for 48 h.

III. SPREAD-PLATE METHOD

This method gives consistently reliable results when bacterial samples are dilute or if the medium being inoculated is highly selective, allowing only a limited number of cells to grow. If a culture has an abundance of bacterial growth, it should be diluted with sterile water prior to spreading, or a different technique should be used. The spread-plate method of isolation is outlined in Figure 8.4.

MATERIALS

Each group should obtain:

A previously diluted mixed broth culture containing:

Escherichia coli

Serratia marcescens

Micrococcus luteus

One nutrient agar plate

Approximately 50 ml of ethyl or isopropyl alcohol in a 250-ml beaker

A large beaker that will completely cover the small beaker and spreading rod

Bent glass spreading rod

One sterile dropper or pipette

Sterile water

Marking pen

Petri plate turntable (optional)

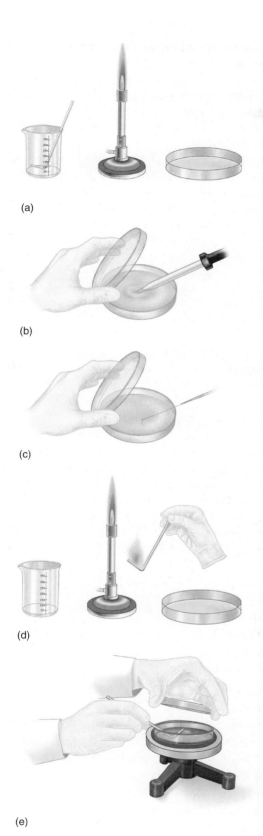

PROCEDURE

A Safety Note: Should the alcohol in the beaker catch fire place the larger beaker over the alcohol containing beaker and spreading rod. When the oxygen is exhausted, the fire will extinguish itself.

1. Place the spreader rod in the beaker, and add enough ethanol to the beaker to cover the lower portion of the rod.

2. Arrange the items on your bench so that the Bunsen burner is between the alcohol and the Petri dish. This reduces the chance of accidentally setting the alcohol aflame.

3. Mark the bottom of your Petri dishes with your name and lab time. Place the label around the periphery so your view of the plate will be as complete as possible after the incubation.

4. Carefully agitate the tube containing the diluted mixed culture until the bacteria is suspended in the media.

5. Place a single drop of sterile water in the center of the plate. Flame your loop, and retrieve a single loopful of the dilute culture. Add this to the drop of water on the plate. Flame the loop prior to placing it aside.

6. Holding the upper end of the spreader rod, remove it from the alcohol and pass it through the flame of the Bunsen burner, allowing the alcohol to ignite. Be sure to keep your hand above the spreader so that the flaming alcohol doesn't run onto your hand. Wait until the alcohol has completely burned off before continuing.

Figure 8.4 (a) When preparing to isolate bacteria using a spread plate, arrange your bench so that the Bunsen burner is between the alcohol and the Petri dish. This minimizes the chances of setting fire to the alcohol. (b) Use a Pasteur pipette to add a single drop of sterile water to the surface of the media. (c) Flame your loop and allow it to cool for about 10 sec, then use it to inoculate the water drop with a loopful of the mixed culture. (d) Remove the glass spreading rod from the alcohol and pass it through the flame, allowing the alcohol to ignite. Remove the rod from the flame and wait for the alcohol to burn off completely. (e) Place the plate on a turntable and rotate the turntable while sliding the spreader back and forth across the surface of the media. If a turntable is not available, the plate can simply be rotated on the benchtop. In either case, finish the inoculation by rotating the plate a final turn while holding the spreader against the edge of the plate. After inoculating the plate, return the spreader to the alcohol; there is no need to flame it again.

7. Place the lower portion of the spreader flat against the agar plate. Rotate the plate using your thumb and middle finger while moving the spreader back and forth. If a turntable is used, gently spin the turntable while moving the spreader back and forth. In both cases, finish by rotating the plate one complete revolution while holding the spreader against the edge of the plate.

8. Return the spreader to the alcohol. There is no need to flame it again.

9. Allow the plate to sit upright for 5 min, then incubate it in an inverted position at 30°C for 48 h.

QUESTIONS—PERIOD ONE

1. What is the importance of generating isolated bacterial colonies?

2. Draw the results you would expect to see in a well-done streak plate.

3. Draw the results you would expect to see in a well-done loop dilution.

4. Draw the results you would expect to see in a well-done spread plate.

5. What is a subculture?

PERIOD TWO—RESULTS
Retrieve your plates from the incubator. To make differentiation easier, each bacterium used in this exercise exhibits a unique color. *Escherichia coli* produces off-white growth, *Micrococcus luteus* produces yellow growth, and *Serratia marcescens* produces red growth. Rarely, if ever, will you encounter a mixed culture with such vibrant colors outside the laboratory, but here the colors will help you to more fully evaluate the success of your separation. Regardless of the method used, a successful outcome will always be marked by many well-isolated colonies. If your plate looks beautiful, all the better, but as long as several well-isolated colonies are present, the plate can be judged a success.

Record the appearance of your plates for each separation technique you employed. Critically analyze your plate(s), rating each on a scale of 1 (unusable) to 10 (Microbiology Hall of Fame). For anything less than a 10, indicate how the plate could have been improved.

Streak plate:

Loop dilution:

Spread plate:

SUBCULTURING BACTERIAL ISOLATES

Although isolated colonies may be used directly from the Petri dishes on which they were isolated, in most cases, it is advisable to transfer, or subculture, an isolated colony to its own container of media. This allows for easier study and organization of bacterial isolates. Today you will subculture each of your isolated bacterial species to nutrient agar slants. This procedure is illustrated in Figure 8.5.

MATERIALS

Each group should obtain:
Bacterial plates containing isolated colonies
Three nutrient agar slants
Bunsen burner
Inoculating needle
Marking pen

PROCEDURE

1. Using any combination of plates, identify one well-isolated colony for each bacterial species. On the bottom of the Petri plate, circle the colony so as to make it easier to find when you are ready to inoculate.
2. Label each agar slant with its respective bacteria, *E. coli*, *S. marcescens*, or *M. luteus*.
3. Flame the inoculating needle completely, including 2–3 cm of the handle, and allow it to cool.
4. Open the Petri dish containing a well isolated *E. coli* colony, and touch the tip of the needle to the colony you selected earlier. Replace the cover on the Petri dish.

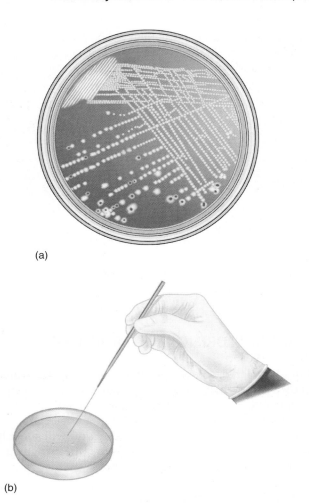

(a)

(b)

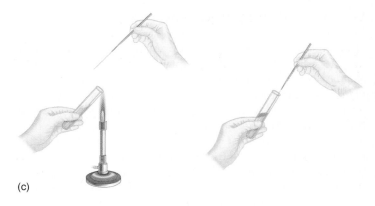

(c)

Figure 8.5 (a) For each species of bacteria, identify a single well-isolated colony. (b) Flame your inoculating needle and allow it to cool. Touch the needle to the isolated colony, picking up a minute amount of bacterial growth. (c) Flame the neck of the nutrient agar tube and inoculate the tube by drawing the needle along the surface of the agar, from bottom to top, moving the needle back and forth as it ascends up the surface of the slant. Flame and recap the tube, and flame the needle before setting it aside. Repeat this process for each of the other two species.

5. Remove the cap from the agar tube labeled *E. coli* and flame the neck of the tube.
6. Inoculate the slant using a fishtail streak. Flame the neck of the tube, and replace the cap. Flame the inoculating needle prior to setting it aside.
7. Repeat steps 4 through 6 for the other two bacteria.
8. Incubate the slants at 30°C for 48 h.

QUESTIONS—PERIOD TWO

1. Sketch the results you would expect to see for three well isolated subcultures.

3. Condensation often gathers in the bottom of agar slants. Why is it important in this exercise to limit condensation on plates but not on slants?

PERIOD THREE
Retrieve your slants from the incubator, and record their appearance, rating them as you did your plates.

2. How would a subculture appear if a colony containing both *S. marcescens* and *M. luteus* was subcultured to a slant?

RESOLUTION OF THE CASE

Neonatal tetanus is caused by toxins produced by *Clostridium tetani,* a bacterium found in high concentration in soil and animal feces. The bacterium is an obligate anaerobe, capable of growth only in the absence of oxygen, which often occurs in injuries where many species of bacteria are present at the site of a wound. In these types of cases, aerobic bacteria deplete the tissues of oxygen, resulting in an anaerobic environment, an ideal condition for the growth of *Clostridium.* Although more than 250,000 deaths a year are attributable to neonatal tetanus worldwide, very few cases occur in the United States as vaccination against tetanus provides both mother and baby ample protection against the bacterium. In this case, however, the mother, because of her philosophical beliefs, had never been vaccinated against tetanus and therefore afforded her baby no protection against the disease. Furthermore, the family lived in a rural area adjacent to a horse pasture, and the family dog often ran between the house and the pasture.

After being delivered by cesarean section, the baby received standard umbilical cord care with isopropyl alcohol prior to being discharged from the hospital at 3 days of age. Once home, the parents applied "Health and Beauty" clay powder to the umbilical cord up to three times a day. The clay was bentonite clay from Death Valley, California, and its manufacturing process did not include sterilization. The clay was originally packaged and sold in 2-lb containers that were divided into smaller containers and distributed to local midwives. The midwives would then further divide the clay and distribute it to their patients with instructions to apply the clay to the umbilical cord with the belief that it would accelerate the drying of the cord.

Isolation of organisms from the umbilical cord yielded pure cultures of the anaerobes *Clostridium perfringens, C. sporogenes,* and *C. tetani* as well as several genera of aerobic (*Staphylococcus, Streptococcus,* and *Bacillus*) bacteria. Presumably, the metabolic activity of the aerobic bacteria created an environment in which *C. tetani* was able to flourish, producing the toxins that led to tetanus. A 10-day course of tetanus immune globulin and antibiotics (penicillin G) resolved the infection, although mechanical ventilation was required for 12 days. The baby was discharged on April 10, 1998, with no apparent neurological sequelae and was developing normally when she was reexamined at age 7 months.

EXERCISE 8 REVIEW QUESTIONS

1. Which separation method is not appropriate for use with cultures containing a great deal of bacterial growth? Why is this method not a good choice for these conditions?

2. Why is agar cooled to 50°C prior to being inoculated with bacteria? What would happen if the agar were significantly warmer or cooler when inoculated?

3. How could you identify a potential contaminant on a streak plate? A pour plate? A spread plate?

4. How would any of the isolation techniques seen in this laboratory be affected by the use of a selective medium?

5. As is often the case in microbiology, many factors contribute to the progression of a case. For each aspect of the case listed, explain its significance.

The mother's vaccination status

Application of Health and Beauty clay powder to the umbilical cord

The horse pasture adjacent to the house

The family dog

Aerobic bacteria at the umbilical cord

C. tetani

REFERENCE

CDC. *Tetanus (Lockjaw) Vaccination.* http://www.cdc.gov/vaccines/vpd-vac/tetanus/default.htm.

Simple Staining, Negative Staining, and Gram Staining

CASE SYNOPSIS

Identification of Bacteria Responsible for the Outbreak of Gastrointestinal Disease

Late in 2006, public health officials in multiple states with the assistance of the Centers for Disease Control and Prevention and the U.S. Food and Drug Administration began investigating a large multistate outbreak of severe gastrointestinal disease. Initial reports of the symptoms associated with the illness (with symptoms primarily of diarrhea and vomiting) and the foods eaten by those who fell ill pointed toward one of four common bacterial species as the responsible agent. Fecal samples of ill persons as well as samples of the epidemiologically implicated foods were collected for analysis. Initial work focused on microscopic examination of the four suspect species: *Bacillus cereus, Staphylococus aureus, Salmonella sp.,* and *Listeria monocytogenes.* Public health officials had been working to identify the source of infections for several months and were hoping that the scope and breadth of the epidemic could be more precisely defined by eliminating cases that were clearly tied to an etiologic agent different from that seen in the majority of cases.

Resolution of the Case appears on page 87

CDC. 2007. Salmonellosis—outbreak investigation, February 2007. www.cdc.gov/ncidod/dbmd/diseaseinfo/salmonellosis_2007/outbreak_notice.htm.

CDC. 2002. Public health dispatch: Outbreak of listeriosis—Northeastern United States, 2002. *Morbidity and Mortality Weekly Report,* 51: 950–951.

CDC. 1997. Outbreak of staphylococcal food poisoning associated with precooked ham—Florida, 1997. *Morbidity and Mortality Weekly Report,* 46: 1189–1191.

CDC. 1994. *Bacillus cereus* food poisoning associated with fried rice at two child day care centers—Virginia, 1993. *Morbidity and Mortality Weekly Report,* 43: 177–178.

STUDENT LEARNING OUTCOMES

After completing this exercise, you should be able to:

1. Prepare a bacterial smear from both solid and liquid cultures.
2. Properly stain bacterial cultures using simple-, negative-, and Gram-staining.

INTRODUCTION

At its most basic, microbiology deals with the differences betweens cells. Microscopic examination is used to determine a cell's size and shape as well as the often characteristic arrangement the cells of a given species assume as they multiply. The initial classification of an isolate is absolutely dependent upon the cells' shape, arrangement, and staining characteristics, although complete identification is not possible from these aspects alone.

Cells of a given species all generally have the same shape, or **morphology.** Bacterial cells are classified as **cocci** (singular, coccus) if they are spherical in shape, rods or **bacilli** (singular, bacillus) if they are elongated, and **spirilla** (singular, spirillum) if they are spiral shaped. Although division based on shape holds true in most instances, variations on this theme do exist. Short curved rods (vibrios), cells intermediate between rods and cocci (coccobacilli), and flexible spiral bacteria (spirochetes) are all seen. **Pleomorphic** bacteria exhibit a variety of shapes, even in the same sample.

Cellular arrangement, the manner in which bacterial cells associate with one another, is a by-product of the manner in which a cell divides and whether the cells stay attached after division. Cocci produce more arrangements than rods. If a coccus divides along a single plane and the cells remain attached after division, a **diplococcus** is formed. If the cells continue to divide in this way without separating, a **streptococcal** arrangement results. If, instead, a second division takes place perpendicularly to the first, a four-cell group called a **tetrad** is produced while a third division (perpendicular to the first two) results in a cuboidal packet of eight cells called a **sarcina.** If the planes of division are not at right angles to one another, an irregular cluster of cells called a **staphylococcal** arrangement results (Figure 9.1). Rod-shaped bacteria divide only along the transverse plane (i.e., the shortest possible distance across the cell), producing the only arrangements seen in the bacilli, **diplobacilli, streptobacilli,** and **palisades** (Figure 9.2). Spiral-shaped bacteria rarely remained attached after division and consequently are seen only as single cells. In most cases, a sample will have multiple arrangements visible, for instance, diplococci, staphylococci, and tetrads. In these cases, the most prevalent

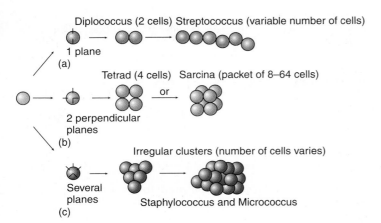

Figure 9.1 Arrangement of cocci resulting from different planes of cell division. (a) Division in one plane produces diplococci and streptococci. (b) Division in two planes at right angles produces tetrads and packets. (c) Division in several planes produces irregular clusters.

arrangement tends to be emphasized because it usually represents the end result of the growing bacterial grouping.

Bacterial cells are essentially transparent, and a careful examination of them involves the same problems as a careful examination of ice cubes in a glass of water. Because the cells and their background do not contrast well, they are difficult to visualize. Staining techniques in the microbiology laboratory have been designed to increase the **contrast** between bacterial cells and their background. These

techniques either obliterate the background, leaving the cell visible as a clear organism against a darkened microscopic field (negative staining) or stain the cell so that it appears as a darkly colored object against a light background (simple staining). Both types of staining are seen in Figure 9.3. Negative staining uses negatively charged dyes such as **nigrosin** or **India ink,** both of which are repelled by the negative charge of the bacterial cell but are attracted to the glass slide. In simple staining, any positively charged, or basic, stain is used to stain the negatively charged bacterial cell. Common basic stains in the microbiology laboratory include **methylene blue, crystal violet,** malachite green, and **safranin** (which is pink). Although simple and negative staining increase the contrast between a cell and its background, making it easier to determine the size, shape, and arrangement of a particular bacterium, they provide little information beyond this. One advantage seen with negative staining is a more accurate determination of the size of a bacterial cell. When preparing a cell for simple staining, heat is applied to the bacterial sample, both to kill the cells and to affix them to the slide. An unwanted consequence of heating is that cells generally shrink. Because negative staining uses no heat, cell shrinkage is minimized or eliminated.

A **differential stain,** as the name implies, allows the discrimination of one cell from another based on differential-staining properties. These stains provide not just morphological information such as size, shape, and arrangement but also give a clue as to the biology of the cell by highlighting structural differences that exist between groups of bacteria. Although several types of differential stains exist, the most widely used, and in fact the most important stain in all of microbiology, is the **Gram stain,** developed in 1884 by the physician Hans Christian Gram. This technique separates bacteria into two groups, Gram positive and Gram negative, based on differences in the structure of the cell wall. The steps

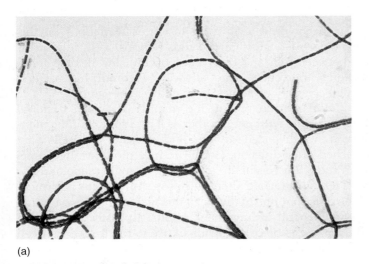

(a)

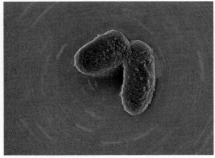

(b)

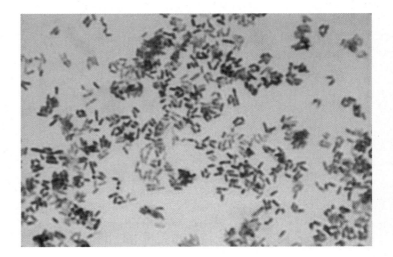

(c)

Figure 9.2 Arrangements of rod-shaped bacteria. (a) *Bacillus anthracis* displays a chain of cells (1000x). (b) "Snapping" of *Corynebacterium diptheriae* (28,000x) cells leads to (c) a palisades arrangement (800x).

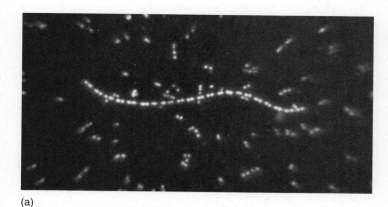

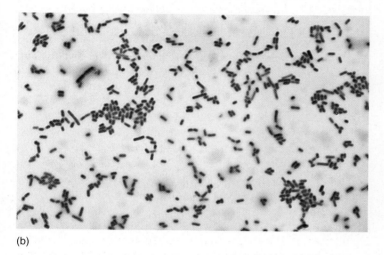

Figure 9.3 (a) In negative staining, the background is stained but the cells are not, resulting in light organisms against a dark field (1000x). (b) In simple staining, bacterial cells are stained directly and appear as dark colored cells against a light background (1000x).

in a Gram stain are outlined here, and the appearance of the cells at each point in the process are illustrated in Figure 9.4:

- Cells are stained with the **primary stain** crystal violet. Both Gram-positive and Gram-negative cells are colored a deep purple at this point.

- Iodine is added as a **mordant,** a chemical that serves to fix a dye in a staining process. In this case, the iodine binds with the crystal violet to create an insoluble complex within the thick peptidoglycan layer of Gram-positive cells.

- Cells are **decolorized** with ethyl alcohol. The alcohol dissolves the lipids that are found in the outer membrane of Gram-negative cells, allowing the crystal violet–iodine complex to escape. At this point, Gram-negative cells are colorless while Gram-positive cells are still dark purple.

- Safranin, a **counterstain,** is then added to increase the contrast of the colorless Gram-negative cells, rendering them pink. The Gram-positive cells are dyed pink as well, but the darker color of the crystal violet masks the lighter color of the safranin.

Reagent	Gram positive	Gram negative
None (Heat fixed cells)		
Crystal violet (30 sec)		
Gram's iodine (1 min)		
Ethyl alcohol (10–20 sec)		
Safranin (1 min)		

Figure 9.4 Color of cells after each step of the Gram-staining procedure.

In the end, those cells staining purple are designated **Gram positive** while those staining pink are deemed **Gram negative.** The ability to reliably Gram stain is absolutely essential to the correct identification of a bacterial species, and reliability comes about only with much practice. In addition, three main factors can affect the outcome of a Gram stain and should be remembered. (1) Young cultures should be used for Gram staining. Gram-positive cultures older than 24 h may sometimes stain Gram negative due to changes in the peptidoglycan layer of the cell wall that accompany aging. (2) Smears must not be too thick. Thick smears can entrap crystal violet so that it is not removed by the alcohol, leading to a false Gram-positive result. (3) Decolorization must be done for an appropriate period of time. Alcohol, if left on too long, will remove the crystal violet from even Gram-positive cells, leaving them pink. Conversely, if cells are not decolorized long enough, Gram-negative cells will be seen as purple, leading to a false impression that the cells are Gram positive.

PRE-LAB QUESTIONS

1. *Define the following terms:*

Morphology

Arrangement

Pleomorphism

Contrast

Mordant

4. *Complete the following table by providing the proper shape for each of the bacterial species listed. A look at the Howcharts in exercise 39 would be helpful.*

Species	Cellular morphology
Salmonella sp.	
Staphylococcus aureus	
Listeria monocytogenes	
Bacillus cereus	

5. *What two staining techniques are appropriate for determining the shape and arrangement of a bacterial species?*

2. *Draw and label the three most common bacterial shapes.*

6. *What stains can be used for negative staining, and why can these same stains not be used for simple staining?*

3. *Draw and label cocci growing as diplococci, tetrads, streptococci, and staphylococci.*

7. *What is the difference between a simple stain and a differential stain?*

PERIOD ONE

The bacteria you will be working with belong to the same genera and have the same characteristics as those seen in the earlier case above but are less virulent.

MATERIALS

Each group should obtain:

Broth cultures of:

Listeria innocua

Salmonella typhimurium

Slant cultures of:

Bacillus megaterium

Staphylococus aureus

Inoculating loop and needle

Methylene blue

Nigrosin and/or India ink

Wash bottle

Bibulous paper

PROCEDURE

Negative Stain

1. Place a small drop of nigrosin or India ink near one end of a clean microscope slide. Use a needle to add a small amount of *B. megaterium* to the ink, and agitate the cells in the dye to break up any clumps.

2. Use a second slide to spread the dye across the bottom slide, as shown in Figure 9.5

3. Allow the slide to air dry, then examine it under oil immersion.

4. Repeat this process on another slide to examine *S. aureus.*

Simple Stain

1. Prepare a smear of *L. innocua* by aseptically adding two loopfuls of the culture to the center of a clean slide. Be sure to flame sterilize your loop every time you enter the culture tube. Spread the culture over the slide, covering an area about the size of a quarter.

2. Allow the smear to air-dry completely. This should take 5–10 min.

3. Heat-fix the smear by passing the slide, with the smear on top, through the flame of your Bunsen burner three times (Figure 9.6).

4. Flood the smear with methylene blue for 1 min (Figure 9.7).

Place a single drop of nigrosin or India ink near one end of a slide. Use a needle to disperse a small amount of organism in the dye.

Place a second slide in front of the drop and move the slide backward until it touches the dye, spreading the dye across the trailing edge of the spreader slide.

Slide the spreader slide forward, dragging the suspension of organisms across the slide. This should result in a smear that is quite thick at one end and very thin at the other.

After allowing the slide to air dry, it may be examined under oil immersion.

Figure 9.5 Negative-staining procedure.

5. Gently wash the smear with water (3–4 sec).

6. Blot the slide with bibulous paper to remove excess water.

7. Examine the slide under oil immersion.

8. Repeat this process on a second slide to examine *S. typhimurium.*

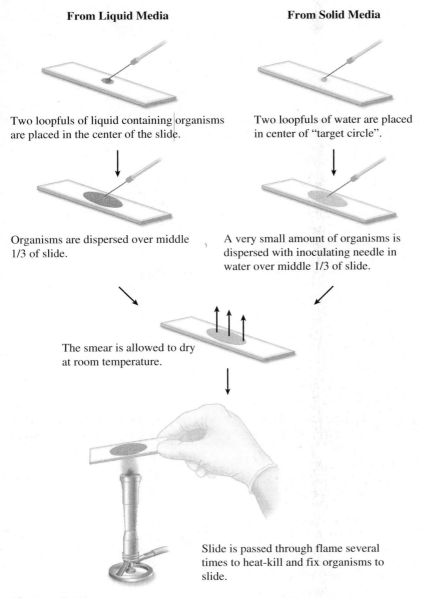

From Liquid Media

Two loopfuls of liquid containing organisms are placed in the center of the slide.

Organisms are dispersed over middle 1/3 of slide.

From Solid Media

Two loopfuls of water are placed in center of "target circle".

A very small amount of organisms is dispersed with inoculating needle in water over middle 1/3 of slide.

The smear is allowed to dry at room temperature.

Slide is passed through flame several times to heat-kill and fix organisms to slide.

Figure 9.6 Preparation and heat fixation of bacterial smears.

(1) A bacterial smear is stained with methylene blue for 1 min.

(2) Stain is briefly washed off slide with water.

(3) Water is carefully blotted off slide with bibulous paper.

Figure 9.7 Simple-staining procedure.

Listeria innocua

Staphylococcus aureus

Results

Make accurate drawings of each specimen, concentrating on its cellular morphology and arrangement.

Salmonella typhimurium

Bacillus megaterium

QUESTIONS—PERIOD ONE

1. Based strictly on cellular morphology, which bacterial species is most easily differentiated from the others? Why?

2. What does the "Gram reaction" of a cell refer to? What color represents Gram positive? Gram negative?

3. Research the bacteria seen in the Case Synopsis, and use the results to complete the table.

Species	Cellular morphology	Gram reaction
Salmonella typhimurium		
Staphylococcus aureus		
Listeria innocua		
Bacillus megaterium		

4. Identify each of the reagents used in a Gram stain and the amount of time that each is usually left in contact with the bacterial smear.

Primary stain

Mordant

Decolorizer

Counterstain

PERIOD TWO

In this period, you will perform Gram stains of each culture that was previously stained using a negative or simple stain.

MATERIALS

Each group should obtain:
Broth cultures of:
 Listeria innocua
 Salmonella typhimurium
Slant cultures of:
 Bacillus megaterium
 Staphylococus aureus
Two unknown cultures
Inoculating loop and needle
Gram-staining kit (crystal violet, Gram's iodine, 95% ethyl alcohol, safranin)
Wash bottle
Bibulous paper
Marking pen

PROCEDURE

Gram Stain

1. Label four slides with the initials of each of the bacterial species used in this exercise.

2. Prepare a heat-fixed smear of *L. innocua* by aseptically adding two loopfuls of the culture to the center of a clean slide. Be sure to flame sterilize your loop every time you enter the culture tube. Spread the culture over the slide, covering an area about the size of a quarter. Prepare a heat-fixed smear of *S. typhimurium* in the same manner.

3. Allow the smear to air-dry completely. This should take 5–10 min.

4. Heat-fix the smear by passing the slide, with the smear on top, through the flame of your Bunsen burner three times. When preparing a heat-fixed smear from bacteria growing as a solid culture, a suspension of cells is created by adding bacteria to a small drop of water on the slide.

5. Prepare a heat-fixed slide of *B. megaterium* by first placing a single loopful of water on a slide.

6. Use a needle to aseptically add a minute amount of bacterial growth to the water. Spread the bacterial suspension over the center of the slide, covering an area about the size of a quarter. Allow the smear to air-dry completely. This should take 5–10 min.

7. Prepare a heat-fixed smear of *S. aureus* in the same manner.

8. Heat-fix the slides by passing it through the flame of your Bunsen burner three times.

9. Flood the slide with crystal violet for 30 sec (Figure 9.8)

10. Briefly (3–4 sec) wash the excess stain from the slide and drain off the excess water.

11. Cover the smear with Gram's iodine for 1 min.

12. Decolorize the smear with ethyl alcohol for no more than 20 sec. Allow the alcohol to flow over the slide until the runoff is colorless. This should take 8–12 sec in most cases.

13. Immediately rinse the slide thoroughly with water. This is necessary to stop the decolorizing effect of the alcohol.

14. Flood the smear with safranin for 1 min.

15. Wash gently with water to remove excess safranin.

16. Blot the slide with bibulous paper to remove excess water.

17. Examine the slide under oil immersion.

18. Repeat the Gram-staining process, being sure to properly prepare the smear, for any unknown organisms.

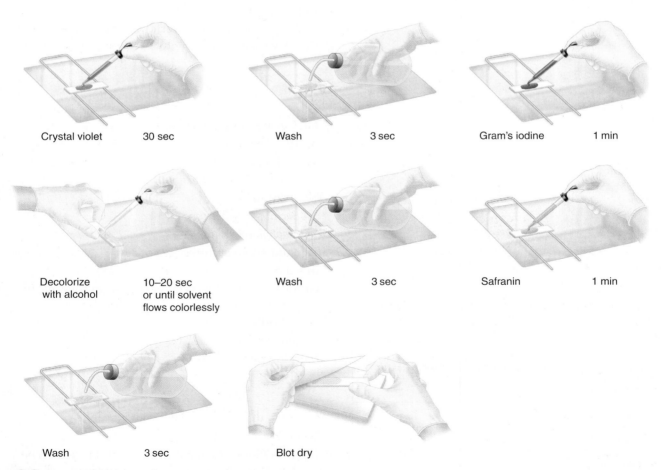

Crystal violet 30 sec	Wash 3 sec	Gram's iodine 1 min
Decolorize with alcohol 10–20 sec or until solvent flows colorlessly	Wash 3 sec	Safranin 1 min
Wash 3 sec	Blot dry	

Figure 9.8 Gram-staining procedure.

RESULTS

Record results for each of your slides, being sure that the drawings accurately represent the cellular morphology, arrangement, and Gram reaction for each sample.

Salmonella typhimurium   *Staphylococcus aureus*

Listeria innocua   *Bacillus megaterium*

Gram stain each of your unknown cultures and describe and sketch your results.

Culture # _____   Culture # _____

Can either unknown culture be conclusively identified as one of the four suspect bacteria based on the tests completed? If so, which one?

RESOLUTION OF THE CASE

Analysis of the suspect food and fecal samples taken from ill individuals implicated *Salmonella* serotype Tennessee as the infectious agent in the great majority of cases. This case was unusual as contaminated peanut butter was found to be the source of the outbreak. A total of 425 cases from 44 states were reported to the CDC. No deaths were attributed to the infection, but 20% of the patients for whom information was available were hospitalized. The FDA is working with peanut butter manufacturers to determine how the contamination could have occurred. Although this particular outbreak was traced to *Salmonella,* previous food poisoning outbreaks have been traced to all four of the bacteria featured in the Case Synopsis.

EXERCISE 9 REVIEW QUESTIONS

1. In this case, which two species could not be separated on the basis of morphology or Gram reaction?

2. If two species of bacteria differ morphologically with regard to size but have the same Gram reaction, which stain could best be used to differentiate the two species? Why?

3. Bacteria within the genus *Bacillus* produce a protective structure called an *endospore,* which can be detected using a special staining technique. How could an endospore stain be used to separate the two indistinguishable species?

4. Although cellular morphology can be determined using a Gram stain, many microbiologists will routinely perform a simple or negative stain on Gram-negative organisms. Why do you think this is?

5. Using the terms *cocci, rod, Gram positive, Gram negative, endospore former,* and *non-endospore former,* complete the accompanying table so that each bacterial species can be properly identified.

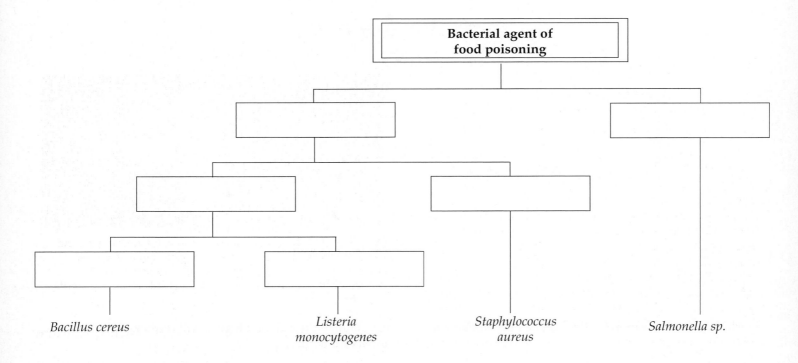

Capsular Staining

CASE SYNOPSIS

Pneumococcal Sepsis after Autosplenectomy—2005

A 15-year-old girl was admitted to the hospital after presenting at the emergency room (ER) in a semiconscious state. Feeling ill was nothing new for this patient—she had a 9-year history of systemic lupus erythematosus (SLE), a condition the ER physicians took into account as they examined her. SLE, sometimes called **lupus**, is an autoimmune disease in which the body produces antibodies against many of its own organs; some organs eventually become damaged or fail to function. The specific symptoms of SLE differ, depending on which organs are affected, but kidney failure, heart problems, lung inflammation, and blood abnormalities are common. The cause of SLE is unknown.

The patient's initial workup revealed abnormally rapid breathing, fever, and low blood pressure. Additionally, her fingers and toes were cold, and she was producing no urine. The ER staff took samples of her blood and cerebrospinal fluid (CSF) and found bacteria in both. Because of the patient's history of SLE, magnetic resonance imaging (MRI) of the abdomen was performed to assess the condition of her organs.

Resolution of the Case appears on page 93

Hühn, R., Schmeling, H., Kunze, C. and Horneff, G. 2005. Pneumococcal sepsis after autosplenectomy in a girl with systemic lupus erythematosus. *Rheumatology*, 44(12): 1586–1588.

STUDENT LEARNING OUTCOMES

After completing this exercise, you should be able to:

1. Perform a capsule stain and differentiate between encapsulated and nonencapsulated bacterial samples.

INTRODUCTION

Bacterial cells are regularly subjected to harsh environmental conditions and are often protected by a coating of macromolecules known as a **glycocalyx.** Although all cells have some sort of protective layer, when the thickness, composition, and organization of this layer reach a certain level of

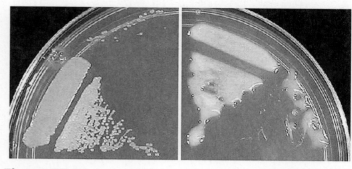

Figure 10.1 Colony appearance of nonencapsulated (left) and encapsulated (right) strains of the same bacterium. The capsule provides bacterial colonies with a mucoid appearance and is responsible for the bacteria's ability to adhere to catheters and other indwelling medical devices.

complexity, it is accorded the status of a cellular structure and referred to as a **capsule** (Figure 10.1).

Bacterial capsules are generally composed of repeating polysaccharide molecules, proteins, or both. The capsule adheres tightly to the cell and generally enhances the pathogenicity of bacteria such as *Streptococcus pneumoniae*, *Haemophilus influenzae*, and *Bacillus anthracis*. These bacteria are able to avoid being engulfed and destroyed by white blood cells known as phagocytes, leaving them free to multiply in the tissues. Capsules also help bacteria resist dehydration, adhere to invasive devices such as catheters, and initiate the formation of biofilms (see Figure 10.1). With few exceptions, when an encapsulated bacteria loses the ability to form capsules, it also loses its pathogenicity.

Because not all species of bacteria form capsules, determining that an unknown bacterial species produces a capsule can be helpful in its identification. Capsular staining is a two-step process. In the first step, the acidic dye nigrosin or India ink is used to obliterate the background. Because these dyes are negatively charged, they will adhere to the glass of the slide but be repelled by the negatively charged bacterial cells, leaving the cells and capsules colorless against a dark background. The smear is allowed to air-dry, but there is no heat fixation because high heat can shrink the cells, resulting in a halo around the cells that could be mistaken for a capsule. In the second step, the basic dye crystal violet is used to stain the bacterial cell. Because the capsule will not stain with either dye, it is visible as a clear halo between the cell and the background. In a properly prepared capsule stain, the background will be gray while the cells themselves are purple, surrounded by a clear halo, which represents the capsule (Figure 10.2). Nonencapsulated cells will still appear as purple cells against a gray background, but the halo will be absent.

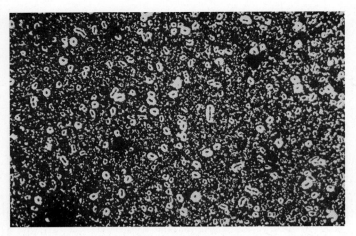

Figure 10.2 Encapsulated bacteria appear as purple cells, surrounded by a clear halo against a gray background (1000x). The clear halo represents the capsule. Note that the capsules have little regular morphology, differing in size and shape from cell to cell.

MATERIALS

Each group should obtain:

Skim milk slant cultures of:

　Klebsiella pneumoniae

　Enterobacter aerogenes

　Alcaligenes denitrificans

Crystal violet stain

Nigrosin or India ink

Inoculating needle

Marking pen

PROCEDURE

1. Label the corner of each slide with the name of the respective bacteria.
2. Place a single drop of nigrosin near one end of a slide.
3. Using an inoculating needle, aseptically transfer a very small amount of bacterial growth from a slant to the nigrosin. Use a second slide to spread the ink suspension over the slide.
4. Allow the slide to air-dry. Do not heat-fix! Heating the slide can lead to shrinkage of the cell, giving the false appearance of a bacterial capsule.
5. Place the slide on a staining rack, and flood the slide with crystal violet. Allow the smear to remain covered for 1 min.
6. Rinse the slide gently with water.
7. Blot dry with bibulous paper.
8. Examine the slide under oil immersion.

RESULTS

Record your results. For each sample, be sure to indicate the cell and the presence or absence of a capsule.

PRE-LAB QUESTIONS

1. *Define the following terms:*

　Biofilm

　Antibodies

　CSF

2. *What is sepsis? How does sepsis figure into this case?*

K. pneumoniae (＿＿＿x)

E. aerogenes (_____x)   *A. denitrificans* (_____x)

RESOLUTION OF THE CASE

The MRI revealed that the lupus had led to the complete destruction of the patient's spleen, a complication called **autosplenectomy** that occurs in approximately 5% of SLE cases. Asplenic individuals have low levels of both immunoglobulin M (a type of antibody) and memory B cells (a type of immune system cell that produces antibodies). Therefore, these patients are at much greater risk of infection by encapsulated bacteria. In this case, ER physicians ordered capsule staining of the bacteria isolated from the patient's blood and CSF. Based in part on the results of the capsule staining, the bacterium isolated from both types of fluid was identified as *Streptococcus pneumoniae,* a heavily encapsulated bacterium commonly encountered in asplenic patients.

It is a serious sign when bacteria are found in the cerebrospinal fluid and blood as these two body compartments are generally off-limits to bacteria and have little or no normal microbial inhabitants, unlike the digestive tract or the respiratory tract. It is more difficult for microbes to enter both of these compartments for several reasons, one being that antibodies can attach to bacteria and prevent them from crossing the boundaries into these areas. Apparently this patient was missing the antibodies that would have acted against the encapsulated bacteria.

The patient was treated for septic shock and respiratory failure for 9 days. Physicians administered dopamine and epinephrine to stabilize her blood pressure, as well as antibiotics to treat the underlying bacterial infection. Artificial ventilation was necessary for the first 4 days of treatment. Prior to being discharged, the patient was injected with pneumococcal vaccine and placed on prophylactic (preventive) penicillin therapy. She fully recovered.

EXERCISE 10 REVIEW QUESTIONS

1. Describe and draw the way in which an encapsulated bacterium would differ from a bacterium without a capsule.

2. Which bacteria in this lab were encapsulated? Which lacked a capsule?

3. Did encapsulated bacteria appear different macroscopically (i.e., on a slant or plate) when compared to bacteria without a capsule?

4. What is meant by the term *prophylactic penicillin therapy?*

Acid-Fast and Endospore Staining

CASE SYNOPSES

Mycobacterium tuberculosis Transmission in a Newborn Nursery and Maternity Ward—New York City, 2003

In September 2003, a female maternity nurse (nurse A) fell ill with what was first diagnosed as asthma. A chest x-ray was interpreted as normal and her symptoms—a productive cough, wheezing, and shortness of breath—were treated with steroids, antihistamines, and a cough suppressant. Despite this treatment, the symptoms persisted, and after approximately 8 weeks, further tests were ordered, including a computed tomography (CT) scan and bronchoscopy. Microscopic examination of sputum samples revealed the presence of *Mycobacterium tuberculosis,* and information from the CT scan was consistent with a diagnosis of tuberculosis (TB). This strain of TB proved unique, matching no other strains in either the New York or national databases. It also was found to be susceptible to each of the first-line anti-TB drugs, and the nurse was successfully treated.

Based on the onset of symptoms, nurse A was deemed to be infectious from September 1 through November 29, 2003. A review of work schedules and hospital records for this time period revealed that nurse A had a total of 1500 contacts, including 32 coworkers, 613 infants in the newborn nursery, and 900 patients in the maternity ward. Over the next 7 months, the Bureau of TB Control worked to get in touch with people who had perhaps come in contact with her. Sputum samples of many of these contacts were examined for the presence of *M. tuberculosis.*

Inhalation Anthrax Associated with Dried Animal Hides—London, 2008

A 35-year-old male presented to a London hospital complaining of difficulty breathing. His symptoms progressed quickly, and he was transferred to the hospital's intensive treatment unit suffering from respiratory failure, which soon progressed to multiple organ failure. A blood culture revealed gram-positive, encapsulated, nonmotile rods preliminarily identified as *Bacillus anthracis.* This is the bacterium that causes the disease anthrax, and it has the ability to survive for long periods of time without water or nutrition. The presence of *B. anthracis* was later confirmed by the Novel and Dangerous Pathogens Division of Britain's Health Protection Agency.

The patient was a drum maker by trade, handcrafting traditional African drums from dried animal hides. This process required him to soak the hides in water for an hour and then scrape the hair off with a razor, thereby releasing large quantities of dust into his studio. Most of the animal skins were goat hides imported from Gambia. Investigators from the Health Protection Agency examined the drum maker's property for the presence of the anthrax bacterium. In his studio, they found endospores of *B. anthracis* on one of five drums and on a few animal skins. No other traces of the bacterium were found at the property.

Resolution of the Case appears on page 101

CDC. 2005. *Mycobacterium tuberculosis* transmission in a newborn nursery and maternity ward—New York City, 2003. *Morbidity and Mortality Weekly Report,* 54: 1280–1283.

Health Protection Agency. 2008. Investigations following a death from anthrax. http://www.hpa.org.uk/NewsCentre/NationalPressReleases/2008PressReleases/081103deathfromanthrax/.

STUDENT LEARNING OUTCOMES

After completing this exercise, you should be able to:

1. Understand the usefulness of acid-fast and endospore staining.
2. Perform an acid-fast stain and differentiate between acid fast and non-acid-fast bacterial samples.
3. Perform an endospore stain and differentiate between endospore-forming and nonendospore-forming bacterial samples.

INTRODUCTION

While differential stains like the Gram stain are often used to assign a bacterial isolate to the most general of categories (Gram positive or negative, and rod or coccus), other differential stains provide a means for a much more specific identification. The physiological symptoms seen in the cases here share some similarities, and in both cases the etiological agent would first be identified as a Gram-positive rod. However, by employing two other differential stains, the organism responsible for each incident can be more precisely classified, and further stains and assays can be intelligently chosen to continue the identification process.

ACID-FAST STAINING

The acid-fast stain is primarily used to detect members of the genus *Mycobacterium,* such as the pathogens *M. tuberculosis* and *M. leprae,* the causative agents of tuberculosis and leprosy, respectively. Other bacterial species, including members of the genus *Nocardia* and even protozoan parasites (*Cryptosporidium* and *Isospora*) show at least some degree of acid-fast behavior. Because so few organisms are acid fast, the stain is generally only used when infection by an acid-fast organism is suspected.

Acid-fast bacteria contain within their cell walls a waxy material called **mycolic acid,** which prevents most stains from penetrating the cell. However, when heat is used to soften the mycolic acid, the primary stain, carbol fuschin, enters the cells and is not removed when decolorized with acid-alcohol (3% hydrochloric acid in 95% ethanol), leaving the cells a deep red color (Figure 11.1). Cells that are non-acid fast are easily decolorized by acid-alcohol, leaving them colorless until the counterstain, methylene blue, is applied. The differences in the appearance of acid-fast and non-acid-fast cells at each step of the staining process are seen in Table 11.1.

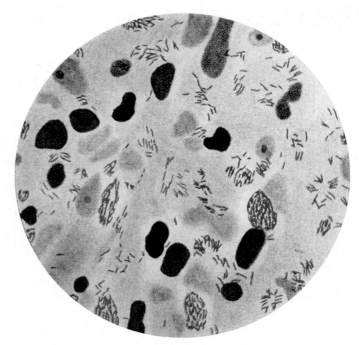

Figure 11.1 A photomicrograph of *Mycobacterium leprae,* the causative agent of leprosy. The red color is characteristic of acid-fast cells (1000x).

TABLE 11.1	Appearance of cells during acid-fast staining	
Step in staining process	Acid-fast bacteria	Non-acid-fast bacteria
Unstained	Clear	Clear
Carbol fuschin + heat	Red	Red
Acid-alcohol	Red	Clear
Methylene blue	Red	Blue

ENDOSPORE STAINING

Some bacteria have the ability to enter a resting stage, producing a tough **endospore** that is highly resistant to heat, cold, chemicals, and other environmental extremes that would kill a vegetative cell. Species within the genera *Bacillus* and *Clostridium* normally exist in a vegetative state, but when certain nutrients, primarily carbon or nitrogen, are depleted, endospore formation is initiated. The process takes 6–8 h, as a vegetative cell first becomes a sporangium with a developing prospore, eventually resulting in the production of a single endospore from each vegetative cell (Figure 11.2). The lifetime of an endospore is essentially

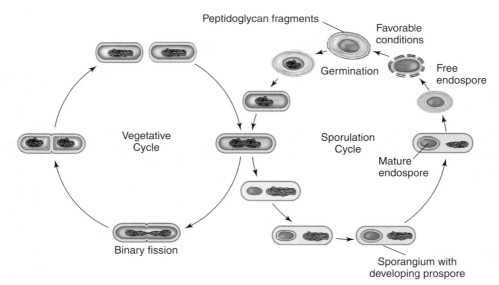

Figure 11.2 Vegetative and sporulation cycles in endospore-forming bacteria.

limitless because the endospore will remain dormant in the environment until conditions improve, at which point the cell reenters the vegetative cycle. Often, the return of optimal growth conditions occurs when an endospore enters the body, as when *Bacillus anthracis* spores are inhaled. Similarly, stepping on a nail covered with *Clostridium tetani* spores can lead to tetanus while *Clostridium botulinum* endospores in food can be a source of botulism.

Because so few medically important species produce endospores, determining that an unknown bacterium is an endospore former can go a long way toward providing its identification. Staining an endospore presents challenges as the protective nature of the endospore prevents the penetration of dye. As with acid-fast staining, heat is used as a mordant to drive the primary stain, malachite green, into the endospore. The cell is then decolorized with water, which removes the green dye from vegetative cells but not from endospores. Finally, safranin is used to counterstain vegetative cells as well as the sporangium (but not the developing spore) of cells that are in the process of sporulation (Figure 11.3). The appearance of cells during each of these stages is summarized in Table 11.2.

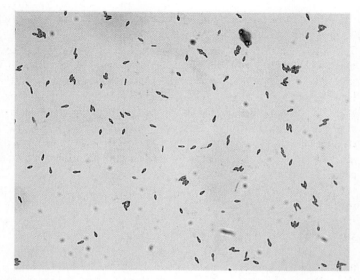

Figure 11.3 Endospore stain of *Clostridium botulinum*. Green staining endospores are formed near one end of the cell, leading to a club-shaped appearance. The pink rods are still in the vegetative cycle (1000x).

TABLE 11.2	Appearance of cells during endospore staining	
	Endospore former	Non-Endospore former
Unstained	Clear	Clear
Malachite green + heat	Green	Green
Water	Green	Clear
Safranin	Green	Pink

PRE-LAB QUESTIONS

1. *Define the following terms:*

Bronchoscopy

Etiological agent

Prophylactic

Differential stain

2. *Name two diseases caused by acid-fast bacteria as well as the species responsible for each.*

3. *Name three bacterial pathogens that form endospores as well the disease associated with each.*

MATERIALS

Each student should obtain:

A nutrient agar slant culture of *Mycobacterium smegmatis*

A nutrient broth culture of *Staphylococcus aureus*

Acid-fast-staining kit (carbol fuschin, acid-alcohol, methylene blue)

Inoculating loop and needle

Marking pen

For Ziehl-Neelsen staining procedure:

Electric hot plate

50-ml beaker

For microwave procedure:

Microwave oven

Empty plastic Petri dish

4. *Complete the table:*

Bacterial species	Gram reaction (+/−)	Gram stain color	Acid fast (+/−)	Acid-fast stain color	Endospore (+/−)	Endospore stain color
Staphylococcus aureus						
Mycobacterium tuberculosis						
Bacillus anthracis						

5. *Why is a Gram stain generally done prior to an acid-fast or endospore stain?*

ACID-FAST STAINING

Additional information about the acid-fast stain as well as an alternate procedure may be found in Exercise 47.

PROCEDURE

Ziehl-Neelsen Staining Procedure

1. Label a slide in the upper left-hand corner for identification.

2. Prepare a mixed smear by transferring two loopfuls of *S. aureus* to a slide, and add to it a very small amount of *M. smegmatis.* The acid-fast bacteria tend to form clumps that will have to be broken apart with the inoculating needle. Air-dry and heat fix the slide.

3. Saturate the smear with carbol fuschin, and steam the slide over boiling water for 5 min. Add additional stain if required to keep the smear from drying out.

4. Allow the slide to cool slightly so that it will not crack. Rinse with acid-alcohol for 8–12 sec.

5. Rinse briefly with water to stop the decolorizing effect of the acid-alcohol.

6. Counterstain with methylene blue for 30 sec.

7. Rinse briefly to remove excess methylene blue.

8. Blot dry with bibulous paper.

9. Examine the stained specimen under oil immersion. Record your observations in the Results section of this exercise.

Microwave Procedure

1. Label a slide in the upper left-hand corner for identification.

2. Prepare a mixed smear by transferring two loopfuls of *S. aureus* to a slide and adding to it a very small amount of *M. smegmatis*. The acid-fast bacteria tend to form clumps that will have to be broken apart with the inoculating needle. Air-dry and heat fix the slide.

3. Place a cut-to-fit piece of paper towel (2 layers of paper) into the bottom of an empty Petri dish. Saturate the toweling with water.

4. Place the heat fixed slide on top of the paper towel, and flood the slide with carbol fuschin.

5. Place the slide in the microwave, and heat for 30 sec at full power.

6. Remove the Petri dish from the microwave, and allow the slide to cool slightly so that it will not crack. Rinse with acid-alcohol for 8–12 sec.

7. Counterstain with methylene blue for 30 sec.

8. Rinse briefly to remove excess methylene blue.

9. Blot dry with bibulous paper.

10. Examine the stained specimen under oil immersion. Record your observations in the Results section of this exercise.

RESULTS

Record your results. Be sure that your drawings indicate both cellular morphology and arrangement while also being accurate with regard to color.

M. Smegmatis and *S. aureus*

ENDOSPORE STAINING
Additional information about the endospore stain may be found in Exercise 46.

MATERIALS

Each student should obtain:
A nutrient agar slant culture of *B. megaterium* (72 h)
Endospore staining kit (malachite green, safranin)
Inoculating loop and needle
Marking pen
For Schaeffer-Fulton staining procedure:
  Electric hot plate
  50-ml beaker
For microwave procedure:
  Microwave oven
  Empty plastic Petri dish

PROCEDURE

Schaeffer-Fulton Staining Procedure

1. Label a slide in the upper left-hand corner for identification and prepare a heat fixed smear of *B. megaterium*.

2. Saturate the smear with malachite green and steam the slide over boiling water for 5 min. Add additional stain if required to keep the smear from drying out.

3. Allow the slide to cool slightly so that it will not crack. Rinse with water for 30 sec.

4. Counterstain with safranin for 30 sec.

5. Rinse briefly to remove excess safranin.

6. Blot dry with bibulous paper.

7. Examine the stained specimen under oil immersion. Record your observations in the Results section of this exercise.

Microwave Procedure

1. Label a slide in the upper left-hand corner for identification and prepare a heat fixed smear of *B. megaterium*.

2. Place a cut-to-fit piece of paper towel (2 layers of paper) into the bottom of an empty Petri dish. Saturate the toweling with water.

3. Place the heat fixed slide on top of the paper towel and flood the slide with malachite green.

4. Place the slide in the microwave, and heat for 30 sec at full power.

5. Remove the Petri dish from the microwave, and allow the slide to cool slightly so that it will not crack. Rinse with water for 30 sec.

6. Counterstain with safranin for 30 sec.

7. Rinse briefly to remove excess safranin.

8. Blot dry with bibulous paper.

9. Examine the stained specimen under oil immersion. Record your observations in the Results section of this exercise.

RESULTS

Record your results. Be sure that your drawings indicate both cellular morphology and arrangement while also being accurate with regard to color.

B. megaterium

POST-LAB QUESTIONS

1. Acid-fast cells often appear clumped together. Speculate on how this may be related to their pathogenicity?

2. As a culture ages, carbon and nitrogen are depleted from the medium. How would an endospore stain of a young (12 h) *Bacillus* culture differ from that of an old (72 h) culture.

3. Anthrax is also known as *Woolsorter's disease*. Consider why this is, and relate your answer to the case seen here.

RESOLUTION OF THE CASES

***Mycobacterium tuberculosis* Transmission in a Newborn Nursery and Maternity Ward—New York City, 2003**

Despite the best efforts of the city and hospital, only 443 of the approximately 1500 possible contacts could be evaluated. Medical exams of 227 infants and 216 adults potentially exposed to nurse A revealed no evidence of TB disease. Of 227 infants screened, 5 tuberculin skin tests (TST) were positive, indicating exposure to *M. tuberculosis* even though no acid-fast rods were detected. The Bureau of TB Control recommended prophylactic treatment with isoniazid for all the patients with a positive TST. Of 807 postpartum patients, 16 (2%) were infected with HIV. It was recommended that these patients all receive isoniazid treatment, regardless of their TST results.

Inhalation Anthrax Associated with Dried Animal Hides—London, 2008

Although rare in the U.K.—and in the United States, for that matter—*B. anthracis* is found throughout much of the world.

Its ability to form endospores and survive harsh environmental conditions (years of heat, cold, and ultraviolet radiation, along with a complete lack of water or nutrients) ensures that many endospores are found in soil. As animals graze, lie, or roll on the ground, some of the endospores are transferred onto their bodies. Livestock in Africa, Asia, and the Middle East account for the majority of anthrax cases seen worldwide.

Despite treatment with rifampin, ciprofloxacin, and clindamycin, as well as with anthrax immunoglobulin, the drum maker died about 2 weeks later. Postexposure prophylaxis was given to eight persons, including the patient's immediate family, the main supplier of the skins, a person who assisted with the drum making, and a hospital worker. This incident was very similar to two 2006 cases in which drum makers in New York City and Scotland contracted anthrax while scraping animal hides for drumheads. In all three cases, the hides were imported from Africa, where anthrax is endemic.

EXERCISE 11 REVIEW QUESTIONS

1. Why was prophylactic tuberculosis treatment recommended for HIV-infected persons as well as infants whose mothers were infected with HIV even if their tuberculin skin tests were negative?

2. Bronchonscopy is, according to the CDC, a cough-inducing procedure and should be avoided in patients suspected of having TB disease. Why is this?

3. One infant received a bacilli Calmette-Guérin vaccination for tuberculosis during a trip to the Dominican Republic. Although the TST of the child was positive when later checked by hospital workers, there was little concern. Why was this?

4. What is meant by the term *secondary contamination?*

5. In the first case, over 1500 contacts of the nurse infected with *M. tuberculosis* were identified, and many of these were tested. In the second case, although inhalation anthrax is the more serious disease, little time was spent identifying contacts of the initial patient. Why were these two cases handled differently?

REFERENCES

CDC. *Anthrax.* http://www.bt.cdc.gov/agent/anthrax.
CDC. *Division of Tuberculosis Elimination (DTBE).* http://www.cdc.gov/tb.

Viable Plate Count

CASE SYNOPSIS

Salmonella Typhimurium Infection Traced to Contaminated Milk

On April 13, 2000, the Centers for Disease Control began an investigation to determine the source of an increased number of cases of infection with *Salmonella enterica* serotype Typhimurium (often referred to simply as *Salmonella* Typhimurium). Health officials in several northeastern states forwarded potential *Salmonella* isolates to the CDC for further testing. Initially, stool samples from 93 persons (76 in Pennsylvania and 17 in New Jersey) yielded *Salmonella* Typhimurium. Further genetic analysis identified 38 isolates that, because of specific similarities in their DNA fingerprint, were classified as outbreak-related strains. Isolates from two dairy cows also matched the outbreak-related strains.

Interviews with patients revealed that nearly all had consumed milk processed at a specific dairy plant and that the odds of becoming infected increased as the volume of milk consumed increased. Other risk factors, including handling reptiles or uncooked chicken, and consuming undercooked eggs or unpasteurized milk, were not associated with the illness. Three employees of the dairy plant reported missing work due to gastrointestinal illness during the outbreak period, and a stool sample from one employee yielded the outbreak-related strain. All three employees reported drinking fully processed milk produced at the plant in the 5 days before the onset of symptoms. The plant in question received raw milk from more than 59 different farms for processing. The pasteurized, finished product was in turn distributed throughout parts of Pennsylvania, Delaware, and New Jersey. The plant had been inspected by the Pennsylvania Department of Agriculture on a regular basis, and the most recent inspection revealed no problems.

Resolution of the Case appears on page 107

STUDENT LEARNING OUTCOMES

After completing this exercise, you should be able to:

1. Determine the number of bacteria in a liquid sample using a viable plate count.
2. Calculate and carry out dilutions of a liquid sample.

INTRODUCTION

As a bodily fluid, milk in the udder of a cow is sterile. During the collection process, however, it is inoculated with the normal microbiota of the cow. Because milk has such a rich composition, being laden with water, proteins, fats, and minerals, microbial growth is a common occurrence. Since milk is such an integral part of so many peoples lives and since its consumption can entail very real risks, the entire production process, from cow to cup, is tightly regulated.

The main safeguard against bacterial infections acquired from milk is **pasteurization.** In this process, heat is used to kill many bacterial species, resulting in a product that, while not sterile, has a reduced **microbial load.** This decrease in bacterial diversity and number results in a product that is safer to drink and has a longer shelf life. Raw milk refers to milk that has not been pasteurized as part of its processing. Dairies that sell raw milk rely on more stringent standards of cow health as well as rigorous collection standards. Despite these efforts, raw milk remains a source of *Salmonella* and *Escherichia coli* outbreaks. For both raw and pasteurized milk, the Food and Drug Administration has set limits on the number of both **coliforms** and total bacteria that may be present in milk before and after processing and/or pasteurization. In-house testing as well as outside inspections are a customary part of the collection routine. Additionally, thorough inspections are also undertaken whenever milk products are thought to be the source of a disease cluster, as was suspected in this case.

PRE-LAB QUESTIONS

1. *Define the following terms:*

   Serotype

   Microbiota

Coliform

Selective media

Differential media

Colony-forming unit (CFU)

2. *If a sample of milk is required to have no more than 20,000 bacterial cells per milliliter, what is the maximum number of cells that are permissible in:*

0.1 ml

0.01 ml

0.001 ml

3. *For each of the media in the accompanying table, indicate what type of bacteria is selected for and how lactose-fermenting bacteria are differentiated from lactose-nonfermenting bacteria.*

PERIOD ONE

Bacterial counts in milk are routinely determined using a viable plate count. In this technique, a diluted volume of milk is spread onto the surface of a plate, and colonies are counted after incubation. The technique used in this exercise is a slightly simplified version of the protocol required by the FDA. Each student team will be responsible for determining the bacterial count of coliforms and total bacteria from one type of milk, either raw or pasteurized.

MATERIALS

Each group should obtain:

Three plates of nutrient agar

Three plates of MacConkey agar

Three tubes each containing 9 ml of sterile water (blanks)

One screw top tube containing pasteurized milk (odd numbered student teams)

One screw top tube containing raw milk (even numbered student teams)

Six serological pipettes, 1 ml

Marking pen

PROCEDURE

1. The dilution and inoculation procedure is illustrated in Figure 12.1.
2. Label the bottom of all six plates and the three water blanks with your name and lab time.
3. Label the water blanks "1:10," "1:1:00," and "1:1000."
4. Vigorously shake your milk sample for 10 sec, making sure that the sample container is completely closed.
5. Within 3 min, aseptically transfer 1 ml of milk to the first (1:10) tube.
6. Pipet up and down several times to mix, and transfer 1 ml of the 1:10 dilution to the 1:100 tube.
7. Pipet up and down several times to mix, and transfer 1 ml of the 1:100 dilution to the 1:1000 tube.
8. Inoculate the nutrient agar plates with the 1:10, 1:100, and 1:1000 dilution samples. Inoculate each of these

Media	Specificity of selection	Color of lactose-fermenting colonies	Color of lactose-nonfermenting colonies
MacConkey agar			
Eosin methylene blue agar			
Hektoen enteric agar			

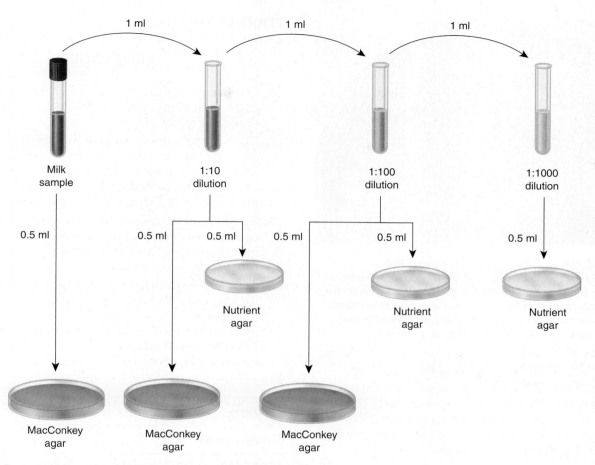

Figure 12.1   Dilution and inoculation procedure for a viable plate count.

plates with 0.5 ml of the appropriate dilution, and spread the inoculum with a sterile glass spreader. Do not invert the plates for at least 10 min.

9. Label the MacConkey agar plates as "undiluted," "1:10," and "1:100." Inoculate the first plate with undiluted milk (shake the milk again if it has been sitting more than 3 min since last being mixed), the second with the 1:10 dilution, and the third with the 1:100 dilution.

10. Incubate the plates at 32°C for 24–48 h. Be sure to properly dispose of all tubes, pipets, and other equipment.

QUESTIONS—PERIOD ONE

1. Why was the milk used for coliform counting diluted to a lesser extent than was the milk used for total bacterial counting?

2. What is the relationship between a cell, a CFU, and a colony?

3. A dilution factor is a number that relates a diluted and undiluted sample. For example, if a 1-ml sample of water is taken from a 1-L (1000 ml) bottle, the dilution factor is 1000. If a plate inoculated with 1 ml of this water produced 72 colonies, then 72 × 1000 or 72,000 CFUs would be present in the 1-L bottle. Based on this information, complete the table:

	Dilution factor	**Organisms in 1-ml milk**
Undiluted milk		
1:10 dilution		
1:100 dilution		
1:1000 dilution		

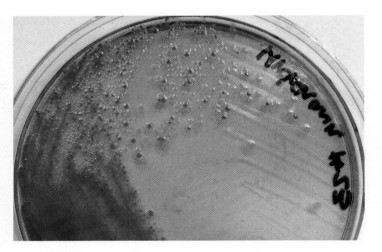

Figure 12.2 When grown on MacConkey agar, lactose-fermenting bacteria should be deep pink or red in color while lactose nonfermenters are white to colorless. Because MacConkey agar selects only Gram-negative bacteria for growth, red colonies can presumptively be considered coliforms (Gram-negative lactose fermenters) while white or clear colonies are considered noncoliforms.

4. FDA standards call for counting only plates with between 25 and 250 colonies per plate. Those plates with fewer than 25 colonies are reported as TFTC (too few to count), and those with more than 250 as TMTC (too many to count). Why do you think the FDA has instituted this recommendation?

PERIOD TWO

PROCEDURE

1. Retrieve your plates from the incubator. The growth of coliforms and noncoliforms on MacConkey agar is illustrated in Figure 12.2.
2. Use a colony counter to count the number of colonies on each plate. Any plate with fewer than 25 colonies should be recorded as TFTC (too few to count) while any plate with more than 250 colonies as TMTC (too many to count). Count colonies of all sizes.
3. Use dilution factors to establish the total number of each type of microorganism per milliliter of milk, using only those plates containing between 25 and 250 colonies. Record your results.
4. Compare the growth between plates inoculated with each type of milk (raw versus pasteurized). Using the morphological appearance of the colonies on each plate, determine the number of **different types** of colonies present on each plate. Review Exercise 40 for help in differentiating colony morphology.

	Distinct colony types
Raw milk	
Pasteurized milk	

Raw milk						
	Type of media used	Undiluted	1:10	1:100	Dilution factor	Organisms/ml
Total bacterial count						
Gram-negative bacteria count						
Coliform count						

Pasteurized Milk						
	Type of media used	1:10	1:100	1:1000	Dilution factor	Organisms/ml
Total bacterial count						
Gram-negative bacteria count						
Coliform count						

QUESTIONS—PERIOD TWO

1. Which milk, raw or pasteurized, had a greater variety (not number) of colonies? Explain this fact.

2. The FDA requires the use of a special type of pipette to inoculate and distribute bacterial cells for counting, rather than the spreading procedure used in this lab. What kind of inaccuracy is almost certainly introduced by the use of the glass spreader in this exercise?

3. Inspection of the dairy plant noted that, among other problems, some processing machines leaked raw milk onto the floor of the plant. Even though this milk was never packaged, why would this be a cause for concern?

RESOLUTION OF THE CASE

Over 200 samples of pasteurized milk from the suspect plant were collected and tested between April and July. Because none of these samples grew *Salmonella,* investigators came to the conclusion that contamination of the milk occurred after pasteurization. Because outbreak-related strains of *Salmonella* were isolated from dairy cows in the region, it was inferred that this strain was circulating among local dairy herds. Furthermore, sanitary conditions at the plant, especially leaks of raw milk, could have led to contamination after pasteurization.

Based on this outbreak, the dairy plant hired an outside consultant to address the FDA's immediate concerns with regard to violations of sanitary standards, including inadequate cooling of milk and excessive condensation, which could have fallen into open containers. Examination of a dozen outbreaks involving pasteurized milk between the years 1960 and 2000 revealed that in the vast majority of cases, the culprit was contamination after pasteurization, not an inadequate pasteurization process. Shortly after this incident, the Pennsylvania Department of Agriculture began to integrate employee training with routine inspections, with the hope that post-pasteurization contamination could be reduced.

EXERCISE 12 REVIEW QUESTION

1. A common method of ensuring pasteurization of milk is to test the activity of an enzyme called *phosphatase*. An adequately pasteurized milk sample will have only a very small amount (<1 µg/ml) of functional enzyme. Explain why this test can be used to determine the adequacy of pasteurization.

REFERENCES

Olsen, S.J., Ying, M., Davis, M.F., Deasy, M., Holland, B., Iampietro, L., Baysinger, C.M., Sassano, F., Polk, L.D., Gormley, B., Hung, M.J., Pilot, K., Orsini, M., Van Duyne, S., Rankin, S., Genese, C., Bresnitz, E.A., Smucker, J., Moll, M. and Sobel, J. 2004. Multi-drug resistant *Salmonella* Typhimurium infection from milk contaminated after pasteurization. *Emerging Infect. Dis.*, 10: 932–935.

U.S. Food and Drug Administration, *The Dangers of Raw Milk: Unpasteurized Milk Can Pose a Serious Health Risk.* http://www.cfsan.fda.gov/~dms/rawmilk.html.

CASE STUDY EXERCISE  13

Cultivation of Anaerobes

CASE SYNOPSIS

Botulism Associated with Commercially Canned Chili Sauce—Texas and Indiana, July 2007

An Indiana husband and wife, both feeling ill, went to separate hospitals to be examined, and several potential diagnoses were suggested. A few days later, on July 11, 2007, both patients were evaluated at the same hospital. There, physicians were able to make a preliminary diagnosis of botulism based on the couple's shared symptoms, which included cranial nerve palsy and descending flaccid paralysis. Both patients required mechanical ventilation. During a search of the couple's refrigerator, local health officials found an unlabeled bag of leftover chili sauce that, when analyzed, contained botulin, the bacterial toxin that causes botulism. Botulin is produced by *Clostridium botulinum,* an endospore-forming bacterium. Because botulism is a reportable disease, the facts of the case were forwarded to the Indiana State Department of Health and the Centers for Disease Control (CDC).

Four days earlier, the CDC had received a similar report from Texas, where two siblings had been diagnosed with botulism, again only after being examined at the same hospital. Both children had eaten Castleberry's Hot Dog Chili Sauce for lunch on June 28. Although the can from this meal had been discarded, another can, bought at the same time, was found in the home.

Based on these two cases, the CDC suspected a common source epidemic.

Resolution of Cases appears on page 115

CDC. 2009. Botulism associated with commercially canned chili sauce—Texas and Indiana, July 2007. *Morbidity and Mortality Monthly Report,* 56: 767–769.

STUDENT LEARNING OUTCOMES

After completing this exercise, you should be able to:

1. Utilize an anaerobe jar.
2. Evaluate the oxygen requirements of a bacterium using fluid thioglycollate media.

INTRODUCTION

One aspect of proper food processing, be it harvesting, cooking, or storage, is the prevention of foodborne illness. Cooking especially is used as a means of killing the bacteria that are assumed to be present on most foods. While nominal temperatures (80°C) are generally enough to kill most vegetative cells, endospores of the species *Clostridium botulinum* are still present even after long exposure to temperatures that would leave foods inedible (i.e., boiling for several hours). Ingested spores are generally destroyed by acid in the stomach or outcompeted for nutrients in the gastrointestinal tract, that is, ingestion of small numbers of endospores is generally not problematic. If, however, the spores are given the chance to germinate in food, the vegetative cells produce **botulin,** the most potent bacterial toxin known. *C. botulinum,* like all members of this genus, is an **obligate anaerobe,** and it is only under completely anaerobic conditions that endospores will germinate. Improper canning of food can lead to viable spores being present in the anaerobic environment created by the canning process. When the cans (or jars) are kept at room temperature, these endospores can germinate and begin to produce botulin. Ingestion of botulin interferes with the release of acetylcholine from motor neurons, resulting in flaccid paralysis (where muscles are unable to contract). Death from botulism is due to paralysis of the respiratory muscles.

Successful culturing of anaerobes requires the creation of an anaerobic environment. One method of doing this is to incubate cultures in an **anaerobe jar,** a container with a tight-fitting lid that prevents oxygen from entering the jar from the outside environment. After inoculated plates are placed in the jar, water is added to a small foil packet, and the lid of the jar is fastened in place (Figure 13.1). The water combines with the chemicals inside the packet to produce hydrogen and carbon dioxide gases. A palladium catalyst, also in the packet, then converts hydrogen and oxygen gas within the jar to water, generating an anaerobic environment. A paper strip impregnated with the dye methylene blue is placed inside the jar as a visual indicator of an anaerobic environment. The blue color normally associated with the dye is only present when oxygen is available to oxidize the dye.

A second method of culturing anaerobic organisms is to use **fluid thioglycollate** media, which permits the growth of a wide variety of bacteria and also allows the determination of the oxygen requirements of an organism. This media, which is a thick broth because of the addition of a small amount of agar, has dissolved oxygen expelled from it during autoclaving.

109

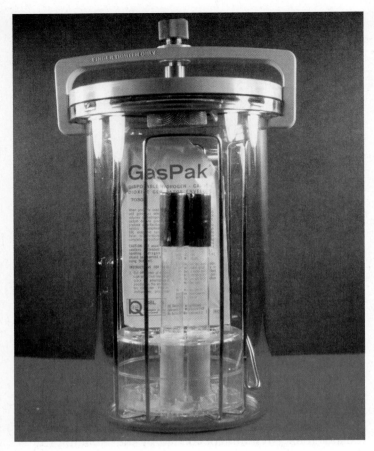

Figure 13.1 Anaerobe jars have a tight-fitting lid that prevents oxygen from entering from the outside. When water is added to the chemical packet inside the jar, a series of reactions take place that result in oxygen within the jar being converted to water, creating an anaerobic environment. A methylene blue indicator strip provides visual assurance that the environment is anaerobic, becoming colorless when no oxygen is available.

During cooling, oxygen diffuses into the media, from top to bottom, resulting in an oxygen gradient within the tube (Figure 13.2). The indicator resazurin shows the location of oxygen in the tube, turning pink where oxygen is present. The medium is inoculated with a vertical stab from top to bottom, ensuring that organisms are initially present throughout the media. After incubation, the position of growth within the media indicates the oxygen requirements of the bacterium.

PRE-LAB QUESTIONS

1. *Define the following terms:*

Endospore

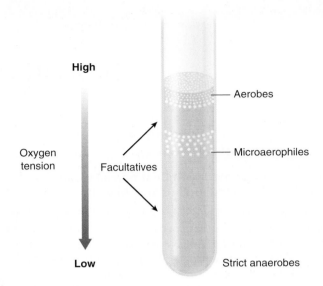

Figure 13.2 Oxygen gradient within a tube of fluid thioglycollate medium.

Obligate anaerobe

Facultative anaerobe

Germination

2. *What is flaccid paralysis?*

3. *How would you characterize* Clostridium botulinum *with regard to its shape and Gram reaction? What other stain would be especially useful in its identification?*

4. *What two techniques are commonly used to culture bacteria such as* Clostridium botulinum?

PERIOD ONE

MATERIALS

Each student should obtain:

Two plates of nutrient agar

Three tubes of fluid thioglycollate media

Broth cultures of:

    Pseudomonas aeruginosa

    Staphylococcus aureus

    Clostridium sporogenes

Access to an anaerobe jar

One gas-generating packet per jar

Inoculating loop and needle

Marking pen

PROCEDURE

Anaerobe Jar

1. Label each nutrient agar plate with your name and lab time.
2. Label each plate as seen in Figure 13.3. Using a loop, inoculate each plate with a single streak of each organism.
3. Invert both plates. Place one plate in the anaerobe jar and the other in a sleeve to be incubated at normal oxygen concentration.
4. When all the plates have been added to the anaerobe jar, prepare it for incubation as follows:
   a. Place a methylene blue strip inside the jar. The strip should be blue due to oxygen in the environment.
   b. Open the corner of the GasPak jar, and pipette 10 ml of water into the packet.
   c. Immediately close the jar, making sure it seals tightly.
5. Incubate the anaerobe jar at 37°C for 24–48 h.

Fluid Thioglycollate Media

1. Examine each of the fluid thioglycollate tubes for the present of a pink color in the upper 1 cm of the tube. This represents the oxygen-containing region of the media; if this color extends further than 2 cm toward the bottom of the tube, the tube should be boiled for 10 min to drive off the excess oxygen and restore the oxygen gradient.
2. Label each tube with the name of the appropriate organism: *P. aeruginosa*, *S. aureus*, or *C. sporogenes*.
3. Use a needle to inoculate each tube with the appropriate organism. Handle and inoculate the tubes gently so as not to disturb the oxygen gradient.
4. Incubate the tubes at 37°C for 24–48 h.

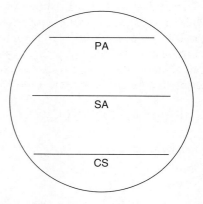

Figure 13.3 Use a loop to inoculate each plate with a single streak of each organism, *Pseudomonas aeruginosa*, *Staphylococcus aureus*, and *Clostridium sporogenes*.

QUESTIONS—PERIOD ONE

1. Why is an anaerobe jar necessary to grow *Clostridium?*

2. How is the environment within the fluid thioglycollate media different from that found within the anaerobe jar?

3. Sketch the appearance of growth on the plate. Assume the plates have been incubated as seen below:

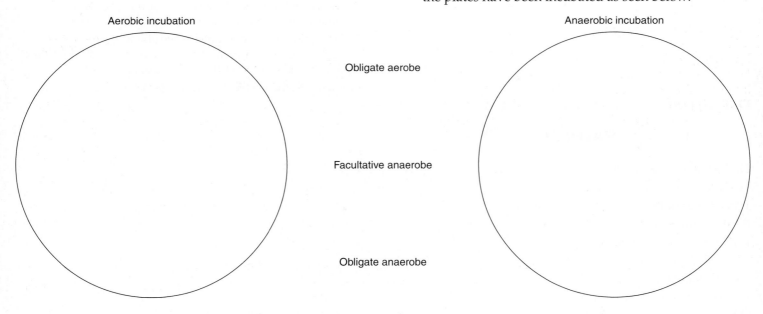

Aerobic incubation

Obligate aerobe

Facultative anaerobe

Obligate anaerobe

Anaerobic incubation

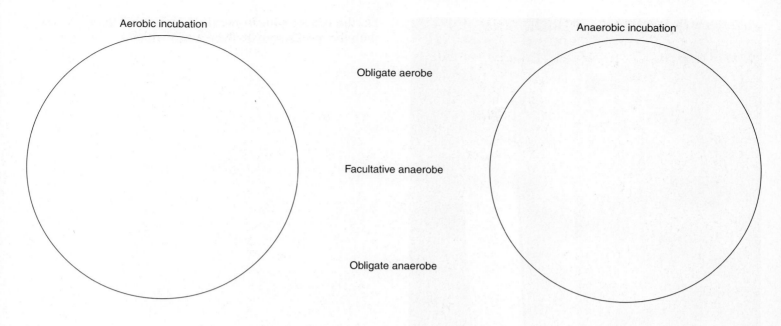

Aerobic incubation

Anaerobic incubation

Obligate aerobe

Facultative anaerobe

Obligate anaerobe

Sketch the appearance of growth in the fluid thioglycollate tubes:

Sketch the appearance of growth in each of your fluid thioglycollate tubes.

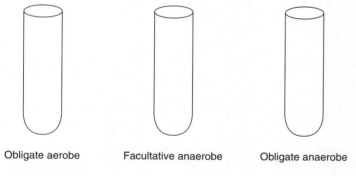

Obligate aerobe   Facultative anaerobe   Obligate anaerobe

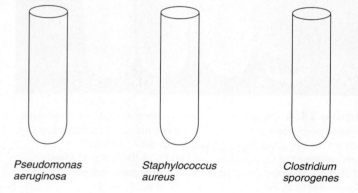

Pseudomonas aeruginosa   *Staphylococcus aureus*   *Clostridium sporogenes*

PERIOD TWO

Retrieve your plates and tubes from the incubator, being careful not to disturb the growth within the tube. Compare your tubes to those seen in Figure 13.4. Sketch the appearance of growth on the plates, being sure to indicate any differences in the amount of growth each organism displays.

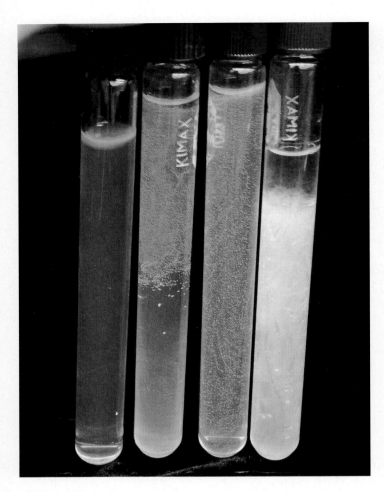

Figure 13.4 Organisms displaying different oxygen requirements in fluid thioglycollate medium. The tube is prepared so that a top-to-bottom oxygen gradient exists within the tube, with high concentrations of O_2 at the top and no O_2 at the bottom. After incubation, the position of growth within the tube indicates the oxygen needs of the bacteria. From left to right: aerobic (*Pseudomonas aeruginosa*), facultative anaerobe (*Staphylococcus aureus*), facultative anaerobe (*Escherichia coli*), and obligate anaerobe (*Clostridium butyricum*).

Do the results seen in the fluid thioglycollate tubes agree with the results seen on the plates? Explain.

RESOLUTION OF THE CASE

The cases of botulism diagnosed in the Indiana couple and the Texas children were of the food-borne variety. This type of botulism is most often associated with home-processed foods because endospores of *Clostridium botulinum* can easily survive normal cooking processes. Furthermore, in an anaerobic environment with low acidity (pH > 4.6) and a temperature greater than 3.9°C (39.0°F), the endospores may germinate and produce botulin.

Commercially processed foods, such as the chili sauce implicated in this case file, are typically processed in a retort, a large chamber where the food is subjected to a combination of steam and high pressure that reliably kills endospores. Put into practice in the 1920s, retort canning has nearly eliminated the risk of acquiring botulism by consuming commercially processed food. Nevertheless, a well-rinsed can of Castleberry's Hot Dog Chili Sauce recovered from the recycling bin of the Indiana couple and the unopened can of chili sauce taken from the Texas house contained production codes indicating that they had been processed at the same facility within a few hours of one another. The knowledge that both cans came from the same location indicated a common source epidemic, in which each occurrence of an illness can be traced to the same source. An investigation of the processing plant was in order.

The U.S. Food and Drug Administration (FDA) initiated an investigation of Castleberry's Food Company on July 17, 2007, after finding that both cans of chili sauce had been manufactured in the same plant and in the same set of retorts. Tests of the facility's canning equipment revealed problems in 2 of the 100 retort cookers in the plant, such that either the temperature or the pressure was not high enough to destroy endospores. Examination of swollen cans in the plant revealed botulin in 16 out of 17 cans.

Based on the inspection results, a recall was issued not only for chili sauce processed at the plant, but also for 90 other products, including chili, meat, chicken, and dog food, that had been processed in the same set of retorts. Castleberry repaired the damaged cookers, installed backup valves to prevent a recurrence, and trained employees to recognize potential problems in the canning process.

EXERCISE 13 REVIEW QUESTIONS

1. Using proper terminology, how would you describe the oxygen requirements of each of the microbes used in this exercise:

   P. aeruginosa

   S. aureus

   C. sporogenes

2. The CDC recommends boiling home-canned foods for 10 min prior to eating. What does this tell you about the botulism toxin?

REFERENCE

CDC. *Botulism.* http://emergency.cdc.gov/agent/botulism/.

Temperature Effects on Bacterial Growth and Survival

CASE SYNOPSIS

Outbreak of Gastroenteritis Associated with Consumption of Alaskan Oysters

A Nevada resident who had just returned home from a cruise that sailed on Prince William Sound in July 2004 was struck by gastrointestinal distress so severe that medical intervention was required. Laboratory tests for this patient indicated the presence of *Vibrio parahaemolyticus,* a pathogenic bacterium known to cause gastroenteritis. Infection usually occurs through consuming raw or undercooked shellfish, particularly oysters. In this case, the patient's illness began 3 days after eating raw oysters obtained from Prince William Sound.

Further investigation by the epidemiology section of the Alaska Division of Public Health revealed that a total of 54 people had developed watery diarrhea, along with various other gastrointestinal symptoms, beginning within 2 days of

consuming raw oysters collected from Alaskan waters. Stool samples provided by eight patients all contained *V. parahaemolyticus.* However, the discovery of this bacterium was puzzling because *V. parahaemolyticus* requires a minimum water temperature of 16.5°C to survive, and the waters of Prince William Sound have historically been colder than that.

Inspectors from the U.S. Food and Drug Administration (FDA) investigated disinfection and food-handling practices aboard the cruise ship and obtained food and water samples for laboratory testing. They also assessed the oyster farm that had supplied the suspect oysters. When samples of oysters, water, and sediment were retrieved for analysis, two water samples, one sediment sample, and six oyster samples were culture-positive for *V. parahaemolyticus.*

Resolution of the Case appears on page 122

State of Alaska Epidemiology Bulletin No. 24. October 26, 2004. Outbreak of *Vibrio parahaemolyticus* gastroenteritis associated with consumption of Alaskan oysters—Summer 2004.

McLaughlin, J.B., DePaola, A., Bopp, C.A., Martinek, K.A., Napolilli, N.P., Allison, C.G., Murray, S.L., Thompson, E.C., Bird, M.M. and Middaugh, J.P. 2005. Outbreak of *Vibrio parahaemolyticus* gastroenteritis associated with Alaskan oysters. *New Engl. J. Med.,* 353: 1463–1470.

STUDENT LEARNING OUTCOMES

After completing this exercise, you should be able to:

1. Explain the effects of temperature on bacterial growth.
2. Determine the optimal growth temperature of a given bacterial species.

INTRODUCTION

Temperature is without a doubt the most widely used method of controlling the growth of microorganisms. One of the purposes of cooking is to kill the many microbes found on raw food, and refrigeration is similarly used to prevent the growth of microbes. Even when temperatures are not artificially maintained, microorganisms are found segregated within the environment, with some growing below 0°C while others are found at temperatures greater than 100°C.

Temperature affects many aspects of an organism's metabolism. Enzymes responsible for the catalysis of biochemical reactions in the cell are generally quite sensitive to temperature and will denature when temperatures climb too high, leaving them unable to function. High temperatures also have the ability to destroy the lipids that make up biological membranes. Low temperatures slow the growth of cells by reducing the rate of chemical reactions in general and also by reducing the fluidity of the cell membrane, decreasing the rate at which nutrients can cross into the cell. As a rule, high temperatures tend to be **microbicidal,** causing irreversible changes in the cell while low temperatures are more often **microbistatic,** with growth rates returning to normal as temperature rises.

For each organism, three **cardinal temperatures** exists. The **minimum temperature** is simply the lowest temperature at which an organism may survive while the **maximum temperature** likewise refers to the highest survivable temperature. Between the two lies the **optimal temperature,** which is defined as the temperature at which an organism shows its greatest rate of growth over time (Figure 14.1). Based on their temperature requirements, microorganisms are generally divided into five groups.

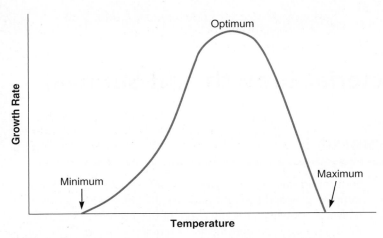

Figure 14.1 The effect of temperature on growth rate. Each bacterial species has a minimum temperature below which the cell will not survive, a maximum temperature, which represents the highest temperature at which growth will continue, and an optimum temperature at which the metabolic activities of the cell proceed at their maximum rate.

Psychrophiles are those organisms adapted to cold temperatures, with minimum and maximum temperatures ranging from −5°C to 20°C, respectively. These bacteria are found in Artic and Antarctic regions where the temperature fluctuates little and cold is a constant. Most bacteria are classified as **mesophiles,** with optimal growth between 15°C and 50°C. **Thermophiles** experience optimal growth between 50°C and 80°C; bacteria in this group may be found in composting organic material and hot springs. Finally, **hyperthermophiles,** or extreme thermophiles, are those hardy bacteria that grow at temperatures above 80°C.

In addition to these designations, bacteria may be classified as **obligate,** or strict, if they are restricted to a specific temperature. For example, the obligate thermophile *Thermus aquaticus* (literally, "hot water") is a bacterium found in hot springs where it thrives at an optimum temperature of 70°C; the species, however, dies below 50°C. In a similar manner, the suffix **troph** is added to a species temperature designation if it grows across a particularly wide temperature range. For instance, many mesophilic bacteria such as *Pseudomonas* and *Campylobacter* are able to grow at 4°C, far below their temperature optimum. These bacteria are referred to as **psychrotrophs** (Figure 14.2).

PRE-LAB QUESTIONS

1. *Define the following terms:*

   Gastroenteritis

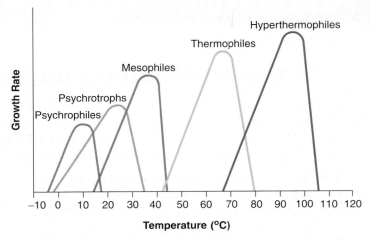

Figure 14.2 Microorganisms can be classified by the range of temperatures at which they flourish. Five groups are generally recognized, the psychrophiles, psychrotrophs, mesophiles, thermophiles, and hyperthermophiles.

   Enzyme

   Denature

   Turbid

2. *What is the difference between microbistatic and microbicidal? Why are high temperatures generally microbicidal and low temperatures generally microbistatic?*

TABLE 14.1   Bacterial assignments

Bacterial species	Student number
Escherichia coli	1, 4, 7, 10, 13, 16, 19, 22, 25
Serratia marcescens	2, 5, 8, 11, 14, 17, 20, 23, 26
Geobacillus stearothermophilus	3, 6, 9, 12, 15, 18, 21, 24, 27

PERIOD ONE

To determine the growth characteristics of an organism, you will inoculate five tubes of nutrient broth with the same organism and then incubate the tubes at temperatures ranging from near freezing (5°C) to far above human body temperature (55°C). In this lab, three different organisms will be examined: *Escherichia coli*, *Serratia marcescens,* and *Geobacillus stearothermophilus*. Each student will be responsible for a single microorganism, as assigned in Table 14.1. A second exercise involves streaking a pair of nutrient agar slants with *Serratia marcescens* to determine the effect of temperature on pigment production in the bacteria.

MATERIALS

Each student should obtain:
Five tubes of nutrient broth
Two nutrient agar slants
Broth cultures of:
   E. coli
   S. marcescens
   G. stearothermophilus
Inoculating loop
Marking pen
Spectrophotometers (optional, only needed for period two)

PROCEDURE

1. Label five tubes of nutrient broth with your name, the organism, and the temperature at which each tube will be incubated (5°C, 25°C, 38°C, 42°C, or 55°C).
2. To transfer approximately the same number of bacteria to each tube, gently mix the culture until it appears uniformly turbid. Repeat this process between each inoculation.
3. Inoculate each tube with a single loopful of your assigned organism.
4. Place each inoculated tube in the proper rack so that each is incubated at its appropriate temperature for 24–48 h.
5. Label two tubes of nutrient agar with your name. Label one tube "25°C" and the other "38°C."

6. Use a single streak to inoculate each tube with *S. marcescens*. Incubate each tube at its appropriate temperature for 24–48 h.

QUESTIONS—PERIOD ONE

1. As bacteria grow in a liquid culture, they cause the media to become cloudy, with the degree of cloudiness being directly proportional to the level of growth. This cloudiness, or turbidity, can be measured using a spectrophotometer, which reports turbidity (or absorbance) on a scale of 0 (clear) to 2 (completely opaque). If a spectrophotometer is not available, turbidity can be estimated visually, usually on a scale of 0 (clear) to 10 (very turbid). The accompanying table provides an example of visual and measured turbidity for a hypothetical bacterium. Based on the data in the table, construct a growth curve similar to that seen in Figure 14.1. What are the minimum, maximum, and optimum temperatures for the organism? Under which temperature group would you classify the organism?

Temperature	Visual appearance of the culture	Measured turbidity
5°C	0	0
10°C	2	0.15
20°C	3	0.25
30°C	5	0.39
40°C	7	0.53
50°C	1	0.02

2. Within a narrow range of temperatures, *S. marcescens* produces a red pigment called prodigiosin. Although the ultimate purpose of the pigment is a mystery, it is known that at least 10 enzymes are required for its production. If one of these enzymes is rendered nonfunctional, how would you expect the appearance of the bacteria to change?

Next, evaluate the growth in each of the five broth tubes. Follow one of the procedures below to evaluate the growth in each tube.

Spectrophotometer Procedure

1. Mix each tube until a uniform turbidity is obtained. If spectrophotometers are available, record the transmittance (%T) and absorbance (Abs) of each tube in the accompanying table.
2. Graph the results (Absorbance vs. Temperature) for your assigned organism. Gather information about other organisms from your classmates, and graph that on the same set of axes.

Visual Procedure (No Spectrophotometer)

1. If a spectrophotometer is not available, rate each tube on a scale of 0 (clear) to 10 (very turbid).
2. Graph the results (Turbidity vs. Temperature) for your assigned organism. Gather information about other organisms from your classmates, and graph that on the same set of axes.

Using the results, sketch a graph of each organism's growth response to changing temperatures. Be sure to label each graph with the name of the organism, the proper temperature, and the units for growth (absorbance, 0–2, or visual reading, 0–10).

PERIOD TWO

Retrieve your tubes from the incubator. First, examine the slants of *S. marcescens,* noting any differences in growth and/or pigment production. Sketch the tubes:

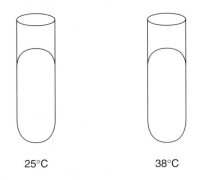

25°C   38°C

	S. marcescens		E. coli		G. stearothermophilus	
Temp (°C)	Visual reading	Spectrophotometer (%T, Abs)	Visual reading	Spectrophotometer (%T, Abs)	Visual reading	Spectrophotometer (%T, Abs)
5						
25						
38						
42						
55						

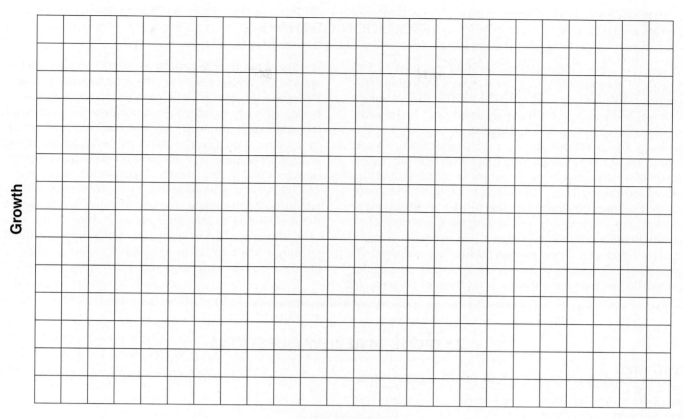

Growth

Temperature

QUESTIONS—PERIOD TWO

1. What are the minimum, maximum, and optimum temperatures for each of the three organisms examined:

	Minimum temperature	Optimum temperature	Maximum temperature
S. marcescens			
E. coli			
G. stearothermophilus			

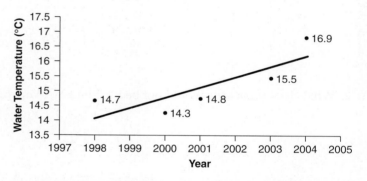

Median (July–August) Water Temperature, by Year—Farm A, Prince William Sound, 1998–2004. (Data compliments of the USDA-funded Molluscan Broodstock Program.)

2. Given that a bacterium will not grow below its minimum temperature, what is the most logical explanation for the finding of *V. parahaemolyticus* in many of the Alaskan samples? Use the accompanying graph to help formulate your answer.

RESOLUTION OF THE CASE

Because it requires a minimum temperature of at least 16.5°C to grow, *V. parahaemolyticus* is considered a mesophile. Historically, Alaskan waters have been too cold for the bacterium to thrive, and people have routinely eaten raw oysters without concern for their safety. But due to the warming of the waters of Prince William Sound, *V. parahaemolyticus* spread from its usual habitat to an area more than 900 kilometers (600 miles) miles further north, where it contaminated the oysters and caused illness. This incident led to the closure of seven oyster farms in Alaska pending further investigation. In addition, as a result of this case, testing for *V. parahaemolyticus* is now routine throughout Alaska, and many oyster farmers have developed a new practice: When sea temperatures approach 15°C, baskets of oysters are lowered to the colder waters 100 feet below the surface to rid them of bacteria.

This case is only one of a number involving illnesses apparently influenced by rising temperatures around the globe. In Sweden, the number of cases of tick-borne encephalitis has increased due to warmer temperatures that facilitated the northward movement of ticks. In other areas, mosquitoes that spread malaria are being found at higher and higher altitudes as warmer summers expand their range upward. The World Health Organization estimated that 154,000 deaths worldwide in the year 2000 could ultimately be attributed to climate change (although only a fraction of these were due to infectious diseases). Not all the news is grim, though: A British study reported a drop in respiratory infections attributable to respiratory syncytial virus, most likely due to warming temperatures.

EXERCISE 14 REVIEW QUESTIONS

1. Name several everyday ways in which temperatures are maintained artificially high or low to inhibit the growth of bacteria.

2. What class of bacteria would not be inhibited by temperatures below 4°C? Above 50°C?

3. Keeping food cold would seem to promote the growth of psychrophilic bacteria. Why are we generally less concerned about this than we are about the growth of mesophilic bacteria?

pH and Microbial Growth

CASE SYNOPSIS

Botilism: Episode Leads to New FDA Regulation

Over 3 days in February 1989, doctors at Kingston City Hospital in New York admitted 3 persons suffering from what was diagnosed as botulism. The index case was an obese 45-year-old man who first displayed bloating, diarrhea, nausea, and vomiting. His symptoms then progressed to include dyspnea, diplopia, slurred speech, a diminished gag reflex, and muscle weakness severe enough that he was unable to stand without assistance. The other two patients exhibited similar but milder symptoms. Administration of antitoxin was used to counteract the effects of the botulism toxin in all three patients; in addition, the index patient required mechanical ventilation. All three patients recovered. The second and third patients were discharged from the hospital after 8 days while the index patient remained for 29 days.

Patient interviews indicated that a dinner at the home of the index patient was the only common source of exposure. Several other people at the dinner did not contract botulism, and foods eaten by these persons were assumed not to be the source of the bacterium. The single food item common to all the suspect persons was garlic bread; the index case had consumed 15 pieces while the other 2 patients had 2 pieces each. The garlic bread was prepared by mixing 2 teaspoons of garlic-in-oil with warm margarine and spreading it onto a piece of pita bread. The bread was then wrapped in tinfoil and heated at 300°F (149°C) for 20 min prior to serving. The garlic itself was processed (in a commercial setting) by mixing chopped garlic with ice water and extra virgin olive oil, without any other additives or heat treatment. The package was labeled with "keep refrigerated" in small type, and the index patient said that the garlic had been refrigerated since the package was opened 6 months prior to the incidence. In that time, only small quantities had been used, and there was no associated illness.

Clostridium botulinum, the causative agent of botulism, forms tough, heat-resistant endospores that may survive normal food preparation processes, such as boiling, that would kill vegetative cells. The endospores themselves are completely inactive; botulism occurs only when the cells reenter the vegetative growth cycle and begin to produce botulinum, a powerful neurotoxin that paralyzes the muscles. Early symptoms of botulism include double vision and difficulty in speaking and swallowing as the muscles of the eyes, tongue, lips, and throat are paralyzed. As the disease progresses, the diaphragm is paralyzed and death by suffocation results. *C. botulinum* is an obligate anaerobe, meaning that endospores will only enter the vegetative cycle when oxygen is excluded. Furthermore, at low pH levels, germination of the spores, resulting in the production of vegetative cells, is completely inhibited.

Laboratory examination revealed the presence of *Clostridium botulinum* cells in the stool samples of all three patients as well as in the leftover garlic. *C. botulinum* toxin was found, at low levels in the serum of two of the patients and also in the leftover garlic.

Resolution of the Case appears on page 128

Morse, D.L., Pickard, L.K., Guzewich, J.J., Devine, B.D. and Shayegani, M. 1990. Garlic-in-oil associated botulism: Episode leads to product modification. *Amer. J. Public Health*, 80: 1372–1373.

STUDENT LEARNING OUTCOMES

After completing this exercise, you should be able to:

1. Explain the effects of pH on bacterial growth.
2. Determine the pH which provides optimal growth of a given bacterial species.

INTRODUCTION

Any solution has within it a number of hydrogen ions, and the pH of a solution is defined as the negative log of the concentration of these ions, in moles per liter. This formula is usually written as:

$$pH = -\log[H^+]$$

For example, pure water has a concentration of 10^{-7} moles/L of hydrogen ions and therefore has a pH of 7.0, which is described as **neutral.** As the concentration of hydrogen ions increases, the pH will drop below 7 and the solution

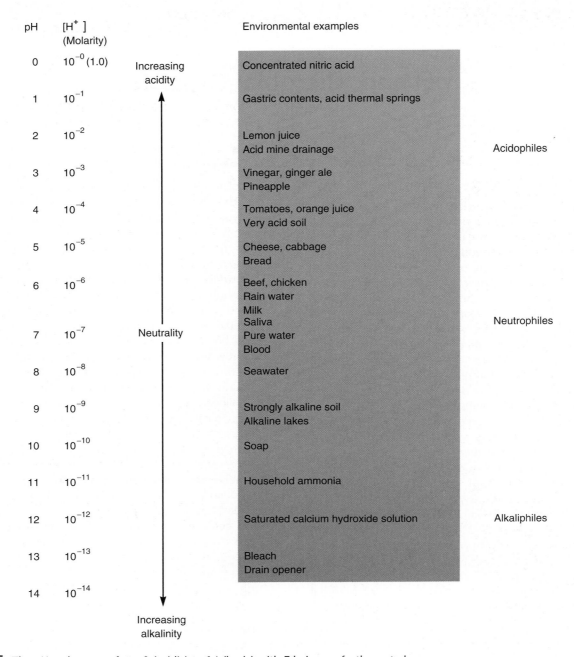

pH	[H⁺] (Molarity)		Environmental examples	
0	10^{-0} (1.0)	Increasing acidity	Concentrated nitric acid	
1	10^{-1}		Gastric contents, acid thermal springs	
2	10^{-2}		Lemon juice Acid mine drainage	Acidophiles
3	10^{-3}		Vinegar, ginger ale Pineapple	
4	10^{-4}		Tomatoes, orange juice Very acid soil	
5	10^{-5}		Cheese, cabbage Bread	
6	10^{-6}		Beef, chicken Rain water Milk Saliva	
7	10^{-7}	Neutrality	Pure water Blood	Neutrophiles
8	10^{-8}		Seawater	
9	10^{-9}		Strongly alkaline soil Alkaline lakes	
10	10^{-10}		Soap	
11	10^{-11}		Household ammonia	
12	10^{-12}		Saturated calcium hydroxide solution	Alkaliphiles
13	10^{-13}		Bleach Drain opener	
14	10^{-14}	Increasing alkalinity		

Figure 15.1 The pH scale ranges from 0 (acidic) to 14 (basic) with 7 being perfectly neutral.

is said to be **acidic.** Conversely, as the number of hydrogen ions in a solution decreases, the pH rises and the solution is deemed **basic.** The pH scale ranges from 0 to 14, with any value below 7 being considered acidic and any value above 7 being considered basic (Figure 15.1).

Microorganisms can be found growing at virtually any pH, although each organism has an optimum pH and is able to grow only within a narrow range on either side of its optimum. The optimum pH of an organism can be used as a means of classification, with organisms described as **neutrophiles** favoring a pH between 5.5 and 8.5, those described as **acidophiles** favoring a pH below 5.5, and those growing best above pH 8.5 being described as **alkaliphiles.** Bacteria

usually maintain an internal environment that is near neutral regardless of external conditions. Most enzymes within the cell are made of protein and are susceptible to damage caused by extremes of pH. When a bacterium is exposed to an environment with a pH outside of its normally acceptable range, internal pH changes can lead to a lethal denaturing of these enzymes. Further damage occurs when the membrane potential of the cell is altered, limiting the production of adenosine triphosphate (ATP). In the laboratory, buffers are added to the media to stabilize the pH within acceptable limits, allowing bacteria to grow well, while in the food industry, the opposite strategy is sometimes employed, with acids being added to foods to prevent the growth of bacteria.

PRE-LAB QUESTIONS

Endospore

1. *Define the following terms:*

Index patient

2. *How are the symptoms in this case—diplopia, dyspnea, slurred speech, a diminished gag reflex, and muscle weakness—consistent with botulism?*

Dyspnea

Diplopia

Buffer

Enzyme

-phile

3. *Based on the optimum pH of the organism, classify each as an acidophile, neutrophile, or alkaliphile:*

Organism	Optimal pH	Classification
Thiobacillus thiooxidans	2.48	
Bacillus acidocaldarius	4.0	
Vibrio cholerae	9.2	
Lactobacillus acidophilus	6.4	
Nitrobacter sp.	8.1	
Escherichia coli	6.5	
Staphylococcus aureus	7.3	
Sulfolobus acidocaldarius	2.5	
Streptococcus pneumoniae	7.8	

PERIOD ONE

To determine the effects of pH on growth, a single organism will be used to inoculate five tubes of nutrient broth that differ only in pH. The tubes will then be incubated at identical temperatures for the same amount of time, and the bacterial growth in each tube will be determined. Because the environment will be the same except for the pH, any differences in the amount of growth can be attributed to the difference in pH. Determine your assigned organism from the table:

Organism	Student number
Escherichia coli	1, 5, 9, 13, 17, 21, 25
Alcaligenes faecalis	2, 6, 10, 14, 18, 22, 26
Lactobacillus acidophilus	3, 7, 11, 15, 19, 23, 27
Staphylococcus aureus	4, 8, 12, 16, 20, 24, 28

MATERIALS

Each student should obtain:
Five tubes of nutrient broth, one each of pH 3, 5, 7, 9, and 11
Broth cultures of one of the following:
  Escherichia coli
  Alcaligenes faecalis
  Lactobacillus acidophilus
  Staphylococcus aureus
Inoculating loop
Marking pen

PROCEDURE

1. Obtain five tubes of pH-adjusted nutrient broth, one each of pH 3, 5, 7, 9, and 11. Label each tube with your name and organism.

2. Inoculate each tube with your assigned organism. To transfer approximately the same number of bacteria to each tube, gently mix the culture until it appears uniformly turbid. Repeat this process between each inoculation. After the five tubes are inoculated, incubate them at 37°C for 24–48 h.

QUESTIONS—PERIOD ONE

1. The pH scale is logarithmic, meaning that a change of one pH unit represents a 10-fold increase or decrease in the hydrogen ion concentration. For example, pH 8 is 10 times more basic than pH 7 and 10 times more acidic than pH 9. Given this information, how many times more acidic (or basic) is:

pH 7 than 9

pH 7 than 11

pH 3 than 11

pH 11 than 5

pH 9 than 5

2. As the bacteria grow in a liquid culture, they cause the media to become cloudy, with the degree of cloudiness being directly proportional to the level of growth. This cloudiness, or turbidity, can be measured on a scale of 0 (clear) to 10 (completely opaque). The table provides an example of visual and measured turbidity for a hypothetical bacterium. Draw a growth curve for this organism. Plot the turbidity (on the vertical axis) versus pH (on the horizontal axis). Identify the minimum, maximum, and optimum pH values for this organism? How would you classify its pH tolerance?

pH	Visual appearance of the culture	Measured turbidity
3	0	0
5	3	0.22
7	4	0.41
9	6	0.57
11	1	0.02

PERIOD TWO
Retrieve your tubes from the incubator. Evaluate the growth in each of the five broth tubes. Follow one of the procedures below to evaluate the growth in each tube.

pH	E. coli		A. faecalis		L. acidophilus	
	Visual reading	Spectrophotometer (%T, Abs)	Visual reading	Spectrophotometer (%T, Abs)	Visual reading	Spectrophotometer (%T, Abs)
3						
5						
7						
9						
11						

Spectrophotometer Procedure
1. Mix each tube until a uniform turbidity is obtained. If spectrophotometers are available, record the transmittance (%T) and absorbance (Abs) of each tube in the table.
2. Graph the results (Absorbance vs. pH) for your assigned organism. Gather information about the other organisms from your classmates, and graph that on the same set of axes.

Visual Procedure (No Spectrophotometer)
1. If a spectrophotometer is not available, rate each tube on a scale of 1 (clear) to 10 (very turbid), after mixing the tube to obtain a uniform turbidity.
2. Graph the results (Turbidity vs. pH) for your assigned organism. Gather information about other organisms from your classmates, and graph that on the same set of axes.

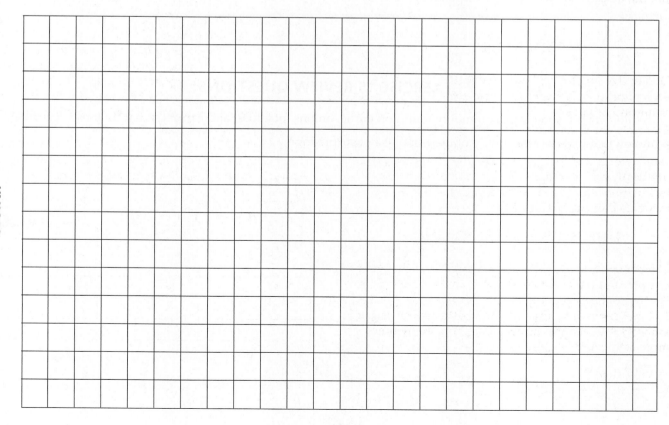

Growth

pH

QUESTIONS—PERIOD TWO

1. What is the minimum, maximum, and optimum pH level for each of the four organisms examined:

	Minimum pH	Optimum pH	Maximum pH
E. coli			
A. faecalis			
L. acidophilus			
S. aureus			

2. Which of the bacteria used in this exercise would grow in processed food that had a pH of 7.0? If the pH of the food was lowered to pH 4.5 during processing, how would that affect each bacterial species?

RESOLUTION OF THE CASE

The pH of the leftover garlic was found to be 5.7, well within the acceptable range for *C. botulinum* growth. This case was but one of a number of episodes of botulism involving vegetable roots or tubers (potatoes, onions, garlic) cooked or stored in oil. If properly refrigerated, these products are safe, but when these products are left unrefrigerated, the bacterium can quickly multiply. Acting in response to this case, the Food and Drug Administration ordered companies to stop producing garlic-in-oil products that are protected only by refrigeration. These products must now be acidified, usually by the addition of phosphoric or citric acid, to prevent the growth of *C. botulinum*. When a product is acidified to a pH of 4.6 or less, according to FDA's Good Manufacturing Practices, inhibition of the growth of *C. botulinum* is assured.

EXERCISE 15 REVIEW QUESTIONS

1. *C. botulinum* endospores are commonly found on garlic, onions, potatoes, and carrots. What do these vegetables have in common that makes the presence of these spores not surprising?

2. *Vibrio cholerae* is the bacterium responsible for cholera, a disease of the gastrointestinal tract. Normally, over one million cells must be ingested to cause infection, but the consumption of antacid significantly lowers this number. Do you think *V. cholerae* is most likely an acidophile, neutrophile, or alkaliphile? Why does antacid lower the infectious dose of this bacterium?

3. The FDA order in this case specifically targeted garlic packed in oil. In what way did the oil help to promote the growth of *C. botulinum*? (Hint: *C. botulinum* is an obligate anaerobe.)

REFERENCE

CDC. *Botulism.* http://emergency.cdc.gov/agent/botulism/.

NOTES

Effects of Osmotic Pressure on Bacterial Growth

CASE SYNOPSIS

Vibrio vulnificus Infection Traced to Sewage Spill—Hawaii, 2006

In 2006, a mortgage broker in his thirties fell into a harbor in Hawaii 6 days after a sewage pipe failure had allowed 48 million gallons of sewage to flow into the harbor water. The man was admitted to the hospital, and within days, he died from massive organ failure caused by septicemia with *Vibrio vulnificus.* You may be familiar with *Vibrio cholerae,* the diarrhea agent. Its relative, *V. vulnificus,* tends to invade breaks in the skin and then proliferate in the bloodstream, often causing massive infections necessitating amputation, and sometimes even leading to death.

The man sickened by the sewage spill had cuts on his feet, providing an obvious portal of entry for water-borne bacteria into his bloodstream. The infection seemed systemic because he had a high fever and symptoms in many areas of his body. For these reasons, the most logical diagnostic procedure was to take a blood sample and perform blood cultures. When this was done, his blood grew *V. vulnificus.*

Bacteria, like other organisms, are acutely sensitive to the concentration of solutes—most commonly salt—in their environment. While many bacteria are restricted to either marine or freshwater habitats, *V. vulnificus* has evolved to thrive in brackish waters, which are part marine and part fresh.

Resolution of Case appears on page 134

The Honolulu Advertiser. June 10, 2006. *Bacteria Presence was not Abnormal.* http://the.honoluluadvertiser.com/article/2006/Jun/10/ln/FP606100345.html.

STUDENT LEARNING OUTCOMES

After completing this exercise, you should be able to:

1. Explain the effects of osmotic pressure on bacterial growth.
2. Determine the salt tolerance of a given bacterial species.

INTRODUCTION

Few environmental factors have as great an effect on an organism as does water. It is by far the most abundant component of cytoplasm and is absolutely required for the biochemical reactions that constitute the metabolism of a cell. Without an adequate supply of water to act as a solvent for chemical reactions and to serve as a source of electrons and hydrogen ions, a cell's metabolism will slow to zero. Molecules of water will always diffuse from areas of low solute concentration, where water molecules are abundant, to areas of high solute concentration, where water is less available. When the diffusion of water molecules occurs across a semipermeable membrane, such as a cell membrane, the process is called **osmosis,** and the pressure created by the flow of water into the cell is known as **osmotic pressure.** In most cases, the internal environment of a cell has a higher concentration of **solutes** (e.g., salts, sugars, and proteins) than is found in the environment outside the cell. In other words, the concentration of water is higher outside the cell than inside, and water will have a tendency to flow into the cell (Figure 16.1). This type of solution is termed **hypotonic** and will lead to an increase in osmotic pressure. Were it not for the rigid cell wall found in bacteria cells, these cells would get larger and larger until they finally lysed.

When the solute concentration outside the cell is the same as that found inside, the solution is considered **isotonic.** In this case, water molecules are crossing the cell membrane in both directions with equal frequency, and there is no net movement of water (see Figure 16.1). Maintaining an isotonic environment is important for cells protected by only a thin cell membrane, such as animal cells. Intravenous saline solution, sports drinks, and media for growing animal cells are all designed to maintain an isotonic environment.

When the concentration of solutes outside the cell is greater than the concentration inside the cell, the solution is said to be **hypertonic.** In this case, the net flow of water will be out of the cell. As water diffuses out, the cell undergoes plasmolysis, a condition where the cell membrane shrinks away from the cell wall and the cytoplasm dehydrates, causing often irreversible damage to the enzymes needed for continued metabolism (see Figure 16.1). Long before anything was known about microorganisms, it was understood that salting or drying food (both of which reduce the amount of available water) would prevent spoilage and that foods high in sugars, such as jellies and chocolate, could be stored without refrigeration. Bacterial cells landing on such a concentrated source of sugar or salt would quickly undergo plasmolysis.

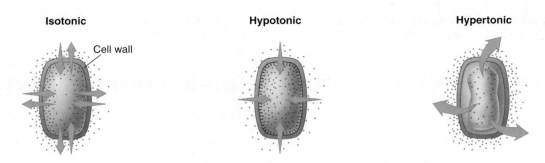

Figure 16.1 Effect of varying solute concentrations on bacterial cells. Under isotonic conditions, the concentration of water inside the cell equals the concentration found outside the cell, and there is no net movement of water. Under hypotonic conditions, the concentration of water is greater outside the cell, resulting in a net movement of water into the cell and an increase in osmotic pressure as the cell membrane is pushed forcefully outward against the cell wall. Under hypertonic conditions, the concentration of water is greater within the cell, and the net movement of water is out of the cell, leading to plasmolysis as the cell membrane shrinks away from the cell wall.

Microorganisms can be grouped according to their ability to grow when exposed to salt concentrations of varying degrees. The majority of bacteria grow best when the salt concentration in the surrounding environment is about 0.9% to 1.0%. Not surprisingly, the human body falls within this range. In contrast, **halophiles** require much higher concentrations of salt to grow optimally, with some obligate halophiles such as *Halococcus* requiring 13% NaCl to grow at all and 25% to grow optimally. Between these two groups exists the facultative halophiles or **halotolerant** bacteria. This group is exemplified by *Staphylococcus aureus*, which can grow in conditions ranging from 1% to 13% NaCl.

When a microorganism is able to grow in high sugar concentrations, it is termed an osmophile. Most osmophiles are yeasts, including members of the genera *Xeromyces* and *Saccharomyces*, both of which can grow on jams and jellies.

PRE-LAB QUESTIONS

1. *Define the following terms:*

Semipermeable membrane

Lysis

Plasmolysis

Halophile

Brackish

2. *What is the relationship between a solute and a solvent?*

3. *Explain why it is important to differentiate movement of water from net movement of water during osmosis?*

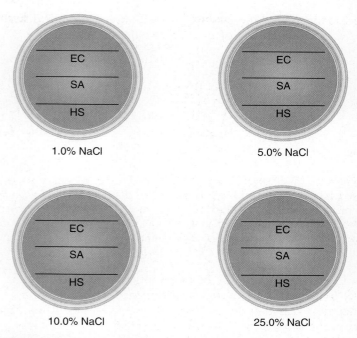

Figure 16.2 Inoculation procedure for NaCl plates.

MATERIALS

Each group should obtain:

One plate of nutrient agar (1.0% NaCl)

One plate of nutrient agar (5.0% NaCl)

One plate of nutrient agar (10.0% NaCl)

One plate of *Halobacterium* agar (25.0% NaCl)

Fresh broth cultures of:

   Escherichia coli

   Staphylococcus aureus

Solid culture (slant or plate) of *Halobacterium salinarium*

Inoculating loop

Marking pen

PERIOD ONE

PROCEDURE

1. Label the bottom of the four nutrient agar plates (one of each salt concentration) as indicated in Figure 16.2.
2. Use a loop to inoculate each plate with a single streak of each bacterium.
3. Incubate all the plates at 37°C for a minimum of 5 days.

QUESTIONS—PERIOD ONE

1. Complete the table.

Cell type	Internal solute concentration	External environment	Solution (hypo-, hyper-, or isotonic)	Osmotic pressure in cell increase (↑) or decrease (↓)	Plasmolysis (+/−)	Lysis (+/−)
Bacterial cell	0.9% NaCl	0.9% NaCl				
Bacterial cell	17.0% NaCl	8.0% NaCl				
Bacterial cell	1.0% NaCl	0.01% NaCl				
Bacterial cell	17.0% NaCl	36.0% NaCl				
Red blood cell	0.9% NaCl	2.0% NaCl				
Red blood cell	1.0% NaCl	0.0% NaCl				

2. How would the growth of a facultative halophile differ from the growth of an obligate halophile on media containing:

    1.0% NaCl

    10% NaCl

    20% NaCl

PERIOD TWO

1. Remove the plates from the incubator, and record the growth of each bacterial species ($+/-$) on each type of media.

Organism	1.0% NaCl	5.0% NaCl	10.0% NaCl	25.0% NaCl
E. coli				
S. aureus				
H. salinarium				

RESOLUTION OF THE CASE

To determine the potential threat to others, a microbiology team from the University of Hawaii was sent to investigate bacteria levels in the harbor water. The team counted the number of *V. vulnificus* in the water and concluded that neither more nor less were present than in other brackish waters around the area. However, the count was conducted 11 days after the end of the sewage spill (and, coincidentally, 11 days after the man had been in the water and became ill). So it is hard to say how high the levels of bacteria were at the time the man was in the water. Nevertheless, the team declared the waters "safe" based on their observation that the bacteria levels were similar to those in other waters in the area.

Climate change seems to be causing the overgrowth of *V. vulnificus*, as well as other bacteria, in waters that previously were less supportive of their growth. At the same time, the bacteria also continue to grow in southern waters. The result was a 51% increase in *Vibrio*-caused wound infections in a recent 6-year period. In fact, more people die annually from *Vibrio* wound infections than from shark attacks.

It seems that a new definition for microbiologically "safe" water may need to be devised.

EXERCISE 16 REVIEW QUESTIONS

1. The incidence of *V. vulnificus* infections is increasing, especially in northern parts of the United States, compared to 10 years ago. Can you think of any reason why this might be happening?

2. Processed ham generally has a salt content of about 5%. When food poisoning is linked to the consumption of processed ham, *Staphylococcus aureus* is almost always responsible. Why are *E. coli* or *H. salinarium* not responsible for a greater percentage of infections?

3. Why is it that honey and corn syrup can be stored at room temperature for long periods of time without fear of bacterial growth?

4. Antibiotics such as penicillin prevent bacterial cells from synthesizing cell wall components, but otherwise they do nothing to kill the cell. How do these antibiotics cause the death of bacterial cells? Why does penicillin not kill animal cells (such as those of a person who takes penicillin)?

REFERENCE

CDC. Vibrio vulnificus: General information. http://www.cdc.gov/nczved/divisions/dfbmd/diseases/vibriov/.

Lethal Effects of Ultraviolet Light

CASE SYNOPSIS

Gastrointestinal Outbreak Traced to Interactive Fountain—New York, March 2006

The popular interactive water fountain, or "sprayground," at Seneca Lake State Park closed for the remainder of the summer on August 17, 2005, after the New York State Health Commissioner announced that an outbreak of cryptosporidiosis had been traced to the park. Cryptosporidiosis results from ingesting water contaminated with the protozoan *Cryptosporidium parvum*. Symptoms include diarrhea, vomiting, fever, and abdominal cramps.

The sprayground itself is an 11,000-square-foot deck with hundreds of individual water jets, spouts, and hoses. After being sprayed into the air, water flows back into a holding tank below the play area before being recycled

through the jets once again. This type of water playground has become immensely popular because even small children can play there without the risk of drowning commonly associated with swimming pools.

Unfortunately, the design of the Seneca Lake water playground allowed contaminants (e.g., vomit, feces, and dirt) to wash into the water-holding tanks supplying it. Testing revealed the presence of *Cryptosporidium* in those water tanks. Even though the water in the tanks was filtered and chlorinated, *Cryptosporidium* is small enough to pass through a filter and is also resistant to chlorine. Until methods could be implemented that would ensure the well-being of patrons, the water park would remain closed.

Resolution of the Case appears on page 141

August 26, 2006. Sprayground at Seneca Lake State Park to reopen. *Ithaca Journal,* vol. 192, p. 1B.

STUDENT LEARNING OUTCOMES

After completing this exercise, you should be able to:

1. Evaluate the effects of ultraviolet light on various species of bacteria.

INTRODUCTION

Electromagnetic radiation is a form of energy. This energy travels in waves and, depending on the wavelength, displays very different properties. The shortest wavelengths (<0.00001–100 nm) include gamma rays and x-rays (Figure 17.1) while radiation with longer wavelengths are classified as ultraviolet light (100–380 nm), visible light (380–750 nm), infrared light (750–1000 nm), and radio waves (>1000 nm). When this energy enters a living organism, it has the ability to cause considerable damage to cells both by interacting with DNA and by breaking down water to form highly reactive free radicals. Although gamma rays and x-rays are very effective for sterilizing thermolabile objects, such as plastic Petri dishes, these high energy (short wavelength) forms of radiation tend to be dangerous to work with. More commonly used is the longer wavelength (lower energy) form of radiation known as **ultraviolet light.** Ultraviolet (UV) light can be broken into three distinct groups,

UV-A (with wavelengths between 315 and 400 nm), UV-B (280–315 nm), and UV-C (100–280 nm). Exposure of bacteria to UV-C for more than a short time results in the formation of pyrimidine dimers (Figure 17.2) as the energy, which peaks at a wavelength of 254 nm, is used to form new covalent bonds between adjacent pyrimidines (thymine-cytosine, cytosine-cytosine, or, most commonly, thymine-thymine). These dimers distort the DNA helix, making replication and transcription difficult; if the dimers are numerous enough, this can result in the death of the cell. The lower energy found in UV light makes it safer to work with but at the same time reduces its effectiveness. While x-rays and gamma rays are able to penetrate solid objects (as anyone who has ever seen an x-ray knows), ultraviolet light is easily blocked and so is useful only for disinfection of surfaces. Because of its combination of effectiveness and relative safety, UV light is commonly used to disinfect laboratory work surfaces as well as air and water where methods such as filtration or chlorination are ineffective or impractical. Although UV light is referred to as relatively safe, this is only in comparison to the more penetrating forms of radiation. Do not expose your skin unnecessarily to UV light and never look directly into the UV light source as eye injury can result.

In this exercise, you will be determining the resistance to ultraviolet light of both endospore-forming and non-endospore-forming bacteria. Recall that endospores are a protective structure formed by some species of bacteria that allow them to survive harsh conditions, including ultraviolet light. Although protozoan cysts are not nearly as

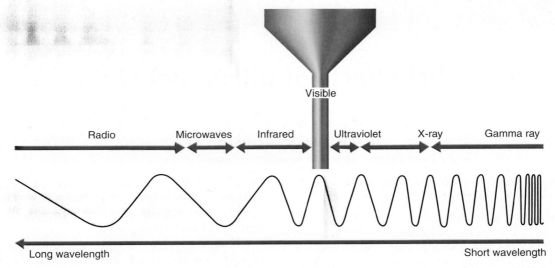

Figure 17.1 The size of its wavelength within the electromagnetic spectrum determines the energy carried by a wave and how it is perceived by humans. Notice that only a small portion of the spectrum appears to us as visible light.

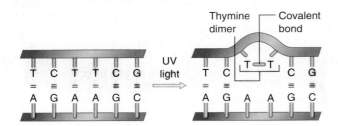

Figure 17.2 Ultraviolet light induces the formation of thymine dimers in DNA. These dimers disrupt the structure of the DNA, interfering with replication and transcription. If enough dimers accumulate, essential genes will not be transcribed or the DNA will not be replicated, leading to the death of the cell.

robust as bacterial endospores, the two structures perform the same task of protecting cells in harsh environments.

PRE-LAB QUESTIONS

1. *Define the following terms:*

   Thermolabile

   Wavelength

   Pyrimidine

2. *In addition to having little penetrating power, the effectiveness of ultraviolet light diminishes greatly as the distance between the light and the surface to be treated increases. How would these two facts affect the design of a water disinfection system?*

3. *Would you expect all microorganisms to be equally susceptible to UV light? What bacterial structure could help to protect a cell against ultraviolet radiation?*

PERIOD ONE

The class will be divided into 10 groups, with each group being responsible for exposing a group of organisms to ultraviolet light for a specified time. Use the table to determine your assignment.

Organism	Group No. 1	Group No. 2	Group No. 3	Group No. 4	Group No. 5	Group No. 6	Group No. 7	Group No. 8	Group No. 9	Group No. 10
Bacillus megaterium (48 h)	15 sec	30 sec	60 sec	90 sec	2 min	5 min	10 min	20 min	30 min	45 min
Staphylococcus aureus (24–48 h)	15 sec	30 sec	60 sec	90 sec	2 min	5 min	10 min	20 min	30 min	45 min

MATERIALS

Each group should obtain:

Two plates of nutrient agar

Broth culture of *S. aureus* containing a swab for spreading

Nutrient broth suspension of *B. megaterium* (48-h culture)* containing a swab for spreading

Two index cards or similar light shields

Laboratory marker

Access to a short wave (~254 nm) ultraviolet lamp

*Broth suspension is prepared by aseptically pouring a tube of nutrient broth into a slant culture, then mixing gently with a sterile swab to distribute the cells within the media.

PROCEDURE

1. Label each nutrient agar plate with your name, lab time, and group number.
2. Label each plate with the name of one of your two organisms. Draw a line across the bottom of each plate dividing it in two. Label one-half "exposed" and the other "control."
3. Use a sterile swab to inoculate each plate. Before removing the swab from the liquid culture, lightly touch it to the inside of the tube to remove excess broth.
4. Swab each plate by spreading the organism over the entire surface of the media. Rotate the plate several times during the spreading process to ensure complete coverage. Repeat this process for the other organism.
5. Depending on the size and style of the UV light, it may be possible to expose more than one plate at a time. If this seems possible, just be sure that each plate is fully exposed to the light. Expose each plate to the ultraviolet light as follows:
   a. Remove the Petri dish cover, and place it face down on a freshly disinfected workbench.
   b. Cover the half of the plate marked "control" with an index card or small metal shield.
   c. Place the plate under the light, and begin timing immediately. When the prescribed time has elapsed, remove the plate from the light and replace the cover.
6. Invert each plate, and incubate at 37°C for 24–48 h.

QUESTIONS—PERIOD ONE

1. Why was half of the plate shielded from the ultraviolet light? How much bacterial growth would you expect on the shielded half of the plate?

2. How much growth would you expect on the side of the plate exposed to the ultraviolet light, and how would you expect this growth to change as the time of exposure to the light increased?

3. What effect does the age of a bacterial culture have on the production of endospores?

PERIOD TWO

Retrieve your plates from the incubator. Compare the growth between the control (protected) side of your plate and the UV-exposed side of the plate. Score the growth on the exposed half of the plate using the following scale:

No growth	−
Slight growth (1–20 colonies)	+
Moderate growth (21–100 colonies)	++
Abundant growth (100–500 colonies)	+++
No reduction in growth compared to control	++++

Record the results of the other groups in the table:

Organism	Group No. 1	Group No. 2	Group No. 3	Group No. 4	Group No. 5	Group No. 6	Group No. 7	Group No. 8	Group No. 9	Group No. 10
B. megaterium (48 h)	15 sec	30 sec	60 sec	90 sec	2 min	5 min	10 min	20 min	30 min	45 min
Growth										
S. aureus	15 sec	30 sec	60 sec	90 sec	2 min	5 min	10 min	20 min	30 min	45 min
Growth										

Graph the survival (growth) of each organism (i.e., the exposed half of the plate) as a function of time of exposure to ultraviolet light (Figure 17.3). Use a different color for each organism.

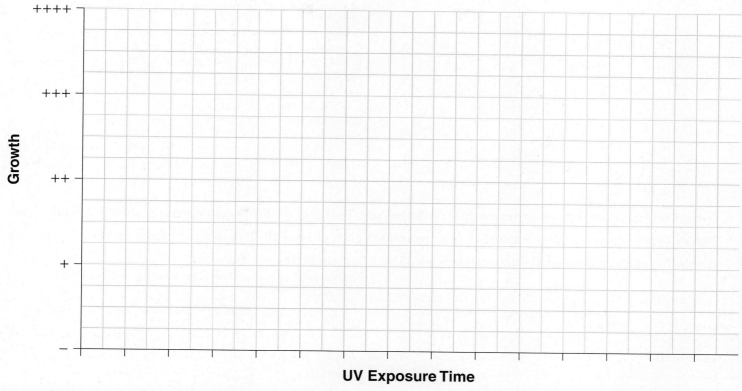

Figure 17.3 The survival of each organism is reported by graphing the number of colonies on the exposed side of the plate (vertical axis) versus the time of exposure to ultraviolet light (horizontal axis), using the scoring table found in the lab.

RESOLUTION OF THE CASE

Estimates of the number of cases of cryptosporidiosis traceable to the Seneca Lake sprayground ranged as high as 3,800 people spread across 37 counties; fortunately, no fatalities were recorded. The total could have been even higher had the park not been shut down in mid-August.

Although *Cryptosporidium* is resistant to chlorine, it is susceptible to ultraviolet (UV) light. Therefore, to prevent a recurrence of cryptosporidiosis, the New York State Department of Health required that recirculated water in interactive fountains be disinfected using UV light. Installation of a donated $65,000 UV system allowed the park to reopen the following summer. Ultraviolet disinfection of water has long been used in Europe, but New York's 2006 ultraviolet disinfection requirements were the first in the United States.

EXERCISE 17 REVIEW QUESTIONS

1. In some laboratories and medical offices, ultraviolet lights are automatically switched on when the last light in the office is switched off. Why are the lights set up in this manner?

2. In this experiment, the effects of UV light on bacteria could have been detected without covering half of the plate. Why was the plate partially covered?

3. Ultraviolet lights lose their effectiveness after prolonged use. Testing of the lights can be accomplished in a manner very similar to the experiment you conducted. If an inoculated plate of susceptible bacteria yielded a great deal of growth despite having been exposed to ultraviolet light for 45 min, how could this be interpreted?

REFERENCE

CDC. *Cryptosporidium* Infection Cryptosporidiosis. http://www.cdc.gov/ncidod/dpd/parasites/cryptosporidiosis/factsht_cryptosporidiosis.htm.

Evaluation of Disinfectants: Use-Dilution Method

CASE SYNOPSIS

Anaphylaxis following Cystoscopy Caused by a High-Level Disinfectant—2004

A urology practice reported nine cases of anaphylactic shock (a severe and sometimes fatal allergic reaction) in patients undergoing cytoscopy. This procedure requires that a cytoscope, a thin tube equipped with a fiberoptic lighting system and telescopic lenses, be passed through the urethra and into the bladder. Cytoscopes are heat-sensitive and are disinfected between uses using a chemical solution. All nine cases of anaphylaxis occurred shortly after the practice switched from a disinfecting solution containing glutaraldehyde to one containing ortho-phthaladehyde in the hope of finding an equally effective chemical that presented fewer workplace hazards.

Resolution of Case appears on page 149

Sokol, W.N. 2004. Nine episodes of anaphylaxis following cystoscopy caused by Cidex OPA (ortho-phthalaldehyde) high-level disinfectant in four patients after cytoscopy. *J. Allergy Clin. Immunol.,* 114(2): 392–397.

STUDENT LEARNING OUTCOMES

After completing this exercise, you should be able to:

1. Understand how the effectiveness of disinfectants is influenced by dilution and time of exposure.
2. Perform and interpret a modified use-dilution test.

INTRODUCTION

Whether in a dentist's office, a physical therapy center, or a hospital, proper control of microorganisms can prevent the spread of infectious agents from person to person. The choice of a chemical disinfection method is based not only on its effectiveness against a wide variety of common microbes but also on its compatibility with laboratory equipment, ability to work even in the presence of contaminating organic material such as blood and protein, user safety, and cost. In addition, all of the above characteristics of a disinfectant are affected by the concentration of the chemical and the length of time microbes are exposed to the chemical.

Complicating matters even further, not all microorganisms are equally susceptible to chemical disinfection, with bacterial endospores generally regarded as most resistant, followed by mycobacteria, Gram-negative bacteria, and finally Gram-positive bacteria. While prions, fungi, and viruses also present risks in the clinical environment, the overwhelming majority of infections are due to bacteria, causing many labs to concentrate on a chemical disinfectant's effectiveness with regard to these organisms. The best disinfectants are those that are broad spectrum, stable, nonstaining, noncorrosive, and relatively safe for users. Choosing a method of chemical disinfection requires balancing all of these factors while also realizing that a single "best" method does not exist for all circumstances.

Chemicals used for disinfection work by causing many small injuries to the cell. Brief exposure to a very dilute disinfectant, for example, may result in the development of a single hole in a cell's membrane. Eventually, the added effects of these insignificant injuries (e.g., many holes in the cell membrane) combine to kill the cell. Two ways to increase the number of these small injuries and, in turn, to increase the effectiveness of a disinfectant are to (1) increase the concentration of the disinfecting agent and (2) increase the amount of time that the chemical is left in contact with the microorganisms. Your goal is to determine both the optimal concentration and minimal amount of time that is needed to achieve complete disinfection.

In order to be labeled as a hospital disinfectant, a chemical must have successfully passed the American Official Analytical Chemist's **use-dilution test**. This test measures how well a disinfectant works against a high microbial load dried onto a hard, nonporous surface such as a benchtop, floor, or piece of heat-sensitive diagnostic equipment. The test involves soaking small stainless steel cylinders in a broth of *Staphylococcus aureus, Salmonella choleraesuis,* or *Pseudomonas aeruginosa.* After soaking for 15 min, the cylinders are removed and placed in an incubator to allow the film of microbes to dry to the cylinder. Each cylinder is then submerged in a test chemical for 10 min, after which it is removed and excess chemical is allowed to drain away. The cylinder is then transferred to a broth tube and allowed to incubate. If live bacteria are present on the cylinder, they will grow and lead to a failing test result. If the broth appears clear after incubation, it can be concluded that no bacteria were able to survive the disinfectant. Today, a modified use-dilution test will be used to evaluate the efficacy of several disinfectants, determining both the most effective concentration that ensures disinfection and the shortest length of time that instruments must remain in contact with the solution to achieve complete disinfection.

PRE-LAB QUESTIONS

1. *Define the following terms:*

Disinfectant

Antiseptic

Sanitization

Broad spectrum

2. *Why would a chemical disinfectant generally be inappropriate for use as an antiseptic?*

3. *Why are chemical disinfectants so often irritating to the skin, eyes, lungs, etc.?*

4. *Why would a Gram-negative cell be more resistant to chemical disinfection than a (vegetative) Gram-positive cell?*

5. *Organisms commonly used to test the effectiveness of disinfectants include* Bacillus megaterium, Mycobacterium bovis, Pseudomonas aeruginosa, *and* Staphylococcus aureus. *Why are these species often used, and (without performing any tests) how would you rank their susceptibility to disinfection?*

TABLE 18.1

Chemical	*Staphylococcus aureus*	*Pseudomonas aeruginosa*
Sodium hypochlorite, 5.25%	Group 1	Group 5
Hydrogen peroxide, 3%	Group 2	Group 6
Amphyl/Lysol (undiluted)	Group 3	Group 7
Ethanol 99%	Group 4	Group 8

TABLE 18.2   **Culture tube preparation**

Nutrient broth tube label	Washer exposed to	Time
Control	No disinfectant	Not applicable
Full strength	Full-strength disinfectant	15 min
1:1	Diluted disinfectant (1:1)	15 min
1:10	Diluted disinfectant (1:10)	15 min
1:50	Diluted disinfectant (1:50)	15 min
1:250	Diluted disinfectant (1:250)	15 min
1 min	Working-strength disinfectant (see Table 18.1)	1 min
5 min	Working-strength disinfectant (see Table 18.1)	5 min
10 min	Working-strength disinfectant (see Table 18.1)	10 min
20 min	Working-strength disinfectant (see Table 18.1)	20 min

PERIOD ONE

The effectiveness of each disinfectant will be tested against *Staphylococcus aureus* or *Pseudomonas aeruginosa*. Quarter-inch stainless steel washers are used as a substrate to which the bacteria will cling. Each student group will be responsible for testing the effect of a single disinfectant on a single organism. The bactericidal effect will be measured at multiple concentrations and exposure times so as to determine the most effective concentration of disinfectant and shortest time of exposure that proves effective. Table 18.1 provides information on the organisms and disinfectants that will be used by each group.

MATERIALS

Each group should obtain:

Ten tubes of nutrient broth

Five empty, sterile test tubes

One empty Petri dish

One piece of sterile filter paper slightly smaller than the Petri dish

50 ml sterile water

Access to broth cultures of *S. aureus* and *P. aeruginosa* containing stainless steel washers

1- and 10-ml pipettes and pipette pumps

Forceps

Marking pen

PROCEDURE

1. Label a plastic Petri dish with the number of your group. Briefly flame a pair of forceps, and use them to transfer a piece of filter paper to the dish.
2. Using flamed forceps, remove 10 washers (one at a time) from the beaker containing your assigned organism. Place each washer on the filter paper, being careful not to roll or scrape the washer across the paper as this will remove many of the organisms.
3. Allow the bacteria to dry by placing the Petri dish in a 37°C incubator for at least 20 min.
4. Study Table 18.2 prior to beginning your inoculations and incubations. Continue to refer to it as you work.

5. Label 10 tubes of nutrient broth with your group number. Set these aside for later inoculation.
6. Using five empty test tubes, prepare serial dilutions of your assigned chemical as follows (Figure 18.1):
   a. Add 10 ml of the concentrated chemical to tube 1. Add 5 ml of sterile water to tube 2 and 8 ml to tubes 3, 4, and 5.
   b. Use a pipette to transfer 5 ml of your chemical from tube 1 to tube 2. Pipette up and down several times to mix.
   c. Using a fresh pipette for each transfer, transfer 2 ml from tube 2 to tube 3, 2 ml from tube 3 to tube 4, and 2 ml from tube 4 to tube 5. Pipette up and down several times to mix after each transfer.

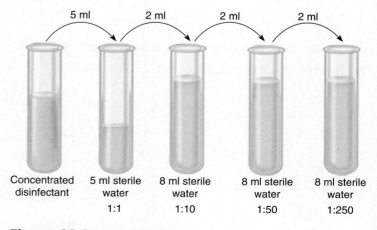

Figure 18.1   Dilution scheme for disinfectants.

d. Your tubes should now contain the following dilutions:

   Tube 1 contains disinfectant in its concentrated form.
   Tube 2 is a 1:1 dilution of tube 1.
   Tube 3 is a 1:10 dilution of tube 1.
   Tube 4 is a 1:50 dilution of tube 1.
   Tube 5 is a 1:250 dilution of tube 1.

7. Retrieve the Petri dish containing your washers from the incubator. Be sure the culture has had time to thoroughly dry to the washers.

8. Remembering to briefly flame your forceps prior to each use, transfer one washer to a tube of nutrient broth. Label the tube "control."

9. After noting the time, transfer one washer to the tube containing full-strength disinfectant. In a similar manner, transfer a single washer to each of the four tubes of diluted disinfectant. Allow the washers to remain in the disinfectant solutions for 15 min.

10. Working with a single washer at a time, use a pair of flamed forceps to remove the washer, shake gently to remove excess disinfectant and place in a tube of nutrient broth.

11. Beginning with the next step, you will need to keep careful track of time. Be sure to note the time that the washers first enter the disinfectant-containing tube.

12. Transfer four washers to the tube containing disinfectant at its most commonly used dilution (full strength for alcohol, bleach, and hydrogen peroxide; 1:100 dilution for Amphyl/Lysol). Arrange the washers in the tube so that they are not stacked on top of one another.

13. After each washer has reached its predetermined exposure time, it should be carefully removed from the disinfectant, carefully shaken to remove excess chemical, and placed into a tube of nutrient broth (Figure 18.2). Label each tube with the proper time.

   • Remove the first washer after 1 min
   • Remove the second washer after 5 min
   • Remove the third washer after 10 min
   • Remove the fourth washer after 20 min

14. Incubate the inoculated tubes at 37°C for 24–48 h.

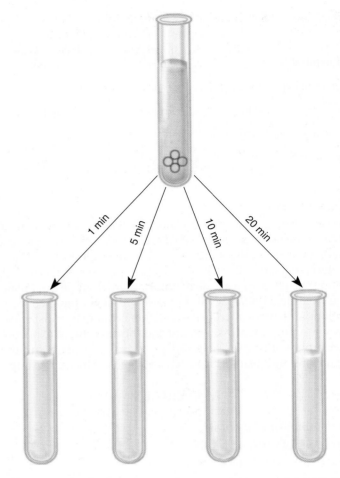

Figure 18.2 Distribution of washers from disinfectant to nutrient broth.

QUESTIONS—PERIOD ONE

1. How is bacterial growth in a broth recognized (in other words, what would the tube look like)?

2. What is the purpose of the control tube in this experiment? What does growth in the control tube indicate? What does a lack of growth indicate?

3. Calculate the final concentration of each of the following dilutions:

100% ethyl alcohol diluted 1:1, 1:2, and 1:10

52,500 ppm sodium hypochlorite (standard household bleach) diluted 1:10, 1:250, and 1:1000

5% iodine diluted 1:1, 1:10, and 1:50

PERIOD TWO

EVALUATION

Retrieve your tubes from the incubator. Recall that if any cells remained viable on the washer after its exposure to the disinfectant, these cells would multiply to produce a turbid culture. A completely clear tube of nutrient broth indicates that the disinfectant was effective in killing all bacterial cells on the washer. Record your results in Tables 18.3–18.6, using + if growth is present and – if no growth is present. Recall that Table 18.2 can help you in differentiating your culture tubes. Gather the results from the other groups in the class.

TABLE 18.3	Optimal concentration of disinfectant (*P. aeruginosa;* growth of control _____)				
Chemical	Full strength	1:1 Dilution	1:10 Dilution	1:50 Dilution	1:250 Dilution
Sodium hypochlorite, 5.25%					
Hydrogen peroxide, 3%					
Amphyl/Lysol (undiluted)					
Ethanol, 99%					

TABLE 18.4	Optimal concentration of disinfectant (*S. aureus;* growth of control _____)				
Chemical	Full strength	1:1 Dilution	1:10 Dilution	1:50 Dilution	1:250 Dilution
Sodium hypochlorite, 5.25%					
Hydrogen peroxide, 3%					
Amphyl/Lysol (undiluted)					
Ethanol, 99%					

TABLE 18.5	Minimum time required for disinfection (*P. aeruginosa;* growth of control _____)			
Chemical	1 min	5 min	10 min	20 min
Sodium hypochlorite, 5.25%				
Hydrogen peroxide, 3%				
Amphyl/Lysol (1:100)				
Ethanol, 99%				

TABLE 18.6	Minimum time required for disinfection (*S. aureus;* growth of control _____)			
Chemical	1 min	5 min	10 min	20 min
Sodium hypochlorite, 5.25%				
Hydrogen peroxide, 3%				
Amphyl/Lysol (1:100)				
Ethanol, 99%				

RESOLUTION OF THE CASE

The urology practice seen in this case originally made the move away from glutaraldehyde-based disinfectants because of the irritating nature of the chemical, which can cause throat and lung irritation, difficulty breathing, nosebleeds, headaches, nausea, and burning of the skin, mucous membranes, and eyes. Even with these harmful qualities, it is commonly used in many laboratory settings to process invasive medical equipment including dialysis apparatus, surgical instruments, bronchoscopes, endoscopes, and ear, nose, and throat instruments. The cases of anaphylactic shock seen here, all of which occurred after switching to a ortho-phthalaldehyde-based disinfectant, were enough to cause the practice to switch back to a glutaraldehyde-based solution. Once the switch was made, no further episodes of anaphylaxis were seen.

EXERCISE 18 REVIEW QUESTIONS

1. Define the terms *microbistatic* and *microbicidal*.

2. The use-dilution test measures only the microbicidal effects of a chemical. Why can it not be used to measure microbistatic effects?

3. What is the difference between a cleaner and a disinfectant? Why do some products advertise the fact that they both "clean and disinfect"?

4. When testing disinfectants in the manner seen here, blood is often included along with the microbial sample. Why?

5. The Environmental Protection Agency mandates that chemical disinfectants state on their label that items to be disinfected must first be cleaned of gross debris. What "debris" would commonly be encountered in a medical or dental environment that would interfere with effective disinfection?

6. What disadvantages could be there be to using stronger than necessary disinfectants?

REFERENCE

Environmental Protection Agency. 2008. *Selected EPA-Registered Disinfectants.* http://epa.gov/oppad001/chemregindex.htm.

Effectiveness of Hand Scrubbing

CASE SYNOPSIS

Puerperal Fever—Vienna, Austria, 1847

In the mid-1800s, the incidence of puerperal fever in many of Austria's maternity ("lying-in") hospitals was quite high. This disease was a serious complication of childbirth, with mortality rates of 25% being not uncommon. Because the disease was thought to be associated with an imbalance of bodily humors (fluids), it was believed that nothing could be done to lower the incidence of the disease.

In 1844, Ignaz Semmelweis was hired as an assistant lecturer in the First Obstetric Division of the Vienna Hospital. He immediately noticed that at 13%, the mortality rate from puerperal fever in the First Obstetric Division was more than six times the rate seen in the Second Obstetric Division, even though both divisions were located in the same building and used the same techniques. The only apparent difference between the two was that doctors were trained in the first division while midwives received their training in the second.

In 1847, Jakob Kolletschka, a physician and friend of Semmelweis became ill after cutting himself with a knife during an autopsy. He died shortly thereafter, and his own autopsy revealed a pathological condition very similar to that seen in the women who had died of puerperal fever. Semmelweis immediately put forth a directive requiring medical students to wash their hands with chlorinated lime (a harsh cleaning agent) between autopsy work and the examination of patients.

Resolution of the Case appears on page 156

STUDENT LEARNING OUTCOMES

After completing this exercise, you should be able to:

1. Differentiate between the resident and transient microbiota of the skin.
2. Perform a viable plate count to measure bacteria removed as a result of hand washing.

INTRODUCTION

Prior to the advent of the germ theory of disease in the mid-1800s, the connection between microorganisms and disease was tenuous at best. With very few exceptions, diseases were thought to result from an imbalance in the body's fluids (whatever that may have meant), as divine retribution for inappropriate behavior, or some other reason that appears nonsensical to our twenty-first century understanding of infectious diseases. So ingrained were these beliefs at the time, however, that the very thought of hand washing as a means of preventing the spread of disease, even among doctors, was rejected as a waste of time.

Today, of course, we take for granted the fact that proper hand washing is the single best method of preventing the spread of infectious disease, and it is understood that everyone, not just healthcare professionals, should wash their hands on a regular basis. In addition to simple hand washing, waterless hand cleaners, antibacterial soaps, and the use of gloves are more common than ever before. The challenge nowadays is less about convincing people of the need to wash their hands and more about educating them on how to do so properly.

Human skin is one of our first lines of defense against infection. As long as it stays intact, the skin is a formidable barrier that prevents the passage of microorganisms to the deeper tissues of the body. Beyond simply acting as a barrier, human skin is also populated by a diverse group of microorganisms, and nowhere is this easier to see than on the hands, which are more likely to come into contact with the environment than any other part of the body. Microorganisms found on the skin may be classified as **residents** or **transients,** depending on how well associated with the skin they are.

The **normal microbiota** of the skin consists of microorganisms that live and multiply in the deeper layers of the epidermis as well as within glands and follicles. These resident microbes are relatively predictable as they consist of those organisms that have become adapted to a very specific environment. The **resident microbiota** also tends to remain quite stable over time as the nature of the skin itself, and hence the environment, is comparatively unchanging. The normal microbiota of the skin can be divided into three primary groups:

Diptheroids The name of this group is indicative of its members morphological similarity to *Corynebacterium diptheriae*. A common member is *Propionibacterium acnes,* an anaerobic bacterium that lives in the hair follicles and breaks down the oily secretions found there.

Staphylococci   Several nonpathogenic members of the genus *Staphylococcus* can be found on the skin. These organisms are generally beneficial as they compete with pathogens for nutrients and produce inhibitory substances, usually preventing pathogens from establishing a presence on the skin. Common inhabitants include the nonpathogenic *S. epidermiditis* as well as, less commonly, *S. aureus*, which can be an opportunistic pathogen.

Yeasts and Fungi   Several species of yeast and fungi are found on the skin. These nonpathogenic organisms generally digest the oily emissions of the secretory glands.

In addition to the normal microbiota, at any given time, the skin harbors a collection of transient bacteria. This population represents those bacteria that are temporarily present on the surface of the skin. They are far more varied in nature, with Gram negative and positive, rods and cocci, and spore formers and vegetative cells all being represented. In contrast to the resident population, the presence of transient bacteria is quite dependent on the hygiene of the person, with most members of the transient population lasting only until the next thorough hand washing.

In this laboratory exercise, the effectiveness of hand scrubbing will be determined. One member of the class will use a sterile brush and water to scrub his or her hands into a sterile basin for 1 min. This will be followed by a 2-min scrub using soap and running water. This process will be repeated until a total of 5 basins of water have been collected, representing the bacteria present on the hands after 0, 3, 6, 9, and 12 min of scrubbing. A second student will be responsible for supplying soap, basins, and brushes as needed. Other students will be responsible for inoculating plates with the wash water to determine the number of bacteria removed during each round of scrubbing.

PRE-LAB QUESTIONS

1. *Define the following terms:*

Resident microbiota

Transient microbiota

Epidermis

Glands

Follicles

2. *If resident and transient bacteria are both found on the hands, why is the transient population generally more varied?*

3. *Would taking up gardening as a hobby be more likely to affect the nature of the resident population or transient population of bacteria on the hands? Why?*

4. *As the hands are being scrubbed and water is collected into separate basins, where would you expect to find the most varied population of bacteria, water collected after 1 min of scrubbing or 10 min?*

MATERIALS

The scrub team (one per class) should obtain:

Five sterile surgical scrub brushes

Five covered flasks, each containing 1000 ml sterile water

Five sterile basins

Antibacterial soap

Hand lotion

Each plating team (five per class) should obtain:

Six veal infusion agar pours

One 1-ml pipette

Six sterile Petri dishes

Marking pen

PERIOD ONE

PROCEDURE

Scrub Team

One person will be designated the scrubber for the entire experiment, and this person should not touch any nonsterile surfaces for the duration of the experiment. The other member of the team will handle all flasks, basins, and soap, and will partially unwrap each scrub brush so it can be picked up by the scrubber without touching the outside of the brush packaging. The scrub procedure is as follows (Figure 19.1):

1. Using a sterile brush, the scrubber should scrub each hand for 30 sec while sterile water is slowly poured over both hands, collecting the wash water in the first sterile basin.
   - While scrubbing, spend an equal amount of time on each hand and be sure to scrub all surfaces, including under the fingernails.
   - Pour any sterile water that remains in the flask at the end of 60 sec into the basin.
   - Mark the basin "A," and notify the first plating team that it is ready.

2. Using the same brush, scrub the hands for 2 min, using antibacterial soap (applied by the other team member) and running water.
   - Be sure to scrub all the surfaces of each hand.
   - Discard the brush.
   - Rinse both hands under running water for 5 sec at the conclusion of the scrub.

3. Repeat the above procedure (steps 1 and 2) for basins B, C, D, and E.
   - Remember to use a new sterile brush for each basin.
   - The final soap scrub (after water for basin E has been collected) may be omitted.
   - When all the scrubbing is completed, the scrubber should dry his or her hands and apply moisturizing lotion.

Plating Teams

The rest of the class should be divided into five plating teams, each of which will work to determine the number of bacterial cells per milliliter in a single basin of water. Proceed as follows:

1. Liquefy six pours of veal infusion agar, and allow to cool to 50°C.

2. Label the bottoms of six empty Petri plates with your group and lab time. Label two of the plates "0.1 ml," two plates "0.2 ml," and two plates "0.4 ml."

3. As soon as you've received your basin, inoculate the plates as follows:
   - Stir the water within the basin for 15 sec, using a sterile pipette, before removing the first sample of wash water.
   - Use the same pipette to inoculate all six plates with the proper volume of water (Figure 19.2).
   - Pour a tube of veal infusion agar, cooled to 50°C, into each plate and swirl gently to distribute the organisms.
   - Allow the plates to solidify and incubate upside down for 24 h at 37°C.

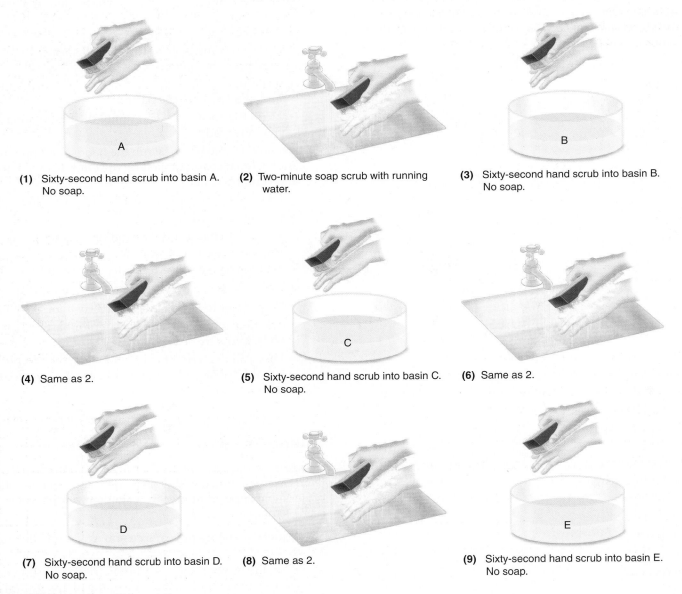

(1) Sixty-second hand scrub into basin A. No soap.

(2) Two-minute soap scrub with running water.

(3) Sixty-second hand scrub into basin B. No soap.

(4) Same as 2.

(5) Sixty-second hand scrub into basin C. No soap.

(6) Same as 2.

(7) Sixty-second hand scrub into basin D. No soap.

(8) Same as 2.

(9) Sixty-second hand scrub into basin E. No soap.

Figure 19.1 Hand-scrubbing routine.

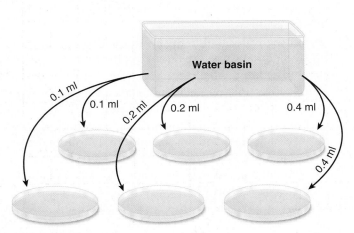

Figure 19.2 Distribute wash water to each of six plates as shown. Stir the wash water for 15 sec prior to beginning the transfer process.

QUESTIONS—PERIOD ONE

1. What is the relationship between a single cell that is removed from the hands during the washing process and a colony on a plate, which may contain 1,000,000 cells?

2. Would you expect resident and transient organisms to be distributed equally among the various basins, or would some basins be more likely to contain transient and some more likely to contain resident bacteria?

PERIOD TWO

1. Retrieve your plates from the incubator. Count the number of colonies on plates that contain between 30 and 300 colonies. Plates with fewer than 30 colonies should be designated as TFTC (too few to count) and those with more than 300 colonies as TMTC (too many to count).

2. Your instructor should have a table similar to Table 19.1 on the board. Record your results on the board.

3. Calculate the number of organisms per milliliter of wash water as follows:
   - Choose a pair of plates (0.1 ml, 0.2 ml, or 0.4 ml) that have between 30 and 300 colonies per plate. If more than one set of plates contain this many colonies, choose the set in which the colony counts are in closest agreement.
   - Count the number of colonies on each plate, and determine the average number of colonies for this pair of plates.
   - Multiply the average number of colonies by the proper dilution factor (i.e., 0.1 ml = 10, 0.2 ml = 5, and 0.4 ml = 2.5) to calculate the total number organisms per milliliter.

4. When all the groups have recorded their numbers, graph the results.

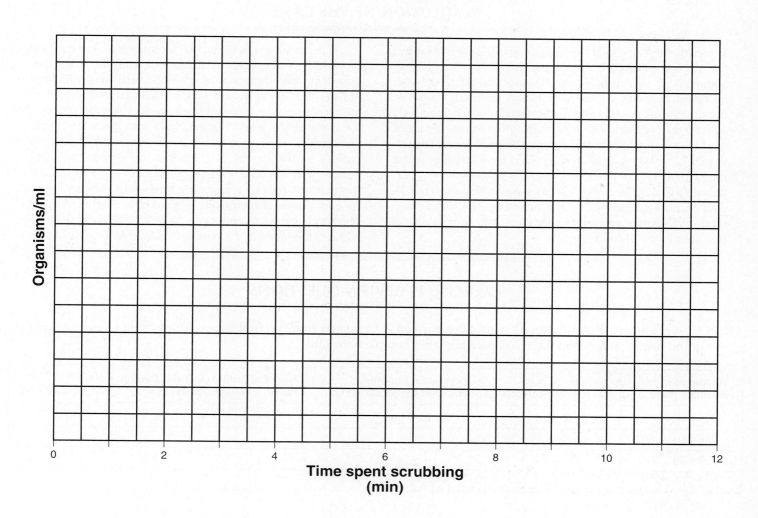

TABLE 19.1

Group	0.1-ml Count Per plate	0.1-ml Count Average	0.2-ml Count Per plate	0.2-ml Count Average	0.4-ml Count Per plate	0.4-ml Count Average	Dilution factor	Organisms/ml
A								
B								
C								
D								
E								

RESOLUTION OF THE CASE

The pathological similarities seen in Jakob Kolletschka and the women who died of puerperal fever were noted by Semmelweis, who postulated a connection between what he called *cadaveric* contamination and puerperal fever. He concluded that doctors and medical students, who routinely worked on cadavers, were transferring the infectious agent to patients whom they later examined. The Second Obstetrics Division, where midwives received their training, had a much lower rate of infection because midwives did not participate in autopsies. After full implementation of Semmelweis' hand-washing policy, the mortality rate in the First Obstetrics division dropped to less than 2%, comparable to that seen in the Second Division.

Unfortunately the lessons of Semmelweis were not well-received by the medical establishment, with the initial reaction being that there was no scientific basis for his findings. The germ theory of disease, which holds that microorganisms are responsible for many diseases, would not be proposed for many more years. Doctors were also quick to criticize Semmelweis' work as it would be an admission of their culpability in the deaths of thousands of women. In 1849, Semmelweis was fired from his position in Vienna for, according to many, political reasons. He traveled to Hungary, where he took charge of the maternity ward of St. Rochus Hospital. The hand- and equipment-washing policies he instituted there reduced the mortality from puerperal fever to less than 1%.

EXERCISE 19 REVIEW QUESTIONS

1. In an experiment similar to the one you just performed, 1 L of water from a scrubbing exercise is used to inoculate plates with the volumes shown in the accompanying table; complete the table.

Volume used to inoculate plate	No. of colonies/plate	Dilution factor (1.0 ml/ volume used to inoculate plate)	No. of organisms in 1 ml (Dilution factor × colonies/plate)	No. of organisms/1.0 L
0.1 ml	14	1.0 ml/0.1 ml = 10	140	140,000
0.2 ml	98			
0.5 ml	240			
1.0 ml	622			

2. The student analyzing the data in the table from question 1 counts only the plates containing 98 colonies and 240 colonies and is commended by her lab instructor for doing exactly the right thing. Why shouldn't the other two plates be counted?

3. In your experiment, was the decline in bacterial cells from one basin to the next (A-E) expected or unexpected? Explain.

4. The 2005 FDA Food Code instructs restaurant workers to wash their hands with soap and water for at least 20 sec prior to handling food. Based on the results of this experiment, is this long enough to remove all transient and resident bacteria from the skin?

5. A correctly done surgical scrub lasts approximately 5 min. Using the information gathered from this exercise, is this enough time to remove all transient and resident microbes from the skin?

6. Regardless of the level of hand washing involved, provide two different reasons why gloves are always worn in the clinical setting?

REFERENCE

Nuland, S.B. 2003. *The Doctors' Plague: Germs, Childbed Fever, and the Strange Story of Ignaz Semmelweis.* New York: W.W. Norton & Co. Ltd. CDC. *Clean Hands Save Lives!* http://www.cdc.gov/cleanhands/.

Antimicrobic Sensitivity Testing: Kirby-Bauer, Tube Dilution, and E-Test Methods

CASE SYNOPSES

Methicillin-Resistant *Staphylococcus aureus* Infections among Tattoo Recipients—Ohio, Kentucky, and Vermont, 2004–2005

From June 2004 to August 2005, health departments in six different communities in Ohio, Kentucky, and Vermont reported clusters of patients whose symptoms were consistent with *Staphylococcus aureus* infection. Initial investigation revealed a total of 34 primary infections, which took the form of boils, folliculitis, erythema, or abscesses, all of which occurred at or near the site of a recent tattoo. Further inquiry yielded 10 secondary cases, which were defined as skin infections consistent with staphylococcal disease occurring in persons who did not receive a tattoo but had been in close contact with someone who had and whose infection

yielded a similar bacterial strain. Characterization of isolates from three of the clusters indicated resistance to oxacillin and erythromycin, necessitating antimicrobial testing.

New Antibiotic Discovered—Germany, 2008

In 2008, scientists at the University of Keil in Germany were investigating defense mechanisms of *Hydra,* a small, freshwater creature famous for regenerating itself when its tissues are severed. During their studies, scientists came across a completely new type of antibiotic that seemed to be active against both Gram-negative and Gram-positive bacteria. Most importantly, it works against some drug-resistant strains of bacteria. They named this protein hydramacin.

Resolution of the Cases appears on page 169

CDC. 2006. Methicillin-resistant *Staphylococcus aureus* skin infections among tattoo recipients—Ohio, Kentucky, and Vermont, 2004–2005. *Morbidity and Mortality Weekly Report,* 55: 677–679.

Jung, S., Dingley, A.J., Augustin, R. et al. 2009. Hydramaccin-1, structure and antibacterial activity of a protein from the basal metazoan *Hydra. J. Biol. Chem.* 284(3): 1896–1905.

STUDENT LEARNING OUTCOMES

After completing this exercise, you should be able to:

1. Determine the susceptibility of a bacterial culture to a variety of antimicrobics.
2. Understand the information provided by each of the major antimicrobic sensitivity-testing methods.

INTRODUCTION

Antibiotics are compounds produced by several species of bacteria and fungi that serve to inhibit the growth of other bacteria. The fungus *Penicillium notatum,* for example, produces **penicillin** while the antibiotic **bacitracin** is manufactured by *Bacillus subtilus.* The discovery of penicillin and its use as a treatment for infection in the 1930s and 1940s changed the way we battled bacterial disease. In the decades that followed, synthetic compounds were formulated that served the same purpose. Together, these antibacterial

compounds, whether natural or man-made, are referred to as **antimicrobics** or **antimicrobials.**

One of the most significant problems associated with infection control in the twenty-first century is the problem of **antibiotic resistance.** Overuse of antimicrobics has led to the selection of resistant strains of bacteria that are immune to these substances, and many bacterial infections now commonly require antimicrobial **susceptibility testing** to identify an effective chemotherapeutic agent. Strains of *Salmonella, Mycobacterium tuberculosis,* and *Staphylococcus aureus* are now frequently encountered that are resistant to one or more antimicrobics, and susceptibility testing is the norm when these species are found.

Susceptibility testing has several goals. Foremost, an antimicrobic that is effective against the pathogenic bacterium must be chosen. The number of available antimicrobics usually makes this relatively simple to achieve, but this is only half the story. Because so many bacteria in the body are beneficial, the ideal chemotherapeutic agent should kill only the pathogen while leaving other bacteria as unaffected as possible, a phenomenon known as **selective toxicity.** The second goal of testing is to identify an appropriate therapeutic dose. Because many antimicrobics are toxic or

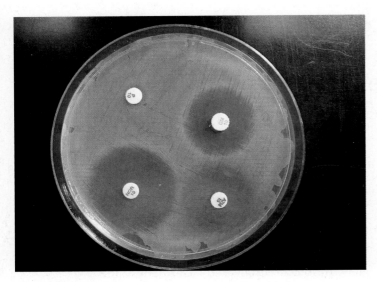

Figure 20.1 In a disc diffusion test like the one seen above, a plate is inoculated with the bacteria to be tested and discs containing precise amounts of antimicrobics are placed on the surface. During incubation, the antimicrobic diffuses from the disc into the media, becoming less concentrated as it spreads. If the bacterium is sensitive to the antimicrobic, a zone of inhibition develops around the disc. The size of the zone can then be used to determine if any of the antimicrobics would be effective chemotherapeutic options.

diffusion test, a bacterial culture is spread over the surface of a plate of media and discs impregnated with various antimicrobics are placed on the plate. As the antimicrobic diffuses through the media, a concentration gradient is created, with high levels of antimicrobics found near the disc and decreasing as the distance from the disc increases. If the bacterium is susceptible to the antimicrobic, an area of no growth, or zone of inhibition, is formed (Figure 20.1). The size of the zone of inhibition is dependent of many variables, including the molecular weight of the antimicrobics (which affects its diffusion), the type of media, and the concentration of the bacterial culture. The **Kirby-Bauer method** of antimicrobic testing is a standardized system in which all of these factors are taken into consideration, allowing results from lab to lab to be compared with one another. Among the variables that are standardized are the type of media used (Mueller-Hinton II) and the depth to which it is poured in the Petri dish (4 mm). The concentration of the bacterial culture is adjusted so that it matches a 0.5 McFarland standard. Each of the 11 McFarland standards represents a specific degree of turbidity that can be used to visually standardize the growth of bacterial cultures. After inoculating the Petri dishes, applying the discs to the media, and incubating overnight, the size of the zone of inhibition can be measured to determine the relative effectiveness of each antimicrobic (see Figure 20.1). To determine the effectiveness of an antimicrobic, the size of the zone of inhibition must be compared to a table that relates the effectiveness of the compound in the laboratory to its effectiveness in a living patient (Table 20.1).

produce unpleasant side effects, especially in large doses, the lowest effective concentration of antibiotic is sought.

Several methods of susceptibility testing exist, each of which has its own advantages and disadvantages. In a disc

TABLE 20.1	Zones of inhibition in the Kirby-Bauer method of antimicrobic sensitivity testing					
			Zone of inhibition		
Antibiotic	**Code**	**Potency**	**Resistant**	**Intermediate**	**Sensitive**
Amikacin	AN-30	30 μg			
Enterobacteriaceae, P. aeruginosa, Acinetobacter, and staphylococci			≤ 14	15–16	≥ 17
Amoxicillin/Clavulinic Acid	AmC-30	20/10 μg			
Enterobacteriaceae			≤ 13	14–17	≥ 18
Staphylococcus spp.			≤ 19	—	≥ 20
Haemophilus spp.			≤ 19	—	≥ 20
Ampicillin	AM-10	10 μg			
Enterobacteriaceae			≤ 13	14–16	≥ 17
Staphylococcus spp.			≤ 28	—	≥ 29
Enterococcus spp.			≤ 16	—	≥ 17
Listeria monocytogenes			≤ 19	—	≥ 20
Haemophilus spp.			≤ 18	19–21	≥ 22
β-hemolytic streptococci			—	—	≥ 24

(continued)

TABLE 20.1	Zones of inhibition in the Kirby-Bauer method of antimicrobic sensitivity testing (continued)					
				Zone of inhibition		
Antibiotic	Code	Potency		Resistant	Intermediate	Sensitive
Azithromycin	AZM-15	15 µg		≤ 13	14–17	≥ 18
Bacitracin	B-10	10 units		≤ 8	9–12	≥ 13
Carbenicillin	CB-100	100 µg				
Enterobacteriaceae and *Acinetobacter*				≤ 19	20–22	≥ 22
P. aeruginosa				≤ 13	14–16	≥ 17
Cefaclor	CEC-30	30 µg				
Enterobacteriaceae and staphylococci				≤ 14	15–17	≥ 18
Haemophilus spp.				≤ 16	17–19	≥ 20
Cefazolin	CZ-30	30 µg				
Enterobacteriaceae and staphylococci				≤ 14	15–17	≥ 18
Ceftriaxone						
Enterobacteriaceae, P. aeruginosa, Acinetobacter and *staphylococci*	CRO-30	30 µg		≤ 13	14–20	≥ 21
Chloramphenicol	C-30	30 µg				
Enterobacteriaceae, P. aeruginosa, Acinetobacter, staphylococci, enterococci, and *V. cholerae*				≤ 12	13–17	≥ 18
Haemophilus spp.				≤ 25	26–28	≥ 29
S. pneumoniae				≤ 20	—	≥ 21
Streptococci				≤ 17	18–20	≥ 21
Ciprofloxacin	CIP-5	5 µg				
Enterobacteriaceae, P. aeruginosa, Acinetobacter, staphylococci, and enterococci				≤ 15	16–20	≥ 21
Haemophilus spp.				—	—	≥ 21
N. gonorrhoeae				≤ 27	28–40	≥ 41
Clarithromycin	CLR-15	15 µg				
Staphylococcus spp.				≤ 13	14–17	≥ 18
Haemophilus spp.				≤ 10	11–12	≥ 13
S. pneumoniae and other streptococci				≤ 16	17–20	≥ 21
Clindamycin	CC-2	2 µg				
Staphylococcus spp.				≤ 14	15–20	≥ 21
S. pneumoniae and other streptococci				≤ 15	16–18	≥ 19
Doxycycline	D-30	30 µg				
Enterobacteriaceae, P. aeruginosa, Acinetobacter, staphylococci, and enterococci				≤ 12	13–15	≥ 13

(continued)

TABLE 20.1	Zones of inhibition in the Kirby-Bauer method of antimicrobic sensitivity testing (continued)					
				Zone of inhibition		
Antibiotic	**Code**	**Potency**	**Resistant**	**Intermediate**	**Sensitive**
Erythromycin	E-15	15 µg			
Staphylococcus spp. and enterococci			≤ 13	14–22	≥ 23
S. pneumoniae and other streptococci			≤ 15	16–20	≥ 21
Gentamicin	GM-120	120 µg			
Enterobacteriaceae, P. aeruginosa, Acinetobacter, and staphylococci			≤ 12	13–14	≥ 15
Imipenem	IPM-10	10 µg			
Enterobacteriaceae, P. aeruginosa, Acinetobacter, and staphylococci			≤ 13	14–15	≥ 16
Haemophilus spp.			—	—	≥ 16
Kanamycin	K-30	30 µm			
Enterobacteriaceae and staphylococci			≤ 13	13–17	≥ 18
Lomefloxacin	LOM-10	10 µg			
Enterobacteriaceae, P. aeruginosa, Acinetobacter, and staphylococci			≤ 18	19–21	≥ 22
Haemophilus spp.			—	—	≥ 22
N. gonorrhoeae			≤ 26	27–37	≥ 38
Loracarbef	LOR-30	30 µg			
Enterobacteriaceae and staphylococci			≤ 14	15–17	≥ 18
Haemophilus spp.			≤ 15	16–18	≥ 19
Meziocillin	MZ-75	75 µg			
Enterobacteriaceae and *Acinetobacter*			≤ 17	18–20	≥ 21
P. aeruginosa			≤ 15	—	≥ 16
Minocycline	MI-30	30 µg			
Enterobacteriaceae, P. aeruginosa, Acinetobacter, staphylococci, and enterococci			≤ 14	15–18	≥ 19
Moxalactam	MOX-30	30 µg			
Enterobacteriaceae, P. aeruginosa, Acinetobacter, and staphylococci			≤ 14	15–22	≥ 23
Nafcillin	NF-1	1 µg			
Staphylococcus aureus			≤ 10	11–12	≥ 13
Nalidixic Acid	NA-30	30 µg			
Enterobacteriaceae			≤ 13	14–18	≥ 19
Neomycin	N-30	30 µg	≤ 12	13–16	≥ 17
Netilmicin	NET-30	30 µg			
Enterobacteriaceae, P. aeruginosa, Acinetobacter, and staphylococci			≤ 12	13–14	≥ 15

(continued)

TABLE 20.1	Zones of inhibition in the Kirby-Bauer method of antimicrobic sensitivity testing (continued)					
				Zone of inhibition		
Antibiotic	Code	Potency	Resistant	Intermediate	Sensitive
Norfloxacin	NCR-10	10 µg			
Enterobacteriaceae, P. aeruginosa, Acinetobacter, staphylococci, and enterococci			≤ 12	13–16	≥ 17
Novobiocin	NB-30	30 µg	≤ 17	18–21	≥ 22
Oxacillin	OX-1	1 µg			
Staphylococcus aureus			≤ 10	11–12	≥ 13
staphylococcus (coagulase negative)			≤ 17	—	≥ 18
Penicillin	P-10	10 units			
Staphylococcus spp.			≤ 28	—	≥ 29
Enterococcus spp.			≤ 14	—	≥ 15
L. monocytogenes			≤ 19	20–27	≥ 28
N. gonorrhoeae			≤ 26	27–46	≥ 47
β-hemolytic streptococci			—	—	≥ 24
Piperacillin	PIP-100	100 µg			
Enterobacteriaceae and *Acinetobacter*			≤ 17	18–20	≥ 21
P. aeruginosa			≤ 17	—	≥ 18
Polymyxin B	PB-300	300 U	≤ 8	9–11	≥ 12
Rifampin	RA-5	5 µg			
Staphylococcus spp., Enterococcus spp., and *Haemophilis spp.*			≤ 16	17–19	≥ 20
S. pneumoniae			≤ 16	17–18	≥ 19
Spectinomycin	SPT-100	100 µg			
N. gonorrhoeae			≤ 14	15–17	≥ 18
Streptomycin	S-300	300 µg			
Enterobacteriaceae			≤ 11	12–14	≥ 15
Sulfisoxazole	G-25	25 µg			
Enterobacteriaceae, P. aeruginosa, Acinetobacter, V. cholerae, and staphylococci			≤ 12	13–16	≥ 17
Tetracycline	Te-30	30 µm			
Enterobacteriaceae, P. aeruginosa, Acinetobacter, V. cholerae, staphylococci and enterococci			≤ 14	15–18	≥ 19
Haemophilus spp.			≤ 25	26–28	≥ 29
N. gonorrhoeae			≤ 30	31–37	≥ 38
S. pneumoniae and other streptococci			≤ 18	19–22	≥ 23
Tobramycin	NN-10	10 µg			
Enterobacteriaceae, P. aeruginosa, Acinetobacter, and staphylococci			≤ 12	13–14	≥ 15

(continued)

TABLE 20.1	Zones of inhibition in the Kirby-Bauer method of antimicrobic sensitivity testing (continued)					
				Zone of inhibition		
Antibiotic	**Code**	**Potency**	**Resistant**	**Intermediate**	**Sensitive**
Trimethoprim	TMP-5	5 µg			
Enterobacteriaceae and staphylococci			≤ 10	11–15	≥ 16
Vancomycin	Va-30	30 µg			
Staphylococcus spp.			—	—	≥ 15
Enterococcus spp.			≤ 14	15–16	≥ 17
S. pneumoniae and other streptococci			—	—	≥ 17

Courtesy of © Becton-Dickinson and Company

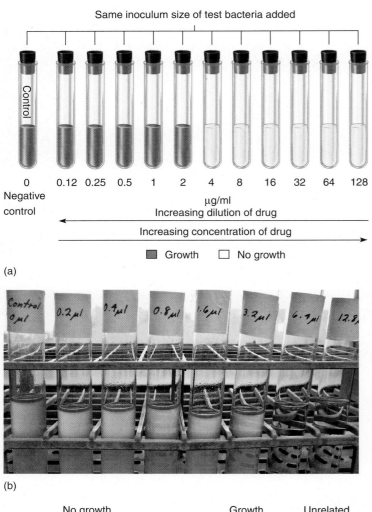

(a)

(b)

(c)

Imagine, for example, that a disc on your plate contained 10 µg of lomefloxacin and the diameter of the zone of inhibition was 17 mm. Consulting the entry for lomefloxacin in Table 20.1 indicates that a zone less than 18 mm in diameter indicates the bacterium is resistant to the antimicrobic; lomefloxacin would therefore be a poor choice for chemotherapy. If a second disc on the same plate containing 10 µg gentamicin produced a zone of 17 mm, this would indicate that the bacterium was sensitive to gentamicin, which would be an effective choice for chemotherapy.

As mentioned before, the effectiveness of a particular antibiotic is only half the story. Determining an appropriate dosage is also important, and the Kirby-Bauer test does not provide this information. To determine what concentration of antimicrobic is appropriate, a tube dilution test may be used. This method of testing tests a single antimicrobic at varying levels of concentration and determines the **minimal inhibitory concentration** (MIC), or lowest level of a chemical that completely inhibits microbial growth (Figure 20.2). The advantage of determining the MIC is balanced by the cost of a much more time-consuming procedure and the inability to test more than a single drug at a time.

A third method, the **E-test,** combines the convenience of the Kirby-Bauer test with the MIC-determining power of a tube dilution test. In this method, the antimicrobics to be tested are impregnated within a thin strip of paper, and the concentration of drug at each point is noted on the strip. The procedure is similar to a disc diffusion test, with the bacterium being

Figure 20.2 Tube dilution test for determining minimal inhibitory concentration (MIC). (a) Moving from left to right, each tube contains twice as much antibiotic as the previous tube in the series. The tubes are inoculated with an identical amount of bacteria and then incubated. The tube on the far left contains no antibiotic and serves as a control for cell viability. The concentration of antibiotic in the first tube that shows no growth (4 µg/ml) is the MIC. (b) A tube dilution test using *E. coli* and tetracycline. (c) A multiwell plate with a set of three tests. Plates such as this are meant to be read automatically and allow the MICs of several drugs to be determined simultaneously.

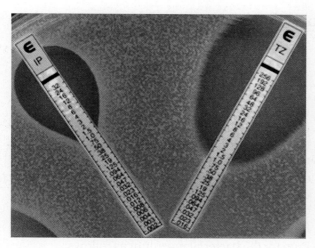

Figure 20.3   A variant of an agar-based method is Etest®, which uses a predefined antibiotic gradient on a calibrated plastic strip to generate a MIC value in μg/mL. The MIC is read where the inhibition ellipse intersects the scale on the plastic strip. Etest is as easy to use as the Kirby Bauer disk test but generates accurate MIC values for a wide range of antimicrobial agents and can be used for many species of bacteria, Mycobacteria and fungi. Etest® and the Etest gradient strip are registered trademarks of AB BIODISK. (IP = imipenem and TZ = ceftazidime).

spread evenly over the surface of the plate. The strips are then applied, and the plate is incubated. The zone of inhibition, rather than being circular is elliptical, and the point at which the bacterial growth touches the strip is the MIC (Figure 20.3).

PRE-LAB QUESTIONS

1. *Define the following terms:*

   Antibiotic resistance

   Broad spectrum

   Narrow spectrum

   Side effects

Abscess

Minimal inhibitory concentration

Minimal bactericidal concentration

2. *Why might microbes produce chemicals that are inhibitory to other microbes?*

3. *Use the characteristics of each of the three major methods of susceptibility testing to complete the table below.*

	Susceptibility testing methods		
	Kirby-Bauer (Y/N)	Tube dilution (Y/N)	E-Test (Y/N)
Determines antimicrobic effectiveness			
Determines MIC			
Can test several antimicrobics simultaneously			

PERIOD ONE

Both *Staphylococcus aureus* and *Escherichia coli* will be tested to determine their susceptibility to several antimicrobics. Odd-numbered students should test *E. coli*, and even numbered students should test *S. aureus*. Use the procedure indicated by your lab instructor.

KIRBY-BAUER METHOD

Materials

Each student should obtain:

One plate of Mueller-Hinton II agar

Broth culture of *S. aureus* or *Escherichia coli* (as appropriate), diluted to match a 0.5 McFarland standard

Sterile swabs

Disk dispenser containing antimicrobic disks

Metric ruler for evaluation of plates

Marking pen

Procedure

1. Label your plate with your name, lab time, and organism.
2. Begin with a fresh broth culture that has been adjusted to match the turbidity of a 0.5 McFarland standard.
3. Using a sterile swab, inoculate a plate of Mueller-Hinton II agar with your assigned bacteria by streaking the surface of the plate three times. Rotate the plate one-third turn between inoculations to ensure complete coverage.
4. Apply antimicrobic disks to the surface of the plate, using either sterile forceps or a dispenser. If using forceps, be sure to space the disks evenly around the plate.
5. Lightly press each disk with a pair of sterile forceps to ensure that it makes complete contact with the agar surface.
6. Incubate the plates for 18–24 h at 35°C. Refrigerate the plates until the lab begins.

TUBE DILUTION METHOD

Materials

Each student should obtain:

Nine tubes each containing 1 ml of Mueller-Hinton II broth

One tube containing 5 ml of Mueller-Hinton II broth

One tube containing 20 ml of Mueller-Hinton II broth

One sterile tube containing 2.0 ml of tetracycline (256 μg/ml)

Ten 1-ml pipettes

Pipette pump

Plate cultures of *S. aureus* or *E. coli* (as appropriate)

Marking pen

Procedure

1. Number nine tubes of Mueller-Hinton II broth (1 ml/tube) 1–9.
2. Using a sterile pipette, add 2.0 ml of tetracycline (256 μg/ml) to tube number 1.
3. Use a sterile pipette to transfer 1 ml from tube 1 to tube 2.
4. Using a new pipette, mix the contents of the second tube, and transfer 1 ml from tube 2 to tube 3.
5. Continue transfers in this way up to tube 8. Be sure to use a fresh pipette for each transfer and to mix the contents of each tube prior to each transfer.
6. Remove 1 ml from tube 8 and discard it.
7. Prepare the bacterial inoculum as follows:
   a. Suspend four to five colonies of the culture to be tested in 5 ml of Mueller-Hinton broth, mixing until a uniform turbidity is obtained.
   b. Transfer 100 μl of this suspension to 20 ml of Mueller-Hinton broth. Mix to ensure even distribution of the bacteria in the media.
8. Add 1.0 ml of the diluted culture to each of the tubes. The concentration of antibiotic is reduced by 50% as it moves from tube 1 to tube 2, tube 2 to tube 3, etc. Tube 9 has no antibiotic and serves as a control. Record the tetracycline concentration on each tube as seen in the table.

Tube no.	Tetracycline concentration
1	128 μg/ml
2	64 μg/ml
3	32 μg/ml
4	16 μg/ml
5	8 μg/ml
6	4 μg/ml
7	2 μg/ml
8	1 μg/ml
9	0 μg/ml

9. Incubate all the tubes at 35°C for 24 h. Refrigerate until the lab begins.

E-TEST METHOD

Materials

Each student should obtain:

One plate of Mueller-Hinton II agar

Broth culture of *S. aureus* or *E. coli* (as appropriate), diluted to match a 0.5 McFarland standard

Sterile swabs

Access to appropriate antibiotic strips (as determined by instructor)

Marking pen

Procedure

1. Label your plate with your name, lab time, and organism.

2. Begin with a fresh broth culture that has been adjusted to match the turbidity of a 0.5 McFarland standard.

3. Using a sterile swab, inoculate a plate of Mueller-Hinton II agar with your assigned bacterium by streaking the surface of the plate several times. Rotate the plate between inoculations to ensure complete coverage. Allow the plate to dry for 10–15 min prior to applying the E-test strips.

4. Using forceps, or a special applicator, apply the strips to the inoculated agar surface with the MIC scale facing upward and the "ε" mark at the rim of the plate (see Figure 20.3). Up to two antimicrobics can be tested on a 90-mm plate and six on a 150-mm plate. Once a strip has touched the agar, do not move it.

5. If large air bubbles appear between the strip and the agar, use forceps to gently press on the strip, moving from low concentration to high. Small bubbles will not affect the results.

6. Incubate at 35°C for 24 h. Refrigerate until the lab begins.

QUESTIONS—PERIOD ONE

1. What causes the zone of inhibition to develop in the Kirby-Bauer test and the E-test?

2. How is the MIC recognized in an E-test?

3. How is the MIC recognized in a tube dilution test?

4. Use Table 20.1 to rate each of the bacteria below as (R)esistant, (I)ntermediate, or (S)ensitive.

Bacterial species	Antimicrobic disk code	Antimicrobic name	Zone diameter	R/I/S
Staphylococcus aureus	LOM-10		22 mm	
Neisseria gonorrhoeae	LOM-10		22 mm	
Streptococcus pneumoniae	C-30		19 mm	
Pseudomonas aeruginosa	C-30		15 mm	

PERIOD TWO
Retrieve your plates and/or tubes from the incubator, and
follow the instructions for the testing method employed.

KIRBY-BAUER METHOD

Measure the zone of inhibition for each antibiotic tested and
record the results.

Bacterial species	Antimicrobic disk code	Antimicrobic name	Zone diameter	R/I/S

TUBE DILUTION METHOD

Sketch the appearance of each tube in your dilution series.
Be sure to indicate the concentration of antimicrobic and the
presence of turbidity ($+/-$) for each tube. Determine the
MIC by identifying the first tube in which no growth
appears.

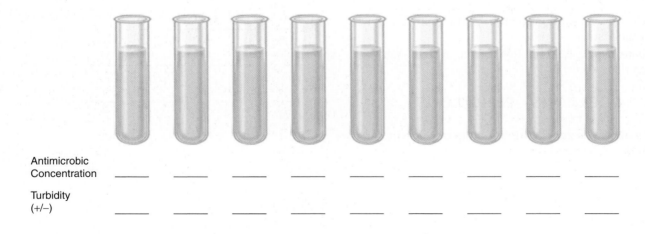

Antimicrobic
Concentration   _____   _____   _____   _____   _____   _____   _____   _____   _____

Turbidity
($+/-$)   _____   _____   _____   _____   _____   _____   _____   _____   _____

Antimicrobic tested_____

Minimal inhibitory concentration_____

E-TEST METHOD

Sketch the appearance of the zone of inhibition surrounding a single E-test strip.

Record the MIC for each of the antimicrobics tested as determined using the E-test.

Antimicrobic	MIC

RESOLUTION OF THE CASES

Methicillin-Resistant *S. aureus* infections among Tattoo Recipients

Antimicrobial susceptibility testing revealed that the isolates were sensitive to several antimicrobics, including trimethoprim-sulfamethoxazole, levofloxacin, and clindamycin. In 40 of the 44 primary and secondary cases, surgical incision and drainage, along with oral antimicrobics, were enough to successfully treat the infection. Four patients had bacteremia and required hospitalization for treatment with intravenous vancomycin. All the patients eventually recovered.

In follow-up interviews, the 34 patients with primary methicillin-resistant *S. aureus* infections identified a total of 13 unlicensed tattooists. Interviews with seven of these tattooists revealed a number of lapses with regard to proper infection-control measures, including not changing gloves between clients, poor hand hygiene, skin antisepsis, and inadequate disinfection of equipment and surfaces. Five patients reported seeing lesions on the hands of tattooists that were consistent with *S. aureus* skin infections, but none of these specimens were cultured.

New Antibiotic Discovered

Most currently used antibiotics are small molecules containing ring structures. But hydramacin is a protein consisting of 60 amino acids, and its structure is unlike that of any known antibiotic, although it is very similar to the proteins found in scorpion venom.

Short protein molecules have been isolated from various animal sources. Hydramacin is a particularly promising one. Of course, there is a long path between the laboratory discovery of a compound that inhibits bacteria and the production of a safe drug derived from that compound. One important phase of testing is to determine whether a compound has toxic effects on the patient. Whether hydramacin, so closely related to scorpion venom, passes this test remains to be seen.

EXERCISE 20 REVIEW QUESTIONS

1. What enables a new drug to be effective against bacteria that are able to resist old drugs?

2. The therapeutic index (TI) is a number that compares the toxic affects of a drug with its MIC using the formula TI = [toxic dose(to humans)]/MIC. If two antimicrobics have the same minimal inhibitory concentration, how could the TI be used to determine which is the better therapeutic choice?

3. Knowing that hydramacin's structure is similar to that of scorpion venom, what aspect of the compound should scientists thoroughly investigate before using it in humans?

REFERENCE

CDC. *Antibiotic/Antimicrobial Resistance.* http://www.cdc.gov/drugresistance/index.html.

Phage Typing of Bacteria

CASE SYNOPSIS

Salmonella Serotype Enteritidis Infections among Workers Producing Poultry Vaccine—Maine, November–December, 2006

On November 25, 2006, a case of salmonellosis in an employee of a facility that produced poultry vaccine was reported to the Maine Department of Health and Human Services (MDHHS). Because a similar case of salmonellosis had been reported 10 days earlier, the MDHHS began an outbreak investigation. Approximately one week prior to the first salmonellosis case, a spill had occurred in a fermentation room at the vaccine production facility, releasing 1–1.5 L of a highly concentrated culture of *Salmonella enterica,* serotype Enteriditis (this bacterium is referred to as SE), that was being used in vaccine production. The room was unoccupied at the time of the spill, and afterward it was cleaned by a worker wearing a biohazard suit, hat, booties,

mask, and gloves using 5% bleach and a commercial disinfectant effective against SE. That worker later reported the first case of salmonellosis.

Following the first two reported cases, the workers in the production area filled out a questionnaire asking about their work routines and whether they had experienced symptoms of salmonellosis (defined as three or more loose, watery stools in a 24-h period since November 1, 2006). Of a total of 26 employees who had been working in the room where the spill occurred, 18 reported illness. No illness was seen in the seven workers who had never entered the room. In addition to the cases from the vaccine facility, seven SE isolates from persons unconnected to the plant were submitted to the MDHHS during that same time period.

Resolution of the Case appears on page 175

CDC. 2007. Salmonella serotype Enteritidis infections among workers producing poultry vaccine—Maine, November–December 2006. *Morbidity and Mortality Weekly Report,* 56: 877–879.

STUDENT LEARNING OUTCOMES

After completing this exercise, you should be able to:

1. Use phage typing to identify the strain of a bacterial species.

INTRODUCTION

Epidemiological investigations depend on being able to link a number of instances of an infection to a common source, whether an item of food, a shared experience, or a common sexual partner. In many cases, simply identifying the bacterial species responsible for an outbreak is enough to formulate an epidemiological explanation. If, for example, a rare organism such as *Vibrio cholera* is isolated in two cases, it is likely that the two are related since so few instances of the disease are seen each year (28 cases in the United States during 2006). For infections such as those caused by *Salmonella,* with 42,000 reported cases in 2006, epidemiological studies require being able to differentiate between two organisms of the same species. One method routinely used to discrimi-

nate between strains of a single bacterial species is **phage typing**.

Bacteriophage, or phage, are viruses that infect bacterial cells, eventually lysing them (Figure 21.1). Each phage has a limited number of bacterial strains that are susceptible to infection by that phage, and these strains therefore constitute the **host range** of that phage. Because the host range is so specific, it is possible to categorize different members of a bacterial species based on their susceptibility to lysis, allowing them to be classified into phage types, or **strains**. In the lab, the bacterium to be tested is inoculated onto a solid media in a Petri dish so that it will grow to completely cover the medium, forming a bacterial lawn. The plate is then marked off into squares and each square is inoculated with a single drop of a suspension containing a different phage to be used in typing. The plate is allowed to incubate for 24 h and is then examined for clear zones within the bacterial lawn (Figure 21.2). These zones are called plaques and represent areas in which the bacterial cells have been infected and, eventually, lysed. When phage typing is performed, the results are usually listed directly after the species name of the organism, such as *Salmonella enteritidis* phage type 4. If a putative connection exists between several isolates and the phage types of the isolates are identical, the chances are great that they are epidemiologically related.

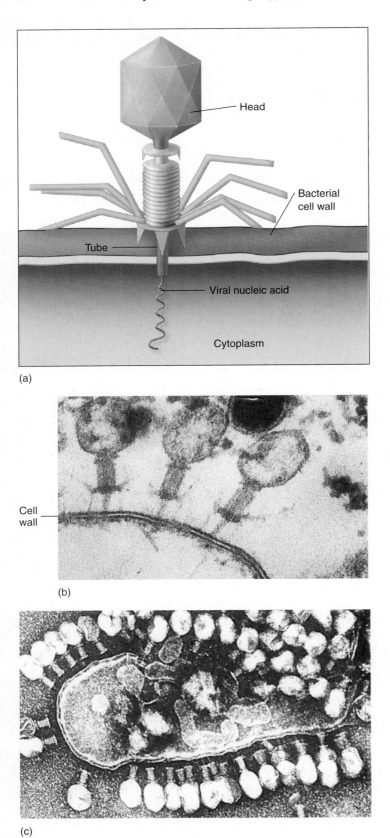

(a)

(b)

(c)

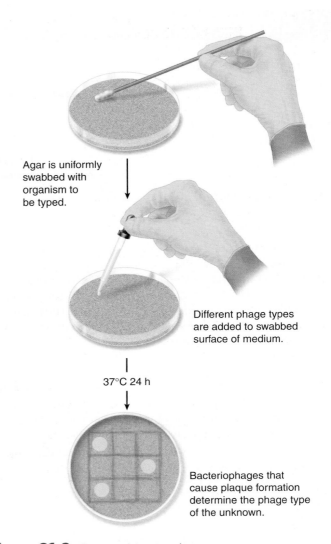

Agar is uniformly swabbed with organism to be typed.

Different phage types are added to swabbed surface of medium.

37°C 24 h

Bacteriophages that cause plaque formation determine the phage type of the unknown.

Figure 21.2 Phage-typing procedure.

Figure 21.1 (a) Typical T-even bacteriophage infecting a bacterial cell. The viral capsid remains outside of the cell as the viruses' genetic information is injected through the cell wall into the cytoplasm of the host bacterium. (b) Electron micrograph of viral infection by bacteriophage. After infection, the virus will enter a replicative phase where hundreds of new viral particles are produced. (c) The bacterial cell will lyse when several hundred viruses have been produced. Each virus is now free to infect a new host cell.

PRE-LAB QUESTIONS

1. *Define the following terms:*

Epidemiology

Bacteriophage

Host range

Plaque

Lysis

2. *What would happen if phage stocks were plated onto the media without any bacterial cells?*

3. *Compare the growth of a bacterial colony with the growth of a bacteriophage plaque.*

PERIOD ONE

In this exercise, strain identification of *Staphylococcus aureus* will be determined by phage typing. The various strains of *S. aureus* have been divided into four lytic groups, as seen in the accompanying table:

Lytic group	Phages in group
I	29, 52, 52A, 79, 80
II	3A, 3B, 3C, 55, 71
III	6, 7, 42E, 47, 53, 54, 75, 77, 83
IV	42D
Not allotted	81, 187

MATERIALS

Each student should obtain:

One plate of trypticase soy agar

Bacteriophage cultures in dropper bottles

Nutrient broth cultures of *S. aureus* with swabs (as many different cultures as are available, one per student)

PROCEDURE

1. Mark the bottom of a Petri dish, with the same number of squares as there are phages to be used. Label each square with the phage type number.
2. Swab the agar completely with the *S. aureus* isolate to be typed.
3. Carefully deposit a single drop of each phage into its respective square.
4. Incubate the plate at 37°C for 24 h.

QUESTIONS—PERIOD ONE

1. Why was it necessary to completely cover the surface of the media with bacteria?

2. What would you expect to see if:
   Your bacterial isolate fell outside the host range of every phage?

   Your isolate was within the range of a single phage?

Your isolate was within the range of three different phages?

PERIOD TWO

1. Remove the plate from the incubator, and record the identities of phages that were able to produce plaques.

2. To what phage types was this strain of *S. aureus* susceptible?

3. To what lytic group does this strain of bacteria belong?

RESOLUTION OF THE CASE

Phage typing of samples revealed that the spilled *Salmonella enterica* Enteritidis culture was phage type 8 (as expected) and that all four employee isolates were also phage type 8. DNA profiling showed no difference between the employee isolates and the community isolates, but phage typing revealed that all seven community isolates were phage type 13A. Based on this information, it was assumed that all of the employees who became ill did so as a result of exposure to a strain of *Salmonella enterica* Enteriditis that was used in vaccine production and that the illnesses seen in the community, despite being caused by the same bacterial species, were unrelated to the incident at the production facility.

Because of this case, the Maine Department of Health and Human Services recommended that the facility implement new practices for handling spills and routinely disinfect work areas. In addition, improved hand-washing practices and the use of personal protection equipment, including gloves, gowns, and face shields, were suggested.

EXERCISE 21 REVIEW QUESTIONS

1. The employee in this case who originally cleaned the culture spill reported having diarrhea for 1 day but taking no time off from work. What is the importance of this fact?

2. The CDC estimates that the 42,000 cases of salmonellosis reported yearly may actually be only 10% of the true number of cases. Why do you think this is?

3. The Food and Drug Administration recently approved a process in which bacteriophage are sprayed onto meat and poultry to make the meat safer to eat. How do you think this process works, and do you feel it is safe for the human who consumes the meat?

REFERENCE

CDC. *Salmonella*. http://www.cdc.gov/salmonella/.

Simulated Epidemic

CDC. 2005. Import-associated measles outbreak—Indiana, May–June 2005. *Morbidity and Mortality Weekly Report,* 54: 1073–1075.

STUDENT LEARNING OUTCOMES

After completing this exercise, you should be able to:

1. Calculate the prevalence and incidence of a disease outbreak.
2. Understand how the source of an outbreak may be determined.

INTRODUCTION

Epidemiology is concerned with the movement of a disease through a population. Rather than focusing attention on a single case, as a medical doctor would, an epidemiologist focuses on the acquisition, frequency, distribution, transmission, and prevention of infectious diseases. Based on the organism being investigated, transmission may occur via inhalation as with tuberculosis and influenza; ingestion as seen with *Salmonella* and *Escherichia coli;* or direct contact with the body or bodily fluids as happens with herpes or hepatitis B. Infection can also be acquired through interaction with healthy people harboring an infectious organism, contact with contaminated items, or the bite of infected insects.

When many people are infected through contact with a single source, such as a contaminated food item, the incident is referred to as a **common source epidemic** while if the disease is passed person-to-person, it is referred to as **propagated transmission epidemic.** Epidemiologists generally work backward from a known case, moving from contact to contact, to determine the first incident of a given outbreak, or **index case.** Identification of the index case allows steps to be taken to break the chain of infection, such as isolating infected individuals and administering antibiotics or vaccines to potentially infected persons.

Epidemiologists also collect statistics on the frequency with which a disease occurs using two common epidemiological measures. The first is **incidence,** which is defined as the number of new cases of a disease occurring among a defined population within a specific period. The equation used to determine incidence is:

$$\text{Incidence} = \frac{\text{no. new cases within defined time period}}{\text{size of at-risk population}} \times K$$

Because the incidence is generally a very small number (e.g., 0.00004), it is customary to multiply the incidence by 10,000 or 100,000 (i.e., the constant K in the equation above) so that the results are more clearly reported. For example, if an incidence of 0.00004 is multiplied by 100,000, it can be reported as 4 cases per 100,000 at-risk individuals rather than 0.00004 cases per person. Because incidence takes into account only new cases, it is an indicator of the spread of a disease. As populations can fluctuate over time, the population size at the midpoint of the period under study is conventionally used.

The second epidemiological measure is **prevalence,** which is defined as the total number of existing cases of disease within a defined population:

$$\text{Prevalance} = \frac{\text{total no. of existing cases}}{\text{total population}} \times K$$

As with incidence, prevalence is generally multiplied by a constant, K, so that it can be reported more clearly. Because prevalence includes both old and new cases, it is an indicator of how widespread a disease has become.

PRE-LAB QUESTIONS:

1. *Define the following terms:*

Otitis media

Carrier

Fomite

Vector

2. *What type of infectious agent is responsible for measles? How is measles spread?*

3. *Does the case summarized here appear to be a common source epidemic or a propagated transmission epidemic? Why?*

4. *Does a common source epidemic always have an index case? Why or why not?*

PERIOD ONE

In today's lab you will determine the index case of a simulated epidemic as well as the incidence and prevalence of the epidemic "disease." Two different procedures are provided: the first involves transmission of a commercial powder that simulates the spread of a microorganism while the second tracks the spread of an actual bacterium from person to person. Use the procedure indicated by your instructor.

TRANSMISSION OF A COMMERCIAL POWDER

Materials

Each student should obtain:

A numbered Petri dish containing a small quantity of a white powder, one of which will be detected later using a UV light and represents the "infectious agent" and the other is undetectable using UV light

Two Kimwipes or tissues

Procedure

1. Label one Kimwipe "round one" and the other "round two."
2. Using your right hand, gather a small amount of powder from your Petri dish and rub the fingers of your hand together to spread the powder across your fingers and palm.

Round One

1. To begin the first round, student number one will shake hands with any other student. When shaking hands, be sure that your palms are fully in contact with one another. The person who was the "shakee" then becomes the "shaker" and shakes the hand of a chosen recipient. Although you will only choose a single person to shake hands with, it is possible that several persons may choose to shake hands with you. The instructor will record the two participants in each handshake. When each person has chosen to shake hands with someone, the round is over.
2. After all the shaking is complete, use your right hand to pick up the tissue labeled "round one." Rub your hand thoroughly on the tissue. Set the tissue aside until after the second round.

Round Two

1. Again beginning with student number one, select a person at random to shake hands with, proceeding as before until everyone has initiated at least one handshake.
2. After all the handshaking is complete, use your right hand to pick up the tissue labeled "round two" and use it to rub your hand.
3. Proceed to the Questions—Period One section.

TRANSMISSION OF LIVING BACTERIUM

Materials

Each student should obtain:

One plate of trypticase soy agar (TSA)

A numbered Petri dish containing a piece of caramel, one of which has been inoculated with *Serratia marcescens* while the others have been moistened with water

Examination gloves

Three sterile cotton swabs

Marking pen

Procedure

1. Divide your trypticase agar plate into thirds. Label the segments "control," "round one," and "round two."
2. Record the number of your caramel on the bottom of your TSA plate.
3. Place three sterile cotton swabs into a container of sterile saline so that a swab can be easily removed with only a single hand.
4. Place a glove on your left hand.
5. Using your right hand, remove a swab from the saline and sample the palm of the gloved hand. Use this swab to inoculate the section of the plate marked "control."
6. Using your gloved hand, pick up the caramel and squeeze it in your hand, rolling it around until the palm and fingers of the glove have fully contacted the candy. Drop the caramel back into the Petri dish. Be sure not to contaminate your gloved hand by touching anything accidentally.

Round One

1. To begin the first round, student number one will shake hands with any other student. When shaking hands, be sure that the palms of the gloves are fully in contact with one another. The person who was the "shakee" then becomes the "shaker" and shakes the hand of a chosen recipient. Although you will only choose a single person to shake hands with, it is possible that several persons may choose to shake hands with you. The instructor will record the two participants in each handshake. When each person has chosen to shake hands with someone, the round is over.
2. After all the shaking is complete, remove a swab from the saline, sample the palm of your glove, and then use the swab to inoculate the section of the plate labeled "round one." Discard the swab in an appropriate container for disposal.

Round Two

1. Again beginning with student number one, select a person at random to shake hands with, proceeding as before until everyone has initiated at least one handshake.
2. After all the handshaking is complete, remove a swab from the saline, sample the palm of your glove, and then use the swab to inoculate the section of the plate labeled "round two." Discard the swab in an appropriate container for disposal.
3. Incubate the plates at 30°C for 48 h.

QUESTIONS—PERIOD ONE

1. Over a 7-month period (October through April), visits for a food-poisoning illness (lasting 3 days or less) and herpes simplex infections (lasting a lifetime) to the only hospital serving a ski resort were as seen in the table. With regard to gastrointestinal illness, which month had the highest incidence?

	Resort population	No. new food-poisoning cases	No. new herpes simplex cases
October	7012	4	2
November	8114	3	4
December	14,763	26	7
January	15,111	31	6
February	16,276	19	7
March	9478	12	2
April	6912	16	0

2. What is the incidence and prevalence of the food-poisoning cases for the entire 7-month period? Why are the values for incidence and prevalence the same?

3. What is the incidence and prevalence of herpes simplex cases for the entire 7-month period? Why are these numbers *not* the same?

4. Which disease represents a (probable) common source epidemic and which a propagated epidemic? Why?

5. In general, how would you determine the identity of the index patient in the herpes outbreak?

PERIOD TWO

TRANSMISSION OF A COMMERCIAL POWDER

1. The powder used to simulate bacteria glows when illuminated with a long-wavelength UV lamp. Use such a lamp to illuminate each of your two tissues in a darkened room. Although these lamps emit very low levels of UV light, it is best not to look directly into the light.

2. For each round of shaking, a glowing tissue is evidence of "bacterial" contamination, and should be recorded as a positive result for that round on the data sheet.

3. Your instructor should have a corresponding table on the board. Record your results on the board, and copy the class results into Table 22.1.

TABLE 22.1	Handshaking results									
	Round 1						Round 2			
Simulated week Number	Shaker (initials)	Result (+/−)	Shakee (initials)	Result (+/−)	Simulated Week Number	Shaker (initials)	Result (+/−)	Shakee (initials)	Result (+/−)	
1					5					
1					5					
1					5					
1					5					
1					5					
1					5					
2					6					
2					6					
2					6					
2					6					
2					6					
2					6					
3					7					
3					7					
3					7					
3					7					
3					7					
3					7					
4					8					
4					8					
4					8					
4					8					
4					8					
4					8					

4. Using the infection and contact information in the table, work backward to find the index patient.

TRANSMISSION OF AN ACTUAL BACTERIUM

1. Retrieve your plate from the incubator and examine each section for the presence of red *S. marcescens* colonies. These colonies represent infection, and should be recorded as + on the data sheet. Other bacterial colonies are most probably not *S. marcescens*, and should be recorded as − on the data sheet.

2. Your instructor should have a corresponding table on the board. Record your results on the board, and copy the class results into Table 22.1.

3. Using the infection and contact information in the table, work backward to find the index patient.

RESOLUTION OF THE CASE

One of the people present at the church gathering on May 15 was a 17-year-old girl who had never been immunized for measles and who had worked from May 4 through 14 as a missionary in Bucharest, Romania, where a large measles outbreak was later reported. The day before the gathering, the girl had returned to the United States, traveling on both international and domestic commercial airliners. When she arrived, she was experiencing fever, cough, conjunctivitis, and coldlike symptoms; family members later recalled that she had exhibited a rash the next day.

Measles is caused by a virus known simply as the measles virus and is spread mainly through droplet contact. Although 30–40 million cases of measles occur annually worldwide, resulting in about 400,000 deaths, widespread vaccination has made measles in the United States nearly unheard of, with only a few cases reported per year. This propagated outbreak was the largest in the United States since 1996. Its severity was due almost entirely to the fact that only one of the 34 people infected had been adequately vaccinated against measles. State and local health departments in Indiana, Ohio, and Illinois worked to control the outbreak through multiple measures, including voluntary isolation of patients, administration of vaccine and immunoglobulin to susceptible contacts, voluntary home quarantine of susceptible individuals who refused vaccination, and alerting hospitals and the media to the measles outbreak.

In the United States, the Advisory Committee on Immunization Practices (ACIP), composed of a group of 15 experts, advises the U.S. Department of Health and Human Services and the CDC on the control of vaccine-preventable diseases. This committee has long advocated that (1) all persons who travel internationally be vaccinated against measles, (2) all school-aged children in the United States receive two doses of measles vaccine, and (3) all hospital workers be fully vaccinated against the disease. Indiana is one of a number of states that allow nonmedical exemption from vaccination for philosophical or religious reasons, although these persons are 22 times more likely to acquire measles than are those who are vaccinated. If the three recommendations of the ACIP had been followed, this outbreak would have been prevented.

Since this outbreak, others have occurred in the United States, including one in 2008 that affected more than 125 people. It seems that, in addition to people who have general philosophical or religious reasons for refusing vaccines, a growing number of individuals believe (incorrectly) that vaccinations are harmful or may cause autism, setting in place the circumstances for more frequent outbreaks. Americans generally enjoy some degree of herd immunity, the phenomenon that protects unvaccinated individuals because almost everyone they come in contact with *has* been immunized and therefore will not transmit the microbe as a result of the high vaccination rate. This is not always the case in third-world countries. Fears about immunization in this country could jeopardize the vaccination coverage rate and, thus, herd immunity.

EXERCISE 22 REVIEW QUESTIONS

1. Who was the index case in the measles epidemic?

2. With regard to the *Serratia* "epidemic"?

Who was the index case?

Who infected you?

Whom did you infect?

3. Use the class data to calculate incidence and prevalence for the *S. marcescens* "epidemic" seen in this exercise. Although the epidemic took only a few minutes to spread throughout the class, the results table for the exercise is divided into 8 weeks so that it is more representative of a true outbreak. For this question, assume that the disease lasts 2 weeks so that people infected in week 1 are not free of the disease until week 4, those infected in week 2 are not healthy until week 5, etc. Also assume that no long-term immunity is generated so that a person can be infected more than once.

Week	Total population	No. new cases*	Total cases†	At-risk population	K	Incidence	Prevalence
1							
2							
3							
4							
5							
6							
7							
8							

* This week.
† This week plus last 2 weeks.

4. What is herd immunity? Why did herd immunity protect the index case from infection with measles in the United States but not in Romania?

REFERENCES

CDC. *Measles Vaccine.* http://www.cdc.gov/vaccines/vpd-vac/measles/default.htm.

Glo Germ. *The #1 Product for Teaching Handwashing, Isolation Techniques, Aseptic Techniques, and General Infection Control.* http://www.glogerm.com.

Morbidity and Mortality Weekly Report

CASE SYNOPSIS

Google Used to Predict Influenza Outbreaks

In 2008, Google launched Google Flu Trends, an application that compiles aggregated data from key word searches for clinical terms, such as **thermometer, chest congestion, muscle aches,** or **flu symptoms.** Google reports the data on a web site, which then provides an early-warning system for the locations of new flu outbreaks. Because the data are collected from searches performed each day, trends in flu symptoms become apparent much more quickly than when they are based on data reported during office visits or in lab reports from physicians around the country. When the CDC compared actual cases over the course of a year with Google's findings, the data from the two sources matched.

Initially, Google was only compiling information about flu trends in the United States and Canada. But after the H1N1 virus appeared in Mexico in 2009, the CDC asked Google to go back and look at Internet searches conducted by people in Mexico during that time. Evaluation of the data indicated that Google detected an uptick several days before the CDC did (Figure 23.1).

Resolution of the Case appears on page 191

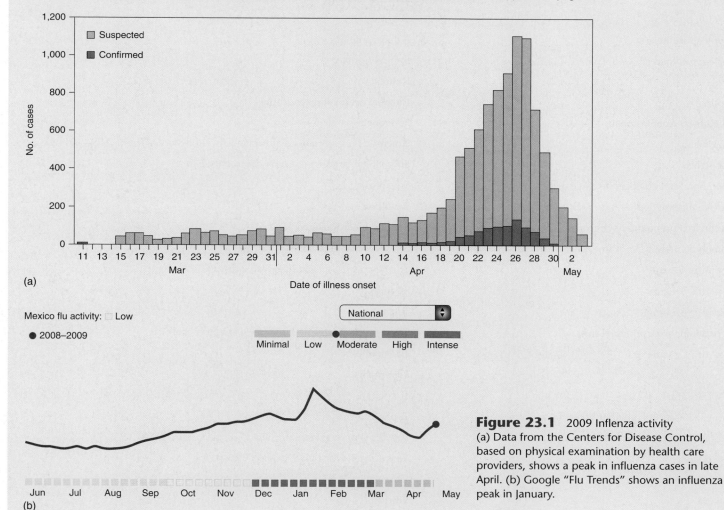

Figure 23.1 2009 Inflenza activity (a) Data from the Centers for Disease Control, based on physical examination by health care providers, shows a peak in influenza cases in late April. (b) Google "Flu Trends" shows an influenza peak in January.

Ginsberg, J. 2008. Detecting influenza epidemics using search engine query data. *Nature,* 457: 1012–1014

STUDENT LEARNING OUTCOMES

After completing this exercise, you should be able to:

1. Calculate the morbidity, mortality, and incidence of a disease outbreak.
2. Access and use the National Notifiable Disease Surveillance System.

INTRODUCTION

Tracking the movement of a disease through a population is the job of **epidemiologists,** who concentrate on the frequency, distribution, and transmission of infectious diseases. The Centers for Disease Control and Prevention (CDC), a government agency headquartered in Atlanta, Georgia, serves as the ultimate storehouse for statistical data related to infectious disease (as well as noninfectious disease and accidents, which are of less interest to microbiologists). Because keeping track

TABLE 23.1   Nationally notifiable infectious diseases

Anthrax	Mumps
Arboviral neuroinvasive and nonneuroinvasive disease	Novel influenza A viral infections
Botulism	Pertussis
Brucellosis	Plague
Chancroid	Paralytic poliomyelitis
Chlamydia trachomatis infection	Nonparalytic poliovirus infection
Cholera	Psittacosis
Coccidiomycosis	Q fever
Cryptosporidiosis	Rabies, animal
Cyclosporidiosis	Rabies, human
Dengue	Rubella
Diptheria	Rubella, congenital syndrome
Erlichiosis/Anaplasmosis	Salmonellosis
Giardiasis	Severe acute respiratory syndrome-associated coronavirus (SARS-CoV)disease
Gonorrhea	Shiga toxin-producing *Escherichia coli* (STEC)
Haemophilus influenza, invasive disease	Shigellosis
Hansen disease (leprosy)	Smallpox
Hantavirus pulmonary syndrome	Spotted fever Rickettsiosis
Hemolytic uremic syndrome, post diarrheal	Streptococcal toxic shock syndrome
Hepatitis A, acute	*Streptococcus pneumoniae,* invasive disease
Hepatitis B, acute	Syphilis
Hepatitis B, chronic	Tetanus
Hepatitis B virus, perinatal infection	Toxic shock syndrome (other than streptococcal)
Hepatitis C, acute	Trichinellosis
Hepatitis C, chronic	Tuberculosis
HIV infection	Tularemia
Influenza-associated pediatric mortality	Typhoid fever
Legionellosis	Vancomycin-intermediate *Staphylococcus aureus* (VISA)
Listeriosis	Vancomycin-resistant *S. aureus* (VRSA)
Lyme disease	Varicella
Malaria	Vibriosis
Measles	Viral Hemorrhagic Fevers
Meningococcal disease	Yellow fever

of all infectious disease would be impossible, the CDC has designated 66 serious diseases as **reportable,** meaning that doctors and hospitals are required to inform the CDC of any cases they encounter (Table 23.1). This information appears each week as part of the CDC publication *Morbidity and Mortality Weekly Report (MMWR).* The CDC then uses this information to track outbreaks, deaths, and even the appearance of never before seen ailments: severe acute respiratory syndrome (SARS) was added to the list in 2003 and potential cases of pandemic influenza H5N1 (bird flu) and H1N1 (swine flu) in 2007 and 2009, respectively.

While the full range of statistical analyses employed by the CDC is beyond the scope of this book, three common epidemiological measurements can be easily calculated. The first of these is **morbidity,** or illness due to a specific disease. Morbidity is calculated by dividing the number of cases of a disease in a given period by the number of people susceptible to that disease:

$$\text{Morbidity} = \frac{\text{no. cases per period}}{\substack{\text{susceptible populaton size} \\ \text{at midpoint of period}}} \times K$$

where K is a factor that varies and serves to assist in more clearly reporting the results. Two aspects of the equation above deserve a little more explanation. First, because population size is constantly changing, it is defined as the number of people present at the midpoint of the time period being studied. Second, because morbidity is generally a very small number, it is commonly multiplied by a power of 10 such as 10,000 or 100,000 (the constant K) so that it can be reported as whole number. For instance, a morbidity rate of 0.000012 can be multiplied by 1,000,000, allowing it to be reported as 12 cases per 1,000,000 persons as opposed to 0.000012 cases per person.

Mortality is the second epidemiological measurement and refers to the number of deaths within a specified period among people having a particular disease. The equation for determining mortality can be written as:

$$\text{Mortality} = \frac{\text{no. disease-related deaths per period}}{\text{no. people with the disease}} \times K$$

The constant K is used just as it was for morbidity, so mortality can be reported as a whole number of cases per, for example, 10,000 or 100,000 people.

Finally, **incidence** compares the number of **new** cases of a disease during a specified period to the size of the susceptible population during that period, as seen in the equation:

$$\text{Incidence} = \frac{\text{no. new cases}}{\substack{\text{susceptible population size} \\ \text{at midpoint of period}}} \times K$$

Just as with morbidity and mortality, the size of the population is determined at the midpoint of the period, and the constant K is used to make the incidence a more easily reportable number.

Morbidity, mortality, and incidence are often combined with other data, such as the geographic location of cases, the sex, race, and behavior of those infected, and even the time of year. Together, this information allows the CDC to track diseases, predict future outbreaks, and devise prevention strategies. In today's lab, you will use the *MMWR* to determine the incidence for a reportable disease.

PRE-LAB QUESTIONS

1. *Define the following terms:*

  Sporadic

  Endemic

  Epidemic

  Pandemic

2. *Use the information in the accompanying table to calculate morbidity, mortality, and incidence for each disease.*

Disease	New cases in January	Deaths from disease	Population on January 15	Morbidity	Mortality	Incidence
Botulism	4	2	714,000			
Cholera	151	37	154,135			
Measles	2,910,000	33,625	6,400,000			

MATERIALS

A computer with Internet access

PROCEDURE

1. Each student will collect data on a different reportable disease based on Table 23.2.
2. Access the CDC web site at http://www.cdc.gov. Select the link for *MMWR*.
3. Click on State Health Statistics and then Morbidity Tables to access the NNDSS (National Notifiable Disease Surveillance System) Interactive Tables.
4. Enter week number 1 of the *previous* year into the selection windows.
5. Click submit, and then click on table II (part 1).

6. If your disease is not listed in part 1, go to part 2 (or 3, 4, . . .) by clicking on the *Next Part* link at the bottom of the page.
7. When you have found your disease, record the number of new cases of the disease for that week in Table 23.3.
8. Return to the NNDSS Interactive Tables by clicking on the link at the bottom of the page.
9. Repeat steps 2 through 7, this time recording the number of new cases occurring in week 3. Continue this process until you've recorded the number of new cases of your disease for all odd-numbered weeks in the year, up to week 51.

RESULTS

1. Calculate the incidence of disease for each period during which morbidity data was collected. Refer to the accompanying table for estimated population size of the United States.

Estimated U.S. population (U.S. Census Bureau)						
2007	2008	2009	2010	2011	2012	2013
300,912,947	303,597,646	306,272,395	308,935,581	311,600,880	314,281,098	316,971,485

2. Graph the incidence rate of the disease you've chosen. The vertical axis should be labeled with the incidence of disease (cases per 10,000 or 100,000 people) while the horizontal axis should indicate each 2-week period in which measurements were taken.

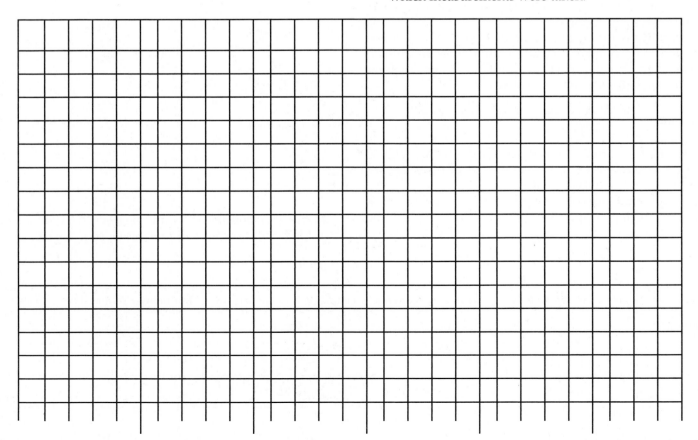

TABLE 23.2	Student assignments: Reportable diseases		
Student number	**Disease**	**Student number**	**Disease**
1	Chlamydia	14	Pertussis
2	Coccidiomycosis	15	Rabies, animal
3	Cryptosporidiosis	16	Spotted fever Rickettsiosis
4	Giardiasis	17	Salmonellosis
5	Gonorrhea	18	Shiga toxin-producing *E. coli* (STEC)
6	Haemophilus influenza, invasive	19	Shigellosis
7	Hepatitis A, acute	20	Streptococcal toxic shock syndrome
8	Hepatitis B, acute	21	*Streptococcus pneumoniae* invasive disease
9	Legionellosis	22	Tetanus
10	Lyme disease	23	Syphilis
11	Malaria	24	Varicella
12	Meningococcal disease	25	West Nile virus disease, all types
13	Mumps		

3. Using the data you've gathered, does the incidence of your disease fluctuate with regard to time of year? If so, is this expected?

4. Could you think of any way in which the incidence of your disease could fluctuate? For example, are persons with certain occupations or living in a particular environment more susceptible?

TABLE 23.3	**Morbidity data for** _____ **(a reportable disease)**			
Week (mm/yy)	No. new cases	U.S. population	K	Incidence

RESOLUTION OF THE CASE

Epidemic tracking based on Internet searches reflects what is called *collective intelligence.* It works because individuals using their personal computers tend to search for terms related to their immediate needs and intentions, and they generally do this before presenting in a doctor's office or emergency room. The methodology used in the Google search was published in the prestigious science journal *Nature,* and another independent study has been published about a similar search analysis conducted by Yahoo, showing that it was effective in predicting flu trends.

Some people worry that data collected from Internet searches may compromise individuals' privacy. However, Google maintains that Flu Trends cannot be used to identify individual users because the data are anonymous and are aggregated before being presented. Another potential drawback is that this data collection method is less likely to be useful for tracking epidemics in societies having a low percentage of computer ownership, namely, developing countries. However, considering the high stakes involved in identifying an epidemic quickly, Internet search term analysis holds great promise for public health. And, unlike most health innovations, it's free!

EXERCISE 23 REVIEW QUESTIONS

1. What is the infectious agent responsible for influenza? How is the disease spread from person to person? What are the symptoms of the disease?

2. Can you think of a downside to tracking influenza using the Internet, as was seen in this case? What if the disease being tracked was HIV or cancer?

3. How might data collected in this way *not* be representative of a particular population?

4. Explain why the two graphs in Figure 23.1 do not show influenza activity peaking at the same time.

5. Some diseases, such as influenza, are considered to be "underreported." What do think this means, and what is the significance of underreported diseases?

REFERENCE

CDC. *Seasonal Influenza.* http://www.cdc.gov/flu.

Bacterial Transformation

CASE SYNOPSIS

Multidrug-Resistant *Acinetobacter* Infections in Soldiers

Acinetobacter baumannii is a Gram-negative bacterium commonly found in soil and water. This bacterium is also frequently associated with nosocomial infections, diseases patients acquire while they are in a hospital. *A. baumannii* tends to thrive in hospital settings because the bacterium is resistant to environmental influences and can survive for months on such objects as faucets, toilets, bedclothes, doorknobs, sinks, and medical equipment. The spread of this bacterium is a concern for all medical facilities, but especially military ones, which have seen increasing numbers of bloodstream infections caused by *A. baumannii*, probably because combat conditions make controlling and treating them more difficult.

Additionally, *A. baumannii* infections have become problematic because increasing numbers of isolates show multiple drug resistance, that is, the bacteria are unfazed by the antibiotics commonly used to treat them. Many of the resistance genes found in *A. baumannii* are similar or identical to those seen in other genera of bacteria, such as *Pseudomonas, Salmonella,* and *Escherichia,* which also commonly occur in health care facilities.

Resolution of the Case appears on page 201

CDC. 2004. *Acinetobacter baumannii* infections among patients at military medical facilities treating injured U.S. service members, 2002–2004. *Morbidity and Mortality Weekly Report,* 53: 1063–1066.

Bio-Rad Laboratories. 2000. Biotechnology Explorer™ pGLO™ Bacterial Transformation Kit (catalog no. 166-0003-EDU). www.explorer.bio-rad.com.

STUDENT LEARNING OUTCOMES

After completing this exercise, you should be able to:

1. Discuss the means by which bacterial cells may obtain new genetic information.
2. Explain the purpose, techniques, and outcome of a bacterial transformation.

INTRODUCTION

Historically, bacterial cells have been described, classified, and categorized by their appearance, or **phenotype,** and these physical characteristics have long been the basis for many laboratory tests. Commonly seen phenotypes include the ability to utilize lactose, produce a cell wall or, as seen here, withstand the effects of a certain drug. A phenotype results when several proteins act together to catalyze a biochemical reaction or contribute to a cellular structure. The information needed to synthesize a protein is in turn encoded in the genes of the organism. An organism's **genotype** refers to the combination of genes it possesses.

New phenotypes can arise when a gene mutates, or changes, resulting in organisms with never before seen abilities. Although mutations are exceedingly rare events, the enormity of most bacterial populations ensures that unique bacteria arise on a regular basis. If a newly acquired mutation provides a cell with a distinct advantage over other cells in a culture, then the mutated cell will eventually come to dominate the environment. As an example, when a random mutation results in a cell gaining the ability to resist the effects of an antibiotic, this cell will have an advantage over other cells in the culture provided the antibiotic is present in the environment. It is in this way that the use of antibiotics does not cause resistant cells to develop but rather provides an environment that allows only these rare cells to grow.

In addition to new genetic variants being created by mutation, novel genes can be acquired from other bacterial species through the process of genetic recombination. Three types of genetic recombination are generally recognized, each of which allows bacterial cells to receive small amounts of DNA from other cells or from the environment. In **conjugation,** DNA is transferred from one living bacterial cell to another with the aid of a bacterial structure known as a **pilus** (singular, **pili**). The second type of recombination, **transduction,** occurs when a bacteriophage transfers a small amount of DNA from a lysed bacterial cell to another cell of the same species. Both conjugation and transduction are commonly implicated in the cell-to-cell transfer of the abilities to resist certain drugs and produce toxins. The third method of recombination is **transformation,** which occurs when a bacterial cell receives small amounts of DNA from the environment. In a famous experiment by Frederick Griffith in the 1920s, the genes responsible for the production of a bacterial capsule were shown to be transferred from dead cells to living ones by this process.

Transformation can also be used in the biotechnology laboratory to add new genes to bacteria, creating recombinant cells that then act as biologic factories, expressing any protein encoded by their newly acquired DNA. Using this method, bacteria have been modified to produce pharmaceutical proteins, enzymes, and hormones easily and cheaply. And, because all organisms use the same genetic code, bacteria can be transformed with DNA from other species, allowing bacterial cells to produce proteins found previously only in cows, pine trees, or humans. In this exercise, *Escherichia coli* cells will be transformed with a gene from the jellyfish *Aequorea victoria*, resulting in bacterial cells that fluoresce under UV light.

Key to the transformation of cells in the laboratory is the use of a small, circular piece of DNA called a **plasmid**. A plasmid is typically a few thousand nucleotides in length and contains several areas of interest, including:

Origin of Replication (ori)   The origin of replication refers to a sequence of DNA that is recognized by the enzymes in the bacterial cell responsible for the replication of DNA. The ori ensures that the plasmid will be replicated each time the bacterial cell divides.

Drug Resistance   At least one gene conferring antibiotic resistance (commonly to ampicillin) is included on most plasmids. The presence of this gene ensures that, when grown on an antibiotic-containing media, cells possessing a plasmid will survive while those lacking a plasmid will die.

Gene of Interest   This is the gene to be expressed, and it can vary depending on the purpose of the transformation.

Promoter   The promoter is a sequence of DNA that lies very close to the gene of interest and acts as a genetic switch, allowing the gene to be transcribed (i.e., used to synthesize RNA) under various circumstances.
Constitutive promoters ensure that a gene is transcribed at all times, while **inducible** promoters allow a gene to be transcribed only when specific conditions exist, such as the presence of a particular nutrient in the environment.

In this exercise, the plasmid being used is named pGLO (Figure 24.1). It contains an ori, the *bla* gene, which provides resistance to ampicillin, and the gene that codes for the production of a green fluorescent protein (GFP). **Transcription** of the GFP gene is controlled by an arabinose promoter.

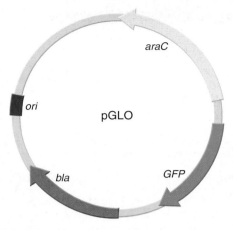

Figure 24.1   The plasmid pGLO has several areas of interest. (1) The *ori* provides a recognition site for DNA polymerase; without it the plasmid would never be replicated. (2) The *GFP* gene provides the code for the green fluorescent protein. Production of this protein will only occur if RNA polymerase is allowed to bind to the gene, which only occurs when arabinose is present. (3) The *araC* gene codes for a DNA-binding protein, which, if arabinose is present, allows RNA polymerase to bind to the *GFP* gene. (4) The *bla* gene codes for a protein that makes the cell resistant to ampicillin. The *bla* gene is under the control of a constitutive promoter, meaning it is always expressed.

When arabinose is present in the environment, it binds to a DNA-binding protein called **araC**. In the presence of arabinose, araC allows the protein RNA polymerase to bind to the promoter and transcribe the genes, which will in turn be translated to provide ampicillin resistance and the green fluorescent protein. If no arabinose is present, the genes will not be transcribed, or translated (Figure 24.2). Table 24.1 summarizes each of the molecules important in this process.

Even though transformation of cells is a natural process, it is also a rare process, occurring very infrequently. The efficiency of the process can be increased dramatically by using cells that have been treated to take up as much DNA as possible; such cells are termed **competent**. The process of making cells competent involves treating them with Ca^{++}, which is thought to neutralize the negative charges on the DNA and the cell membrane, making it easier for DNA to enter

TABLE 24.1	Important components of the transformation process
Name	**Function**
Green fluorescent protein (GFP)	Protein fluoresces under ultraviolet light. Used to determine if transformed gene is being transcribed.
Plasmid (pGLO)	Vector used to introduce the GFP gene into *E. coli*.
DNA-binding protein (araC)	Regulates the arabinose promoter. If arabinose is not present, araC prevents RNA polymerase from binding to the arabinose promoter. In the pGLO plasmid, araC is used to regulate transcription of the GFP gene.
β-lactamase gene (*bla*)	Produces β-lactamase, which hydrolyzes antibiotics containing a β-lactam ring, including ampicillin. Used to select for transformed cells.
Origin of replication (ori)	Allows the plasmid to be recognized (and replicated) by DNA polymerase within the cell.

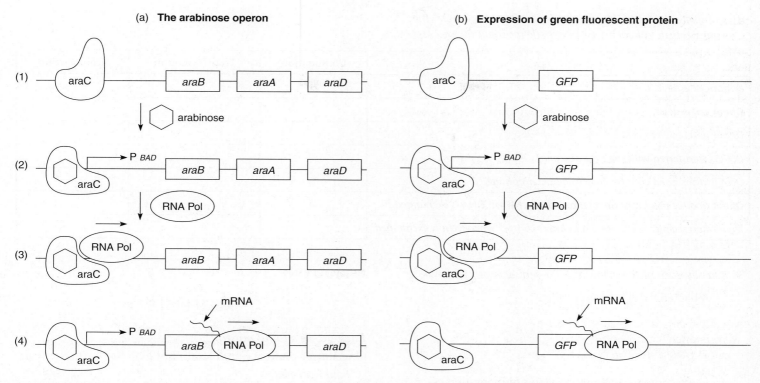

(a) The arabinose operon

(b) Expression of green fluorescent protein

Figure 24.2 (a) The digestion of the sugar arabinose requires the production of three proteins that are only expressed when arabinose is present in the environment. Expression of all three proteins is under control of a single promoter, or genetic switch. This grouping of genes together with a common promoter is known as an operon. (1) When arabinose is not present in the environment, a DNA-binding protein called araC obstructs the promoter, not allowing RNA polymerase to bind to the DNA. (2) When arabinose is present, it binds to araC, changing the conformation of the protein. (3) The new conformation of araC allows RNA polymerase to bind to the promoter. (4) The RNA polymerase is able to transcribe the genes in the operon. The mRNA produced will be translated by ribosomes in the cell, producing the enzymes needed to digest arabinose. (b) Regulation of green fluorescent protein (GFP) is obtained by linking the *GFP* gene to the arabinose operon in place of the genes needed for arabinose digestion. In this way, when arabinose is present in the environment, RNA polymerase will transcribe the *GFP* gene, leading to production of the green fluorescent protein.

the cell. The cells are also subjected to a brief heat shock that helps increase the permeability of the membrane to DNA. Even under these idealized conditions, less than one cell in a thousand will be transformed.

After transformation, the cells are plated onto media containing ampicillin, an antibiotic that acts as a selective agent, allowing only those cells containing plasmids to grow. The media will also, in some cases, contain arabinose, the inducer molecule that interacts with the DNA-binding protein, resulting in transcription of the gene of interest. This exercise is based on the Biotechnology Explorer™ pGLO™ Bacterial Transformation Kit produced by Bio-Rad Laboratories.

PRE-LAB QUESTIONS

1. *Define the following terms:*

Nosocomial infection

Phenotype

Mutation

Plasmid

Competent

2. *Complete the following table by indicating with an "X" the characteristics of each type of genetic recombination.*

	Conjugation	Transformation	Transduction
Donor cell is alive.			
Donor cell is dead.			
Donor and recipient cells are the same species.			
DNA is transferred with the help of a pilus.			
DNA is transferred with the help of a bacteriophage.			
Donor and recipient cells are in direct contact when DNA is exchanged.			
Donor and recipient cells are not in direct contact when DNA is exchanged.			

3. *Classify each of the following as a genotype or phenotype*

A bacterial cell is Gram positive.

E. coli contains a gene that allows fermentation of lactose.

E. coli ferments lactose.

S. aureus is resistant to methicillin.

Micrococcus luteus colonies are yellow.

PERIOD ONE

MATERIALS

Each group should obtain:
One starter plate of *E. coli* strain K-12
  Four agar plates
  One LB (Luria-Bertani)
  Two LB/amp (LB/ampicillin)
  One LB/amp/ara (LB/ampicillin/arabinose)
1-ml transformation solution
1-ml nutrient broth
1–20 µl, 20–200 µl and 100–1000 µl, pipettors, or five
  volume-calibrated Pasteur pipets
Two 2-ml microcentrifuge tubes and foam holder
Ice bath (crushed ice in a beaker or foam cup)
Marking pen

Every group should have access to the following:
Rehydrated pGLO plasmid DNA
42°C water bath
37°C incubator
Long-wavelength UV lamp

PROCEDURE

1. Mark the bottom of each Petri dish with the media it
   contains (LB, LB/amp, or LB/amp/ara) and the
   number of your group.

2. Label two microcentrifuge tubes with your group number. Label one tube +pGLO and another −pGLO.

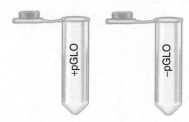

3. Using a sterile transfer pipet, add 250 μl transformation solution to each tube, and place the tubes on ice.

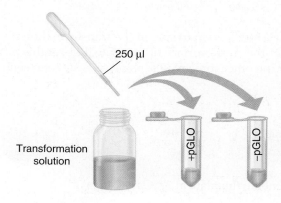

4. Using a sterile loop, transfer an entire colony of *E. coli* to the + pGLO tube. Use the loop to completely disperse the cells in the transformation solution. Place the tube back on ice. Repeat this process for the −pGLO tube, closing the lid of the tube before returning it to the ice.

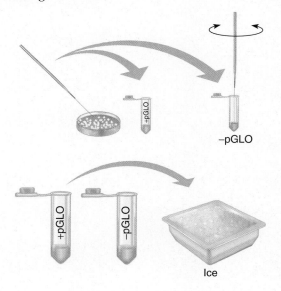

5. Use a supplied sterile loop to transfer a loopful of the pGLO DNA solution to the +pGLO tube. Mix the DNA solution with the cell suspension in the tube. Close the lid of the tube and return it to the ice. Dispose of the pipet.

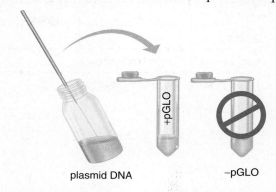

6. Incubate the tubes on ice for 10 min. If the tubes are in a tube holder, be sure that they make good contact with the ice.

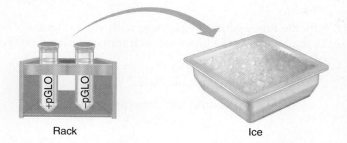

7. Transfer from tubes from the ice bath to a water bath set at 42°C. Allow the tubes to remain in the water for **exactly 50 sec** and then return them to the ice. For best results, the temperature change from cold to hot and back to cold must be as rapid and complete as possible, so work quickly and be sure the tubes are fully immersed in both the ice and water.

8. Continue incubating the tubes on the ice for a total of 2 min.

9. Using a sterile pipet, add 250 µl LB broth to the +pGLO tube and close the top. Repeat this process with a second pipet for the other tube.

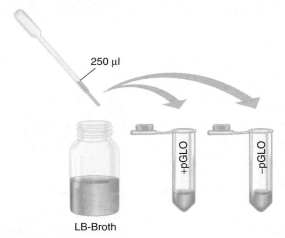

LB-Broth

10. Allow both tubes to incubate at room temperature for 10 min.

11. Mix the tubes gently by tapping with your finger. Using a new sterile pipet for each transfer, inoculate the plates as follows:

    − pGLO 200 µl onto LB plate

    − pGLO 200 µl onto LB/amp plate

    + pGLO 200 µl onto LB/amp plate

    + pGLO 200 µl onto LB/amp/ara plate

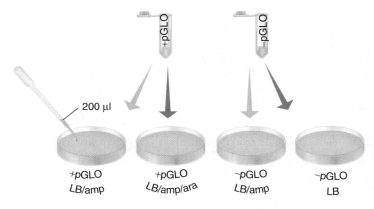

+pGLO	+pGLO	−pGLO	−pGLO
LB/amp	LB/amp/ara	LB/amp	LB

12. Use a sterile loop to spread the transformation mixture over the entire surface of the plate. Repeat this process for the other three plates, using a sterile loop each time.

13. Incubate the plate at 37°C for 24 h.

QUESTIONS—PERIOD ONE

1. Complete the table by indicating with an "X" the
characteristics you would expect to see on each of your
plates.

	−pGLO in LB	−pGLO in LB/amp	+pGLO in LB/amp	+pGLO in LB/amp/ara
Growth of bacteria				
Bacteria resist ampicillin				
Bacterial lawn				
Isolated bacterial colonies				
Colonies fluoresce under UV light				

2. What would be indicated by each of the following
results:

   No bacteria grew on the LB plate.

3. What is the purpose of the arabinose in the LB/amp/
ara plates? What would happen if arabinose was
inadvertently left out of the medium?

   Control bacteria (not transformed with pGLO) grew on
the LB/amp plate.

4. In what way does the LB/amp/ara act as a selective
media? How does it act as a differential media?

PERIOD TWO

1. Remove the plates from the incubator and examine each plate for bacterial growth. Sketch each plate.

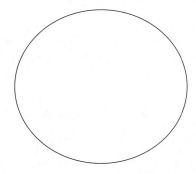

−pGLO on LB

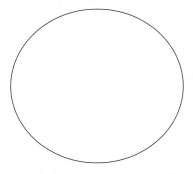

−pGLO on LB/amp

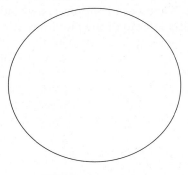

+pGLO on LB/amp

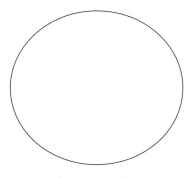

+pGLO on LB/amp/ara

2. Remove the cover from each of the four plates and examine the colonies with a UV light. What **two** new traits have the bacteria gained?

3. Record the number of colonies present on each of the two transformed (+pGLO) plates.

TRANSFORMATION EFFICIENCY

Transformation efficiency is commonly used to describe how successfully DNA molecules enter bacterial cells. It is expressed as the number of transformed cells per microgram (μg) of DNA. As an example, if 0.25 μg of DNA were spread on the LB/amp/ara plate and this resulted in a total of 115 colonies, then the transformation efficiency would be 115 colonies/0.25 μg or 460 colonies/μg DNA.

The amount of DNA used in the transformation process can be calculated by multiplying the volume of DNA used by its concentration. The loop used to originally transfer DNA to the transformation solution was calibrated to contain 10 μl, and the DNA was at a concentration of 0.08 μg/μl.

Total quantity of pGLO DNA used = _____μg

Only a fraction of this DNA was actually applied to the plate. The fraction applied to the plate is determined by dividing the volume of transformant solution spread on the plate by the total volume of the solution. Two hundred microliters of transformant solution was applied to the plate out of a total of 510 μl (250 μl transformant solution + 250 μl LB + 10 μl pGLO), meaning that the fraction of DNA spread on the plate was 39.2%.

The amount of DNA present on a single plate can be calculated by multiplying the total quantity of DNA by the fraction used per plate.

pGLO DNA used per plate =

Finally, the transformation efficiency is calculated by dividing the total number of transformed cells (recognizable as colonies after incubation) by the total amount of DNA used and is expressed as transformants/μg.

Transformation efficiency = _____transformants/μg

RESOLUTION OF THE CASE

Had the resistance genes found in *A. baumannii* been unique (i.e., significantly different from resistance genes seen in other species), this would have suggested that they had first arisen within *A. baumannii* with no genetic material contributed by other sources. Instead, the fact that many of the resistance genes were previously known to exist in other bacterial species indicated that they had probably been transferred to *A. baumannii* from one of these species. The fact that all three types of genetic recombination—conjugation, transformation, and transduction—occur in *A. baumannii* also supports this hypothesis. Since these other species are commonly found in health care facilities, the facilities themselves are the most likely setting for the transfer of genetic material between other bacterial species and *A. baumannii*.

The multidrug resistance of *A. baumannii* is a serious concern in both military and civilian medical facilities and has led to several changes in their practices. Laboratory characterization of patient and environmental samples is in progress to determine the prominence of *A. baumannii* in soil samples and in medical facilities. Care facilities have increased surveillance for patients colonized with *A. baumannii*. Clinicians have had to treat infections most carefully and to consider new combinations of antimicrobials. Infection control measures have been introduced or revised, and the use of alcohol-based hand sanitizers has been implemented. Finally, researchers and the pharmaceutical industry are continuing to search for new and innovative ways to treat all infectious diseases.

EXERCISE 24 REVIEW QUESTIONS

1. What evidence is there that the resistance genes found in *A. baumannii* were transferred from another bacterium?

2. How is long-term antibiotic therapy connected to the emergence of antibiotic-resistant bacterial species?

The Ames Test

CASE SYNOPSIS

West Nile Virus Update—United States, January 1–December 31, 2009

West Nile Virus (WNV) was first observed in the United States in 1999 and is now considered to be a well-established, seasonal epidemic in North America. Peak season for the disease is late summer and early fall when the mosquitoes that carry the virus are most active. Although most people infected with West Nile Virus show few if any symptoms, about 1 in 150 will develop a severe, life-threatening illness.

Surveillance data for WNV for the period January 1 through December 31, 2009 revealed a total 720 cases of WNV illness spread across 38 states; 32 of these cases were fatal. Additionally, 116 people were thought to have become infected with WNV through blood donations contaminated with the virus. Because the virus is most often spread by mosquitoes, the Centers for Disease Control and Prevention recommends that mosquito repellents containing N,N-diethyl-meta-toluamide (DEET) be applied to the skin to reduce the number of mosquito bites. Some people have questioned the safety of DEET, fearing the compound may cause mutations in the DNA, which could potentially lead to cancer.

Resolution of the Case appears on page 207

CDC. *West Nile virus* http://www.cdc.gov/ncidod/dvbid/westnile/index.htm.

STUDENT LEARNING OUTCOMES

After completing this exercise, you should be able to:

1. Explain how an Ames test can be used to evaluate the mutagenicity of a chemical compound.
2. Perform an Ames test and evaluate the results.

INTRODUCTION

The safety of chemicals we put in and on our bodies is, rightfully, of great importance. Sunscreens, insecticides, cosmetics, food preservatives, medications, and the like all must be tested to ensure their safety. One area of particular concern is whether a chemical interacts with DNA to produce a change in the sequence of DNA known as a **mutation.** This is especially important as most chemicals that cause cancer also induce mutations. In the past, small animals were inoculated with suspect chemicals and then monitored for the development of tumors, a slow and expensive process. The Ames test is an assay that detects chemical **mutagens** using bacteria, rather than animals, producing results faster, cheaper, and without the ethical qualms of animal testing.

Bacteria that are able to synthesize all of the biochemicals needed for their own growth are termed prototrophs. These bacteria contain hundreds of enzymatic pathways, each working to synthesize a single biochemical molecule. If a mutation in the gene for an enzyme results in a bacterial strain being unable to synthesize a needed molecule, such as an amino acid, then that strain is referred to as an auxotroph (with regard to that molecule). The missing molecule must be added to the media if an auxotroph is to grow (Figure 25.1). The Ames test measures the ability of a chemical to induce **back mutations,** or **reversions,** of an auxotroph to its original prototrophic state in essence correcting the mutated gene by inducing a second mutation that restores the original, functional gene sequence. The greater the number of revertant colonies, the more mutagenic the test substance must be.

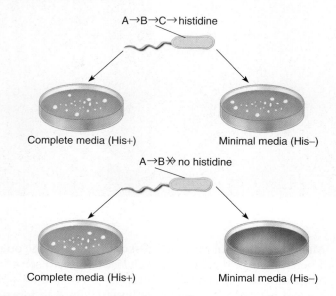

Figure 25.1 A histidine prototroph (top) is able to synthesize histamine and is able to grow on both complete media (containing histidine) and incomplete media (without histidine). A histidine auxotroph (bottom) is unable to synthesize histidine and is only able to grow when histidine is present in the media.

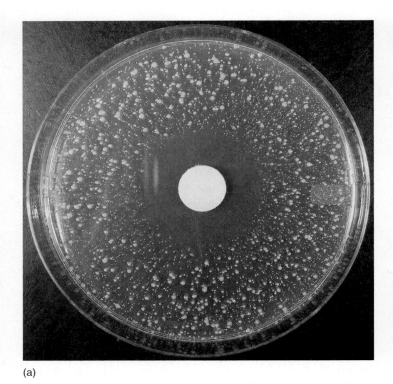

(a)

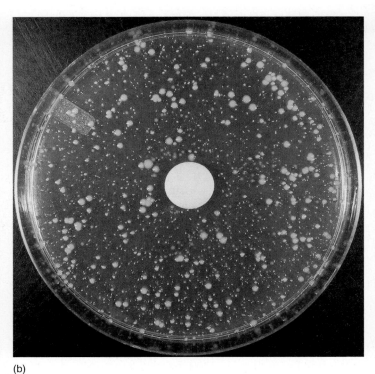

(b)

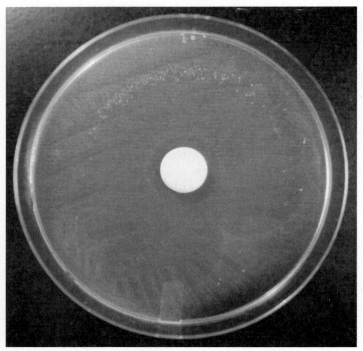

(c)

Figure 25.2 (a) A *Salmonella enterica* histidine auxotroph is grown on a minimal medium. The disk in the center of the plate is saturated with a chemical being tested for mutagenic properties. Colonies visible on the plate represent those cells that have undergone back mutations (either spontaneously or in response to the chemical) in the histidine synthesis gene, giving them the ability to grow on minimal media. (b) A control plate of minimal media in which the test substance is sterile water. Because water is nonmutagenic, all of the colonies here represent spontaneous mutations. (c) A *Serovar typhimurium* histidine auxotroph growing on complete medium and exposed to the test substance shows a zone of inhibition around the disk. Because the bacterium should grow well on complete media, the zone of inhibition indicates the toxicity of the test substance.

minimal media, which contains no histidine, and once it solidifies, a paper disk impregnated with the chemical to be tested is placed on top of the agar. The first few generations of growth produce a very faint bacterial lawn; once the histidine is exhausted, the bacteria should stop growing. Any colonies seen on the plate after incubation are the result of reversions that have resulted in a cell able to synthesize its own histidine. The more mutagenic the test substance, the greater the number of revertant colonies the test plate should show (Figure 25.2).

Besides the test plate, two controls are included as part of the Ames test. The first is a plate exactly like the test plate, the only difference being that the filter paper disk is saturated with sterile water (which is known to be nonmutagenic). Colonies found on this plate are the result of spontaneous mutations and are not caused by the test substance. The second control reveals toxicity of the test compound and consists of a plate of complete agar

Many variations of the Ames test exist, and the one outlined here is one of the simplest. *Salmonella enterica* serovar Typhimurium, a bacterium auxotrophic for histidine (an amino acid needed for protein synthesis) is used to inoculate a tube of top agar containing a minute amount of histidine and biotin. The histidine allows the bacteria to complete several rounds of replication, which is essential if mutations are to be created. The top agar is poured onto a plate of inoculated with *Salmonella* and treated with the mutagenic substance. A zone of inhibition around the disk indicates that bacteria are unable to grow, and therefore unable to mutate, in the presence of the compound (see Figure 25.2).

In this lab, you will run an Ames test along with controls for spontaneous mutations and toxicity, using a chemical of your own choosing. Chemicals most likely to be mutagenic will often have organic ingredients such as those found in cleaning solvents, painting supplies, or hair dyes. Be sure that the compound you choose to test does not contain protein. Exercise caution when handling these chemicals and when transporting them to the lab.

PRE-LAB QUESTIONS

1. *Define the following terms:*

Mutation

Mutagen

Carcinogen

Spontaneous mutation

Induced mutation

2. *The top agar used in the Ames test contains a small amount of histidine. What would happen if histidine was not included in the medium?*

3. *In what way is a back mutation different from most other mutations?*

4. *The negative control in the Ames test is a minimal medium plate in which sterile water is the only compound used to saturate the paper disk. What would you use if you wanted a positive control?*

PERIOD ONE

Each student will perform an Ames test on a single compound and will also complete a toxicity plate for the same compound as well as a negative control.

MATERIALS

Each student should obtain:

Two minimal media agar plates

One complete media agar plate

Three tubes of top agar

One tube of sterile water

Sterile Pasteur pipettes

Forceps along with a small beaker of alcohol for flaming

Beaker and hot plate

Three serological pipettes (1 ml)

Three sterile filter paper disks

Compound to be tested

Nutrient broth suspension of *Salmonella enterica* serovar Typhimurium TA 1538

PROCEDURE

1. Label one minimal media plate "experimental" and a second plate "negative control." Label a complete media plate "toxicity." Also label each plate with your name and lab time.

2. Liquefy four tubes of top agar and cool to 50°C. The next few steps must be accomplished quickly! Organize your work area so that inoculation, mixing, and pouring of the top agar can be accomplished in less than 20 sec.

3. Using a 1-ml serological pipette, inoculate a tube of top agar with 0.1 ml of *S. enterica* serovar Typhimurium TA 1538.

4. Vortex the tube for 3 sec at low speed to distribute the organism throughout the top agar. Alternately, the tube may be rolled vigorously between the hands for several seconds.

5. Pour the inoculated top agar onto the plate of minimal agar labeled "experimental."

6. Repeat steps 3, 4, and 5, for each of the other two plates.

7. Flame a pair of forceps by dipping the end in alcohol and passing through the flame of a Bunsen burner to burn off the alcohol.

8. Use the forceps to grasp a sterile filter paper, and submerge it approximately halfway into the experimental compound. The disk should be saturated with the test substance but not dripping.

9. Place the disk on the center of the experimental plate, and press lightly with the forceps to ensure that the disk is in contact with the agar.

10. Using a disk dipped in sterile water, repeat steps 7, 8, and 9 with the plate labeled "negative control."

11. Using a disk dipped in the experimental compound, repeat steps 7, 8, and 9 with the plate (containing complete media) labeled "toxicity."

12. Invert each plate and incubate at 37°C for 24–48 h.

QUESTIONS—PERIOD ONE

1. What is the purpose of the negative control plate? How should it be interpreted?

2. Should the numerous small colonies on the minimal media plate be counted as revertants? Why do they stop growing?

3. If the experimental plate contains 285 colonies and the negative control plate contains 241 colonies, how many revertants can be attributed to the test substance? What do the rest of the colonies represent?

4. How would you recognize if your test compound was toxic to *S. enterica* serovar Typhimurium? How could a compound's toxicity interfere with a determination of its mutagenicity?

PERIOD TWO

Retrieve your plates from the incubator. Count the number of large colonies on the minimal media plates labeled "experimental" and "negative control," and record the results in Table 25.1. Count only the larger colonies, ignoring the numerous tiny background colonies. Using the complete media plate (labeled "toxicity"), measure any zone of inhibition produced by the test substance. Measure the zone in millimeters from the edge of the disk to the edge of growth, and record the results in Table 25.1. Subtract the number of revertant colonies on the control plate from the number seen on the experimental plate to determine if the test substance has any mutagenic effect.

TABLE 25.1	Ames test data			
Plate	Type of media	Test substance	No. revertant colonies	Size of inhibition zone
Experimental				
Negative control				
Toxicity				

RESOLUTION OF THE CASE

The West Nile Virus home page at the CDC recommends using an insect repellent containing DEET and says that questions about the safety of the repellent are one of the five common myths associated with WNV. Based on Ames tests done by the Agency for Toxic Substances & Disease Registry, a 50% DEET solution has been shown to have no mutagenic effects in *Salmonella enterica* serovar Typhimurium.

EXERCISE 25 REVIEW QUESTIONS

1. The strain of bacteria used in the Ames test has a mutation in the uvrB gene, inactivating a gene involved in mutation repair. Why is this important?

2. In another version of the Ames test, liver enzymes are included along with the test substance to more closely simulate what happens to the test compound as it passes through the body. How does the presence of liver enzymes increase the reliability of the test?

DNA Extraction from Bacterial Cells

CASE SYNOPSIS

At the Limits of Science: 9/11 ID Effort Comes to an End

When the twin towers of the World Trade Center fell on the morning of September 11, 2001, every family affected by the attack asked the same question: "Did my loved one survive?" Weeks later, when the relative had not returned home, the answer should have been obvious. But it was only natural to hope that he or she might still walk in the door, especially since the devastation had left little concrete evidence. Starting in 2002, the New York City medical examiner's office began attempting to identify the human remains using tried-and-true methods of genetic analysis. However, these methods depend on relatively long pieces of undamaged human DNA, and little DNA of that description survived the catastrophic collapse and burn of the Trade Center. Frustrated with the slow pace of identification, officials decided to try two promising, but relatively unproven, experimental techniques.

Faced with a dozen refrigerated semitrailers filled with human remains and at least 1,700 families hoping for closure, the medical examiner suggested two methods that rely on very small pieces of DNA: mini short tandem (DNA) repeats (mini-STR) and single nucleotide polymorphism (SNP). Although the New York State Health Department refused to certify these experimental methods, fearing incorrect identification, the city medical examiner's office pushed ahead.

Both methods involve genetically testing a portion of human remains and then comparing the results with DNA taken from tissue samples, such as hair and skin, known to belong to the victim. As part of the investigation, family members provided toothbrushes, combs, razors, or other personal effects that contained the DNA of a person thought to have perished in the World Trade Center. SNP analysis is based on the fact that every person has between 3 million and 10 million variants in his or her 3 billion bases of DNA. Therefore, the chances of finding a variation in a small piece of DNA are relatively good. Mini-STR relies on a much smaller number of specific chromosomal regions—commonly 13—that typically show variability from person to person; in most cases, this is enough to conclusively identify human remains.

Resolution of the Case appears on page 212

The New York Times. April 3, 2005. At Limits of Science, 9/11 ID Effort Comes to End. p. 34.

STUDENT LEARNING OUTCOMES

After completing this exercise, you should be able to:

1. Understand the difference between genotype and phenotype.
2. Explain the general principles involved in isolating DNA.

INTRODUCTION

While DNA analysis is used, with ever increasing frequency, to identify human remains after disasters, the same technology is commonly employed to identify microorganisms. Whether in a hospital laboratory dealing with a single patient or an epidemiology lab dealing with thousands, the ability to identify organisms based solely on their DNA has become in many cases the primary means of microbial identification.

After all, the physical appearance of an organism is determined by its DNA. Characteristics such as the ability to ferment a sugar, grow under anaerobic conditions, or produce the thick peptidoglycan layer typical of a Gram-positive bacterium can all be traced to the DNA carried within a cell. These characteristics, or **phenotypes,** are the common targets of the physiological and biochemical tests used in the lab to identify the bacterial species responsible for an illness or disease outbreak. When even more specificity is needed, however, genetic testing is often employed. These tests focus on differences in the **genotype,** or underlying genetic code, of an organism. In many cases, two bacteria that appear identical based on laboratory tests may nevertheless contain several differences in their genetic code. These differences may be discovered using a variety of techniques, including DNA sequencing, polymerase chain reaction (PCR) analysis, DNA fingerprinting, and other assays beyond the scope of this book. In this way, an outbreak of *Escherichia coli* in one area of the country can be linked to an outbreak in another area of the country if the two strains show the same genetic profile. Conversely, what appears to be a single large outbreak may in fact be the result of several smaller outbreaks that unfortunately occur at the same time and place. This information is crucial when trying to determine the extent of a food recall or the spread of a disease.

Although the specific type of genetic analysis used will vary based on the information required, all analyses begin with the isolation of DNA from the bacteria of interest. In this exercise, DNA will be extracted from *E. coli* using a protocol that is rapid and does not involve the use of phenol or chloroform (both carcinogens) to separate cellular proteins from the DNA. As a result, the DNA produced by this method is usually contaminated with a small amount of protein that could interfere with later tests. The isolation of DNA in this simplified protocol can be summarized in four steps:

1. The bacterial cells are centrifuged to concentrate them.
2. A detergent is used to break down the cell wall and cell membrane, releasing the DNA.
3. A protease, along with heat, is used to denature and digest proteins within the cell.
4. Cold alcohol is used to precipitate the DNA from the solution, allowing it to be spooled onto a glass or metal rod.

PRE-LAB QUESTIONS

1. *Define the following terms:*

   Genotype

   Phenotype

MATERIALS

Each group should obtain:

One 15-ml polypropylene tube containing a 7-ml overnight culture of *E. coli* in Luria-Bertani medium

Ice bath (beaker filled with crushed ice)

Vortex mixer

Tris-EDTA (TE) buffer (10mM tris, 1 mM EDTA, pH 8.0)

10% sodium dodecyl sulfate (SDS) solution

20 mg/ml proteinase K solution

1.0 M sodium acetate solution, pH 5.1

95% isopropyl alcohol, stored on ice

Hooked glass rod (e.g., Pasteur pipette with a tip bent by heating in the flame of a Bunsen burner)

Each group should have access to:

65°C water bath with a submerged test tube rack

100–1000 μl pipettor and tips

PROCEDURE

This procedure is illustrated in Figure 26.1.

1. Mark the culture tube with the name of your group.
2. Centrifuge the tube in a tabletop centrifuge at top speed for 10 min to pellet the cells.
3. Decant the supernatant, being careful not to disturb the pellet. Remember to dispose of the supernatant properly.
4. Resuspend the pellet in 750 μl TE buffer.
5. Add 300 μl of 10% SDS solution.
6. Add 50 μl of 20 mg/ml proteinase K solution.
7. Mix the tube by *gently* vortexing every few seconds for 5 min. Alternatively, swirl the tube every few seconds to gently mix the contents.
8. Incubate the tube in the 65°C water bath for 5 min.
9. Add 400 μl of 1 M sodium acetate solution and mix gently.
10. Place the tube on ice for 5 min.
11. Slowly pour 3 ml of ice-cold isopropyl alcohol down the side of the tube, layering the alcohol atop the aqueous mixture in the tube.
12. Use the glass rod to gently stir the interface of the two layers in the tube. As you mix, the DNA will thicken into a mucousy layer and adhere to the rod (Figure 26.2).
13. Transfer the DNA to a microcentrifuge tube.

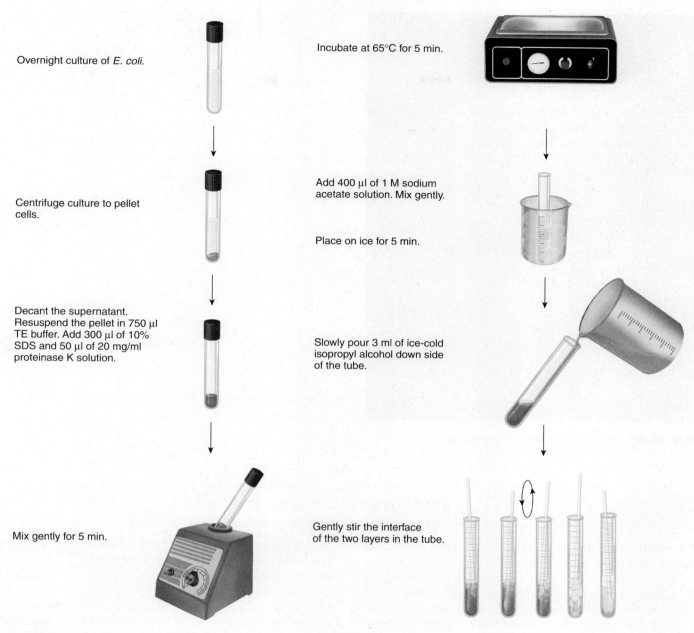

Overnight culture of *E. coli.*

Centrifuge culture to pellet cells.

Decant the supernatant. Resuspend the pellet in 750 μl TE buffer. Add 300 μl of 10% SDS and 50 μl of 20 mg/ml proteinase K solution.

Mix gently for 5 min.

Incubate at 65°C for 5 min.

Add 400 μl of 1 M sodium acetate solution. Mix gently.

Place on ice for 5 min.

Slowly pour 3 ml of ice-cold isopropyl alcohol down side of the tube.

Gently stir the interface of the two layers in the tube.

Figure 26.1 DNA extraction procedure.

Figure 26.2   DNA being spooled onto a glass rod.

RESOLUTION OF THE CASE

As investigators at the World Trade Center spent 2 years and millions of dollars working with the new genetic analysis techniques, they encountered many cases suggestive of a DNA match, but in the end, they were able to make only a few unequivocal identifications. As of 2008, fewer than half of the 2,749 victims of the World Trade Center had been positively identified. As the researchers said, "We hit the lim-

its of science." For now, they have cataloged their data and preserved the tissue samples, and are waiting until technology improves enough to identify the rest of the victims. Once that happens, acquiring the needed tissue samples from family members and creating a master database for potentially millions of individual fragments from thousands of victims will be a massive project.

EXERCISE 26 REVIEW QUESTION

1. What aspect of the structure of DNA allows the molecule to be "spooled" around a glass rod?

DNA Profiling

CASE SYNOPSIS

Multistate Outbreak of *Salmonella* Infections Associated with Peanut Butter and Peanut Butter-Containing Products—United States, 2008–2009

A group of scientists at the Centers for Disease Control (CDC) noted 13 cases of *Salmonella enterica* infection in sick people in a dozen states during November 2008. The typical symptoms of salmonellosis (infection with *Salmonella*) include vomiting and diarrhea, and may result from ingesting any of more than 1,500 different strains, or unique subspecies, of *S. enterica.* The differences that exist between the various types of *Salmonella* have led to *S. enterica* being subdivided into many strains, or serotypes, based on variations in their major surface components. In fact, *Salmonella* strains are often identified by their genus, species, and serotype, such as *S. enterica* Typhimurium or *S. enterica* serotype Tennessee.

Two weeks after the first cluster of infections, a similar outbreak of 27 cases of the disease, spread across 14 states, was found to be caused by the same strain of the organism seen in the first outbreak. By February 2009, 682 people from 46 states and Canada had become infected, 9 had died, a large corporation had filed for bankruptcy, and several criminal investigations had begun.

PulseNet is a branch of the CDC that seeks to identify food-borne disease clusters by carefully studying the bacterial isolates thought to be the source of an outbreak. Usually this means obtaining DNA profiles, sometimes called fingerprints, of each bacterium and using that information to compare *isolates* (isolated strains of bacteria) from different outbreaks. Because the profiles from the two outbreak strains in this case were similar to one another—but also different from any fingerprint within the PulseNet database—CDC scientists initiated an epidemiological investigation.

S. enterica was identified in unopened 5-pound containers of King Nut peanut butter in Minnesota and Connecticut, in the peanut butter factory, and in bacteria isolated from the patients. At the time, King Nut peanut butter was manufactured by the Peanut Corporation of America (PCA) in Blakely, Georgia, and sold to schools, hospitals, restaurants, cafeterias, and other large institutions rather than directly to consumers. Examination of the bacteria revealed several different *S. enterica* strains, but only a few of them were linked to the illnesses.

Resolution of the Case appears on page 221

CDC. 2009. Multistate outbreak of *Salmonella* infections associated with peanut butter and peanut butter-containing products—United States, 2008–2009. *Morbidity and Mortality Weekly Report,* 58: 85–90.

STUDENT LEARNING OUTCOMES

After completing this exercise, you should be able to:

1. Explain how differences in DNA sequence can be used to differentiate closely related organisms from one another.
2. Generate and interpret an agarose gel displaying a DNA profile.

INTRODUCTION

Identification of any organism has always relied on physical characteristics, or **phenotypes,** and the process is largely the same whether identifying a felon through his blood type and hair color, an animal through its paw prints, or a bacterial species by its Gram reaction and ability to ferment lactose. Over the last several years, there have been an increasing number of occasions where DNA alone was used to establish the identity of a person, animal, or bacterium. This is possible, in short, because all of the physical characteristics usually relied upon for identification are the result of sequences of DNA (genes) in the cell. A man with type A blood possesses the genes for the A antigen, while a Gram-positive bacterium has within it the genes responsible for the Gram-positive cell wall. Comparison of unknown DNA to that of a known sample allows for identification in much the same way that the fingerprints of an unknown criminal can be matched to those of a known suspect.

DNA itself is an extremely long molecule consisting of millions (in prokaryotic cells) or billions (in eukaryotic cells) of nucleotides (Figure 27.1). Each of the four

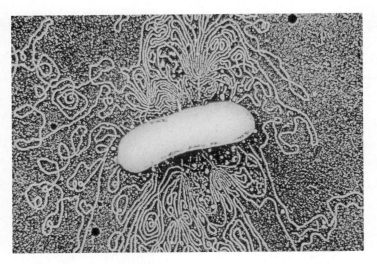

Figure 27.1 An *Escherichia coli* cell that has lysed and disgorged its single, long DNA molecule.

nucleotides of DNA consists of a phosphate group, a sugar (deoxyribose), and one of four possible bases: adenine (A), thymine (T), guanine (G), or cytosine (C). While the phosphate and sugar molecules comprise the backbone of the DNA strand, the bases extend out, away from the strand. Each sugar molecule is chemically bonded to two phosphate molecules, with one bond extending from the 3' (three prime) carbon of the sugar and the other from the 5' (five prime) carbon; this bonding specifies the orientation of the DNA strand. A second strand of DNA lies parallel to the first with weak hydrogen bonds between the nucleotides holding one strand to another in a specific manner. Adenine always bonds to thymine, with two hydrogen bonds holding the nucleotides together, while cytosine and guanine always bond and are held together with three hydrogen bonds. The orientation of the two strands is such that they are running in opposite directions, with one strand running 5' to 3' while the other runs 3' to 5', an arrangement referred to as antiparallel (Figure 27.2). Finally, the two strands coil around each other to form a double helix, a configuration that provides the molecule with its greatest stability.

When examining DNA, the sequence of nucleotides, especially among closely related organisms, is remarkably similar, with only one or two nucleotide differences per thousand separating individuals of the same species. One way to evaluate DNA molecules would be to identify the sequence of nucleotides over a long span and compare this sequence between organisms, a relatively arduous and expensive task. A simpler method depends on the use of **restriction enzymes** that recognize very specific sequences of DNA.

These enzymes, more properly known as **restriction endonucleases,** have a specific four to twelve nucleotide

recognition sequence (six nucleotide is by far the most common) and cut DNA whenever this sequence is encountered (Figure 27.3). Restriction endonucleases are produced by many different species of bacteria where their purpose, within the cell, is to destroy the DNA of infecting bacteriophage by cutting it into small pieces. In the laboratory, the same enzymes can be used to cut DNA every time a restriction site appears. If even a single nucleotide within the recognition sequence is incorrect, the restriction enzyme will not cut the DNA at that site. In this way, single nucleotide changes in the sequence of DNA can be detected without having to determine the exact sequence of long segments of DNA.

Unfortunately, a tube containing many small pieces of DNA looks no different than a tube containing a single large piece of DNA. To evaluate the action of a restriction enzyme on a DNA strand, the pieces of DNA are separated using a procedure called **agarose gel electrophoresis.** In this technique, an electric field is used to force DNA through a gel made of agarose, a more refined version of the agar used to solidify media. DNA, being negatively charged, is attracted to the positive pole of an electric field while the agarose acts as a molecular filter, allowing smaller pieces of DNA to pass through the gel faster than larger pieces. The end result is the separation of DNA fragments, with smaller pieces closer to the positive pole of the gel and larger pieces nearer the negative end (Figure 27.4).

The combination of restriction endonuclease digestion followed by electrophoresis of the DNA sample yields a specific set of DNA fragments. Though most of the fragments produced through restriction digestion will be common to a number of related samples and will be of little use, some will be rare and may be used as an identifying feature (Figure 27.4). These atypical DNA strands are known as restriction fragment length polymorphisms (RFLPs), or more commonly, DNA profiles.

DNA profiles are commonly used to identify crime suspects from small amounts of blood, while disaster victims whose bodies are badly damaged are often identified in the same manner. One of the most powerful uses of the technique however has been the tracking of microbial epidemics. By determining the particular RFLP pattern of a bacterial isolate responsible for an illness, scientists can identify the same bacterium when it is encountered again. In this way, bacteria responsible for an outbreak can be differentiated from unconnected isolates.

In this exercise, DNA profiling will be used to identify the bacterial isolate responsible for a disease outbreak. Samples representing DNA from several different bacterial isolates will be digested with restriction endonucleases, and the products of digestion separated using agarose gel electrophoresis. The gels will then be stained, to visualize the DNA, and examined to determine which bacterial isolate matches the outbreak strain.

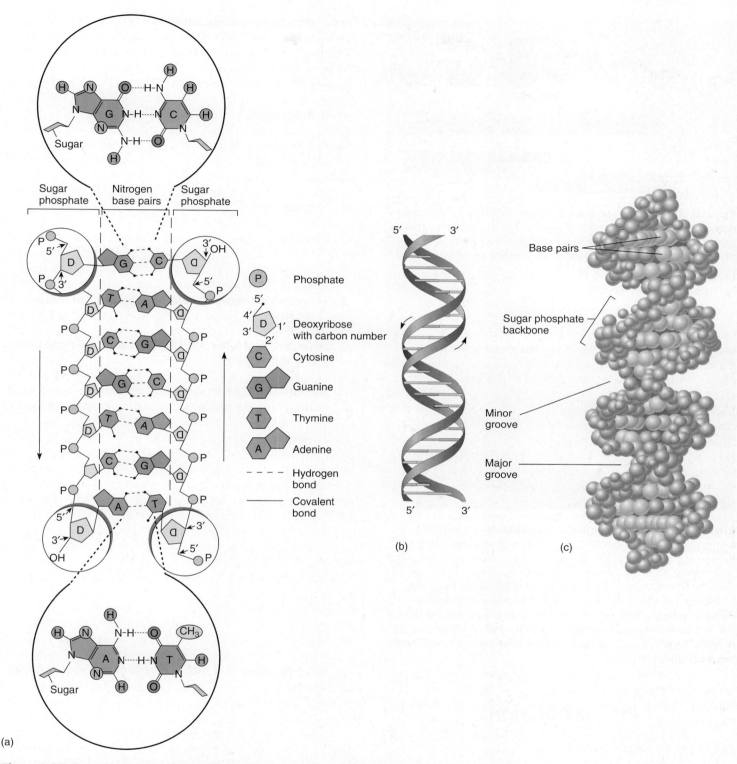

(a)

(b)

(c)

Figure 27.2 Structure of the DNA molecule. (a) The two strands are antiparallel, running next to each other but in opposite directions so that, when read top to bottom, one strand is oriented 5' to 3' while the other is 3' to 5'. For purposes of clarity, the molecule has been shown without its natural helical twist. (b) A skeletal model of DNA highlighting the antiparallel nature of the molecule and the base-pairing arrangement. (c) A space-filling model that more accurately depicts the three dimensional structure of DNA.

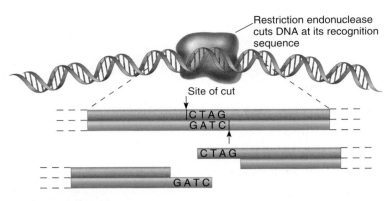

Figure 27.3 Restriction endonucleases cut DNA each time they encounter a specific recognition sequence. These sequences generally consist of four to twelve nucleotides arranged in a palindrome, such that the sequence of the two strands are identical when each is read in the same direction (i.e., 5' to 3').

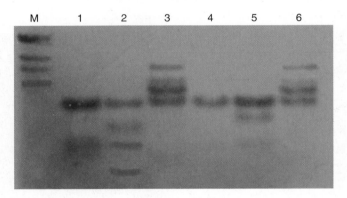

Figure 27.4 Agarose gel electrophoresis. Smaller DNA fragments move more quickly than larger fragments, appearing further from the wells at the top of the gel. Note that some fragments are common to all the samples while some are unique. In this gel, two of the samples have the same pattern of restriction fragments, indicating that their DNA sequences are identical in the regions recognized by the restriction enzyme that was used to cut the DNA. The first lane contains a collection of DNA fragments whose sizes are precisely known that can be used to estimate the size of fragments in the other lanes.

PRE-LAB QUESTIONS

1. *Define the following terms:*

Phenotype

Restriction endonuclease

Antiparallel

Gene

2. *The restriction enzyme* BamHI *has a recognition sequence of* $\frac{G{\downarrow}GATTC}{CCTAG{\uparrow}G}$ *and cuts as shown by the arrows. How many pieces of DNA would result from cutting the following DNA strand with* BamHI?

ATGCAGTGGATCCTGACGATCAGATGGATCCTTAGACATAGACAGTGAAGCTTCGG
TACGTCACCTAGGACTGCTAGTCTACCTAGGAATCTGTATCTGTCACTTCGAAGCC

Write the sequence of each of the pieces produced by BamHI *digestion.*

3. *The restriction enzyme* HindIII *has a recognition sequence of* $\frac{A^{\downarrow}AGCTT}{TTCGA_{\uparrow}A}$ *and cuts as shown by the arrows. How many pieces of DNA would result from cutting the DNA strand above with both* HindIII *and* BamHI?

Write the sequence of each of the pieces produced by double digestion with both BamHI *and* HindIII.

Underline the fragment of DNA that would be found closest to the positive end of the gel after electrophoresis, and circle the fragment that would be closest to the negative end.

PERIOD ONE

MATERIALS

Each group should obtain:

One tube (80 µl) EcoRI/PstI enzyme mix (this should be placed on ice immediately)

Six microcentrifuge tubes

Micropipettor (2–20 µl) and tips (one rack)

Small ice bath (a beaker filled with ice is adequate)

Marking pen

Floating tube holder

Every group should have access to the following:

Bacterial DNA (outbreak strain) in 2x restriction enzyme buffer

Isolate #1 DNA in 2x restriction enzyme buffer

Isolate #2 DNA in 2x restriction enzyme buffer

Isolate #3 DNA in 2x restriction enzyme buffer

Isolate #4 DNA in 2x restriction enzyme buffer

Isolate #5 DNA in 2x restriction enzyme buffer

Microcentrifuge

37°C water bath

PROCEDURE

1. Mark each of the six microcentrifuge tubes with your group number and the contents of the tube (outbreak, isolate #1, isolate #2, etc.)

2. Using a separate tip for each addition, add 10 µl of the appropriate DNA mixture to each tube. The DNA may be dispensed onto the inner wall of the tube, it does not need to be placed at the bottom of the tube.

3. Using a new tip for each transfer, add 10 µl of the *EcoRI/PstI* restriction endonuclease mixture to each tube. Again the mixture may be placed on the inside of the tube.

4. Ensure each tube is properly capped and place all six tubes into a centrifuge in a balanced manner (each tube should have another tube directly opposite it in the rotor). Centrifuge the tubes for 1 sec. This will ensure that the DNA sample and the enzyme mixture combine at the bottom of the tube. In the absence of a centrifuge, each tube may be tapped several times on the benchtop to combine the reagents in the bottom of the tube.

5. Place all six tubes in a floating tube rack, and place the rack in the 37°C water bath. Allow the digest to proceed for 90 min. Store the tubes in the refrigerator until the next class.

QUESTIONS—PERIOD ONE

1. In what way have the DNA samples in the tubes changed during the 90 min. incubation period? How will this change be visualized?

PERIOD TWO

MATERIALS

Each group should obtain:

Agarose gel electrophoresis chamber, gel mold, comb, and
   power supply

1% agarose gel, 50 ml in 1x TAE (Tris-Acetate-EDTA) buffer
   (large gels may require a greater volume)

400 ml 1x TAE buffer

Digested DNA samples

DNA size standards (*HindIII* digest of λ DNA), 10 μl

DNA sample loading dye, 80 μl

Micropipettor (2–20 μl) and tips (one rack)

Small ice bath (a beaker filled with ice is adequate)

Marking pen

Agarose gel staining solution 1x strength

Gel staining tray

Every group should have access to the following:

Microcentrifuge

65°C water bath

Rocking platform

PROCEDURE

1. Seal the ends of the gel mold in preparation for pouring the gel. Check with your instructor on the proper way to seal the gel molds you will be using.

2. If melted agarose has already been prepared for you, skip to step 3. Otherwise proceed as follows:
   - Add 0.5 g agarose to 50 ml 1x TAE buffer in a 250 ml flask. Plug the opening of the flask loosely with a foam plug.
   - Place the flask in the microwave, and heat at full power for 2 min.
   - While wearing heat resistant gloves or oven mitts, retrieve the flask from the microwave, keeping it pointed away from you. Hot agarose commonly super-heats, and when disturbed, it may rapidly boil and spray out of the flask. Swirl the flask gently and ensure that all of the agarose has melted. There should be no tiny translucent particles floating in the melted agar.
   - If any particles are seen in the agar, microwave an additional 30 sec, swirl, and observe again. Generally two or three additional 30 sec cycles are required to completely melt the agar.

3. Ensure that the agar is between 50°C and 60°C. Generally, if you can comfortably hold the flask in your hand, it is between 50°C and 60°C. Pour the melted agarose into the gel mold to a depth of 5–8 mm. Place the comb in the slot near one end of the gel so that the teeth of the comb are submerged in the agarose.

4. Allow the gel to solidify for 10–20 min. The agarose will become translucent as it solidifies.

5. Place the gel in the electrophoresis chamber with the wells nearer the cathode (negative, or black electrode).

6. Pour electrophoresis buffer into the chamber until the top of the gel is 2–3 mm below the surface of the buffer.

7. Carefully remove the comb from the gel by pulling straight up.

8. Using a new tip for each addition, add 5 μl gel loading dye to each tube. Once again, the dye may be dispensed on the wall of the tube and the tube briefly centrifuged or tapped on the benchtop.

9. Obtain a tube containing DNA marker from your instructor. This marker is a known piece of DNA—the genome of a virus called Lambda (λ)—that has been digested by the restriction enzyme *HindIII*. The sizes of the DNA fragments created by this digestion are precisely known and can be used as a standard of comparison.

10. Incubate the sample and marker tubes in a 65°C water bath for 5 min., and then place the tubes quickly on ice.

11. Using a separate pipet tip for each sample, load each DNA digest into a separate well of the agarose gel as follows:

Lane 1:	HindIII marker	10 μl
Lane 2:	Outbreak DNA	20 μl
Lane 3:	Isoalte #1 DNA	20 μl
Lane 4:	Isolate #2 DNA	20 μl
Lane 5:	Isolate #3 DNA	20 μl
Lane 6:	Isolate #4 DNA	20 μl
Lane 7:	Isolate #5 DNA	20 μl

12. Place the lid on the electrophoresis chamber and connect the electrical leads to the power supply,

ensuring that black is connected to black (−) and red is connected to red (+).

13. Turn on the power supply and set it to 100 V. You should see tiny bubbles rising from the electrodes in the electrophoresis chamber, indicating that electricity is indeed flowing through the electrophoretic apparatus.

14. After 45–60 min., turn off the power supply, remove the cover from the gel chamber, and carefully remove the gel mold and gel from the chamber (be careful, the gel will be slippery).

15. Slide the gel off the gel mold and into a small tray containing 1x gel stain. Place the tray on a rocking platform and allow the gel to gently rock in the stain overnight (between 8 and 48 h).

QUESTIONS—PERIOD TWO

1. Why are the DNA fragments attracted to the positive pole of the gel?

2. How does the size of a DNA fragment relate to its speed of passage through the gel?

3. The λ DNA used as a size marker in the gel is a linear piece of DNA 48,502 base pairs (bp) in length. The restriction enzyme *HindIII* has seven recognition sites along the λ DNA, the position of each site is indicated on the map below.

23130 bp	25157 bp	27479 bp	36895 bp	37459 bp	37584 bp	44141 bp

Keeping in mind that the first (leftmost) base in the map above is at 0 bp. and the full length of the λ DNA is 48,502 bp, determine the sizes of the fragments produced by digestion of λ DNA with *HindIII*.

4. Which of the fragments seen in the previous answer would be located closest to the negative end of the gel after electrophoresis? Which would be located closest to the positive end?

5. One of the DNA fragments produced when λ DNA is cut with *HindIII* is 6,557 bp in length. Estimate the size of a band that traveled slightly further through the gel than the 6,557-bp band.

relationship between the size of a DNA fragment and the distance it travels through the gel. Once this graph, or standard curve, is produced, it can be used to determine the sizes of all of the other unknown fragments of DNA on the gel by measuring how far they have traveled. A complicating factor when constructing a standard curve is that there is not a linear relationship between size and distance traveled; in other words a 2,000 base pair fragment does not travel half as far as a 1,000 base pair fragment. Rather, a logarithmic relationship exists between size and migration distance, which can be easily dealt with by using semilogarithmic graph paper to construct the standard curve. This paper has graphing lines along the vertical axis that are spaced in a way that eliminates the need to deal with logarithms. Because this is semi-log paper, only the vertical axis is logarithmic while numbers on the horizontal axis are evenly spaced.

Each group should obtain or have access to:

A small plastic ruler

An agarose gel transilluminator (helpful, but not necessary)

Digital camera

PERIOD THREE

EVALUATION

Although proving that two isolates are genetically identical is complicated, discovering that they possess the same restriction fragment pattern is persuasive evidence. This genetic evidence can be used in conjunction with other facts (e.g., eating the same food, drinking the same water) to create a link between individual cases in an outbreak.

The first step in analyzing an electrophoretic gel is accomplished by determining the sizes of the DNA fragments produced by restriction enzyme digestion of the isolate and comparing them to the fragments produced by other bacterial samples. Determination of fragment sizes requires that the DNA be stained to make it visible, and the distance of each bands migration be measured and plotted on a graph.

The fragments in the DNA marker lane are all of known sizes and can be used to construct a graph displaying the

PROCEDURE

1. Carefully rinse any excess stain from your gel, and place the gel on the transilluminator. If a transilluminator is not available, a white piece of paper will work almost as well.

2. It is far easier to deal with a photo of the gel than with the gel itself. Take a photo of your gel and print it (it does not need to be exactly the same size as the gel itself).

3. Use a ruler to measure the distance traveled by each band in the marker lane, and record that distance in Table 27.1. For consistency, measure each distance from the top of the well to the center of the band.

4. Repeat this process for each of the other samples (remember, not all of your samples will have the same number of bands). Record the information in Table 27.1.

| | **TABLE 27.1** | DNA restriction fragments: size versus migrating distance |

Band	**λ/HindIII size marker**		**Isolate**		**Patient #1**		**Patient #2**		**Patient #3**		**Patient #4**		**Patient #5**	
	Distance (mm)	Size (bp)	Distance (mm)	Size (bp)	Distance (mm)	Size (bp)	Distance (mm)	Size (bp)	Distance (mm)	Size (bp)	Distance (mm)	Size (bp)	Distance (mm)	Size (bp)
1		23,130												
2		9,416												
3		6,557												
4		4,361												
5		2,322												
6		2,027												

5. Construct a standard curve for the λ/*HindIII* marker. For each of the six DNA fragments, place a point at the intersection of the fragments size and its migration distance. Draw a straight "best fit" line through the points, and continue the line all the way to the bottom right of the graph.

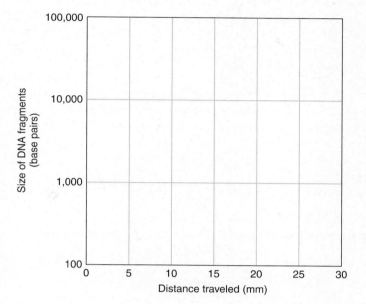

6. To determine the size of an unknown fragment, work backward, beginning with the distance that the fragment migrated. Move up from that point until you meet the standard curve. Now move horizontally

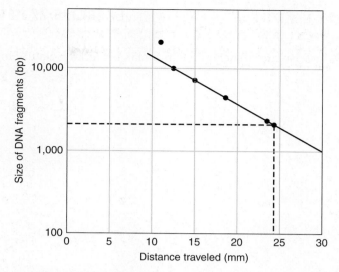

Figure 27.5 **Determining the size of an unknown DNA fragment.** After measuring the distance traveled through the gel by the fragment of interest, move upward from that point on the horizontal axis until you intersect the standard curve. Move horizontally to the left until you intersect the vertical axis. The point of intersection is the size of your unknown fragment. Note that the largest fragment of DNA used as a size marker is typically not used to construct the standrad curve.

to the left until you intersect the vertical axis. This allows you to determine the size of the unknown fragment (Figure 27.5). Repeat this process for all the unknown fragments. Record the sizes of each fragment in Table 27.1.

RESOLUTION OF THE CASE

In this case, *S. enterica* Typhimurium was identified as the outbreak strain and was found in peanut products manufactured in the PCA plant as well as in ill persons—and even in a tanker truck that had been used to transport peanut paste. Complicating matters was the fact that other companies had used the peanut paste to manufacture food items; at last count, the paste had been traced to over 3,000 peanut-containing products, including peanut butter crackers and dog biscuits. Two other *S. enterica* strains, Mbandaka and Senftenberg, were discovered in cracks in the concrete floor of the PCA processing plant, and a third variant, Tennessee,

was found in peanut butter in the factory. Comparison of DNA from these three strains with DNA from strains isolated from ill individuals revealed that none of these strains were linked to any illness.

On January 28, 2009, PCA announced a voluntary recall of all peanuts and peanut-containing products processed in its Georgia facility since January 1, 2007. Records indicated the company had knowingly shipped peanut butter containing *Salmonella* at least 12 times in the previous 2 years, and a criminal inquiry was begun that same month. PCA filed for bankruptcy on February 13, 2009.

EXERCISE 27 REVIEW QUESTIONS

1. Which isolates matched the outbreak bacterial isolate? How do you know?

The following questions refer to the gel seen in Figure 27.4.

2. The DNA marker in the left-hand lane is the same λ/*HindIII* marker used in your gels. Please label each marker band with its proper size.

3. The two smallest fragments produced when λ DNA is digested with *HindIII* are 564 bp and 125 bp. Why aren't these fragments seen on the gel?

4. The largest DNA fragment in sample #1 is seen in every other sample as well, while the largest fragment in sample #3 is seen in only one other lane. Which of these fragments are more useful for identification? Why?

5. In the case discussed in this lab, three different *Salmonella* variants, Typhimurium, Mbandaka, Senftenberg, and Tennessee were found in the PCA plant. Would you expect profiles of the three strains to have many bands in common or to be very different from one another?

6. It was stated in the case that although the Mbandaka, Seftenberg, and Tennessee strains were found within the PCA plant, none were linked to any illness. How could a DNA profile be used to show that a bacterial isolate is *not* linked to an illness?

7. As part of a DNA profiling exercise, your partner places his gel into the electrophoresis chamber with sample wells close to the positive electrode. He then loads the gel, and connects the electrodes to the power supply (positive to positive and negative to negative). He is just about to run the gel when you noticed what he's done wrong.

   a. What has he done wrong?

   b. What will happen if the gel is run as it is currently set up?

   c. Can this situation be fixed without disturbing the gel?

Measures of Water Quality: Most Probable Number Procedure

CASE SYNOPSIS

E. coli Contamination of Water Supply—Frazier Park, California, 2007

On May 3, 2007, analysis of a water sample collected from a municipal water system serving a small community tested positive for the presence of coliform bacteria and *E. coli,* indicating fecal contamination of the water supply. The sample was one of many collected as part of regularly scheduled testing of the water supply and was drawn from a testing station near one of several main storage tanks. Based upon the initial positive test, established procedure dictated that additional samples be collected and tested, and that "Boil Water" notices be distributed to homes and businesses that receive water from the system. As the name implies, these notices inform people of the need to boil water prior to using it for drinking, brushing one's teeth, or washing dishes. Until the source of contamination was identified, water from the system remained potentially unsafe without being boiled first.

Resolution of the Case appears on page 231

Hedlund, P. May 18, 2007. Thristy field mouse was source of LOW water woes. *Mountain Enterprise,* 41(37): 1.

STUDENT LEARNING OUTCOMES

After completing this exercise, you should be able to:

1. Carry out water testing using the most probable number technique and evaluate the results.

INTRODUCTION

Ensuring access to safe water is one of the hallmarks of modern civilization. In addition to harmless microorganisms picked up from air and soil, water can also acquire pathogens that are of great concern. Although these pathogens may include protozoans (*Giardia* and *Cryptosporidium*), bacteria (*Campylobacter, Salmonella,* and *Shigella*), and viruses (Hepatitis A and Norwalk virus), water testing is usually focused on detecting fecal contamination. High levels of fecal contamination indicate the presence of not just fecal bacteria but point toward other microbial pathogens being present as well. Water sources such as wells, reservoirs, and storage tanks are generally tested for the presence of **indicator bacteria,** species that are part of the normal intestinal microbiota flora of birds and mammals and that are easily detected using common laboratory methods.

The most useful of the indicator organisms are the **coliforms,** which are defined as Gram-negative bacteria that ferment lactose, producing both acid and gas. These bacteria survive in natural waters but will not multiply there, meaning that high numbers of coliforms indicate recent fecal contamination. The single most widely used indicator bacterium is *Escherichia coli,* and water testing commonly seeks to report both total coliforms and *E. coli* densities.

The **most probable number** (MPN) procedure is used to determine the number of coliforms in a sample of water. Through the use of three different types of media, both the number of total coliforms and the number of *E. coli* in a sample can be determined. All the broths include an inverted Durham tube to trap any gas produced as a result of fermentation. After incubation, the presence of gas within the tubes indicates a positive reaction. The media are as follows:

Lauryl tryptose broth (LTB)   Selective for coliforms. Lactose is present as a fermentable carbohydrate, and lauryl sulfate selects for coliforms. Tubes containing gas bubbles within their Durham tubes are considered to be presumptively positive for coliforms.

Brilliant green bile (BGB) broth   Contains lactose for fermentation and 2% bile to inhibit noncoliforms. Production of gas is considered to be evidence of coliforms.

***E. coli* (EC) broth**   Contains lactose for fermentation and bile salts to select for coliforms. When incubated at 45.5°C, the medium is selective for *E. coli,* with a bubble in the Durham tube again serving to indicate gas production.

The procedure begins when three sets of LTB broth are inoculated. Each set contains five tubes. Each tube in the first set receives 1.0 ml of the original water sample. Each tube in the second set receives 1 ml of a 10-fold dilution while each tube in the third set receives 1 ml of a 100-ml dilution, as seen in Table 28.1.

After inoculation, the LTB tubes are incubated at 35°C for 48 h and then examined for gas production (Figure 28.1). A loop is used to subculture presumptively positive cultures

TABLE 28.1	Inoculation of LTB tubes		
	Water volume/tube	Dilution factor	Equivalent volume of original water sample
Set A	1 ml	1	1 ml
Set B	1 ml	10^{-1}	0.1 ml
Set C	1 ml	10^{-2}	0.01 ml

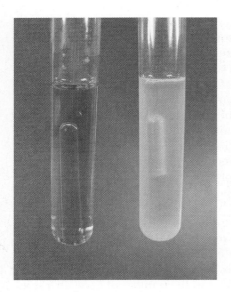

Figure 28.1 Laurel tryptose broth. The presence of a bubble in the Durham tube indicates a positive reaction.

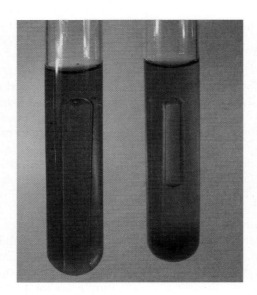

Figure 28.2 Brilliant green lactose broth. The presence of a bubble in the Durham tube indicates a positive reaction.

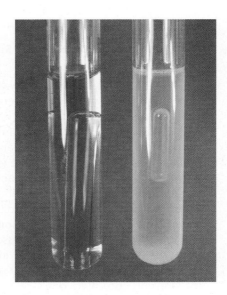

Figure 28.3 EC broth. The presence of a bubble in the Durham tube indicates a positive reaction.

to BGB broths and EC broths, which are incubated for 48 h at 35°C and 45°C, respectively. After incubation, the number of positive BGB tubes (Figure 28.2) and EC tubes (Figure 28.3) are recorded. These numbers are then used, along with a statistical table, to determine the most probable number of both coliforms in general and *E. coli* specifically.

PRE-LAB QUESTIONS

1. *Define the following terms:*

   Coliform

   Lactose

Indicator bacteria

2. *What is the advantage of knowing the number of indicator organisms as opposed just to being aware of their presence in a water sample?*

3. *Based on the information in the following table, classify each of the following organisms as a coliform or noncoliform.*

	Gram reaction	Lactose fermentation to produce gas	Coliform/ noncoliform
Enterococcus faecalis	+	–	
Escherichia coli	–	+	
Proteus mirablis	–	–	
Enterobacter aerogenes	–	+	

4. *A 100-ml water sample (bottle A) was diluted as follows: 1 ml was added to 99 ml of sterile water in bottle B; 1 ml from bottle B was added to 99 ml of sterile water in bottle C; 1 ml from bottle C was added to 99 ml sterile water in bottle D; and 1 ml from bottle D was added to 99 ml of sterile water in bottle E. One ml from bottle D was used to inoculate a plate of nutrient agar. After incubation, 31 colonies were counted on the plate.*

How many cells were in bottle D?

How many colonies would be expected if 1 ml from bottle B was plated in a similar manner?

How many cells were present in the 100 ml of bottle A?

MATERIALS

Each group should obtain:
Nine tubes of LTB
Up to nine tubes of BGB broth
Up to nine tubes EC broth
100-ml water dilution bottle
Two 9-ml dilution tubes
Water sample to be tested (100 ml)
Marking pen

PERIOD ONE

PROCEDURE

This procedure is illustrated in Figure 28.4. Serial dilution of the original water sample will be used to prepare it for testing. The first sample will be undiluted water (10^0), the second will be a 1:10 dilution (10^{-1}), and the third will be a 1:100 dilution (10^{-2}). Proceed as follows:

1. Label the undiluted water sample "A."
2. Aseptically add 9.0 ml of sterile water to each dilution tube. Label one tube "B" and the other "C."
3. Add 1.0 ml of the water sample to tube B. Pipette up and down several times to mix the contents of the tube.
4. Using a new pipette, transfer 1.0 ml from tube B to tube C. Pipette several times to mix.
5. Place 9 tubes of LTB in a test tube rack, arranged in groups of three.
6. Label each of the first three tubes "10^0," each of the second three "10^{-1}," and each of the final three "10^{-2}."
7. Add 1 ml of the undiluted water (sample A) to each of the tubes labeled "10^0." Pipette several times to mix.
8. Using a new pipette, add 1 ml of the water sample labeled "B" to each of the tubes labeled "10^{-1}." Pipette several times to mix.
9. Again using a new pipette, add 1 ml of the water sample labeled "C" to each of the tubes labeled "10^{-2}." Pipette several times to mix.
10. Incubate the tubes at 35°C for 48 h.

QUESTIONS—PERIOD ONE

1. What is the purpose of the small Durham tube in each LTB tube?

Create a serial dilution
of your water sample
by transferring 1 ml
of the undiluted sample
to tube B and 1 ml from
tube B to tube C.

Add 1 ml of the
undiluted water
(sample A) to each
of the tubes labeled
10^0, 1 ml of sample
labeled 10^{-1}, and
1 ml of sample C to
each of the tubes
labeled 10^{-2}. Pipette
several times to mix.

Incubate the tubes at
35°C for 48 h.

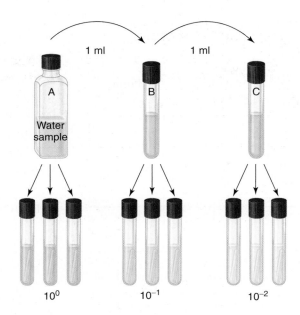

Figure 28.4 Preparation of tubes for most probable number test.

2. Sketch what you would expect to see when examining an LTB tube displaying a positive reaction and one displaying a negative reaction.

3. What does a positive result in this test indicate about the microorganisms present in the water sample?

4. Bacteria displaying a positive result after incubation in LTB broth could presumptively be classified as

_____ while those displaying a negative

result could be classified as _____.

5. If, after incubation, two of the three tubes labeled "10^{-1}" show a positive result, would you expect to have more or less than two "10^{-2}" tubes with positive results? Why?

TABLE 28.2	Results for most probable number procedure		
	No. positive results		
	Set A (10^0)	**Set B (10^{-1})**	**Set C (10^{-2})**
LTB			
BGB broth			
EC broth			

QUESTIONS—PERIOD TWO

1. What is the significance of an LTB tube that shows bacterial growth but no gas within the Durham tube?

PERIOD TWO

1. Remove the broths from the incubator, and examine all the tubes in each group for the presence of gas. A gas bubble occupying more than 10% of the Durham tube should be recorded as a positive result while the lack of a gas bubble should be recorded as a negative result. Record the results in Table 28.2.

2. For each positive lauryl tryptose broth, label one BGB broth and one EC broth with the letter corresponding to the LTB that will be used for inoculation. As an example, if three LTB tubes from sample A are positive, then three BGB and three EC broths should be labeled with an "A."

3. Use a loop to inoculate one BGB broth and one EC broth from each positive LT broth. Be sure that each tube is clearly labeled with respect to the LTB tube from which it was inoculated (Figure 28.5).

4. Incubate the BGB tubes at 37°C and the EC tubes at 45.5°C for 48 h.

2. What is indicated by a positive result in BGB broth that has been incubated at 37°C?

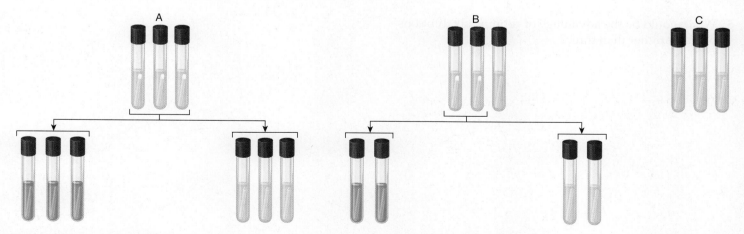

Figure 28.5 Each positive tube of laurel tryptose broth (those containing a gas bubble within the Durham tube) is used to inoculate a tube of brilliant green bile broth and EC broth. Tubes displaying a negative reaction may be discarded. Incubate the BGB tubes at 37°C and the EC tubes at 45.5°C for 48 h.

3. What is indicated by a positive result in EC broth that has been incubated at 45.5°C?

PERIOD THREE

1. Remove the tubes from the incubator, and examine each tube for the presence of gas. Record your results in Table 28.2.

2. Calculate the MPN of coliforms by referring to Table 28.3. Determine the *presumptive* number of coliforms using the results from the LTB tubes. For example, if you had gas in two of the 10^0 tubes, one of the 10^{-1} tubes, and one of the 10^{-2} tubes, your results would be read as "2-1-1." Table 28.3 indicates that such a sample is statistically estimated to have 20 coliforms/10 ml of water. The last two columns in the table indicate that you could be 95% certain that the number of coliforms was between 7 and 89. In the same manner, determine the confirmed number of coliforms using the results of the BGB tubes and determine the *E. coli* density using the EC tubes.

3. Record the final results for your sample:

Total coliform MPN/10 ml (BGB data) _____

E. coli MPN/10 ml (EC data) _____

4. It is actually more common to run an MPN test with six dilutions of water rather than the three used in this exercise. Use this information to complete the following table:

	LTB tube set					
	A	**B**	**C**	**D**	**E**	**F**
Volume of water from previous tube	10 ml	1 ml	1 ml			
Volume of sterile water	0 ml	9 ml	9 ml			
Dilution factor	10^0	10^{-1}	10^{-2}	10^{-3}		
Volume of original water sample in 1 ml (1 ml × dilution factor)	1 ml	0.1 ml	0.01 ml	0.001 ml	0.0001 ml	

5. What would be the advantage of running six dilutions of water rather than three?

TABLE 28.3	Most probable number (MPN) determination form for multiple tube test					
No. tubes (of three) giving a positive reaction			MPN of Coliforms/10 ml	95% confidence limits		
Set A (10⁰ Dilution)	Set B (10⁻¹ Dilution)	Set C (10⁻² Dilution)			Lower	Upper
0	0	1	3		< 0.5	9
0	1	0	3		< 0.5	13
1	0	0	4		< 0.5	20
1	0	1	7		1	21
1	1	0	7		1	23
1	1	1	11		3	36
1	2	0	11		3	36
2	0	0	9		1	36
2	0	1	14		3	37
2	1	0	15		3	44
2	1	1	20		7	89
2	2	0	21		4	47
2	2	1	28		10	150
3	0	0	23		4	120
3	0	1	39		7	130
3	0	2	64		15	380
3	1	0	43		7	210
3	1	1	75		14	230
3	1	2	120		30	380
3	2	0	93		15	380
3	2	1	150		30	440
3	2	2	210		35	470
3	3	0	240		36	1300
3	3	1	460		71	2400
3	3	2	1100		150	4800

(*Source:* Eaton, A.D., Clesceri, L.S., and Rice, E.W. (eds.). *Standard Methods for the Examination of Water and Wastewater, 12th ed.* New York: The American Public Health Association, Inc.)

RESOLUTION OF THE CASE

Over the next several days, additional tests continued to reveal the presence of *E. coli* in the water system, always in close proximity to the same storage tank. Draining and inspection of the tank revealed a dead field mouse that had made its way to the top of the tank and then slipped through a 3/8-in. opening along the edge of a hatch. Although the opening had existed for several years, it was thought that the installation of a permanent ladder to the tank just a few months prior to the incident probably allowed the mouse to gain access to the top of the tank. The mouse was removed, the tank cleaned and disinfected, and water within the distribution system was disinfected with high levels of chlorine. After two consecutive series of tests revealed no *E. coli* in the system, the "Boil Water" order was lifted on May 11.

EXERCISE 28 REVIEW QUESTIONS

1. If all nine of the LTB tubes showed a positive result after incubation, what would this indicate and what should be done?

2. If none of the nine LTB tubes showed a positive result, what would this indicate?

3. If positive results for gas were found in several of the LTB tubes but no gas was detected in the BGB tubes, predict what results you would see in the EC tubes. How would you interpret these results?

Measures of Water Quality: Membrane Filtration Method

CASE SYNOPSIS

Fecal Contamination of Airline Drinking Water—2005

In the United States, public water systems provide the water that is loaded onto aircraft, just as they provide water for homes and businesses. Depending on the airport, water is delivered either through a system of pipes or via a hose from a water tanker. Regulation of water on aircraft is a complex subject, with the Environmental Protection Agency (EPA) regulating the systems that supply water to the aircraft and drinking water once aboard the plane, the Food and Drug Administration is charged with ensuring the safety of culinary water, for example, ice, and the Federal Aviation Administration oversees all aspects of the aircraft itself, including the *potable* water system. Further complicating the matter, aircraft that travel internationally may board water from foreign sources that are not subject to EPA drinking water standards. In the United States, the EPA is primarily responsible for the safety of drinking water aboard aircraft.

During August and September 2004, the EPA collected water samples from galley water taps, water fountains, and lavatory faucets of 158 randomly selected passenger aircraft at 7 U.S. airports. Testing of this water revealed the presence of *coliform* bacteria on 20 planes (12.7%), with two of the aircraft also testing positive for the presence of *Escherichia coli*. A second round of sampling in November and December 2004 retrieved water samples from galley water taps and lavatory faucets in 169 aircraft. Testing revealed 29 samples (17.2%) that were positive for the presence of coliforms, with none testing positive for *E. coli*. The airplanes tested included both U.S. and foreign flag aircraft.

Resolution of the Case appears on page 238

U.S. Environmental Protection Agency. October 19, 2005. *EPA Announces Drinking Water Agreements with 24 Domestic Airlines.* http://yosemite.epa.gov/opa/admpress.nsf/a4a961970f783d3a85257359003d480d/0b6df456b61c61fa8525709f0064c1c7!OpenDocument.

STUDENT LEARNING OUTCOMES

After completing this exercise, you should be able to:

1. Carry out water testing using the membrane filtration method and evaluate the results.

INTRODUCTION

Ensuring access to safe water is one of the hallmarks of modern civilization. In addition to harmless microorganisms picked up from air and soil, water can also acquire pathogens that are of great concern. Although these pathogens may include protozoans (*Giardia* and *Cryptosporidium*), bacteria (*Campylobacter, Salmonella,* and *Shigella*), and viruses (Hepatitis A and Norwalk virus), water testing is usually focused on detecting fecal contamination. High levels of fecal contamination indicate the presence of not just fecal bacteria, but point toward other microbial pathogens being present as well. Water sources such as wells, reservoirs, and storage tanks are routinely tested for water safety. Changes in lifestyles and technology have complicated this problem as smaller water tanks, holding anywhere from just a few to thousands of gallons, can be found everywhere from community festivals and soccer games to cruise ships and airliners. Water from these sources is generally tested for the presence of indicator bacteria, species that are part of the normal intestinal flora of birds and mammals and that are easily detected using common laboratory methods.

The most useful of the indicator organisms are the coliforms, which are defined as Gram-negative bacteria that ferment lactose, producing both acid and gas. These bacteria survive in natural waters but will not generally multiply there, meaning that high numbers of coliforms indicate recent fecal contamination. The single most widely used indicator bacterium is *E. coli,* and water testing commonly seeks to report both total coliforms and *E. coli* densities.

The **membrane filtration method** is a rapid procedure that is used to test water for the presence of coliforms. In this method, a fixed volume of water is passed through a thin filter that traps the cells in its matrix. The filter is then transferred to a Petri dish containing Endo agar, a medium that is both selective (for Gram-negative organisms) and differential (for lactose fermentation). The selective and differential aspects of the media together allow for the identification of both coliform and non-coliform bacteria. Endo agar contains lactose as well as a pH indicator. After incubation, lactose-fermenting bacteria will appear pink or red due to the production of acid from lactose fermentation while some bacteria that produce large amounts of acid, such as *E. coli,* have a metallic sheen that aids in their recognition. Non-lactose fermenters (noncoliforms) produce no acid from lactose and appear colorless to slightly pink. The differential aspect of the medium allows the determination of both coliform and non-coliform bacteria.

To arrive at a statistically significant result, the filter should trap between 20 and 80 coliforms, and no more than 200 total bacterial cells. To end up with this many cells, the amount of water filtered through the membrane can be adjusted, based on how many bacteria are assumed to be in the water. More water would be filtered from "clean" sources, such as a kitchen tap or swimming pool, while less would be required from "dirty" sources, such as sewage, to achieve the same number of trapped cells. If only a small volume of water is to be filtered, it is customary to dilute the sample with sterile water so that the sample is large enough to spread evenly across the surface of the filter. Because the added water is sterile, it contains no bacterial cells and is not included when calculating the number of cells per milliliter; only the amount of the original sample is used in this calculation.

PRE-LAB QUESTIONS

1. *Define the following terms:*

Coliform

Lactose

Indicator organism

Potable water

Pathogen

2. *Explain the following sentence: "All* E. coli *are coliforms, but not all coliforms are* E. coli.*"*

3. *Based on the information provided about each organism, complete the following table:*

	Gram reaction	Lactose fermentation to produce gas	Coliform/noncoliform
Enterococcus faecalis	+	−	
Escherichia coli	−	+	
Proteus mirablis	−	−	
Enterobacter aerogenes	−	+	

4. *A 100-ml water sample (bottle A) was diluted as follows: 1 ml was added to 99 ml of sterile water in bottle B; 1 ml from bottle B was added to 99 ml of sterile water in bottle C; 1 ml from bottle C was added to 99 ml sterile water in bottle D; and 1 ml from bottle D was added to 99 ml of sterile water in bottle E. Fifty ml from bottle D was filtered and the filter was placed on Endo agar. After incubation, 74 colonies were counted on the plate.*

How many cells were in bottle A? In bottle D?

How many colonies would be expected if 1 ml from bottle B was plated in a similar manner?

MATERIALS

Each group should obtain:

One sterile membrane filter (0.45 μm pore size)

One Endo agar plate

One sterile membrane filter assembly

One vacuum assembly

100-ml water dilution bottle

Water sample to be tested (100 ml)

Up to four 99.0-ml dilution tubes (for some groups only)

1.0-ml pipettes and pipettor (for some groups only)

Marking pen

PROCEDURE

Prior to period one, do the following:

1. As close to your lab time as possible, obtain your water sample using a 100-ml dilution bottle. Tightly cap the bottle, and clean the outside with household disinfectant.

2. If the sample must be held for more than a day before the lab, loosen the cap slightly and store the bottle at 4°C (refrigerator temperature) until your lab period.

PERIOD ONE

Depending on the water sample chosen for testing, dilution of the sample may be necessary. The guidelines below are estimates of the amount of water needed to produce statistically reliable results.

Tap or bottled water	100 ml
Filtered swimming pool water	100 ml
Lake or spring	10–100 ml
Beach water	0.1–10 ml
River water	0.001–1.0 ml
Raw sewage	0.0001–0.1 ml

1. Use Table 29.1 to record the degree to which your sample was (or was not) diluted.

2. Assemble the filter assembly as noted below and illustrated in Figure 29.1.

   • Use alcohol flamed forceps to place a membrane filter between the two halves of the filter housing.

   • Clamp the two halves of the housing together.

   • Be sure the assembly is clamped or held securely in place.

3. Pour 100 ml of the appropriate dilution of your sample into the upper portion of the filter apparatus (Figure 29.2).

TABLE 29.1	Dilution of water samples for testing			
	Tube A: 100-ml sample	Tube B: 99 ml water + 1 ml from tube A	Tube C: 99 ml water + 1 ml from tube B	Tube D: 99 ml water + 1 ml from tube C
Volume of original sample in tube	100 ml	1.0 ml	0.01 ml	0.0001 ml
Dilution factor	1	100	10,000	1,000,000
Indicate dilution used for testing				

4. Turn on the vacuum pump, and draw the sample through the filter into the flask below.
5. Turn off the vacuum pump, and allow the suction to dissipate.
6. Disassemble the filter apparatus.
7. Using alcohol flamed forceps, transfer the membrane filter to the Endo agar plate. Allow the filter to roll onto the agar so that air bubbles do not become trapped between the agar and filter.
8. Allow the plate to remain right side up for 3–5 min, giving the filter time to adhere to the plate.
9. Invert the plate, and incubate at 37°C for 24–48 h.

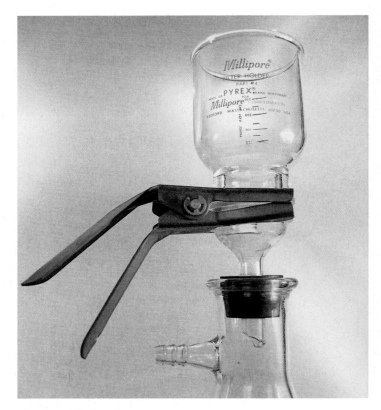

Figure 29.1 Membrane filter assembly.

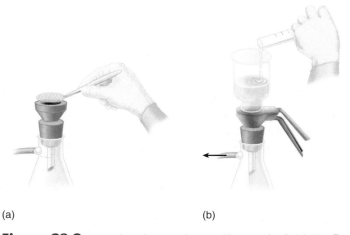
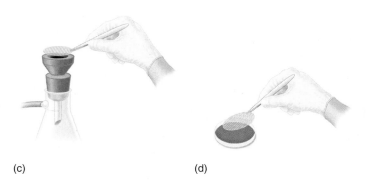

(a)   (b)   (c)   (d)

Figure 29.2 Procedure for membrance filter method. (a) Use flamed forceps to place a sterile membrane grid side up on the filter holder. (b) After assembling the filter, 100 ml of the appropriate dilution of a water sample is poured into the filter apparatus. Vacuum is applied to filter the sample. (c) When the vacuum dissipates, disassemble the filter, and carefully remove the filter using a pair of sterile forceps. (d) Keeping the grid side up, roll the filter onto a plate of Endo agar. Keep the plate upright for 5 min to allow the membrane to adhere to the plate. Invert the plate and incubate at 37°C for 24–48 h.

QUESTIONS—PERIOD ONE

1. In this assay, why is Endo agar used?

2. Based on the information provided about each organism, complete the following table:

	Gram reaction	Lactose fermentation to produce gas	Growth on Endo agar	Color on Endo agar	Coliform/ noncoliform
E. faecalis	+	–			
E. coli	–	+			
P. mirablis	–	–			
E. aerogenes	–	+			

3. A 100-ml water sample taken from a small pond is divided in half, and each half is filtered through a membrane. The first membrane is placed on a plate of Endo agar while the second is placed on a plate of nutrient agar. Sketch the results you would expect for each plate, assuming that the amount of water used was enough to produce a statistically acceptable number of colonies.

4. In what way would a dark red colony growing on an Endo agar plate be interpreted differently than a light pink colony?

PERIOD TWO

1. Remove the plates from the incubator, and count the colonies that are dark pink to red, purple, have a metallic sheen, or have a black spot in the center.

2. Calculate the number of coliforms per 100 ml of your original sample as follows:

   Coliforms/100 ml =
   (coliform colonies on plate) × (dilution factor)

3. Record both your own and class results.

4. Potable water must have less than 1 coliform/100 ml. In the right-most column of the table, indicate whether the water samples are potable.

Sample	Number of colonies/plate	Dilution factor	Colonies/100 ml	Potable or nonpotable

RESOLUTION OF THE CASE

In early 2005, the EPA issued a press release detailing the results of their investigation. The release stated that passengers with compromised immune systems may consider requesting only canned or bottled beverages, and may want to decline coffee or tea unless made with bottled water. U.S. flag aircraft that tested positive for coliforms were disinfected and then retested to ensure that disinfection was successful. Foreign carriers having airplanes that tested positive for coliforms were notified of the results and advised to disinfect and retest the aircraft.

Shortly thereafter, an agreement was reached between the EPA and 24 domestic airlines to ensure the safety of the water onboard aircraft. Included in this agreement was a requirement that airlines do the following:

- Undertake regular monitoring of aircraft water systems
- Regularly disinfect water systems and water transfer equipment
- Notify the public and take immediate corrective action when a total coliform positive result is encountered
- Conduct a study of possible sources of contamination and other aspects of water practices, and supply this information to the EPA

EXERCISE 29 REVIEW QUESTIONS

1. Is a water sample that produces no colonies on Endo agar sterile? Explain?

2. Membrane filters are available with pores of many sizes. What would be the effect of using a filter with a pore size of 0.8 μm as opposed to the 0.45-μm filter that was used?

3. Boiling is a generally accepted method for destroying coliforms in drinking water. Given this fact, why do you think the EPA warned passengers against coffee or tea unless bottled water was used?

REFERENCE

EPA. *Aircraft Drinking Water.* http://water.epa.gov/lawsregs/rulesregs/sdwa/airlinewater/index.cfm.

Measures of Milk Quality: Methylene Blue Reductase Test

CASE SYNOPSIS

Salmonella Typhimurium Infection Associated with Raw Milk and Cheeses Consumption—Pennsylvania, 2007

In February 2007, two cases of laboratory-confirmed infection with *Salmonella enterica* serotype Typhimurium were reported to the Pennsylvania Department of Health (PDH). Both patients reported consuming unpasteurized, or raw, milk from the same dairy, located in York County, Pennsylvania (dairy A). At the same time, the Pennsylvania Department of Agriculture (PDA) received several reports of diarrheal illness associated with consumption of raw milk from the same dairy. (In Pennsylvania, raw-milk sales are regulated by the Pennsylvania Department of Agriculture, which issues permits to dairies that adhere to regulations concerning milk sanitation and also displays public notices explaining the potential hazards of consuming raw milk.) On February 26, the PDH and PDA initiated an investigation to determine the source of the outbreak as well as how many cases could be traced to the initial source. Samples taken from the raw-milk bulk tank at the dairy yielded *Salmonella* Typhimurium that was genetically identical to that seen in the patients. Stool samples of patients and family members were also tested for the presence of the pathogen and food histories were obtained for each patient. By July 14, a total of 29 cases of diarrheal illness caused by *Salmonella* Typhimurium and associated with consumption of raw milk from the same dairy were identified, with the cases grouped into three distinct time periods (Figure 30.1).

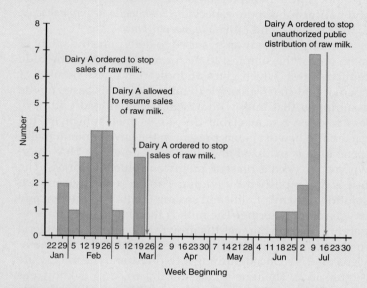

Figure 30.1 Number of cases of diarrheal illness caused by outbreak strain of *Salmonella* Typhimurium, by week of onset—Pennsylvania, 2007.

Resolution of the Case appears on page 244

CDC. 2007. *Salmonella* Typhimurium infection associated with raw milk and cheeses consumption—Pennsylvania, 2007. *Morbidity and Mortality Weekly Report*, 56: 1161–1164.

STUDENT LEARNING OUTCOMES

After completing this exercise, you should be able to:

1. Perform the methylene blue reductase test and evaluate the results.
2. Understand the use and importance of pasteurization.

INTRODUCTION

Prior to the adoption of widespread **pasteurization** of milk in the United States (circa 1938), approximately 25% of foodborne and waterborne outbreaks of disease were attributable to milk. By 2001, less than 1% of foodborne and waterborne outbreaks were associated with milk. According to the Centers for Disease Control, between 1998 and 2001, a total of 45 outbreaks attributable to unpasteurized milk (or cheese made from unpasteurized milk) were responsible for 1007 illnesses, 104 hospitalizations, and 2 deaths. Because the majority of foodborne illness goes unreported, the actual number of cases most likely far exceeds these numbers.

Because some people consume raw milk for convenience, taste preference, or supposed health benefits (which are not supported by scientific evidence), at least 27 states allow some form of raw-milk sales to the public. Human pathogens are often shed in the feces of cows and can contaminate milk when they are present in or on the udders; hygienic milking practices can reduce, but not eliminate, this risk. Pasteurization decreases the number of pathogenic organisms and is a reliable method of ensuring the safety of milk. In states that regulate the sale of raw milk, the milk is tested on a regular basis for the presence of total and **coliform** bacteria, especially *Salmonella, Campylobacter, E. coli* O157:H7, and *Listeria monocytogenes*.

Among the many assays used to determine the safety of milk, one of the simplest to perform is the **methylene blue reductase test,** which gives an indication of the number of microorganisms in a sample of milk. The test is based on the assumption that milk with a high bacterial load will have less free oxygen available due to its metabolism by the bacteria. When oxygen is present in the environment, methylene blue is oxidized and displays its familiar blue color; when no oxygen is present, methylene blue is colorless. The test is accomplished by adding 1 ml of a methylene blue solution to 10 ml of milk and sealing the tube to prevent any oxygen from entering. The milk is then incubated at a temperature that encourages the growth of bacteria. As bacteria in the milk multiply, oxygen is consumed and the methylene blue is reduced, turning it colorless and the milk white (Figure 30.2). The time it takes for this to occur is known as the methylene blue reduction time (MBRT), and the shorter the time, the greater the number of bacteria present in the milk. High-quality milk has an MBRT of at least 6 h while an MBRT of 30 min indicates milk of very poor quality.

Figure 30.2 Methylene blue reductase test. The tube on the left illustrates what can be expected when the test has just begun. The blue color indicates that dissolved oxygen is present in the milk. The lack of blue color in the middle tube indicates that oxygen has been depleted due to bacterial metabolism. The right-hand tube shows the endpoint of the reaction, when 80% of the tube has turned white.

Serotype

PRE-LAB QUESTIONS

1. *Define the following terms:*

Coliform

Pasteurization

Raw milk

2. *How is the reduction of methylene blue recognized?*

MATERIALS

Each student or group should obtain or have access to:

Three samples of pasteurized milk

One sample of raw milk

Four sterile test tubes with screw caps or rubber stoppers

E. coli broth culture

Methylene blue solution

Five 1-ml pipettes and pipettor

Four 10.0-ml pipettes and pipettor

Water bath set at 35°C containing a test tube rack

Marking pen

PROCEDURE

1. Prepare four tubes with your name and lab time. Label one tube "negative control," the second "positive control," the third "pasteurized milk," and the fourth "raw milk." Using a new pipette for each addition and aseptic technique throughout, prepare each tube as seen in the table.

2. Tightly seal each tube, and gently invert the tubes three times to mix the contents.

3. Place the tube labeled "negative control" in the refrigerator and the other three tubes in the water bath.

4. After 5 min incubation, remove each tube from the water bath, and invert once to mix. Return the tubes to the water bath. Do the same to the negative control, returning it to the refrigerator after mixing.

5. Record the time in Table 30.1.

6. At 30-min intervals, carefully remove the tubes containing raw milk and the positive control from the water bath to check for reduction of the methylene blue. When 80% or more of the milk in the tube has turned white, the endpoint of reduction has been reached. Continue the incubation for 8 h or until both the raw milk and positive controls have reached an endpoint. Record these times in Table 30.1.

7. Determine the MBRT for the positive control and the raw-milk tube. Assess the quality of each tube as follows:

   Excellent: MBRT of 8 h or more

   Good: MBRT of 6–8 h

   Fair: MBRT of 2–6 h

   Poor: MBRT less than 2 h

Tube	Milk contents	Methylene blue	*E. coli*
Negative control	10 ml pasteurized milk	1 ml	—
Positive control	9 ml pasteurized milk	1 ml	1 ml
Pasteurized milk	10 ml pasteurized milk	1 ml	—
Raw milk	10 ml raw milk	1 ml	—

TABLE 30.1	Methylene blue reductase results			
Tube	Start time	End time	MBRT	Milk quality
Negative control				
Positive control				
Pasteurized milk				
Raw milk				

RESOLUTION OF THE CASE

Of the 29 incidents of infection with *Salmonella* Typhimurium, the first cluster consisted of 15 cases with dates of onset between February 3 and March 5. Samples of raw milk collected from a bulk milk tank at dairy A (on February 20) and from the home of an ill person (on February 28) yielded the outbreak strain of *Salmonella* Typhimurium. On March 2, the PDH ordered the dairy to stop selling raw milk and advised the public not to consume raw milk or raw-milk products from dairy A. The dairy was allowed to resume sales after two consecutive samplings of raw milk from the milk tanks tested negative for the bacterium. A second outbreak occurred in late March and was linked to cheese made from raw milk purchased from dairy A. The dairy was again ordered to cease raw-milk sales, and its state-issued permit to sell raw milk was suspended. The final cluster of cases occurred during the summer of 2007. Of the 11 cases seen, 10 occurred among people living close to dairy A. Further investigation revealed that the dairy had, despite its suspended permit, been selling raw milk. Analysis of the milk from a bulk milk tank and the home of an ill person revealed the presence of the outbreak-related strain of *Salmonella*

Typhimurium. The dairy was again ordered to halt distribution of raw milk, and its raw-milk permit was subsequently revoked.

Eight inspections of the dairy by the PDA during the first four months of 2007 revealed improper cleaning of milking equipment, insufficient supervision of workers, illnesses in lactating cows, and bird and rodent infestation, any of which could have contributed to the outbreaks. One of the most sobering aspects of this case is that it happened despite the fact that Pennsylvania has stringent standards regulating raw-milk production. Twice monthly milk testing, annual PDA inspections, and annual herd skin testing for *Mycobacterium bovis* and *Brucella.* Even more telling is the fact that this outbreak was not an isolated incident. During roughly the same period, three clusters of illness from *Campylobacter* were associated with consumption of raw milk from three different dairies, and an additional three dairies were involved in a recall of raw milk after *Listeria monocytogenes* was found in milk samples, all in Pennsylvania.

EXERCISE 30 REVIEW QUESTIONS

1. Is milk with a short MBRT always unsafe to drink?

2. Why must the milk samples be sealed to prevent oxygen from entering the tubes?

3. What would be the result of a methylene blue reductase test if the milk was contaminated with a high number of obligate psychrophiles or thermophiles?

REFERENCE

CDC. *Salmonella.* http://www.cdc.gov/salmonella/.

NOTES

Bacterial Counts of Food

CASE SYNOPSES

Food Poisoning among Inmates at a County Jail—Wisconsin, August, 2008

One August morning in 2008, a large proportion of the inmates at a Wisconsin county jail awoke complaining of nausea, vomiting, and diarrhea. The local health department suspected an outbreak of foodborne illness and, along with the Wisconsin Division of Public Health, initiated an investigation.

Because of the strict routine and controlled environment of prison life, it was relatively easy to find out what the inmates had eaten in the past 24 hours and how their food had been prepared. A written questionnaire distributed to the inmates revealed 194 probable cases of food intoxication. Four respondents commented on the unusual taste of the casserole they had eaten the night before, which contained macaroni, ground beef, ground turkey, frozen vegetables, and gravy. Stool samples were obtained from six symptomatic inmates and cultured for the presence of pathogenic bacteria.

Resolution of the Case appears on page 250

CDC. 2009. *Clostridium perfringens* infection among inmates at a county jail—Wisconsin, August 2008. *Morbidity Mortality Weekly Report,* 58: 138–141.

STUDENT LEARNING OUTCOMES

After completing this exercise, you should be able to:

1. Understand how bacterial contamination of food can be quantified.
2. Evaluate the results of a standard plate count.

INTRODUCTION

Food often serves as a common vehicle for the spread of pathogens. The Centers for Disease Control estimates that 76 million people get sick, more than 300,000 are hospitalized, and 5000 Americans die each year from foodborne illness. These illnesses often arise when bacteria or their toxins are incorporated into food during growth or processing. For example, vegetables are often associated with endospores of the soil-dwelling bacterium *Clostridium botulinum.* Improper processing of food containing these endospores can create the anaerobic environment needed for the spores to germinate and produce botulism toxin. Beef and poultry are both commonly contaminated when the digestive tract of the animal ruptures during processing, releasing bacteria that are part of the normal flora of the gastrointestinal tract of these animals. *Escherichia coli, Campylobacter jejuni,* and *Salmonella sp.* are all commonly spread in this manner. Vegetables have also been contaminated when irrigation water became contaminated with animal feces.

Preventing foodborne bacterial illness takes many forms. Food is collected and animals are butchered as aseptically as possible to prevent incorporation of bacteria. Processing of foods may include washing, pasteurization, boiling, desiccation, or the addition of chemical preservatives such as citric acid. Proper storage of foods is used to minimize bacterial incorporation or growth; preventing contact with animals such as rats and flies and using temperature, either hot or cold, are also employed to inhibit microbial growth. Finally, proper food handling prevents incorporation of bacteria during preparation. Together, these methods help to keep the food supply safe in the vast majority of instances.

Overseeing all aspects of food collection, handling, and preparations are inspectors from state and federal agencies whose job it is to determine the source of an outbreak like the one seen in the case synopsis. One of the most straightforward ways of detecting the presence of bacteria in food is to use a standard plate count, in a manner similar to that seen in water or milk testing. The only additional step that needs to be taken is to suspend the food in sterile water prior to plating. In this lab, you will measure the bacterial levels of several different foods using a **standard plate count.**

PRE-LAB QUESTIONS

1. *Define the following terms:*

   Common vehicle

Anaerobic

Normal microbiota

Dilution factor

Colony forming unit (CFU)

2. *Briefly describe how a standard plate count works.*

PERIOD ONE

MATERIALS

Each student should obtain:
Three Petri plates
Three tubes of plate count agar
Two 1.0-ml pipettes and pipettor
One 99-ml sterile water blank

Everyone should have access to:
Food blender
One sterile blender jar for each type of food
Balance, along with sterile weighing papers
180-ml sterile water blanks
Samples of ground beef and fresh and frozen vegetables
Marking pen

PROCEDURE

1. Melt three tubes of plate count agar in a beaker of boiling water and hold at 50°C.
2. Label three Petri dishes with your name and lab time. Label one plate "A," one plate "B," and one plate "C."
3. Using sterile weighing paper, aseptically weigh 20 g of the food to be tested.
4. Transfer the food to a sterile blender jar, and add 180 ml of sterile water (Figure 31.1).
5. Blend the food for 5 min.
6. Use a 1-ml pipette to dispense 0.1 ml to plate A.
7. Use the same pipette to dispense 1.0 ml to the 99-ml water blank.
8. Tightly close the blank and shake vigorously for 15 sec.
9. Use a new pipette to dispense 1.0 ml to plate B and 0.1 ml to plate C.
10. Aseptically add one tube of agar to each plate, swirling gently to ensure that the agar covers the plate.
11. When the agar has solidified, invert the plates and incubate at 35°C for 24–48 h.

QUESTIONS—PERIOD ONE

1. If 20 g of meat is blended along with 180 ml of water, this represents a _____ :10 dilution of the meat.

3. FDA standards call for counting only plates containing between 25 and 250 colonies per plate. Those plates with fewer than 25 colonies are reported as TFTC (too few to count) and those with more than 250 as TMTC (too many to count). Why do you think the FDA has instituted this recommendation?

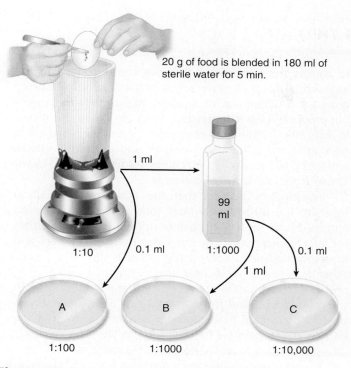

20 g of food is blended in 180 ml of sterile water for 5 min.

1 ml

99 ml

1 ml

1:10   0.1 ml   1:1000   0.1 ml

1 ml

A   B   C

1:100   1:1000   1:10,000

Figure 31.1 Dilution procedure for bacterial counts of food.

PERIOD TWO

PROCEDURE

1. Retrieve your plates from the incubator.
2. Use a colony counter to count the number of colonies on each plate. Any plate with fewer than 25 colonies should be recorded as TFTC while any plate with more than 250 colonies as TMTC. Count colonies of all sizes.
3. Use dilution factors to establish the total number of each type of microorganism per gram of food, using only those plates containing between 25 and 250 colonies. Record your results in Table 31.1.

2. If 1 ml of the suspension above is used to inoculate a plate and 122 colonies are seen on the plate after incubation, how many bacterial cells are present in 1 g of meat?

TABLE 31.1	Results					
	Food	1:10	1:100	1:1000	Dilution factor (Best plate)	Bacterial cells/gram
Colony count						
Colony count						
Colony count						

RESOLUTION OF THE CASE

Biochemical tests of the prisoners' stool samples were negative for *Salmonella*, *Shigella*, *Campylobacter*, and *Escherichia coli* O157:H7. However, *Clostridium perfringens* enterotoxin was present in all six samples. *C. perfringens* is found in soil and also commonly inhabits the intestinal tracts of mammals, including humans. In addition, it is a frequent contaminant of meats and gravies and is usually associated with inadequate heating and cooling during the cooking process. When food products contaminated with *C. perfringens* are allowed to remain at temperatures between 40°C and 50°C (104°F and 122°F), enterotoxin-producing vegetative cells are rapidly produced; illness results from the enterotoxin's action on the small intestine. *C. perfringens* is responsible for an estimated 250,000 cases of diarrhea annually in the United States.

In instances where the number of bacteria in a sample is expected to be especially large, as would be the case with a fecal sample, many types of specialized media may be used to narrow the possibilities. Selective media contain inhibitory substances that allow only a single type of microbe to grow, while differential media allow most organisms to grow but produce visible differences among the various microbes. In this case, samples of the casserole the prisoners had eaten were analyzed using both selective and differential media and found to contain 43,000 colony-forming units (CFUs) of *C. perfringens* per gram of casserole.

Investigators learned that the company distributing meals to the jail routinely froze food that was not served and held it for up to 72 h before using it to prepare dishes for later consumption. In this case, the ground beef and macaroni had been cooked the previous day, and several other food items were near their expiration dates. Also, proper documentation of the cooling temperatures for both the ground beef and the macaroni was unavailable. Investigators concluded that improper handling of food in the kitchen was responsible for the prisoners' illness.

EXERCISE 31 REVIEW QUESTIONS

1. Ten grams of ground beef are added to 90 ml of sterile water. After blending, 0.1 ml is used to inoculate a plate of general-purpose media. After a 24-h incubation, a total of 73 colonies are present on the plate.

    How many CFUs were present on the plate?

    How many CFUs are present in 1 g of the ground beef?

2. If an investigation is only concerned with the contamination of food by coliforms, such as *E. coli*, what media could be used for plating, and how would coliforms be recognized?

3. Carrots are commonly associated with *C. botulinum* endospores. Why do you think this is?

4. It is generally assumed that culture-confirmed cases of *Salmonella* infection represent only about 10% of the actual number of infections. Why do you think this is?

5. What staining technique could be especially helpful in the identification of *Clostridium* species.

Epidemiology of Gastrointestinal Illness: Differentiation of Enterobacteriaceae

CASE SYNOPSIS

Gastroenteritis among Evacuees from Hurricane Katrina—Houston Texas, 2005

Following Hurricane Katrina in August 2005, relief agencies provided food and shelter to an estimated 240,000 of the region's residents in a variety of locations. Approximately 24,000 evacuees were temporarily housed in the Reliant Park Sports and Convention Center in Houston, Texas, which was renamed Reliant City for the time being. A medical clinic was set up to serve the immediate needs of the residents. Over the next several weeks, 1,169 individuals visited the clinic exhibiting symptoms of acute gastroenteritis, specifically diarrhea, vomiting, or both.

Besides providing shelter to the evacuees displaced by Hurricane Katrina, Reliant City housed numerous staff members and volunteers who also required cots, bedding, food, water, toilets, and shower facilities. Soon these workers, along with police officers and others having direct contact with the shelter residents, were reporting gastrointestinal symptoms similar to those of the patients who had presented at the clinic. This secondary spread, presumably by person-to-person contact or fomite transmission, indicated a causative agent with a very low infectious dose (I.D.).

Resolution of the Case appears on page 260

CDC. 2005. Norovirus outbreak among evacuees from Hurricane Katrina—Houston, Texas, September 2005. *Morbidity and Mortality Weekly Report*, 54(40): 1016–1018.

STUDENT LEARNING OUTCOMES

After completing this exercise, you should be able to:

1. Understand the processes used to culture, isolate, and identify common enteric bacteria.
2. Isolate and identify two unknown enteric bacterial species.

INTRODUCTION

The colon contains a diverse population of bacteria, many of which have a **commensal** or **mutualistic** relationship with their human hosts. Folic acid and vitamin K are both synthesized by enteric bacteria, and the competition for nutrients initiated by this large resident population helps prevent colonization of the gut by pathogenic organisms. The normal population found in the large intestine encompasses several genera of **obligate anaerobes,** such as *Bacteroides, Bifidobacterium,* and *Clostridium,* as well as a wide variety of facultative anaerobes that are of greater medical importance because of their ability to survive and spread outside the body. This latter group includes the primary pathogens *E. coli, Salmonella,* and *Shigella* as well as several opportunists. Many gastrointestinal diseases can be traced back to the consumption of food or water carrying these microorganisms, often a result of fecal contamination.

Routine testing for the presence of enteric bacteria is carried out not only by medical laboratories but by the food-processing industry, municipal water districts, and government agencies tasked with ensuring the safety of the nation's food supply. The enormity of the microbial population found in the large intestine makes the isolation of individual pathogens difficult, and a variety of differential and selective media have been developed to make this task easier.

Two bacteria belonging to different groups (species, genera, phyla, etc.) must differ in one or more characteristics. After all, if the two were identical in every way, they would be the same bacteria. The key to successful identification relies upon selecting tests based upon their ability to distinguish between two otherwise similar groups. Table 32.1 lists the characteristics of some of the most common enteric genera; all are Gram-negative rods, are oxidase negative, and ferment glucose to produce acid.

TABLE 32.1	Characteristics of selected Gram-negative rods						
	Gram-negative rod	Oxidase	Ferment glucose to produce acid	Ferment lactose to produce acid	H$_2$S produced	Indole	Motility	Urease
Citrobacter	+	−	+	+	+	+	+	*
Escherichia	+	−	+	+	−	+	+	−
Enterobacter	+	−	+	+	−	−	+	−
Klebsiella	+	−	+	+	−	−	−	*
Shigella	+	−	+	−	−	*	−	−
Salmonella	+	−	+	−	+	−	+	−
Proteus	+	−	+	−	+	*	+	+

*No result is given for a test if different species within the genus react differently.

Because results in the first three columns don't differ from genus to genus, they cannot be used to differentiate one genus from another. They can, however, be used to differentiate the family Enterobacteriaceae as a group from other groups of bacteria that do not share these characteristics. The Pseudomonads, for instance, ferment neither glucose nor lactose while members of the genera *Bacillus* and *Staphylococcus* are both Gram positive. The key characteristics for distinguishing members of the Enterobacteriaceae appear in the last five columns. Note that the characteristics seen here differ among the various genera and are therefore quite useful in differentiating between one group and another.

Microbiological media described as **selective** contains a substance that inhibits the growth of most bacteria, allowing only one type to grow, while **differential media** allows all microbes to grow but each will take on a different visual appearance based on its unique physiology. The enteric bacteria may be easily discriminated using only a few well-chosen types of media with selective and differential properties. These include:

MacConkey Agar   This medium contains crystal violet and bile salts, which select against Gram-positive organisms, allowing Gram-negative cells to be isolated from a complex mixture. The media is also differential in that lactose-fermenting bacteria will appear pink while nonlactose-fermenting species will be colorless (Figure 32.1).

Kligler's Iron Agar (KIA)   Central to the identification of many members of the Enterobacteriaceae, KIA can detect three primary characteristics of a bacterium: the ability to ferment the sugars lactose and glucose to produce acid, the production of gas from the fermentation of sugars, and the production of hydrogen sulfide, H$_2$S (Figure 32.2).

Sulfur Indole Motility (SIM) Media   Combining three tests needed to distinguish many members of the Enterobacteriace, SIM is a semisolid medium that allows the detection of bacterial motility while also indicating whether or not hydrogen sulfide (H$_2$S) gas is produced. The gas combines with iron in the media to form a black precipitate. In addition, after incubation, the addition of Kovac's reagent allows the detection of indole, which is produced in some bacteria when the amino acid tryptophan is broken down (Figure 32.3).

Urea Broth   The hydrolysis of urea results in the production of carbon dioxide and ammonia, the latter rapidly increasing the pH of the media. This alkalinization causes the pH indicator in the media to turn from peach to pink, indicating a positive reaction (Figure 32.4).

In this lab, you will isolate two enteric bacteria from a mixed culture and determine the genus of each using selective and differential media.

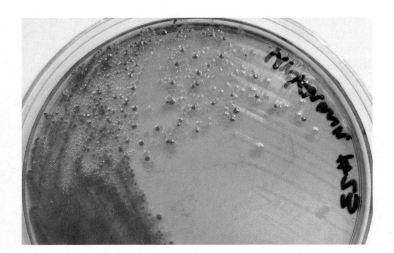

Figure 32.1 MacConkey agar contains crystal violet and bile salts, which select against Gram-positive organisms, allowing Gram-negative cells to be isolated from a complex mixture. The media also contains lactose along with the pH indicator neutral red, which turns red as the pH drops below 6.8. Lactose-fermenting bacteria produce lactic acid, lowering the pH of the media and producing pink colonies. Nonlactose fermenters produce no acid and consequently produce clear colonies.

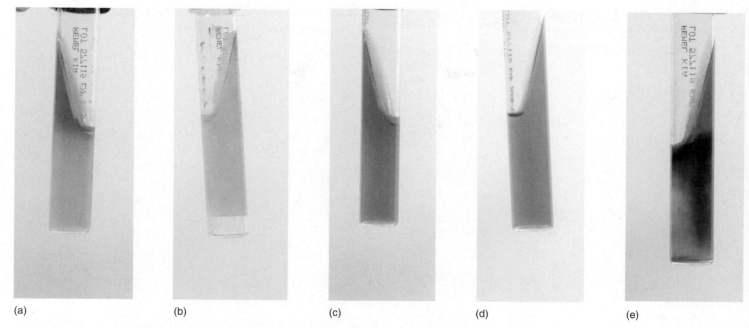

(a) (b) (c) (d) (e)

Figure 32.2 Interpretation of Kligler's iron agar. In addition to beef extract, yeast extract, and peptone, Kligler's iron agar contains glucose and lactose as well as a pH indicator to signal the production of acid resulting from their fermentation. (a) An organism that produces a red slant and yellow butt after 24 h of incubation indicates the fermentation of glucose alone. (b) An organism that ferments both glucose and lactose will produce a yellow slant and yellow butt. In this case, the production of gas has lifted the media from the bottom of the tube. (c) A slight reddening of the slant and no change in the butt indicates no fermentation. Peptone has been catabolized aerobically, producing alkaline end products and causing the minor color change within the slant. (d) An uninoculated control. (e) A black precipitate indicates sulfur reduction. This reaction only occurs in acid conditions, so fermentation of one or both sugars is occurring even if the color of the butt cannot be seen.

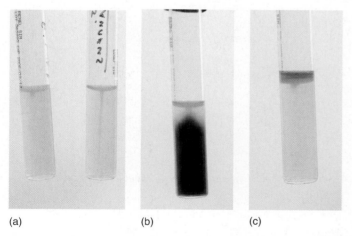

(a) (b) (c)

Figure 32.3 Interpretation of SIM media. (a) Motility should be evaluated first. SIM media contains enough agar (0.4%) to retard nonmotile bacteria while allowing motile species freedom of movement. Motility is detected as turbidity radiating from the central stab line. (b) Reduction of sulfur appears as a black precipitate after incubation. (c) Bacteria that possess the enzyme tryptophanase catabolize the amino acid tryptophan to produce indole. Detection of indole is accomplished by layering the tube with 3–4 drops of Kovac's reagent and noting the appearance of a deep red ring in the reagent layer. A red ring indicates an indole positive reaction while no change indicates that the sample is indole negative.

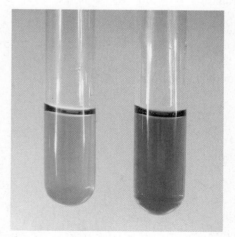

Figure 32.4 Interpretation of urea medium. The enzymatic hydrolysis of urea by the enzyme urease results in the production of ammonia and carbon dioxide. Ammonia drives the pH of the media higher and is detected by including in the media the pH indicator phenol red. A lack of nutrients and a strong buffer system in the media inhibit urea hydrolysis of all but the strongest urease producers, *Proteus, Providencia,* and *Morganella.* Organisms that rapidly hydrolyze urea will turn the media pink after 24 h of incubation (right). The media remains peach colored after incubation with urease-negative organisms (left).

PRE-LAB QUESTIONS

1. *Define the following terms:*

   Enteric

   Ferment

   Glucose

   Lactose

   Commensal

   Mutualistic

   Resident population

   Zoonotic

   Fomite

2. *What is the difference between an obligate anaerobe and a facultative anaerobe?*

3. *What is the difference between a primary pathogen and an opportunistic pathogen?*

4. *What are selective and differential media and how do they differ?*

5. *Using Table 32.1 as a guide, fill in the middle column in the accompanying table with a single characteristic that would allow you to distinguish between the two organisms presented in the first column. Fill in the right-hand column in a similar fashion, indicating what media could be used to differentiate the organisms in each pair.*

Organisms	Characteristic used to differentiate organisms	Media used
Salmonella **vs.** *Shigella*		
Enterobacter **vs.** *Escherichia*		
Klebsiella **vs.** *Enterobacter*		
Klebsiella **vs.** *Shigella*		
Proteus **vs.** *Citrobacter*		

PERIOD ONE

MATERIALS

Each student should obtain:

One broth containing a mixed culture of two enteric
  bacteria
One MacConkey agar plate
Marking pen

PROCEDURE

1. Use an inoculating loop to streak your unknown broth onto a MacConkey agar plate.
2. Incubate the plate at 35°C for 24 h.

QUESTIONS—PERIOD ONE

1. Using the characteristics seen in Table 32.1, label each of the boxes in Figure 32.5 with the appropriate genus.
2. How do lactose-fermenting bacterial species differ from non-lactose-fermenting species on MacConkey agar?

3. Of the organisms seen in Table 32.1, which ones will produce pink colonies on MacConkey media? Which will produce colorless colonies?

PERIOD TWO

MATERIALS

Each student should obtain:

Two tubes each of Kligler's iron agar, SIM media, and urea
  broth
Marking pen

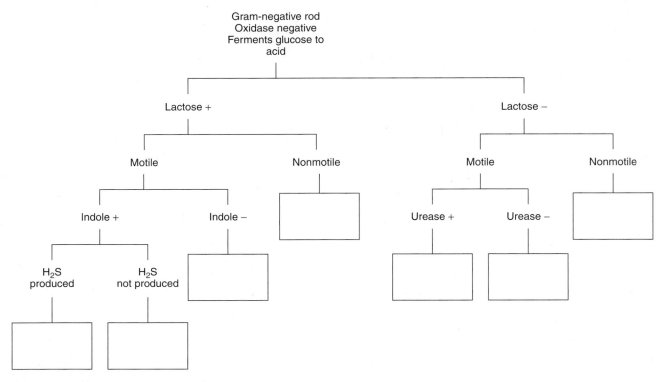

Figure 32.5 Separation outline for selected Gram-negative rods.

PROCEDURE

1. Retrieve your plate from the incubator.
2. Identify a well-isolated lactose-fermenting colony (pink) and a well-isolated nonlactose fermenter. Circle these colonies on the bottom of the plate to identify them.

You will be inoculating two identical sets of media, one set from a lactose-fermenting colony and a second set from a non-lactose-fermenting colony (six tubes total). For each set of inoculations, be sure to use the *same, well-isolated* colony (Figure 32.6). Use a needle to inoculate each set of media as follows:

3. **KIA**   Streak the surface of the slant and then stab into the butt of the agar. Be sure to stab in the center

and to not allow your needle to touch the side of the tube.

4. **SIM medium**   Stab the agar approximately three-fourths of the way to the bottom of the tube.
5. **Urea broth**   Inoculate using the same colony used for the KIA and the SIM agar.
6. Incubate your samples at 37°C for 24 h.

QUESTIONS—PERIOD TWO

1. Complete the table by providing the color of the slant and butt and indicating how you would visually determine the "reduction of sulfur" (i.e., H_2S production) and the production of gas.

Kliglers iron agar results	
Visual result (color of slant and butt)	**Interpretation**
	Glucose and lactose fermentation
	Glucose fermentation
	Reduction of sulfur
	Production of gas

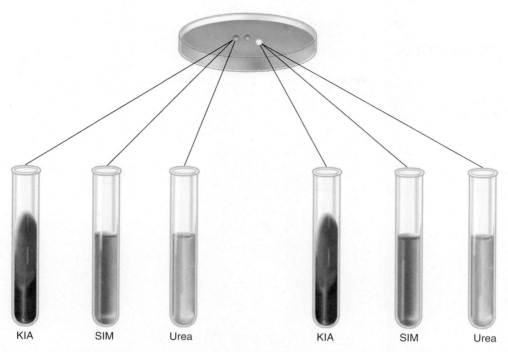

Figure 32.6 Inoculation procedure. Using a single lactose-fermenting colony, inoculate Kligler's iron agar (KIA), sulfur, indole, motility (SIM), and urea media. Do the same with a non-lactose-fermenting colony.

2. Sketch the results you would expect to see for SIM tubes displaying a positive and negative reaction for motility, H₂S production, and indole production.

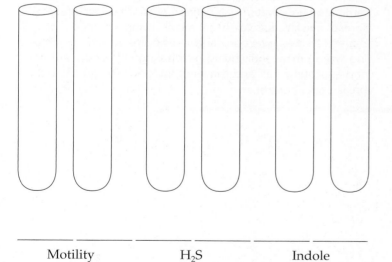

Motility   H₂S   Indole

3. Of the characteristics identified using SIM media, which would be most useful in differentiating:

   Citrobacter and *Escherichia*

   Escherichia and *Enterobacter*

   Enterobacter and *Klebsiella*

4. Urea hydrolysis is most helpful in distinguishing *Proteus* from which other genera of Enterobacteriaceae?

PERIOD THREE

EVALUATION

Retrieve your tubes from the incubator, and evaluate each isolate. Identify each sample using Figure 32.5, and provide the genus of each of your isolates:

_____

_____

RESOLUTION OF THE CASE

Initial laboratory testing for bacterial species most commonly suspected in cases of acute gastroenteritis—*Salmonella*, *Shigella*, *E. coli*, and *Campylobacter*—was negative. Similarly, none of the most common parasitic enteropathogens—*Cryptosporidium*, rotavirus, and adenovirus—were found. However, testing of stool samples or rectal swabs from 44 of the symptomatic patients identified norovirus, a frequent cause of gastroenteritis outbreaks in the United States, in 22 of these samples. Norovirus is highly contagious (I.D. < 100 organisms) and is easily spread from person to person and by contact with contaminated materials. The typical incubation period is 24–48 hours, and the resulting symptoms persist for 12–60 hours. Such outbreaks are frequently caused by contaminated food or water and can also be associated with crowded living conditions, such as those at Reliant City.

It is likely that one or more individuals were infected with the norovirus when they arrived at the shelter. Although the source of the initial infection is unknown, contact with contaminated floodwaters is a definite possibility. The infection spread quickly due to the crowded living conditions and shared facilities. Infection control measures, including isolating symptomatic individuals, distributing gel hand sanitizer, and educating staff and evacuees, quickly brought the outbreak under control.

EXERCISE 32 REVIEW QUESTIONS

1. How are gastrointestinal infections of the type seen in this case commonly spread?

2. Norovirus is a common cause of gastrointestinal infections on cruise ships, seemingly as far away from a refugee center as one could possibly get. With regard to transmission of norovirus, in what way are a cruise ship and Reliant City similar?

3. Young children are commonly stricken by gastrointestinal disease. What behaviors seen in toddlers would increase their risk of infection?

4. What procedures, if any, should be implemented to prevent similar outbreaks in the future?

5. Define the following terms, and indicate where in this case each was possibly encountered:

   Fomite

   Common vehicle

   Carrier

   Vector

   Direct transmission

   Indirect transmission

REFERENCE

CDC. *E. coli.* http://www.cdc.gov/ecoli/.

Isolation and Identification of Staphylococci

CASE SYNOPSIS

Methicillin-Resistant *Staphylococcus aureus* Skin Infections from an Elephant Calf—San Diego, California, 2008

Over the past several years, methicillin-resistant *Staphylococcus aureus* (MRSA) has become infamous as the cause of skin infections among football players, wrestlers, fencers, and other athletes who share equipment or engage in contact sports. However, humans are not the only victims of MRSA. On January 29, 2008, the San Diego Zoo reported a MRSA outbreak involving a newborn African elephant and three of its human caretakers. The humans exhibited cutaneous pustules that were laboratory confirmed as MRSA infections. An investigation was initiated to determine the course and scope of the outbreak.

After the elephant calf was born at the zoo on November 28, 2007, its mother was unable to provide enough milk, so the calf was separated from her on December 24. A variety of zoo caretakers—nursery staff, elephant keepers, nutritionists, veterinarians, and veterinary technicians—bottle-fed the calf and played with it daily. These activities undoubtedly helped spread the bacterium because *S. aureus* is most commonly spread through contact with skin lesions. Some of the caretakers would also lie alongside the calf and blow into its trunk to encourage it to drink from a bottle. Because the calf was still not receiving sufficient nutrients through bottle feeding, surgeons inserted a central feeding line into a vein on the calf's neck. Three days later, cellulitis developed at the surgical site, and shortly thereafter pustules appeared on the calf's leg and elbow. Samples from all three locations were laboratory confirmed as MRSA on January 26, 2008. Treatment with topical, oral, and intravenous antibiotics successfully resolved the infection, but the calf failed to thrive and was euthanized on February 4.

Resolution of the Case appears on page 270

CDC. 2009. Methicillin-resistant *Staphylococcus aureus* skin infections from an elephant calf—San Diego, California, 2008. *Morbidity and Mortality Weekly Report*, 58: 194–198.

STUDENT LEARNING OUTCOMES

After completing this exercise, you should be able to:

1. Understand the processes used to culture, isolate, and identify bacteria in the genus *Staphylococcus*.
2. Isolate and identify an unknown staphylococcal species.

INTRODUCTION

Staphylococcus aureus is a Gram-positive, coccus-shaped bacterium commonly found in the nose of just under a third of the healthy population (Figure 33.1). Although no harm comes to persons who are colonized with *S. aureus*, the same bacteria is responsible for a variety of infections, from minor skin lesions such as pimples and boils to serious wound infections, osteomyelitis, pneumonia, and endocarditis. *S. aureus* has been a troublesome **nosocomial** opportunist for decades and remains a common cause of infection in hospitals, nursing homes, and dialysis centers. Methicillin-resistant *S. aureus* is a strain of the bacterium that is resistant to the beta-lactam

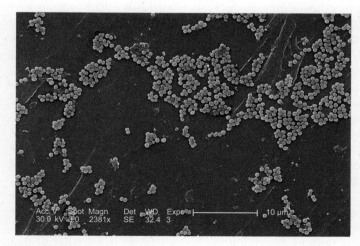

Figure 33.1 Methicillin-resistant *Stapylococcus aureus* (MRSA) magnified 2381x displaying the irregular arrangement of the cells that is one of the characteristic of the genus.

antibiotic methicillin as well as to other antibiotics of the same class (oxacillin, penicillin, and amoxicillin). Infection with MRSA severely limits treatment options. While 25% to 30% of the population is colonized with traditional *S. aureus*, about 1% carries MRSA.

TABLE 33.1	Differentiation of staphylococcal species		
	S. aureus	*S. epidermidis*	*S. saprophyticus*
Growth on media containing 7.5% NaCl	+	+	+
Beta hemolysis	+	−	−
Mannitol fermentation	+	−	(+)*
Coagulation of plasma	+	−	−
Reaction to novobiocin	Sensitive	Sensitive	Resistant

*(+) = most strains are positive.

Several characteristics of the genus *Staphylococcus* help to explain its virulence. All members of the genus are highly resistant to drying, extremes in pH, and high temperatures, allowing them to remain viable on surfaces for long periods of time. Once in the body, an array of toxins and enzymes help to promote invasion and growth of the bacteria. For many years, infections with MRSA were limited almost entirely to healthcare facilities. Over the past decade, however, a greater number of cases have been community acquired, with infections associated with schools, prisons, and even shared athletic equipment. This trend has become so pronounced that infections are now routinely identified as being healthcare-associated (HA) or community-associated (CA) MRSA, with the latter designation referring to infection in persons who have not been hospitalized or had a medical procedure within the last year. Statistics for 2005 indicate that MRSA was responsible for 94,000 life-threatening infections and nearly 19,000 deaths in the United States. Approximately 85% of these infections were associated with the healthcare setting while 15% were community associated.

Although *S. aureus* is the principle pathogen of the genus, *S. epidermidis* is a common cause of nosocomial infections in immunocompromised patients, and *S. saprophyticus* is an occasional cause of urinary tract infections. While more than 30 species of *Staphylococcus* exist, very few of them are of clinical importance. The characteristics of the three most medically important species are seen in Table 33.1.

Isolation and identification of staphylococcal species are accomplished by using selective media to isolate *Staphylococcus* from a complex mixture and then using a variety of differential tests to determine the species of an isolate. Recall that **selective media** contains a substance that inhibits the growth of most bacteria, allowing only one type to grow, while **differential media** allows all microbes to grow, but each will take on a different visual appearance based on its unique physiology. The various staphylococcal species may be easily discriminated using only a few well chosen types of media with selective and differential properties. These include:

m-Staphylococcus Broth This medium contains 7.5% sodium chloride, making it highly selective for staphylococci.

Blood Agar (trypticase soy agar with 5% sheep's blood) This differential medium allows for the detection of **hemolysis,** or the lysing of red blood cells by bacterial toxins. *S. aureus* produces α-*toxin* that causes the red blood cells in the immediate vicinity of a bacterial colony to lyse (Figure 33.2).

Mannitol Salt Agar This media is both selective and differential, containing 7.5% NaCl, mannitol, and phenol red indicator. The high salt content inhibits most bacteria other than staphylococcus. Fermentation of the mannitol by *S. aureus* and some strains of *S. saprophyticus* produces acid, lowering the pH of the media and causing the phenol red indicator to turn from red to yellow (Figure 33.3).

Once a presumptive staphylococcal species has been isolated, a single biochemical test can be used to distinguish *S. aureus* from other species of the genus. Production of the enzyme coagulase is present in 97% of *S. aureus* isolates and allows these isolates to coagulate blood plasma (Figure 33.4). In this lab, you will isolate staphylococcal species from your

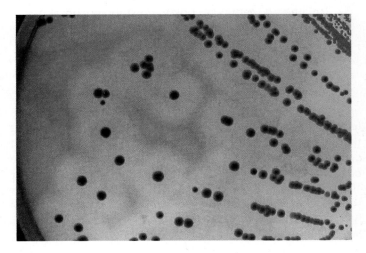

Figure 33.2 *Staphylococcus aureus* displaying beta hemolysis on blood agar. The lightened areas around the colonies are the result of α-toxin, a hemolytic toxin that lyses red blood cells.

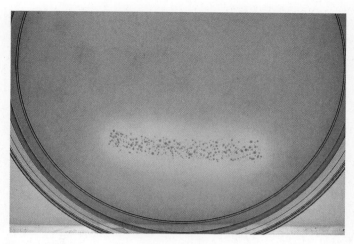

Figure 33.3 *Staphylococcus aureus* growing on mannitol salt agar. The high concentration of NaCl in the media (7.5%) inhibits the growth of bacteria other than *Staphylococcus*. The yellow color indicates that mannitol in the media has been fermented, producing acid. The fermentation of mannitol serves to differentiate many species within the genus.

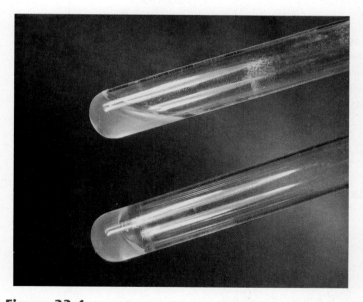

Figure 33.4 A positive coagulase reaction is indicated by any degree of coagulation, or thickening, of blood plasma. Even tubes in which the media has thickened without becoming completely solid should be interpreted as a positive reaction. The top tube illustrates a negative result while the bottom displays a positive result. The coagulase test is used to distinguish *Staphylococcus aureus* (positive) from other species of *Staphylococcus* (negative).

nose, throat, and a fomite and presumptively identify each isolate as *S. aureus* or one of the less pathogenic members of the genus. The procedure for this lab is outlined in Figure 33.5.

PRE-LAB QUESTIONS

1. *Define the following terms:*

   Colonize

   Ferment

   Fomite

   Coagulase

2. *What are pustules, vesicles, and boils, and where on the body would they be found?*

3. *What is an opportunistic pathogen? What is a nosocomial infection? How are the two related?*

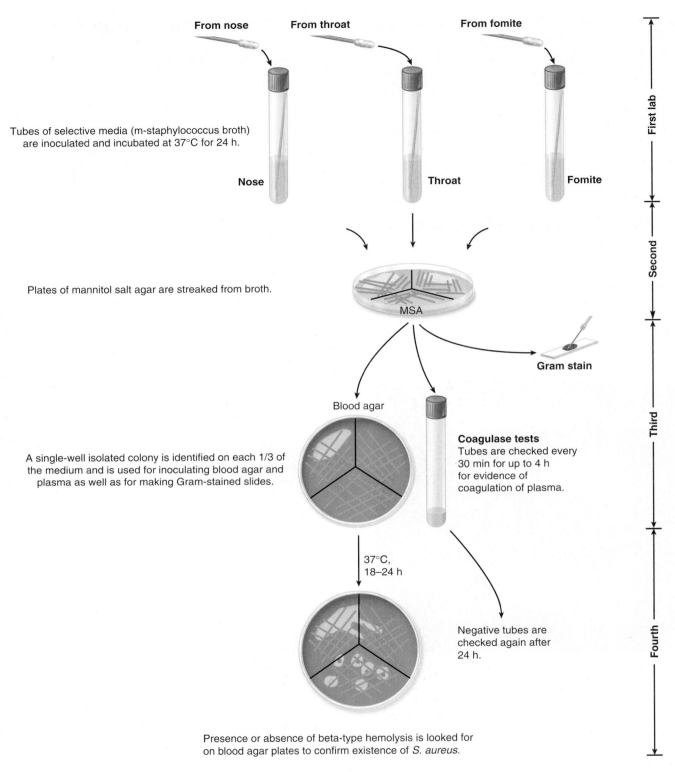

From nose

From throat

From fomite

Tubes of selective media (m-staphylococcus broth)
are inoculated and incubated at 37°C for 24 h.

Nose

Throat

Fomite

First lab

Plates of mannitol salt agar are streaked from broth.

MSA

Second

Gram stain

Blood agar

A single-well isolated colony is identified on each 1/3 of
the medium and is used for inoculating blood agar and
plasma as well as for making Gram-stained slides.

Coagulase tests
Tubes are checked every
30 min for up to 4 h
for evidence of
coagulation of plasma.

Third

37°C,
18–24 h

Negative tubes are
checked again after
24 h.

Fourth

Presence or absence of beta-type hemolysis is looked for
on blood agar plates to confirm existence of *S. aureus*.

Figure 33.5 Procedure for presumptive identification of staphylococci.

4. *Explain the selective and differential properties of mannitol salt agar.*

QUESTIONS—PERIOD ONE

QUESTIONS—PERIOD ONE

1. Why is a high-salt media used during the initial isolation of staphylococci?

5. *Using Table 33.1 as a guide, fill in the middle column in the accompanying table with a single characteristic that would allow you to distinguish between the two species presented in the first column. Fill in the right-hand column in a similar fashion, indicating what media or test could be used to differentiate the organisms in each pair.*

Organisms	Characteristic used to differentiate organisms	Media or test used
S. aureus **vs.** S. epidermidis		
S. epidermidis **vs.** S. saprophyticus		
S. aureus **vs.** S. saprophyticus		

PERIOD ONE

MATERIALS

Each student should obtain:

Three tubes of m-staphylococcus broth
Three sterile swabs
One tongue depressor
Marking pen

PROCEDURE

1. Label the three tubes of m-staphylococcus broth "nose," "throat," and "fomite." Also label each with your name and lab time.

2. Moisten a swab by immersing it partially in the tube of broth labeled nose. Use this tube to swab the nasal membrane just inside your nostril. Place the entire swab into the nose tube.

3. Using a similarly moistened swab, swab the surface of a fomite and place the swab into the fomite tube.

4. Using the tongue depressor to hold down the tongue, use the final swab to sample the back of your lab partner's throat. Avoid touching the tongue or cheeks with the swab. Place the swab into the throat tube. Dispose of the tongue depressor in a hard-sided biohazard container.

5. Incubate the broths at 37°C for 24 h.

2. Would m-staphylococcus broth be an appropriate choice for isolation of bacteria other than *Staphylococcus* from a complex mixture?

PERIOD TWO

MATERIALS

Each student should obtain:
One mannitol salt agar plate

PROCEDURE

1. Retrieve your tubes of m-staphylococcus broth from the incubator.
2. Use a marker to divide a plate of mannitol salt agar into thirds. Label each third with "nose," "throat," or "fomite."
3. Remove a single loopful of bacteria from the nose tube, and streak it onto the agar, being sure to use no more than a third of the plate. A three quadrant streak (as seen in figure 33.5) should be adequate.
4. Repeat this process for the throat and fomite.
5. Incubate the plates for 24–48 h at 37°C.

QUESTIONS—PERIOD TWO

1. What is the function of each of the following components found in mannitol salt agar?

   7.5% NaCl

   Phenol red

2. In mannitol salt agar, indicate with a + or − the conclusion you would draw from each of the following results:

Visual result	Interpretation	
	Growth in 7.5% NaCl	Fermentation of mannitol
No growth, agar is pink		
Colony growth, agar is pink		
Colony growth, agar is yellow		

PERIOD THREE

MATERIALS

Each student should obtain:
One blood agar plate
Three small test tubes containing 0.5 ml of rabbit plasma
Marking pen

PROCEDURE

1. Retrieve your mannitol salt agar plate from the incubator. Examine the plate, noting if the media has turned yellow around any colonies. Record your results on the data sheet at the end of the exercise.
2. Circle a single colony in each of the three sections of your mannitol salt agar plate. The circled colony will be used to perform a Gram stain as well to inoculate a blood agar plate and a tube of plasma.
3. Use a maker to divide a blood agar plate into three sections and label each section "nose," "throat," or "fomite."
4. Use the circled nose colony to inoculate a small area on a blood agar plate. A sterile loop can than be used to streak the organism over one-third of the plate, in a manner similar to that used to inoculate the mannitol salt agar plate earlier.
5. Prepare a Gram-stained slide from the same colony. Record the results of the Gram stain on the data sheet.
6. Finally, use a loop to transfer all that remains of the circled colony to a tube of rabbit plasma. Cover the tube with a piece of labeling tape or parafilm, and label it appropriately.
7. Repeat the process of blood agar inoculation, Gram staining, and plasma inoculation for the other two circled colonies.
8. Incubate the blood agar plate at 37°C for 24 h.
9. Incubate the coagulase tubes at 37°C for 4 h. Check for coagulation of the plasma every hour for 4 h (see Figure 33.4). If no coagulation has occurred, continue incubating for a total of 24 h. If no coagulation has occurred after 24 h, the tubes may be considered negative for the presence of coagulase. Record the results for each isolate on the data sheet.

QUESTION—PERIOD THREE

1. Describe the difference in appearance between alpha and beta hemolysis. What type of hemolysis is displayed by *S. aureus*?

PERIOD FOUR

EVALUATION

Retrieve your tubes and plates from the incubator, and evaluate any hemolysis and/or coagulase activity. Record this information on the data sheet. Identify each sample as being *S. aureus* or *Staphylococcus sp.* using Table 33.1.

Data Sheet: Isolation and Identification of Staphylococci

Growth in m-staphylococcal broth (+/−):

Nose

Throat

Fomite (_____)

Gram stain:

	Nose	Throat	Fomite

Culture/biochemical results:

	Growth on mannitol salt agar (+/−)	Fermentation of mannitol (+/−)	Coagulation of plasma (+/−)	Beta hemolysis on blood agar (+/−)
Nose				
Throat				
Fomite				

Identification:

Nose _____

Throat _____

Fomite _____

RESOLUTION OF THE CASE

To determine the scope of this community-acquired MRSA outbreak, investigators obtained rectal and trunk cultures from the other 11 African elephants at the zoo and nasal cultures from 53 of 55 elephant caretakers. While no other elephants tested positive for MRSA, epidemiological investigation identified two caretakers as carriers and revealed a total of 20 suspected (based on signs and symptoms alone) or confirmed (through laboratory testing) MRSA infections among the caretakers. The strain was identified as MRSA USA 300, the most common strain implicated in CA-MRSA infection. This strain was identical to that isolated from the elephant calf and from the wounds on three of the caretakers.

Examination of work records showed that the calf had become infected with MRSA after being exposed to caretakers carrying the same strain. Investigators surmised that transmission then occurred from the calf to other human caretakers because the activities involving direct contact between the calf and a caretaker were those that most often resulted in infection. It was also noted that veterinary staff workers were more likely to wear personal protective equipment and were thus less likely to be infected than were the nursery staff or the elephant keepers.

EXERCISE 33 REVIEW QUESTIONS

1. How are *S. aureus* infections commonly spread, and what steps could a healthcare worker take to limit the spread?

2. Evidence of MRSA infection of the elephant calf was initially seen at the surgical site as well as the calf's knee and elbow (which don't differ substantially from your own knee and elbow). Why were these three sites the first to show evidence of infection?

3. Most MRSA lesions found on animal keepers were found on the hands, forearms, and wrists. Explain why this is as well as what it tells you about the use of personal protective equipment.

4. HA-MSRA infections are generally resistant to many classes of antimicrobics while CA-MRSA infections are usually resistant to only one or two antimicrobial classes. Why do you think this is?

5. When identifying staphylococcal species, sensitivity to the antimicrobic novobiocin can be used to differentiate *S. epidermidis* from *S. saprophyticus* (details of this procedure can be found in Exercise 92). This, however, was not done during this exercise and is not done in many clinical laboratories once *S. aureus* has been ruled out as the infectious agent. Explain how sensitivity is assessed, and why the identification is sometimes left uncompleted.

REFERENCE

CDC. *MRSA infections.* http://www.cdc.gov/mrsa/index.html.

Isolation and Identification of Streptococci

CASE SYNOPSIS

Invasive *Streptococcus pyogenes* after Allograft Implantation—Colorado, 2003

In September 2003, a 17-year-old male was admitted to a Colorado hospital after presenting at the emergency room with pain and erythema at the incision site of a recent surgical procedure, as well as a fever of 39°C and chills. Six days earlier, the patient had undergone elective surgery to repair an anterior cruciate ligament using a hemipatellar tendon allograft. In this type of surgery, a tendon removed from a cadaveric donor is used to replace all or part of the damaged ligament.

Surgical exploration of the wound revealed extensive tissue damage necessitating removal of the allograft and fasciotomy of the affected thigh. Cultures of blood, wound aspirate, and the removed allograft all revealed the presence of group A *Streptococcus*. The patient was treated with antibiotics along with aspiration of fluid from the affected leg. After 17 days, the patient was discharged and completed a course of intravenous antibiotics at home.

Resolution of the Case appears on page 282

CDC. 2003. Invasive *Streptococcus pyogenes* after allograft implantation—Colorado, 2003. *Morbidity and Mortality Weekly Report*, 52: 1173–1176.

STUDENT LEARNING OUTCOMES

After completing this exercise, you should be able to:

1. Understand the processes used to culture, isolate, and identify bacteria in the genus *Streptococcus*.
2. Isolate and identify an unknown streptococcal species.

INTRODUCTION

The genus *Streptococcus* comprises many species of Gram-positive, spherical bacteria that typically grow in pairs or chains (Figure 34.1). They are easily distinguished from *Staphylococcus* species both by their arrangement and by the fact that they don't produce the enzyme catalase. Although the most common infection that can be traced to the genus is the ubiquitous strep throat (caused by *Streptococcus pyogenes*), other species within the genus are responsible for at

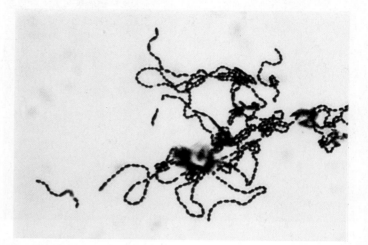

Figure 34.1 Chains of streptococci. Chains only appear in liquid media and become longer as the culture ages.

least some cases of meningitis, pneumonia, endocarditis, erysipelas, and necrotizing fasciitis, or flesh-eating bacterial infections. In addition, many species of streptococci are human commensals and can be found residing in the mouth, on the skin, and throughout the upper respiratory tract.

The identification and classification of streptococcal species has a long history. As far back as 1919, the ability of certain species to cause the hemolysis of red blood cells when grown on blood agar was seen as an important differential characteristic, and this means of differentiation is still used today. Some species are able to reduce the iron in hemoglobin, resulting in a green halo around these colonies referred to as **alpha (α) hemolysis** while others completely destroy the blood cells around them, resulting in a clear zone known as **beta (β) hemolysis** (Figure 34.2). An absence of hemolysis is often referred to as gamma (γ) hemolysis.

In 1933, Rebecca Lancefield classified streptococcal species using carbohydrate antigens found on the cell surface of many species of the genus and designated them group A, group B, etc. Her original system has since been expanded to group V, but the system is not inclusive as several groups do not possess the carbohydrate antigens necessary for classification. Identification of medically important streptococcal species in the laboratory is usually based on a combination of hemolysis along with a small number of biochemical and physiological characteristics. Faster immunological methods are also used, with the most common being the rapid strep test, which is used in physicians' offices to diagnose infection with group A strep in just a few minutes.

Beta-hemolytic streptococci that are human pathogens are found in Lancefield groups A and B (and to a lesser

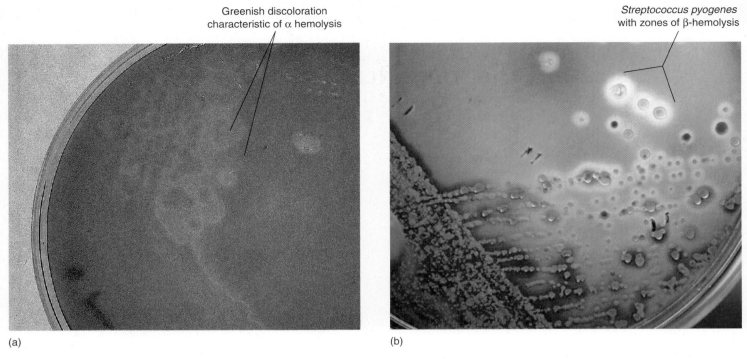

Greenish discoloration
characteristic of α hemolysis

Streptococcus pyogenes
with zones of β-hemolysis

(a)   (b)

Figure 34.2 (a) Alpha (α) hemolysis is marked by a greenish halo surrounding bacterial colonies such as those of the viridans bacteria. The color is a result of reduction of the hemoglobin in the red blood cells of the media. (b) Beta (β) hemolysis is marked by the complete destruction of the red blood cells in the media and leads to a clear halo surrounding the colonies of beta-hemolytic bacteria, such as *Streptococcus pyogenes.*

extent C). **Group A** strep is the causative agent of several diseases including strep throat, rheumatic fever, scarlet fever, and necrotizing fasciitis. Although several species display group A antigens, almost all human infections in this group are due to infection with *S. pyogenes.* Group B strep generally refers to *S. agalactiae,* an infrequent cause of meningitis in newborns and the elderly. Occasional colonization of the female reproductive tract is also seen. Group C strep includes *S. equi* along with the subspecies *S. equisimillis* and *S. zooepidemicus,* all of which are responsible for infections in a variety of mammals including cattle, horses, and moose.

Alpha-hemolytic streptococci include *S. pneumoniae,* a potentially serious pathogen that can cause not just pneumonia but also meningitis and otitis media. The *viridans streptococci* include several species (*S. mitis, S. mutans, S. milleri, S. salivarius,* and *S. sanguis*) that are common microbiota of the mouth. Members of the viridans group that enter the bloodstream can potentially lead to tooth abscesses, meningitis, or endocarditis. Entry into the bloodstream is often facilitated by dental work that causes small cuts in the soft tissue of the mouth. Neither *Streptococcus pneumoniae* nor the viridans group possesses the carbohydrate receptors used in Lancefield classification.

Lancefield **group D** contains several species of bacteria that have since been reclassified and placed in the genus *Enterococcus.* Chief among these are *E. faecalis* and *E. faecium,* both of which are common inhabitants of the large intestine. *S. bovis* is a nonenterococcus that colonizes animals, and occasionally humans, but is only rarely linked to disease.

The sheer number of potentially pathogenic streptococcal species ensures that their isolation and identification are

common practices in the microbiology laboratory. In this exercise, you will use common biochemical and physiological tests to identify a streptococcal isolate. The key to successful identification of streptococcal species is to select tests that will distinguish between two otherwise similar groups. Table 34.1 lists the characteristics of some of the most commonly encountered streptococcal groups.

Initial characterization of streptococcal species is based on the type of hemolysis exhibited. For example, an isolate that displays beta hemolysis may be *Streptococcus pyogenes* while alpha hemolysis may indicate *Steptococcus bovis* or *Enterococcus faecium.* After that, several types of **differential media** and biochemical tests are used to presumptively identify an isolate, with different media being used depending on the type of hemolysis. Final confirmation in the laboratory often relies on **serological tests** that determine whether or not specific antigens are present on the cells of an isolate.

Differential media and tests used for the identification of alpha-hemolytic streptococcal species include:

Optochin Sensitivity   *S. pneumoniae* is sensitive to this antibiotic while members of the viridans group (*S. mitis, S. mutans,* and *S. salivarius*) are resistant (Figure 34.3).

Bile Esculin Hydrolysis   All members of group D (*E. faecalis, E. faecium,* and *S. bovis*) are positive for this test (Figure 34.4).

Salt Tolerance   Group D enterococci will grow in the presence of 6.5% NaCl while other group D organisms will not.

TABLE 34.1	Characteristics of selected streptococcal species							
	Lancefield group	Hemolysis	Bacitracin sensitivity	CAMP reaction	SXT sensitivity	Bile esculin hydrolysis	Tolerance to 6.5% NaCl	Optochin sensitivity
S. pyogenes	A	β	S	−	R	−	−	R
S. agalactiae	B	β	R	+	R	−	−	R
S. pneumoniae	None	α	R	−	*	−	−	S
S. equi	C	β	R	−	S	−	−	R
S. equisimilis	C	β	R	−	S	−	−	S
E. faecalis	D	β	R	−	R	+	+	R
E. faecium	D	α	R	−	R	+	+	R
S. bovis	D	α †	R	−	*	+	−	R
S. mitis	None	α †	R	−	S	−	−	R
S. salivarius	None	α †	R	−	S	−	−	R
S. mutans	None	None	*	−	S	−	−	R

*No result is given for a particular characteristic if members of the group commonly differ in their reactions.
†Weak alpha hemolysis.

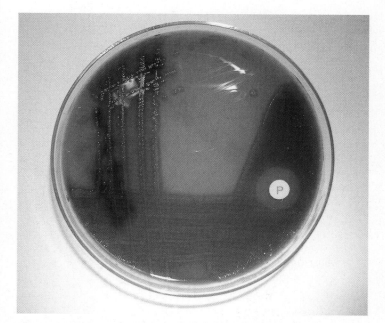

Figure 34.3 Sensitivity to optochin is indicated by a zone of inhibition (an area of no growth) surrounding the disk.

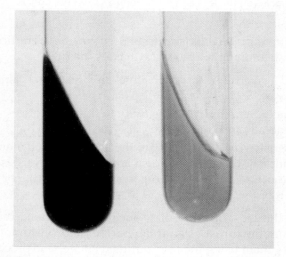

Figure 34.4 Bile esculin hydrolysis. Group D streptococci and enterococci are able to hydrolyze esculin in the presence of bile. The end product of the reaction, esculetin, reacts with iron in the media to produce a dark brown color. After inoculation, the media is incubated for up to 72 h. If the slant is more than 50% darkened at any point, a positive result is recorded (left). Slants that remain light in color are negative (right).

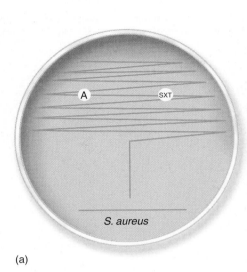

(a)

(b)

Figure 34.5 CAMP reaction, bacitracin sensitivity, and SXT sensitivity. (a) Inoculation of a blood agar plate for CAMP testing is accomplished by heavily streaking approximately 40% of the surface of the plate with the unknown bacterium. A single line of the unknown is then brought down to a point about two-thirds of the way across the plate. A loopful of *Staphylococcus aureus* is then streaked perpendicularly to the unknown streak so that a gap of 1 cm separates the two organisms. Bacitracin and SXT differentiation disks are placed within the heavily streaked area, as shown. (b) Interpretation of the CAMP test is completed by observing the area of hemolysis between the two organisms. If no hemolysis occurs in the 1 cm gap between the unknown bacterium and *S. aureus,* the test is negative. If an arrowhead-shaped area of hemolysis appears between the two bacteria, indicating a synergistic effect between the hemolysins of each species, the CAMP test is considered to be positive. Sensitivity to bacitracin and SXT can be determined by noting the presence or absence of a zone of inhibition. In this case, the unknown bacterium is resistant to bacitracin (on the left) and sensitive to SXT (on the right).

Differential media and tests used for the identification of beta-hemolytic streptococcal species include:

CAMP Test   If this test is positive, the organism is almost always *S. agalactiae* (Figure 34.5).

Bacitracin Sensitivity   Sensitivity to this antibiotic generally indicates *S. pyogenes* (see Figure 34.5).

SXT Sensitivity   Sensitivity to the antibiotic SXT, along with resistance to bacitracin and beta hemolysis, is indicative of a group C organism such as *S. equi* or *S. equisimilis* (see Figure 34.5).

In this lab, you will isolate one or two streptococcal species from the throat of your lab partner and presumptively identify them using the tests outlined above. The procedure for this lab is outlined in Figure 34.6.

Erythema

Allograft

Fasciotomy

PRE-LAB QUESTIONS

1. *Define the following terms:*

Colonize

Hemolysis

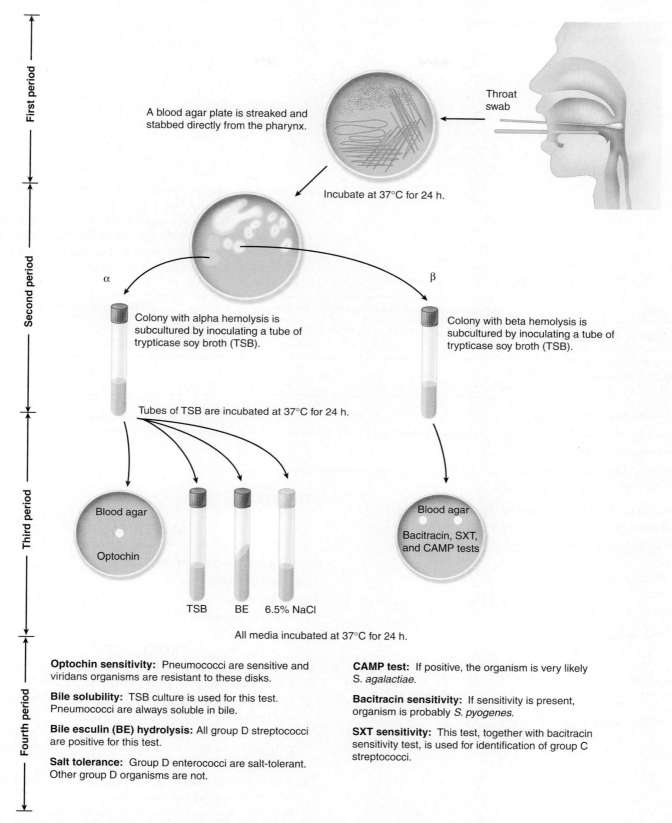

A blood agar plate is streaked and stabbed directly from the pharynx.

Throat swab

Incubate at 37°C for 24 h.

α

Colony with alpha hemolysis is subcultured by inoculating a tube of trypticase soy broth (TSB).

β

Colony with beta hemolysis is subcultured by inoculating a tube of trypticase soy broth (TSB).

Tubes of TSB are incubated at 37°C for 24 h.

Blood agar

Optochin

TSB   BE   6.5% NaCl

Blood agar

Bacitracin, SXT, and CAMP tests

All media incubated at 37°C for 24 h.

Optochin sensitivity: Pneumococci are sensitive and viridans organisms are resistant to these disks.

Bile solubility: TSB culture is used for this test. Pneumococci are always soluble in bile.

Bile esculin (BE) hydrolysis: All group D streptococci are positive for this test.

Salt tolerance: Group D enterococci are salt-tolerant. Other group D organisms are not.

CAMP test: If positive, the organism is very likely S. agalactiae.

Bacitracin sensitivity: If sensitivity is present, organism is probably S. pyogenes.

SXT sensitivity: This test, together with bacitracin sensitivity test, is used for identification of group C streptococci.

First period — Second period — Third period — Fourth period

Figure 34.6 Procedure for streptococci isolation and identification.

2. *What does the enzyme catalase do, and why is its function important to the cell?*

Erysipelas

Otitis media

4. *Using Table 34.1 as a guide, fill in the middle column in the accompanying table with a single characteristic that would allow you to distinguish between the two species presented in the first column. Fill in the right-hand column in a similar fashion, indicating what media or test could be used to differentiate the organisms in each pair.*

Organisms	Characteristic used to differentiate organisms	Media or test used
S. pyogenes vs. *S. agalactiae*		
S. equi vs. *S. pyogenes*		
E. faecalis vs. *E. faecium*		
S. pneumoniae vs. *S. salivarius*		

3. *In what part of the body would each of the following infections be found?*

Meningitis

Pneumonia

Endocarditis

PERIOD ONE

MATERIALS

Each student should obtain:
One plate of blood agar
One sterile swab
One tongue depressor

PROCEDURE

1. Label the plate with your name and lab time.
2. Using the tongue depressor to hold down the tongue, use the swab to sample your lab partner's throat. Sample the back of the throat along with any white patches on or near the tonsils. Avoid touching the cheeks or tongue.
3. Roll the swab over an area approximately equal to one-fifth of the surface of the blood agar plate. Be sure that the entire surface of the swab contacts the agar.
4. Exchange plates with your lab partner so that you are each manipulating your own bacteria.
5. Using a loop, streak out the bacteria as seen in Figure 34.7.

Figure 34.7 Procedure for inoculation of blood agar plate.

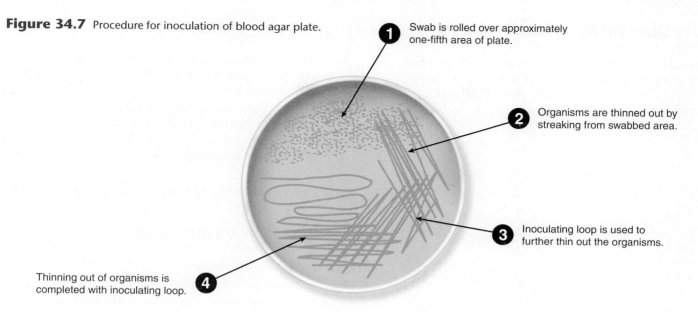

1 Swab is rolled over approximately one-fifth area of plate.

2 Organisms are thinned out by streaking from swabbed area.

3 Inoculating loop is used to further thin out the organisms.

4 Thinning out of organisms is completed with inoculating loop.

6. Incubate the plate at 37°C for 24 h.
7. Be sure to dispose of the swab and tongue depressor in a hard-sided biohazard container.

QUESTIONS—PERIOD ONE

1. Draw and label an area of a blood agar plate in which alpha hemolysis is occurring and an area where beta hemolysis is occurring.

2. If multiple colonies are seen on your blood agar plate after incubations and all of them exhibit alpha hemolysis, are they all members of the same species? Why or why not?

PERIOD TWO

MATERIALS

Each student should obtain:
Two tubes of trypticase soy broth (TSB)

PROCEDURE

1. Retrieve your blood agar plate from the incubator.
2. Using Figure 34.2 as a guide, examine the plate for well-isolated colonies displaying alpha or beta hemolysis. While you will undoubtedly encounter alpha-hemolytic colonies, you most likely will not see any beta hemolysis as beta-hemolytic streptococci are rarely isolated from healthy individuals.
3. Choose two, well-isolated colonies to inoculate tubes of TSB. If you have a colony that displays beta hemolysis, use it as one of your samples, otherwise pick two alpha-hemolytic colonies that differ in some way (size, shape, or degree of hemolysis). Be sure to label your tubes for later identification (alpha-1 or beta-1).
4. Incubate the tubes for 24–48 hours at 37°C.

QUESTIONS—PERIOD TWO

1. What physiological tests are necessary to differentiate the groups of alpha-hemolytic streptococci from one another? Which groups contain alpha-hemolytic streptococci?

2. What physiological tests are necessary to differentiate the groups of beta-hemolytic streptococci from one another? Which groups contain beta-hemolytic streptococci?

PERIOD THREE

MATERIALS

Each student should obtain:

Trypticase soy broth tubes from the previous period
Gram-staining kit
Marking pen

PROCEDURE

1. Make a Gram-stained slide for each of your two isolates, and examine them at 100x. Do they appear to be pure cultures?
2. Draw the Gram-stain results for each isolate on the data sheet in PERIOD FOUR. Follow the appropriate procedure depending on the type of hemolysis displayed by your isolate.

INOCULATION WITH ALPHA-HEMOLYTIC ISOLATES

MATERIALS

Each student should obtain:

One blood agar plate (for up to four unknowns)
One 6.5% sodium chloride broth per unknown (i.e., heart infusion broth supplemented with NaCl)
One bile esculin slant per unknown
One optochin disk per unknown
Candle jar or CO_2 incubator
Inoculating loop
Marking pen

PROCEDURE

1. Mark the bottom of a blood agar plate to divide it into halves, thirds, or quarters, depending on the number of alpha-hemolytic isolates to be tested. Label each section with the same label that is on the trypticase soy broth tube used for inoculation.
2. Use a loop to completely streak the surface of the agar within each section using the appropriate isolate. Place an optochin disk in the center of each area and press down slightly on the disk to ensure that it adheres to the media.
3. Using a loop, inoculate the sodium chloride broth and streak the surface of the bile esculin slant.
4. Incubate the blood agar plates in a candle jar or CO_2 incubator for 24 h at 37°C. All other media may be incubated aerobically at 37°C. The bile esculin slants should be incubated for 48 h and the sodium chloride broth for up to 72 h.

INOCULATION WITH BETA-HEMOLYTIC ISOLATES

MATERIALS

Each student should obtain:

One blood agar plate per unknown
One bacitracin differential disk

One SXT sensitivity disk

One broth culture of *S. aureus* (subspecies *aureus*)

Dispenser or forceps for manipulating antimicrobic disks

Inoculating loop

Marking pen

PROCEDURE

1. Mark the bottom of a blood agar plate with your name, lab time, and the identifying information on the trypticase soy broth tube containing your beta-hemolytic isolate.

2. Using Figure 34.5 as a guide, inoculate the blood agar plate with your isolate and with *S. aureus*. Note that a small (1 cm) gap should remain between the streak of the unknown organism and the *S. aureus* streak.

3. Using a dispenser or forceps, place a bacitracin disk and an SXT disk on the heavily inoculated areas of the plate, as seen in the figure. Press each disk gently to ensure that it adheres to the media.

4. Incubate the blood agar plates aerobically for 24 h at 37°C.

QUESTIONS—PERIOD THREE

1. How would tolerance of 6.5% NaCl be recognized?

2. How is sensitivity or resistance to an antibiotic in a disk diffusion test recognized? Sketch what you would expect to see if an isolate was sensitive to bacitracin and resistant to SXT.

3. How is a bile esculin tube interpreted?

PERIOD FOUR

EVALUATION

Retrieve your tubes and plates from the incubator.

- For alpha-hemolytic isolates, evaluate sensitivity to optochin and bile esculin hydrolysis using Figures 34.3 and 34.4, respectively. Tolerance to 6.5% NaCl is positive if the medium is turbid and negative if it is clear. Record your results on the accompanying data sheet and use them to identify the streptococcal group to which your isolate belongs based on Table 34.1.

- For beta-hemolytic isolates, evaluate CAMP activity, bacitracin resistance, and SXT resistance using Figure 34.5. Record your results on the data sheet and use them to identify the streptococcal group to which your isolate belongs based on Table 34.1.

Data Sheet: Isolation and Identification of Streptococci

Gram stain:

Isolate 1   Isolate 2

Cultural/biochemical results:

	Hemolysis	Optochin sensitivity	Bile esculin hydrolysis	Tolerance to 6.5% NaCl	CAMP	Bacitracin sensitivity	SXT sensitivity
Unknown 1							
Unknown 2							

Group identification:

Unknown 1 _____

Unknown 2 _____

RESOLUTION OF THE CASE

The donor that supplied the allograft in this case also supplied tendon allografts for five other patients, none of whom encountered similar complications. All the remaining allografts from this donor were either recalled or put on hold. Records revealed that cultures of the donor tissue yielded group A *Streptococcus* (GAS) prior to processing, but after treatment with an antimicrobial solution, all cultures were negative for GAS. GAS was also isolated from a specimen of the donor's blood, which had been stored prior to recovery and processing of tissue. Genetic analysis revealed that strains of GAS detected in the donor's blood and tissue, as well as in the recipient after implantation, were both rare and identical to one another.

This case demonstrated the apparent ability of group A *Streptococcus* to survive standard tissue-processing treatment and to remain undetectable using common methods after treatment. Based on this case, the American Association of Tissue Banks proposed sterilizing or discarding certain tissues if specific organisms, including GAS, are detected.

EXERCISE 34 REVIEW QUESTIONS

1. In a case similar to the one seen here, antimicrobial treatment did not eradicate *Clostridium sordellii,* and postprocessing cultures failed to detect contamination with the organism, resulting in the death of a recipient of an allograft. Based on the characteristics of the genus *Clostridium,* answer the following:
   a. Why do you think antimicrobial treatment was ineffective in killing the *C. sordellii?*

   b. Why was *C. sordellii* not detected prior to implantation but was able to grow once it was implanted?

2. The CDC says that postsurgical GAS infection should be regarded as a *sentinel event* that should lead to an epidemiological investigation within the hospital. What is a sentinel event?

REFERENCE

CDC. *Group A Streptococcal Disease.* http://www.cdc.gov/ncidod/dbmd/diseaseinfo/groupastreptococcal_g.htm.

CASE STUDY EXERCISE 35

Blood Typing

CASE SYNOPSIS

Transfusion Reaction Leads to Death due to ABO Incompatibility—Florida, 2008

A 67-year-old patient at a Florida hospital was given a blood transfusion as part of a minor surgical procedure. Shortly after receiving the donated blood, the patient experienced a hemolytic transfusion reaction and died that same day, December 29, 2007.

Hemolytic transfusion reactions are responsible for the majority of transfusion-related deaths. This type of reaction occurs when antibodies in the recipient's blood recognize red blood cells in the incorrectly matched donated blood, eventually resulting in the destruction of those red blood cells. By comparison, if the donated blood is correctly matched to the blood type of the recipient, the donated blood is perceived as the patient's own and no reaction occurs.

In the Florida case, the medical team had anticipated the potential need for blood during or after the surgery, and the patient's blood had been typed 11 days earlier in his hospital room. Initial investigation into the case indicated that the donated blood matched the patient's blood type. Another possibility investigators considered was that other, rarer red blood cell antigens, such as MN or Kell, could have been incompatible between the donor and the recipient, leading to the fatal transfusion reaction. However, this also turned out to be unrelated to the fatal transfusion reaction.

Resolution of the Case appears on page 289

Bert Fish Medical Center Clinical Laboratory. 2008. Statement of deficiencies and plan of correction. Florida Agency for Health Care Administration, St. Petersburg, FL.

STUDENT LEARNING OUTCOMES

After completing this exercise, you should be able to:

1. Understand the theoretical basis for the A/B/O blood groups.
2. Determine the A/B/O and Rh group for an unknown blood sample.

INTRODUCTION

Human red blood cells have on their surface several different antigens, which have historically been used to divide blood into four major groups (many minor groups exist as well). The antigens themselves are sugars, which in turn are attached to a second sugar embedded in the cell membrane of **erythrocytes.** The two most common antigens are referred to as A or B, and everyone inherits a single allele (alternative version of the antigen producing gene) from each of their parents, for a total of two alleles. An additional possibility occurs when no sugar at all is attached to the transmembrane protein, resulting in the lack of an antigen, which is referred to as O. In genetic terms, the A and B antigens are codominant, meaning that if a person possesses the gene for the antigen, it is fully expressed on the surface of the cell. At the same time, A and B are both dominant to O, meaning that an O phenotype is only seen when there is no A or B antigen present. This combination of A, B, and O antigens results in four possible blood types depending on the particular combination of alleles present (Figure 35.1). Two A alleles or an A and O

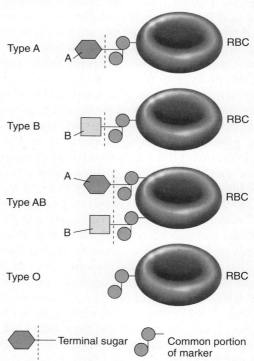

Figure 35.1 The molecular basis for the A/B/O blood groups. The designation of a blood sample as A, B, AB, or O indicates the type of terminal sugar found on that cell, either A or B. The presence of one sugar does not effect the presence of another sugar, so erythrocytes possessing both A and B sugars are designated AB. The lack of a terminal sugar results in the blood being designated as type O.

TABLE 35.1	Characteristics of ABO Blood Groups		
Genotype	Antigen present on erythrocyte membrane	Blood type	Antibodies present in plasma
AA, AO	A	A	Anti-B
BB, BO	B	B	Anti-A
AB	A and B	AB	Neither anti-A nor anti-B
OO	Neither A nor B	None	Anti-A and Anti-B

result in blood type A; two B alleles or a B and an O result in type B; a combination of one A and one B result in type AB, and two O alleles give rise to type O (Table 35.1).

Under most circumstances, the immune system produces **antibodies** against an antigen only after first encountering it. As an example, antibodies against Varicella Zoster, the virus responsible for chickenpox, are only produced after the initial encounter with the virus, either through infection or vaccination. When the virus is encountered a second, third, or fourth time, the rapid response of these antibodies ensures that viral particles are quickly marked for destruction.

An exception to the normal response of the immune system occurs when the antigen involved is a red blood cell. A person's blood serum typically contains preformed antibodies to any blood cell displaying an antigen different than the body's own blood type, even though previous contact with the blood has never occurred. Reaction between these preformed antibodies and the antigens present on "foreign" red blood cells (from donated blood) can lead to an immediate immune response known as a **transfusion reaction.** In the most serious type of transfusion reaction, massive and immediate hemolysis occurs as donated red blood cells react with antibodies and are marked for destruction by the complement system. The cell debris from the lysed cells quickly clogs the **glomeruli** (filtering apparatus) of the kidney, resulting in failure of the organ. Death is a common result.

Precise identification of blood type enables blood transfusions to be safely performed. The key to a safe transfusion is to ensure that the recipient does not produce antibodies against the donated blood, and this can most easily be accomplished through accurate blood typing. Recall that a person will typically have antibodies against those blood antigens that are not normally present in her or his body. This means that Type A blood contains antibodies against the B antigen (anti-B) found on type B and type AB blood cells. Type B blood contains antibodies against the A antigens (anti-A) found on type A and type AB blood. Type O blood contains antigens against both type A and type B antigens and will therefore react with type A, type B, and type AB blood. Type AB blood produces neither anti-A nor anti-B antibodies and will not react with blood from any of the major blood groups (Table 35.1).

An additional major blood type is represented by the Rh antigen, so named because the red blood cell antigen was first discovered in **Rh**esus monkeys. Instead of a letter designation such as A or B, the presence or absence of the Rh antigen is indicated by a plus or minus sign after the other major blood antigens. For example A+, B+, and AB+ all indicate the presence of the Rh antigen while A−, B−, and AB− indicate its absence. Like the A and B antigens, the gene for the Rh antigen is expressed in a codominant manner, meaning that if the gene is present, the antigen will be expressed on the surface of the red blood cell. Unlike the A and B antigens, antibodies to the Rh antigen are only produced after the body has first encountered the Rh antigen and been sensitized. Because of this, transfusion reactions occur only after the second (and subsequent) encounter of Rh-positive blood by an Rh-negative recipient.

Typing of blood is accomplished by mixing a drop of blood with purified antibodies against one of the three major antigens. If the antigens on the red blood cells are recognized by the purified antibodies, a visible clumping will occur as the red blood cells are bound together, a process known as **agglutination.** For example, if anti-B antibodies cause the agglutination of a blood sample, the cells in the sample must possess B antigens. If a separate drop of the same blood does not agglutinate when combined with anti-A antibodies, then the blood cells must not possess A antigens. Finally, if agglutination is seen when a third drop of the blood is mixed with anti-Rh antibodies, then it can be determined that the blood possesses the Rh antigen. Such a blood sample would be identified as B+ (Figure 35.2).

PRE-LAB QUESTIONS

1. *Define the following terms:*

    Antibody

    Antigen

2. *What feature of the red blood cell is responsible for its blood type?*

3. *What type(s) of blood could a person with type A+ blood receive? Why?*

4. *Blood from a person with B- blood could be safely transfused to persons with what blood type? Why?*

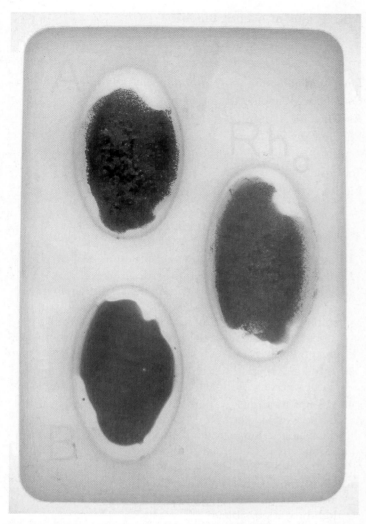

Figure 35.2 Blood Typing. Each well of the blood-typing slide contains a drop of the blood to be tested along with a single type of antibody: anti-A, anti-B, or anti-Rh. If the red blood cells express a cell surface antigen recognized by the antibodies in the well, agglutination of the cells occurs. Here, because agglutination has occured in the "A" well (containing anti-A antibody) on the upper left and the "Rh" well (containing anti-Rh antibody) seen on the right-hand side of the slide, the blood can be designated as A+.

Erythrocyte

Agglutination

PERIOD ONE

In this exercise, you will use purified antibodies to determine the blood type of either your own blood or aseptic blood supplied by your instructor. Follow the procedure for the proper exercise below.

Because of the risk of exposure to transmissible bloodborne pathogens such as HIV or hepatitis, blood and blood products require the utmost care in handling. Gloves must be worn whenever you are working with blood, all contaminated materials must be treated as biohazard waste and disposed of properly, and all sharp tools, such as lancets, should be disposed of in a hardsided biohazard container. Except in the case of extreme emergency (potentially lethal bleeding), you should never touch the blood of another person. If you feel there is any reason you should not participate in this exercise, inform your instructor.

ABO Typing of Human Blood

MATERIALS

Each student should obtain or have access to:
One disposable blood-typing slide
Purified antisera
   Anti-A, anti-B, and anti-Rh
Disposable blood lancets
Alcohol swabs
Toothpicks
Small adhesive bandages
Lightbox for viewing the agglutination reactions

PROCEDURE

1. Obtain a disposable blood-typing slide.
2. Place a single drop of anti-A serum into the well labeled "A." Replace the cap on the serum vial before continuing so as to prevent cross-contamination. In the same way, add a drop of anti-B serum to the well labeled "B," and a single drop of anti-Rh (anti-D) serum to the well labeled "Rh."
3. Scrub the tip of the middle finger with an alcohol pad.
4. Carefully remove a sterile lancet from its package. If the packaging of the lancet looks damaged in any way, it should not be used.
5. Press the middle finger approximately 1 cm from the fingertip with the thumb of the same hand and puncture the tip of the finger with the lancet. Discard the lancet into a hard-sided biohazard container. Never reuse a lancet.
6. Apply gentle pressure to the finger to force out a drop of blood into each of the three wells on the blood-typing slide.
7. Wipe any excess blood from your fingertip using an alcohol pad. If the bleeding continues for more than a minute or two, cover the wound with an adhesive bandage.
8. Using a toothpick, carefully mix the blood and antisera in the well labeled "Rh." Dispose of the toothpick in a hard-sided biohazard container. Repeat this process, using a fresh toothpick each time for the blood and serum in the wells labeled "A" and "B."
9. Gently agitate the blood-typing slide by sliding it back and forth on the lightbox or by rocking the entire lightbox, if possible.
10. Examine the slide for any agglutination reactions, using Figure 35.2 as a guide. Positive A and B reactions should be visible in 1 min or less while positive Rh reactions may require up to 5 min.
11. Sketch the results of your blood-typing reaction, and provide an interpretation in the space provided. Dispose of all materials in a hard-sided biohazard container.

ABO Typing of Aseptic Blood

MATERIALS

Each student should obtain or have access to:
One disposable blood-typing slide
Aseptic blood sample
Purified antisera
   Anti-A, anti-B, and anti-Rh
Toothpicks
Lightbox for the viewing of the agglutination reactions

PROCEDURE

1. Obtain a disposable blood-typing slide.
2. Place a single drop of anti-A serum into the well labeled "A." Replace the cap on the serum vial before continuing so as to prevent cross-contamination. In the same way, add a drop of anti-B serum to the well labeled "B," and a single drop of anti-Rh (anti-D) serum to the well labeled "Rh."
3. Transfer a single drop of blood from one of the sample bottles to each of the wells labeled "A," "B," and "Rh."
4. Using a toothpick, carefully mix the blood and antisera in the well labeled "Rh." Dispose of the toothpick in a hard-sided biohazard container. Repeat this process, using a fresh toothpick each time for the blood and serum in the wells labeled "A" and "B."
5. Gently agitate the blood-typing slide by sliding it back and forth on the lightbox or by rocking the entire lightbox, if possible.
6. Examine the slide for any agglutination reactions, using Figure 35.2 as a guide. Positive A and B reactions should be visible in 1 min or less while positive Rh reactions may require up to 5 min.
7. Sketch the results of the blood-typing reaction and provide an interpretation in the space provided. Dispose of all materials in a hard-sided biohazard container.

Sketch the results of your blood typing reaction in the space below.

Sample Identification (student name or sample number)

Established blood type: _____

RESOLUTION OF THE CASE

As the investigation into the elderly Florida man's death continued, the blame for the incident was eventually placed on poor record keeping.

According to the hospital policy in place at the time, phlebotomists—healthcare workers trained to draw blood—were to use at least two separate patient identifiers, such as the patient's name and his or her medical records number, when drawing blood and when labeling the collection container. Furthermore, the phlebotomist was supposed to handwrite the medical records number from the patient's armband on the blood collection tube. In this case, the staff person who drew the blood did not transcribe the medical records number from the patient's armband to the blood tube label. Although the name on the tube was that of the patient, the blood within the tube was actually that of his hospital roommate. When the patient was transfused with blood that had been typed and cross-matched based on this mistaken identification, the transfusion reaction ensued.

In the United States, 54 people died from hemolytic transfusion reactions in 2008. Administrative or clerical mistakes are the most common cause of this type of reaction, but adherence to strict record-keeping procedures has helped decrease the incidence. Blood is now commonly labeled with a bar code that can be quickly and accurately scanned, reducing mistakes due to poor handwriting. In some hospitals, radio-frequency identification, or RFID, is used to manage the blood supply. RFID technology relies on small electronic chips to identify each bag of donated blood, while a similar chip embedded in the patient's armband broadcasts the patient's blood type. Medical personnel are alerted to a potential transfusion reaction by an audible alarm that sounds if the chips in the patient's armband and on the blood bag do not match.

EXERCISE 35 REVIEW QUESTIONS

1. The term "universal donor" is applied to persons whose blood can be safely transfused into another person without fear of a transfusion reaction involving the A, B, O, or Rh antigens in the recipient. The term "universal recipient" applies to those persons who may receive blood from any donor without fear of a transfusion reaction. What are the blood types of the universal donor and universal recipient, and why is this so?

2. A condition known as hemolytic disease of the newborn results when anti-Rh antibodies produced by a pregnant Rh- mother cross the placenta and attack the blood cells of a developing Rh+ fetus, causing fetal anemia. This Rh incompatibility is never seen in a first pregnancy. Why is this?

3. The human A and B antigens are terminal sugars attached to another sugar (known as the H antigen) that is in turn attached to the surface of the red blood cell. People displaying the Bombay phenotype produce no H antigen and therefore have no way to attach either A or B antigens to their red blood cells even though they may produce the antigens. When typing blood from a person with the Bombay phenotype, what blood type would be indicated? Why?

4. If the RFID technology discussed earlier had been used in the case seen here, would it have prevented the transfusion reaction seen in this case?

McCall, R.E., and Tankersley, C.M. 2008. *Phlebotomy Essentials, 4th ed.,* chapter 6. Maryland: Lippincott Williams and Wilkins.

Differential White Blood Cell Count

CASE SYNOPSIS

Screening for Parasitic Infection of Refugees—United States, 2008

War, famine, and political repression displace millions of people around the world. Some families are internally displaced, meaning that they must leave their homes but can still remain within their country; others become refugees, migrating to another country to find peace or safety. The United States receives a large share of these refugees. The largest numbers come from the Near East (especially Iraq and Iran) and southern Asia. Many also emigrate from eastern Asia, especially Burma, and from Africa, particularly Somalia and Sudan. In 2008, the United States received 60,191 refugees.

When refugees arrive in another country, they need housing, food, and medical attention. Since many refugees come from areas having high rates of diseases that are not common in the United States, healthcare workers follow a set of guidelines in order to provide the needed care. One of the first tests run is a CBC, or complete blood count, in which a quantity of blood is drawn and analyzed. One type of white blood cell, the eosinophil, is a particularly useful diagnostic tool for the refugee population. An elevated eosinophil count often means the patient has a worm or parasite infection.

When screening refugees newly arrived in the United States, the Centers for Disease Control and Prevention recommends the use of the accompanying flowchart. The first

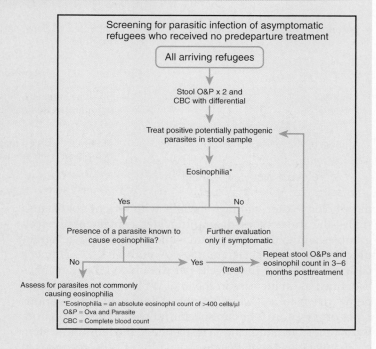

tests recommended are a CBC and an examination for ova and parasites (O&P) in the stool. Eosinophilia is defined as the presence of more than 400 eosinophils in 1 μl of blood.

Resolution of the Case appears on page 295

CDC. 2010. Domestic intestinal parasite guidelines. http://www.cdc.gov/immigrantrefugeehealth/guidelines/domestic/intestinal-parasites-domestic.html.

STUDENT LEARNING OUTCOMES

After completing this exercise, you should be able to:

1. Understand the significance of a differential white blood cell count.
2. Use a prepared slide to perform and evaluate a differential white blood cell count.

INTRODUCTION

Whole blood is composed of several different types of blood cells suspended within a complex solution called **plasma.** Red blood cells are the most common cellular component and are responsible for the transport of oxygen and carbon dioxide between the lungs and the tissues, while the far less numerous white blood cells—or **leukocytes**—are in charge of the body's immune reactions. Being able to identify and count each of the different leukocytes found in a blood sample provides a glimpse into the health of a patient and indicates whether other tests may be called for.

The five major types of white blood cells (WBCs) are divided into two groups based on their microscopic

TABLE 36.1	Characteristics of leukocytes			
Cell type	Prevalence in circulation	Appearance in stained smears (Wright's or Giemsa stain)	Diameter	Function
Granulocytes				
Neutrophil	55%–65%	Multilobed nucleus; light purple granules	12 μm–15 μm	Phagocytosis of (primarily) bacteria
Eosinophil	1%–3%	Bilobed nucleus; red granules	12 μm–15 μm	Immune response against worms and other large eukaryotic parasites
Basophil	0.5%–1%	Indented nucleus; dark purple granules	12 μm–15 μm	Involved in the inflammatory response
Agranulocytes				
Lymphocyte	20%–35%	Spherical nucleus, very little cytoplasm visible; no granules	7 μm–12 μm (occasionally larger)	Specific immune response
Monocyte	3%–7%	Horseshoe-shaped nucleus, cytoplasm easily seen; no granules	12 μm–20 μm	Phagocytosis as mature macrophages

morphology. After staining—usually with Wright's or Giemsa stain—the three types of **granulocytes** contain easily seen granules within their cytoplasm while the two types of **agranulocytes** have no such granules. A normal WBC count is between 5,000 and 10,000 cells per μl of blood. **Leukocytosis** is defined as an increase in the total WBC count while **leucopenia** refers to a lowered count. Additionally, the various types of WBCs are found in specific ratios, and deviation from those ratios often indicates a challenge to the immune system by an infectious agent. Because each type of WBC responds in a different way, the nature of the infectious agent can often be deduced from changes in the ratios of the various WBCs.

Each of the three types of granulocytes are about 12–15 μm in diameter, about twice the size of a red blood cell (Table 36.1). **Neutrophils** are the most numerous, comprising 55% to 65% of total WBCs and contain small, light purple granules within their cytoplasm. As they mature, their nucleus becomes segmented into multiple lobes (Figure 36.1), causing this type of cell to also be referred to as a **seg.** Immature neutrophils do not show this segmentation and are often called **bands.** An increase in neutrophils, and especially an increase in immature band forms, indicates systemic bacterial infection.

Eosinophils comprise 1% to 3% of WBCs and are recognized by their red cytoplasmic granules and bilobed (horseshoe-shaped) nucleus (Figure 36.2). An increase in eosinophils is an indication that the body is fighting an infection caused by a fungus, worm, or other large, eukaryotic parasite.

Basophils are the least prevalent WBC, comprising only 0.5% to 1% of total WBCs. Recognized by their dark purple granules and constricted nucleus (Figure 36.3) basophils are involved in the inflammatory response and allergies.

Agranulocytes include monocytes and macrophages, neither of which displays prominent cytoplasmic granules.

Monocytes (Figure 36.4) are generally 12–20 μm in diameter, have an indented or horseshoe-shaped nucleus, and normally comprise 3% to 7% of total circulating WBCs. Monocytes will eventually leave the circulatory system and mature to become macrophages, one of the body's main phagocytic cells. **Lymphocytes** (Figure 36.5) are involved in the body's specific immune reactions and are divided functionally into T-cells, which are responsible for cell-mediated immunity, and B-cells, which are responsible for humoral immunity. Lymphocytes are generally about the same size as red blood cells (7.5 μm) although larger cells (up to 18 μm) are occasionally seen.

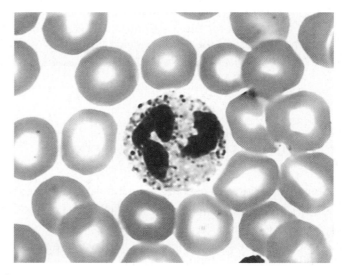

Figure 36.1 Neutrophils are the most common granulocytic white blood cell, comprising approximately 55% to 65% of WBCs. Mature neutrophils are recognized by their multilobed nucleus, and increased numbers of neutrophils is a common sign of systemic bacterial infection (1600x).

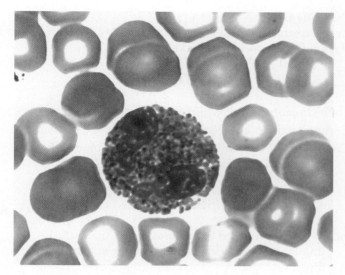

Figure 36.2 Eosinophils make up 1% to 3% of white blood cells and characteristically display red granules and a bilobed nucleus. Increased eosinophils typically indicate infection with large eukaryotic pathogens (1250x).

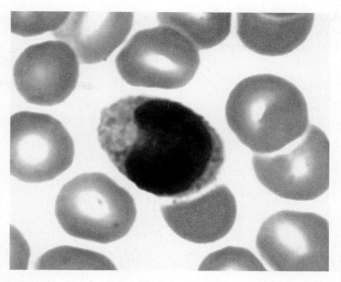

Figure 36.4 Monocytes are the largest WBCs and have no visible granules within the cytoplasm, much of which is visible in the cell. They make up about 3% to 7% of circulating WBCs and will eventually leave the circulatory system to mature into macrophages (1600x).

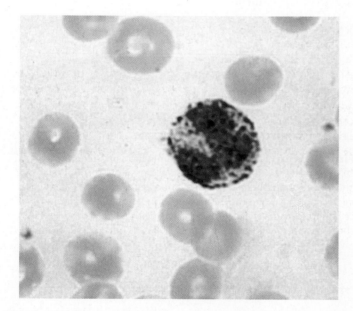

Figure 36.3 Basophils are the least common granulocyte, making up 0.5% to 1.0% of circulating WBCs. Large blue-black granules are present in the cytoplasm, and the nucleus is usually constricted (320x).

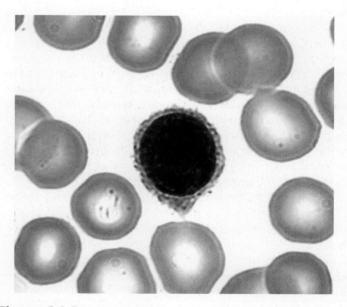

Figure 36.5 Lymphocytes appear as small spherical cells, comparable in size to red blood cells. They have a round, dark-staining nucleus, which fills most of the cell. Lymphocytes comprise between 20% and 35% of WBCs and are divided functionally into T-cells and B-cells although no morphological differences exist between the two types of cells. T-cells are responsible for cell-mediated immunity while B-cells are responsible for humoral immunity (1600x).

PRE-LAB QUESTIONS

1. *Define the following terms:*

Leukocyte

Phagocytosis

2. *In the flowchart seen in the beginning of this exercise, why is the number of eosinophils in a blood sample, as opposed to other WBCs, important?*

3. *Why don't healthcare providers test refugees for very specific diseases rather than for increased eosinophils?*

PERIOD ONE

In this exercise, you will be performing a differential blood cell count of a prepared blood smear.

MATERIALS

Each student should obtain or have access to:

Commercially prepared normal human blood smears

Commercially prepared abnormal human blood smears (eosinophilia or neutrophilia)

PROCEDURE

1. Obtain a prepared blood smear and, using the oil immersion lens, locate an area of the smear where the blood cells are close together but not overlapping.
2. Scan the slide using the pattern seen in Figure 36.6. Talley, in Table 36.2, the first 100 white blood cells you see.
3. Calculate the percentage of each WBC and compare to accepted values.

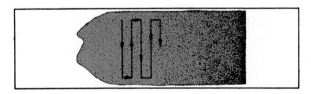

Figure 36.6 Nonoverlapping examination of a blood smear. Follow a path like the one shown to ensure that you count each cell only once. Choose a recognizable blood cell at the right side of your microscopic field and move the slide, counting as you go, until the cell disappears off of the left side of the field. Repeat this process, choosing a new cell each time, until 100 cells have been counted.

TABLE 36.2	Differential WBC count				
Cell type	Number counted	Percentage $\dfrac{\textit{number of cell type}}{\textit{total cells counted}} \times 100$	Normal limits	Within normal limits (Y/N)	Diagnosis (normal, eosinophilia, neutrophila, etc)

RESOLUTION OF THE CASE

Very often, the first step in diagnosis is to detect the body's reaction to an infection. A signal that something is going on often points the physician in the right direction, saving the time, expense, discomfort, and inconvenience involved in undergoing multiple tests. Especially for refugees who are asymptomatic, multiple tests for every possible infection are not called for. Rather, the most common infections can be diagnosed based on a CBC, which will show the body's reaction to any infection, plus a stool test, which may yield the parasites or ova themselves.

Of course, a CBC can yield vital diagnostic information about any person. For example, eosinophilia often occurs in the United States in people who become infected with *Trichinella*. This infection, which was formerly associated with consuming undercooked pork, is now most often caused by eating bear meat.

EXERCISE 36 REVIEW QUESTIONS

1. Why is the number of band neutrophils an important indicator of infection?

2. A relatively common infection of people who hunt and eat wild game, particularly wild boars and bears, is trichinellosis. Research trichinellosis and decide what type of WBC would be found in increased numbers during this infection?

3. What does the screening chart instruct physicians to do once an eosinophilia-causing infection is treated? Why?

Slide Agglutination

CASE SYNOPSIS

Leptospirosis Infection—Hawaii, 2005

Sometimes dedication to your job can get you in trouble. In 2004, a 56-year-old genetics professor at the University of Hawaii in Oahu was determined to continue working in his lab even though a local stream had overflowed and the campus was flooded. For 4 days, he slogged through standing water in his lab to keep his research going.

Some time afterward, the professor developed blisters on his feet. A few days later, he started having flulike symptoms—fever and chills, followed by nausea and vomiting. He began to feel better, but then developed another phase of illness that featured tremors, impaired balance, and colored illusions appearing before his eyes. Based on

these signs and symptoms, physicians suspected that the dedicated genetics professors was suffering from leptospirosis, a bacterial infection caused by *Leptospira interrogans*, which is usually transmitted by direct or indirect contact with animal urine. It is considered the most common zoonosis in the United States, with cows, sheep, deer, and pigs being the animals most frequently infected. The bacterium is a spirochete that is thought to enter the bloodstream through minute breaks in the skin. Physical contact with water potentially contaminated with animal urine—swamps, rivers, floods, and the like—is a known risk factor for contracting the disease.

Resolution of the Case appears on page 299

Gaynor, Kate, et al. Leptospirosis on Oahu: An outbreak associated with flooding of a university campus. *American Journal of Tropical Medicine and Hygiene,* 76(5): 882–885.

STUDENT LEARNING OUTCOMES

After completing this exercise, you should be able to:

1. Understand how a slide agglutination test can be used to detect both antigens and antibodies.
2. Use a slide agglutination test to identify a bacterial isolate.

INTRODUCTION

When the immune system of a mammal detects a foreign molecule, one of the outcomes is the production of **antibodies,** proteins found floating free within the bloodstream or attached to the surface of certain white blood cells. Antibodies recognize and bind to specific molecules called **antigens**—typically a cell, virus, or protein—marking the foreign object for destruction in a number of ways. While the full workings of the immune system are extraordinarily complex, the production of specific antibodies is a straightforward sign that someone has been exposed to a particular antigen. Furthermore, if the concentration of a specific antibody in the bloodstream, called the **titer,** increases over a period of days, it generally means that the infection is

ongoing, with ever-increasing numbers of antibodies being produced. If the antibody titer stays the same over an extended period, it indicates a past exposure that has since been resolved. The union of antibody and antigen results in **agglutination,** as the antibody molecules bind large numbers of antigens to one another, forming a visible clump. Because antibodies are collected from blood serum, such assays are known as **serological** tests.

Serological tests are often used to identify an infection. For example, if *Leptospira* infection is suspected, antibodies against *Leptospira* should be present in the bloodstream. If a small amount of a patient's blood is mixed with purified *Leptospira* cells and an agglutination reaction occurs, then the blood contains anti-*Leptospira* antibodies and the patient has been exposed to the *Leptospira* bacterium (Figure 37.1a). If no agglutination reaction occurs, no anti-*Leptospira* antibodies are present in the blood. The same type of antibody/antigen reaction can be used to identify an unknown microbial species just as easily. In this case, an unknown microbe can be combined with a known antibody; if an agglutination reaction occurs, then the antibody and antigen recognize one another (Figure 37.1b).

The ability of an antibody to recognize only one specific antigen, while not binding to other similar antigens, is a measure of the **specificity** of a serological test. Typically, serological tests have a high degree of specificity and are able to distinguish among various strains, or **serotypes,** of a bacterial species.

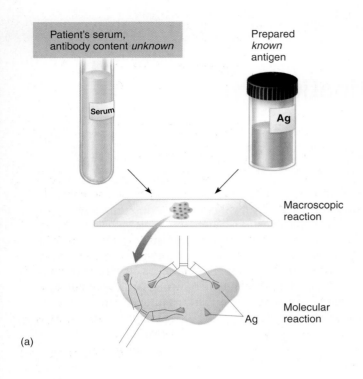

(a)

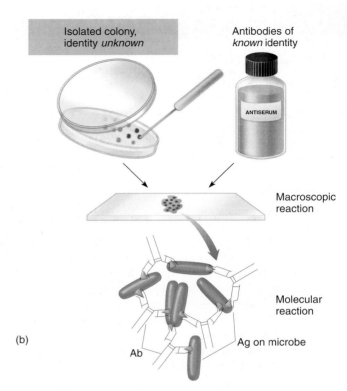

(b)

Figure 37.1 (a) When serological testing is used to diagnose a disease, a blood sample is combined with a known antigen. If the blood contains antibodies specific to the antigen, agglutination, or a similar reaction, will be visible. This indicates that the patient has been exposed to the microorganism in question and has developed antibodies against it. (b) When serological testing is used to identify an unknown microorganism, known antibodies are combined with the microbe and a visible reaction serves as evidence of the identity of the unknown.

One of the simplest types of serological tests is a **slide agglutination test** in which antibody and antigen—one known, the other unknown—are combined on a microscope slide and the presence or absence of agglutination is noted. In this exercise, you will use antibodies to determine if an unknown sample contains *Salmonella*, a genus far easier to cultivate than *Leptospira*.

PRE-LAB QUESTIONS

1. *Define the following terms:*

Agglutination

Titer

2. *Explain the difference between an antigen and an antibody.*

PERIOD ONE

MATERIALS

Each student should obtain or have access to:

Two numbered dropper bottles containing unknown suspensions of inactivated bacteria, one of which is *Salmonella* [*Salmonella* somatic (O) antigen Group B] and the other *Proteus* [*Proteus* OXK antigen]

Labeled control bottle containing *Salmonella* suspension
    [*Salmonella* somatic (O) antigen Group B]

Salmonella O antiserum, poly A-I

Microscope slide

Toothpicks

Marking pen

PROCEDURE

1. Mark your slide to separate it into three equal areas. Label the areas "A," "B," and "C" (Figure 37.2).

2. Place a drop of the first unknown into the area labeled "A."

3. Place a drop of the second unknown into the area labeled "B."

4. Place a drop of the *Salmonella* control suspension into the area labeled "C."

5. Place a drop of *Salmonella* O antiserum next to, but not touching, each of the three drops on the slide.

6. Using a different toothpick for each of the three reactions, mix the two drops together thoroughly.

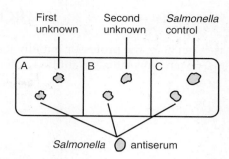

Figure 37.2 Divide your slide into three sections. Each section will contain one drop of a bacterial suspension and one drop of *Salmonella* O antiserum.

7. Allow the slide to sit for 3 to 4 min, and then look for any agglutination. The area marked "C" on the slide contains *Salmonella* cells and *Salmonella* anti-serum and will provide a positive control for the appearance of agglutination.

8. Record your results, indicating which of your unknown bacterial suspensions contained *Salmonella*.

9. Discard the toothpicks into a hard-sided biohazard container.

RESOLUTION OF THE CASE

Once the professor had been definitively diagnosed with leptospirosis, health officials issued an Internet survey of the campus to try to determine if any of the hundreds of students and staff who had helped with flood cleanup had been affected. In the end, they diagnosed only one other case, although 90 people did report experiencing a febrile illness within 30 days of the flood.

One problem with leptospirosis is that the early signs are no different from those of other flulike illnesses, so affected people do not necessarily present to the healthcare system.

Certain occupations predispose people to this disease, namely ones that put people in touch with animal urine. This includes veterinarians, meat packers, and farmers but this case presented a unique situation leading to the exposure of a profession that would not normally be at risk in a location that would not normally be considered risky. It was only because of the work of a particularly astute physician taking a thorough history that leptospirosis was considered as a diagnosis. Sometimes, dedication to your job can get you out of trouble.

EXERCISE 37 REVIEW QUESTIONS

1. In the case seen here, the professor initially tested negative for leptospirosis even though he was already showing symptoms. Seven days later, his antibody titer was much higher. Explain why this occurred.

2. How would you interpret a slide agglutination test for leptospirosis that was positive and whose titer did not change over the course of several weeks?

REFERENCE

CDC. *Leptospirosis.* http://www.cdc.gov/nczved/divisions/dfbmd/diseases/leptospirosis/.

Enzyme-Linked Immunosorbent Assay (ELISA)

CASE SYNOPSIS

Hepatitis C Virus Transmission at an Outpatient Hemodialysis Unit—New York, 2001–2008

Hepatitis C is a chronic liver infection that can be either silent (with no noticeable symptoms) or debilitating. Either way, 80% of infected persons experience continuing liver destruction. Chronic hepatitis C infection is the leading cause of liver transplants in the United States. The virus that causes it is bloodborne, and therefore patients who undergo frequent procedures involving transfer of blood are particularly susceptible to infection. Kidney dialysis patients belong to this group. In 2008, a for-profit hemodialysis facility in New York was shut down after nine of its patients were confirmed as having become infected with hepatitis C while undergoing hemodialysis treatments there between 2001 and 2008.

When the investigation was conducted in 2008, investigators found that 20 of the facility's 162 patients had been documented with hepatitis C infection at the time they began their association with the clinic. All the current patients were then offered hepatitis C testing to determine how many had acquired hepatitis C during the time they were receiving treatment at the clinic. They were considered positive if an enzyme-linked immunosorbent assay (ELISA) showed the presence of antibodies to the hepatitis C virus.

Resolution of the Case appears on page 306

CDC. 2009. Hepatitis C virus transmission at an outpatient hemodialysis unit—New York, 2001–2008. *Morbidity and Mortality Weekly Report,* 58: 189–194.
Bio-Rad Laboratories. 2010. Biotechnology Explorer™ ELISA Immuno Explorer Kit (catalog no. **166-2400EDU**). www.explorer.bio-rad.com.

STUDENT LEARNING OUTCOMES

After completing this exercise, you should be able to:

1. Understand how an ELISA can be used to detect both antigens and antibodies.
2. Use an ELISA to detect the presence of antibodies in a serum sample.

INTRODUCTION

The immune system serves the purpose of protecting our health by detecting and reacting to the presence of foreign molecules within the body. One way in which this is done is through the production of **antibodies,** proteins that have the ability to bind to any molecule deemed foreign to the body. These foreign molecules, or **antigens,** are often part of the surface structure of cells or viruses, and the binding of antibodies to the antigens leads to their inactivation or eventual destruction by cells of the immune system.

While the intricacies of the immune system are still being unraveled, the reaction between an antibody and an antigen is a comparatively simple and well understood process that need not even take place within the body. Antibodies have the capability to detect and bind to specific antigens in an artificial environment, such as a test tube, serving as an exquisitely precise system capable of detecting cells, viruses, and small molecules like sugars and proteins. Assays that utilize antibodies are called **serological** tests, as the antibodies central to the test are isolated from blood serum. Serological tests are accurate, simple to perform, and relatively inexpensive, making them not only useful in medical laboratories, but also in kits to be used in the home. Home pregnancy tests use antibodies to detect the presence of human chorionic gonadotropin, a protein found only in the urine of pregnant women while home-based tests to detect marijuana use and HIV infection work in much the same manner.

Antibodies are formed from four protein chains: two identical heavy chains and two identical light chains, held together in a Y shape (Figure 38.1). The key to an antibody's function can be found in the two upper tips of the Y where the antibody's structure differs from antibody to antibody, allowing each antibody to recognize a different antigen (antigen is actually a contraction of the words "antibody generator" and is used to describe any molecule that causes the immune system to produce antibodies). Antibodies can be "raised" to recognize virtually any antigen quite easily. The antigen is first injected into an animal (typically a rabbit) where the immune system recognizes the antigen as foreign and produces antibodies against it. After a few weeks, great quantities of antibodies can be extracted from the blood of the animal.

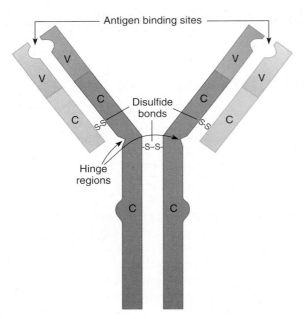

Figure 38.1. Structure of an antibody molecule. A typical antibody molecule is composed of four protein chains, two identical heavy chains and two identical light chains. Each chain has a constant (C) region, which differs little from one antibody to the next, and a variable (V) region, which differs greatly between antibodies. Chemical linkages called disulfide bonds help the four chains to assemble into the shape of a "Y." Foreign antigens bind to the tips of the Y, where the variable regions of a light chain and a heavy chain come together.

Because the binding of antibody to antigen is not visible without the use of an electron microscope, antibodies are often linked to dyes, enzymes, or radioactive isotopes so the endpoint of a reaction can be more easily determined. One of the most commonly used tests of this type is the **enzyme-linked immunosorbent assay,** commonly known by its acronym, **ELISA** (the same test is sometimes referred to as an enzyme immunoassay, or EIA). In this type of test, a color change announces the binding of antibody to antigen. Two types of ELISA tests are commonly seen. In a **capture ELISA,** antibodies are used to detect the presence of a specific antigen, as when measles virus is detected in the blood. The second type of test, known as an **indirect ELISA,** detects the presence of antibodies in the blood stream, which occurs when someone has been exposed to a pathogen, such as the hepatitis C virus. An indirect ELISA is useful in determining the amount of antibody in a patient's blood, information that can be used to differentiate between a new and chronic infection. New infections typically have a high level of antibodies as the immune system works to defend the body against the pathogen; older, chronic infections are generally marked by a far lower antibody level. Additionally, an indirect ELISA is more reliable when the levels of a virus in the serum are unknown. Both types of ELISAs are summarized in Figure 38.2.

In this exercise, simulated blood serum will be screened for the presence of antibodies to hepatitis C using an indirect

ELISA. You will also run two control reactions to provide a visual example of both positive and negative results. All the samples will be run in triplicate.

PRE-LAB QUESTIONS

1. *Define the following terms:*

Antigen

Antibody

Enzyme

Negative control

Positive control

2. *Serological tests can be used to detect HIV infection long before symptoms appear. Why is this important?*

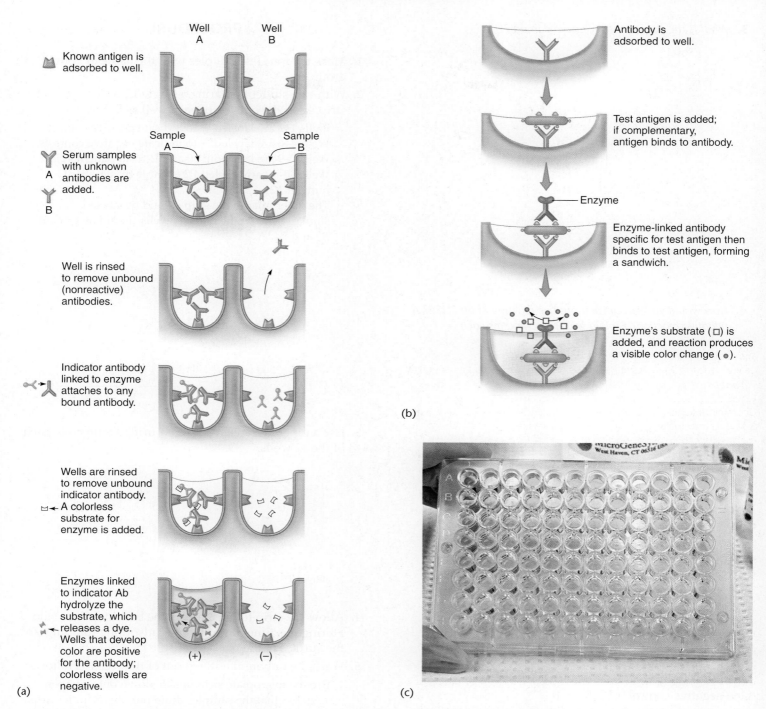

Known antigen is adsorbed to well.

Well A   Well B

Serum samples with unknown antibodies are added.

Sample A   Sample B

Well is rinsed to remove unbound (nonreactive) antibodies.

Indicator antibody linked to enzyme attaches to any bound antibody.

Wells are rinsed to remove unbound indicator antibody. A colorless substrate for enzyme is added.

Enzymes linked to indicator Ab hydrolyze the substrate, which releases a dye. Wells that develop color are positive for the antibody; colorless wells are negative.

(+)   (−)

(a)

Antibody is adsorbed to well.

Test antigen is added; if complementary, antigen binds to antibody.

Enzyme

Enzyme-linked antibody specific for test antigen then binds to test antigen, forming a sandwich.

Enzyme's substrate (□) is added, and reaction produces a visible color change (●).

(b)

(c)

Figure 38.2 (a) An indirect ELISA tests for the presence of a specific antibody in the blood, such as one specific to hepatitis C (typically referred to as anti-hepatitis C). Presence of the antibody indicates that the patient has been exposed to, and most likely infected by, the pathogen in question. (b) A capture ELISA tests for the presence of an antigen by using antibodies known to bind to a specific cell or virus. Both hantavirus and measles virus are detected using this type of test. (c) A microtiter ELISA plate with 96 wells for HIV antibodies. Colored wells indicate a positive reaction.

3. *What is the purpose of the enzyme in an ELISA?*

4. *How would you recognize a positive reaction in an ELISA? A negative reaction? Explain your reasoning for each.*

MATERIALS

Each group (four students) should obtain:

Four serum test samples (0.25 ml each)
One positive control tube (0.5 ml)
One negative control tube (0.5 ml)
Secondary antibody (1.5 ml)
Enzyme substrate (1.5 ml)
Two 12-well microplate strips
Disposable plastic transfer pipet
Beaker containing 80 ml wash buffer
Several paper towels
Marking pen
50 μl fixed volume pipettor or 20–200 μl adjustable volume
    pipettor and tips (~20 per group)

PROCEDURE

1. Mark the four test samples with the name of your group.
2. With two students sharing a single 12-well microplate strip, label the outside of each well as follows:

    Wells 1–3 are labeled + to indicate positive controls.
    Wells 4–6 are labeled − to indicate negative controls.
    Wells 7–9 are labeled P1 (sample from patient #1).
    Wells 10–12 are labeled P2 (sample from patient number #2).
    The other microplate strip should be labeled similarly except the final six wells should be labeled P3 and P4.

3. Use a pipet to transfer 50 μl of purified antigen to each of the 12 wells.

Purified
antigen

4. Allow the strip to incubate on the benchtop for 5 min. During this time, the antigen is adsorbing (binding) to the surface of the wells.
5. Wash the unbound antigen out of the wells as follows:

    Tip the microplate strip upside down onto a paper towel so that the samples drain out, then tap the strip a few times upside down on the paper towel. Discard the paper towel.
    Use a transfer pipet filled with wash buffer to fill each well with wash buffer, being careful not to allow the wash buffer to overflow the well. Retain the pipet as it will be used for all the wash steps.
    Tip the microplate strip upside down onto a few paper towels so that the wash buffer drains out, then tap the strip a few times upside down on the paper towels. Discard the paper towels.

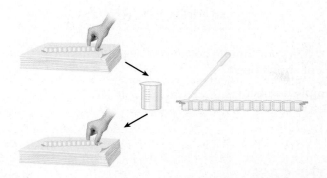

6. Repeat the wash step for a total of two washes.

7. Use a fresh pipet tip to transfer 50 µl of the positive control into each of the three "+" wells.

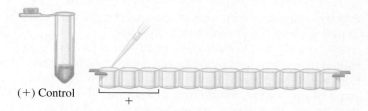

(+) Control

8. Use a fresh pipet tip to transfer 50 µl of the negative control into each of the three "−" wells.

(−) Control

9. Use a fresh pipet tip to transfer 50 µl of the serum test sample into each of the three wells for that patient. Recall that each student should have their own test serum tube.

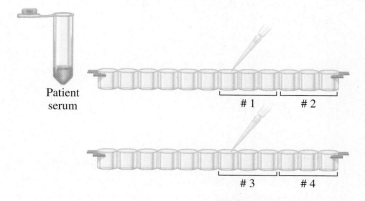

Patient serum

1   # 2

3   # 4

10. Allow the strip to incubate on the benchtop for 5 min. During this time, serum antibodies are binding to the antigen in the wells.

11. Wash the unbound primary antibody out of the well exactly as was done in step 5. Repeat this step for a total of two washes.

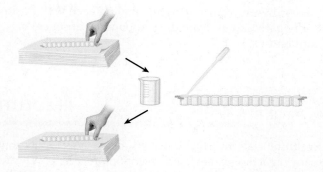

12. Use a fresh pipet tip to transfer 50 µl of secondary antibody into all 12 wells of the microplate strip.

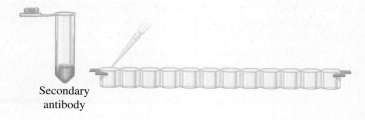

Secondary antibody

13. Allow the strip to incubate on the benchtop for 5 min. During this time, the secondary antibody is binding to the primary antibody.

14. Wash the unbound secondary antibody out of the well exactly as was done in step 5. Repeat this step twice, for a total of three washes.

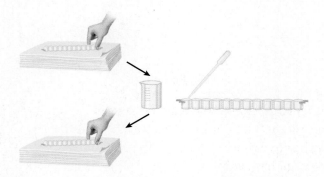

15. Use a fresh pipet tip to transfer 50 µl of enzyme substrate into all 12 wells of the microplate strip.

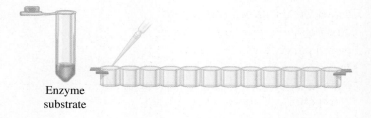

Enzyme substrate

16. Allow the strip to incubate on the benchtop for 5 min. During this time, the enzyme horseradish peroxidase is catalyzing a reaction that results in the formation of a colored product.

17. Record your results at right, labeling each well positive or negative to indicate the presence (or absence) of antibodies.

Was your sample positive or negative for the presence of antibodies?

Sample number _____

Presence of antibodies _____

RESOLUTION OF THE CASE

Health officials investigating the hemodialysis facility found many breaches in infection prevention procedures. The facility served between 70 and 100 patients per day at 30 dialysis stations. Relatively little time was allotted between patients, and the equipment at each station was usually cleaned with a single, bleach-soaked gauze pad. After disinfection, visible blood remained on the equipment, dialysis chairs, and floor. Staff did not don gloves or wash hands between patients. Since a significant (if small) proportion of the patients using the facility had come in with hepatitis C and there were obvious breakdowns in infection control,

investigators felt the new infections most likely came from the patients and not from the staff. This was confirmed when the RNA of viruses from some of the nine newly infected patients was sequenced and matched with sequences from the already-infected patients.

Eventually the hemodialysis facility in New York was shut down permanently. While not all of the new infections could be traced to poor infection control practices, some of them were. In addition to losing its certificate to operate, the facility paid a civil settlement to the state of New York.

EXERCISE 38 REVIEW QUESTIONS

1. What is a false positive reaction? A false negative? Provide at least two ways that each could arise in an ELISA.

2. Why is washing the wells, especially after the addition of the secondary antibody, so critical to the success of an ELISA.

3. Explain how an indirect ELISA test would be affected if the patient whose blood was being tested was severely immunocompromised and produced very few antibodies. How could the test be changed to account for this circumstance?

NOTES

Identification of Bacterial Unknowns

CASE SYNOPSIS

Respiratory Disease Strikes Legionnaires Convention—Philadelphia, 1976

On August 4, 1976, epidemiologists from the United States Public Health Service and the Pennsylvania State Health Department began investigating an outbreak of a mysterious illness believed responsible for a number of deaths among people who attended an American Legion convention in Philadelphia from July 21–24. According to Dr. Leonard Bachman, the Pennsylvania Health Secretary, there was no immediate need for concern, with a spokesman for the secretary telling the Associated Press, "It doesn't seem to be related to food poisoning. They have flu symptoms. It looks like flu." The following day, Dr. David J. Sencer, director of the Centers for Disease Control (CDC) testified before a Senate committee that while the mystery disease had not yet been identified, his scientists had eliminated plague and several other exotic ailments from consideration. He also indicated that the disease, which by this time had taken the lives of 23 persons in Pennsylvania, was diminishing rapidly and did not appear to be contagious.

While investigating the American Legion outbreak, federal and Pennsylvania health officials took note of two other similar outbreaks of respiratory illness. In one, 3 members of the Independent Order of Odd Fellows died and 16 became ill with respiratory symptoms after 1,500 of them met in Philadelphia from September 14–20, 1974. The Odd Fellows stayed at the same hotels the Legionnaires did during their convention. In the second case, 94 people developed pneumonia during July and August of 1965 at St. Elizabeth's Hospital in Washington D.C. The death rate in that outbreak was 17 percent, the same as in the American Legion case. The cause of the St. Elizabeth's outbreak was never discovered.

As the summer of 1976 gave way to fall, many people were concerned that the mystery of the so-called legionnaires' disease, which by this time had killed 27 persons and sickened 128, remained unexplained. One of the most thorough epidemiological investigations in history had eliminated a number of possible factors, but investigators were no closer to finding the cause of the disease than they had been when the first person became ill 3 months earlier. An unfortunate postscript was added to the story when, on November 10, after a long decline in revenues brought about by the public's hesitancy to stay in the hotel so closely linked to a deadly outbreak of disease, the 72-year-old Bellevue Stratford hotel closed.

On January 19, 1977, Dr. David Sencer, director of the newly reorganized CDC in Atlanta, announced that the scientists at the CDC had discovered the organism responsible for the outbreak in Philadelphia. The same bacterium also appears to have been the cause of the earlier outbreak of fatal pneumonia among patients at St. Elizabeth's Hospital in Washington D.C. in 1965. "The present findings provide very strong evidence that the two epidemics were caused by the bacterium," said a report on the discovery released by the CDC. In their haste to announce their findings, scientists at the CDC left unanswered the question of whether the bacterium responsible for the outbreak was something entirely new to the scientific community or some obscure organism that had previously been identified but not linked to pneumonia in humans.

According to the CDC, identification of the bacterium is the first step in determining its source and the manner in which it was transmitted to humans in the Philadelphia and Washington D.C. outbreaks, details that so far remain unknown.

Resolution of the Case appears on pages 326–327

INTRODUCTION

It is ironic when you consider that science education, with microbiology being no exception, is generally built around memorizing and working with facts that are already known and theories established long ago, yet science itself is focused on uncovering unknown information and explaining unsolved mysteries. The period between 1850 and 1950 is often referred to as the "Golden Age of Microbiology" because the etiologic agents of so many diseases were discovered. Even here in the twenty-first century, with the causes of most infectious diseases well known, the last 35 years have seen the "discovery" of Legionnaires' disease, HIV/AIDS, SARS, and bird flu. In each of these cases, microbiologists, using skills similar to those you've acquired over the past few weeks or months, have isolated and characterized the microbe responsible.

Beginning now, you will be asked to do what a trained microbiologist does on a daily basis: isolate, characterize, and identify unknown bacteria. The identification of unknowns in the microbiology lab is as old as the teaching of the subject itself. Just as old is the worry and stress that often goes along with the assignment. But, by relying on the skills you've developed in the laboratory and taking things one step at a time, identification of your unknowns will seem more like a mystery to be solved and less like a punishment to be endured.

The key to a successful identification is to carefully assemble a list of your unknowns "vital statistics." For instance, correctly identifying your unknown as a Gram-positive organism eliminates all Gram-negative organisms from consideration; knowing that your unknown is a coccus eliminates all rods, etc. By employing a well-thought-out series of biochemical and physiological tests along with macroscopic and microscopic observations of your bacteria, all but one bacterial species will be eliminated. These tests are the heart of any identification and can be broken into three general categories:

Morphological Characteristics   These include the physical characteristics of your unknown on both macroscopic (colony) and microscopic (cellular) levels. Colony size, shape, and pigment along with cell shape and size, motility, and reaction to the Gram, acid fast, and endospore stains may all play a role in the identification.

Cultural Characteristics   Generally referring to the growth exhibited by your organism, cultural characteristics include growth on both liquid and solid media as well as oxygen usage.

Biochemical and Physiological Assays   These include a wide array of tests, each of which evaluates a bacterium for the presence of a specific enzymatic reaction. In most cases, inoculation of differential media followed by incubation is used to detect the presence or absence of a specific enzyme. In some other tests, the presence of the enzyme is detected chemically, without incubation, leading to an immediate result.

Meticulous technique is required as you complete each test as is careful recordkeeping and intelligent interpretation of each test. Do not rely on your memory. Record the results of each test as it is completed along with any difficulties in interpretation or equivocal results. If a medium is supposed to look yellow or red but instead looks orange, write it down. If a culture seems motile at 25°C but nonmotile at 37°C, write it down. These details may seem insignificant at the time but may be immeasurably helpful as you get closer to the identification of your unknown.

To help you in identifying your unknown bacterium, several flowcharts are provided in this exercise. Beginning with the Gram reaction and shape of your unknown bacteria, the charts will help you in selecting proper tests and media. The results of each test will point you toward an ever decreasing number of possible choices, ending when all but a single microbe has been eliminated from consideration. Congratulations, you've just identified your unknown!

Almost.

Although it would seem that you're done at this point, a few tasks still remain. First, a little background information on your presumptive species will go a long way toward ensuring the accuracy of your identification. The closest that microbiology comes to an "official" classification system is *Bergey's Manual of Systematic Bacteriology*. This multivolume set of reference books is a compendium of information on thousands of different bacterial species. Unfortunately, two aspects of Bergey's limit its usefulness. First, its sheer size can make it difficult to navigate under the best of circumstances, and second, a major reorganization of the manual is currently underway. The newer versions of the manual utilize an entirely different form of hierarchal organization, and many laboratories have a combination of old and new manuals in stock. The easiest way to sidestep this problem is to not worry about where your particular unknown falls in the great scheme of things but rather to simply use the index in Bergey's to find the section dealing with your microbe.

Here is where science gives way to art and experience. Once you've found your unknown, learn something about your bacteria: do your test results seem to agree with what you've read? If not, should any tests be repeated? Should new tests be run? Note that a positive result for any test only means that 90% or more of strains test positive, not every last one. The same can be said for negative results, with 90% of strains testing negative. Keeping this in mind, you can see that your results really point toward a probability of a species, and your analysis of these results and their limitations is an equally important consideration. For instance, more than 90% of dogs have four legs but your experience would still allow you to recognize a three-legged dog as a member of the species. If the vast majority of your tests agree with what is expected for your species, you should be confident in your identification.

What if you're wrong? All is not lost. Accurate identification of a bacterial unknown is difficult under the best of circumstance and it is not unusual to slightly (or even greatly) miss the mark. If your identification is not correct, several possibilities exist:

- Eliminate the most obvious errors by rechecking slides and test results and running once more through the identification charts. You would not be the first person to mistake a short rod for a cocci or to go left on a flowchart instead of right.

- Your organism could be providing you with a false-negative reaction or, less likely, a false-positive reaction on a test. Rerunning the test and/or checking the flowchart to see where you would end up if the reaction was reversed could help.

- Your culture may be contaminated. If your culture has become contaminated at some point, none of your tests should be relied on. A Gram stain can be used to determine if you are still working with a pure culture.

- You may have been given the wrong unknown! Although professional microbiologists like to think of themselves as infallible (as well as good looking and possessed of a great sense of humor), it would not be the first time that an unknown had been incorrectly labeled. Check with your instructor if you feel this may truly be the case.

PRE-LAB QUESTIONS

1. *Define the following terms:*

   Pure culture

   Etiologic agent

   Morphological characteristic

2. *What is meant by a* false-positive result *and a* false-negative result?

PERIOD ONE

MATERIALS

Each student should obtain:

Two plates of trypticase soy agar

One numbered unknown culture (a mixed culture containing two unknown species)

Marking pen

PROCEDURE

1. Label your plates with your name and lab time.
2. Streak your unknown culture onto each plate.
3. Incubate one plate at 25°C and one plate at 37°C for 24–48 h.

QUESTIONS—PERIOD ONE

1. If one of your unknowns grew best at 25°C and one grew best at 37°C, how would you recognize this?

2. Provide several ways in which two unknowns could be distinguished from one another on a streak plate.

PERIOD TWO

MATERIALS

Each student should obtain:
Two tubes of trypticase soy broth
Four tubes of trypticase soy agar

PROCEDURE

1. Retrieve your trypticase soy agar plates from the incubator.
2. Check each plate for isolated colonies that have different morphologies. Look at colony size, shape, and pigment color to make your decision.
3. Label one tube of broth with your name and lab time as well as the number of your first unknown. Label the agar slants in the same manner. Also, label one slant "working stock" and one slant "reserve stock." Do the same for the other three tubes of media, being sure to label them with the number of your other unknown.
4. If you have achieved isolation of your two bacterial species, subculture an isolated colony to one trypticase soy broth and two trypticase soy agar tubes (Figure 39.1). Do the same for the other unknown.

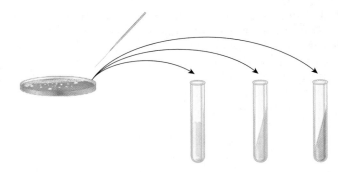

Figure 39.1 Subculturing of an isolated bacterial colony. Using a needle, subculture a single, well-isolated colony to a tube of trypticase soy broth and two trypticase agar slants. Inoculate the slants with a single streak along the agar from bottom to top.

5. For each unknown species, examine the streak plates to determine if it grows better at 25°C or 37°C. The optimum growth of a bacterial species is generally marked by more rapid growth. Remember that your two unknowns may have different temperature optima.
6. Incubate each culture at its optimum temperature for 24–48 h.

QUESTIONS—PERIOD TWO

1. Identify each of the following bacteria using the test results in the accompanying table.

Characteristic	Unknown 1	Unknown 2	Unknown 3
Gram reaction	+	+	−
Shape	Rod	Cocci	Rod
Endospore stain	+	−	−
Growth in fluid thioglycollate media	Aerobe	Facultative anaerobe	Facultative anaerobe
Mannitol	−	+	+
Nitrate	−	−	+
Other information	Voges-Proskauer + Catalase +	Catalase + Maltose + Coagulase − Nitrate + Glucose + Nonmotile Urease +	Facultative anaerobe No capsule
Identity of unknown			

2. For each of the unknowns above, were any tests unnecessary to correctly identify the unknown?

3. Why would you be more likely to encounter a false-positive result rather than a false-negative result?

PERIOD THREE

PROCEDURE

1. Retrieve your tubes from the incubator, and check each for the presence of growth.

2. The broth tube and agar slant marked "working stock" should be used for further testing. The slant marked "reserve stock" should be refrigerated and used only if problems such as contamination are noted with the working stock.

3. All cultures should be stored in the refrigerator from this point forward.

4. Collect information about your unknown by running any necessary tests. Use the flowcharts in this exercise to determine the identification of your unknown species. Note that these charts represent just over 200 species, a small fraction of all bacteria.

5. The first test to be performed is the Gram Stain. Based on the results of this test, begin with the flowcharts in Section I (Gram-Positive Bacilli), Section II (Gram-Positive Cocci), or Section III (Gram-Negative Bacilli and Cocci).

SECTION I: GRAM-POSITIVE BACILLI

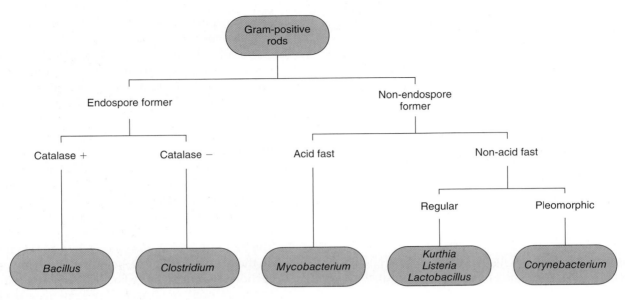

Figure 39.2 Separation outline for Gram-positive rods.

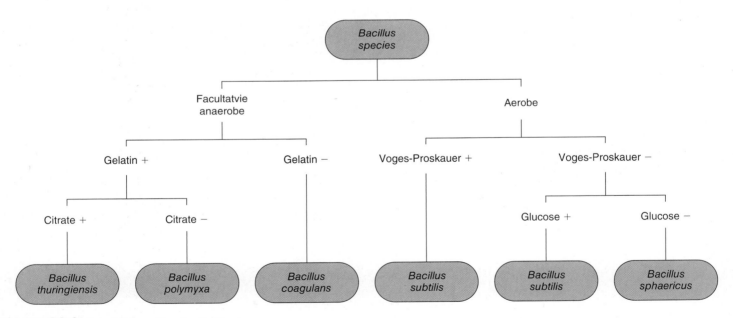

Figure 39.3 Separation outline for *Bacillus*.

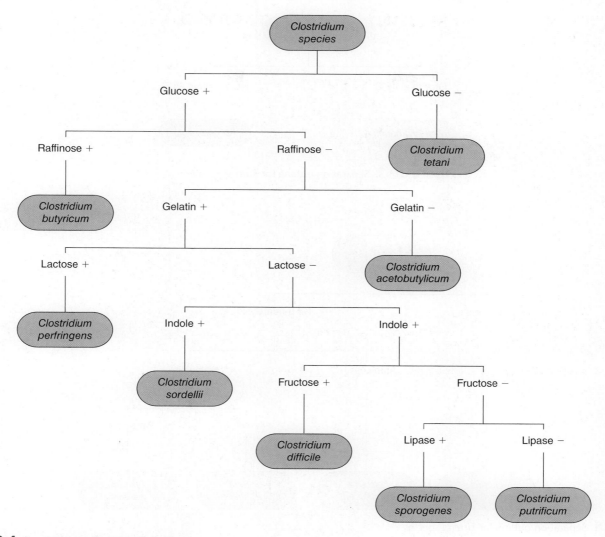

Figure 39.4 Separation outline for *Clostridium*.

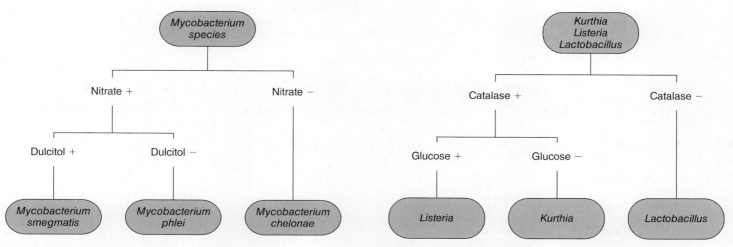

Figure 39.5 Separation outline for *Mycobacterium*.

Figure 39.6 Separation outline for *Kurthia, Listeria, and Lactobacillus* genera.

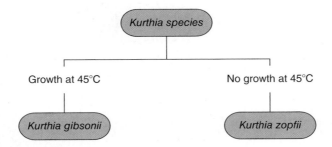

Figure 39.7 Separation outline for *Kurthia* species.

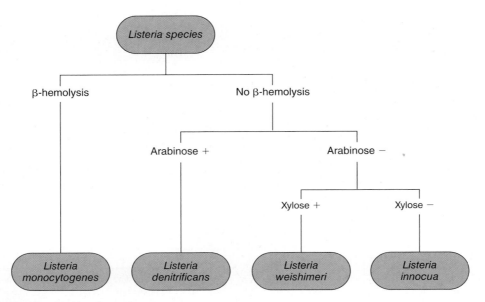

Figure 39.8 Separation outline for *Listeria* species.

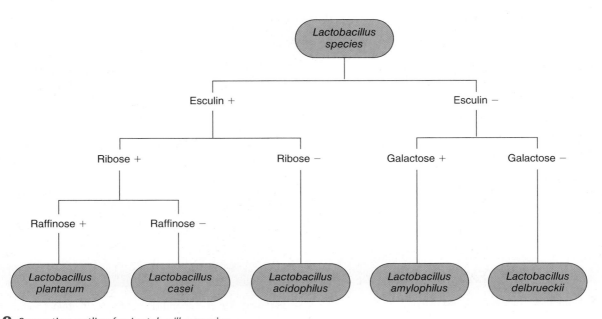

Figure 39.9 Separation outline for *Lactobacillus* species.

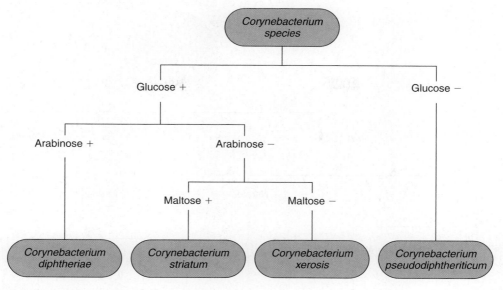

Figure 39.10 Separation outline for *Corynebacterium* species.

SECTION II: GRAM-POSITIVE COCCI

If your Gram stain indicates a Gram-positive coccus, begin the identification process with figure 39.11. Once you've identified the genus of your unknown, continue to figures 39.12 through 39.17 (as appropriate) to complete the identification process.

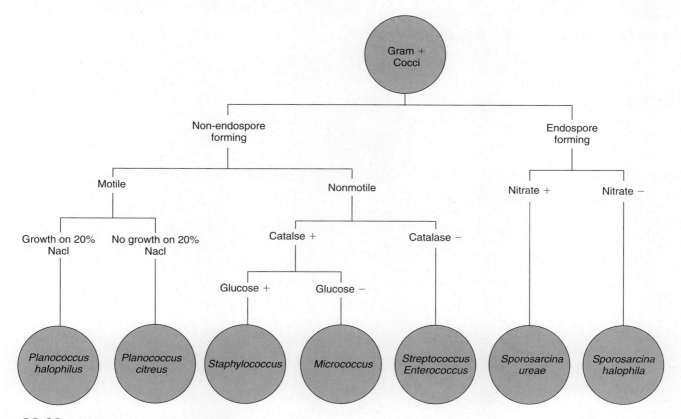

Figure 39.11 Separation outline for Gram-positive cocci.

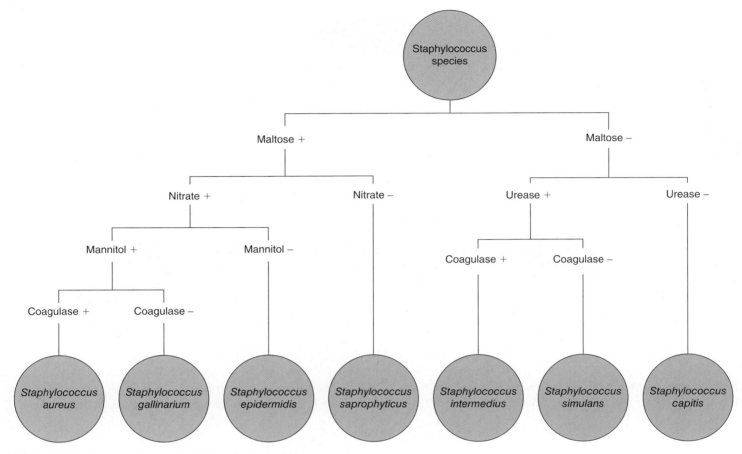

Figure 39.12 Separation outline for *Staphylococcus* species.

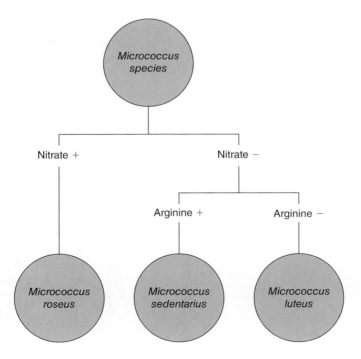

Figure 39.13 Separation outline for *Micrococcus* species.

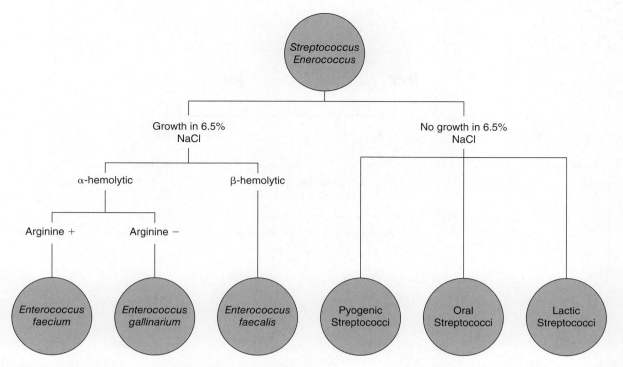

Figure 39.14   Separation outline for *Streptococcus* and *Enterococcus*.

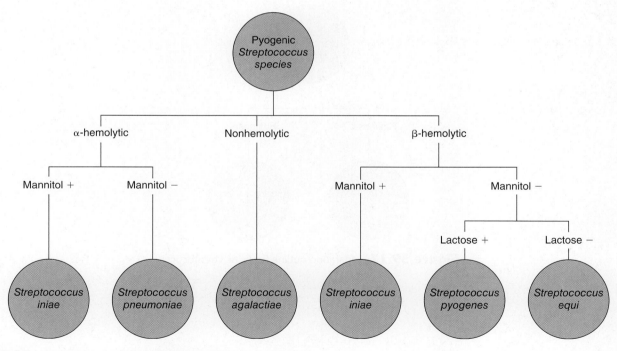

Figure 39.15   Separation outline for pyogenic streptococci.

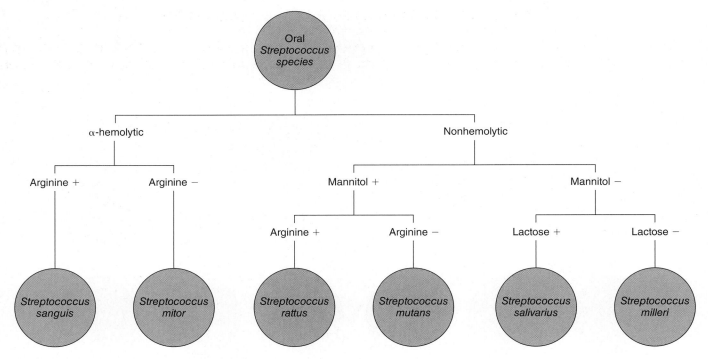

Figure 39.16 Separation outline for oral streptococci.

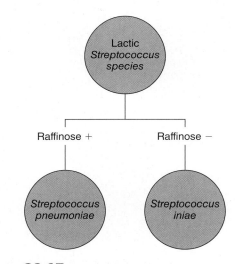

Figure 39.17 Separation outline for lactic streptococci.

SECTION III: GRAM-NEGATIVE BACILLI AND COCCI

If your Gram stain indicates a Gram-negative organism, begin the identification process with figure 39.18. Once you've identified the genus of your unknown, continue to figures 39.19 through 39.32 (as appropriate) to complete the identification process.

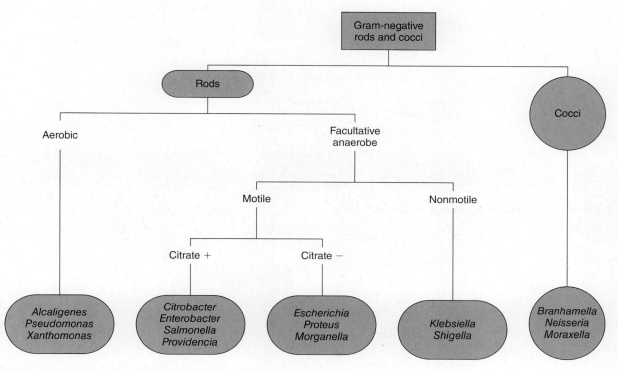

Figure 39.18 Separation outline for Gram-negative rods and cocci.

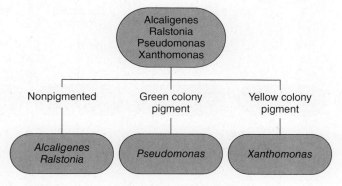

Figure 39.19 Separation outline for *Alcaligenes, Ralstonia, Pseudomonas,* and *Xanthomonas.*

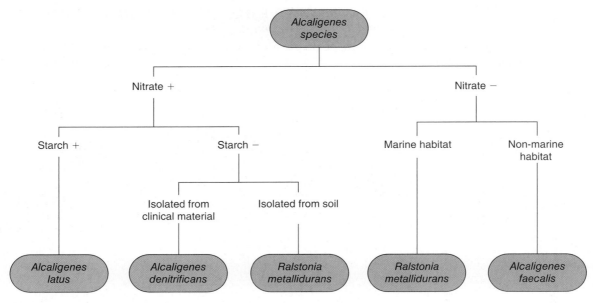

Figure 39.20 Separation outline for *Alcaligenes* species.

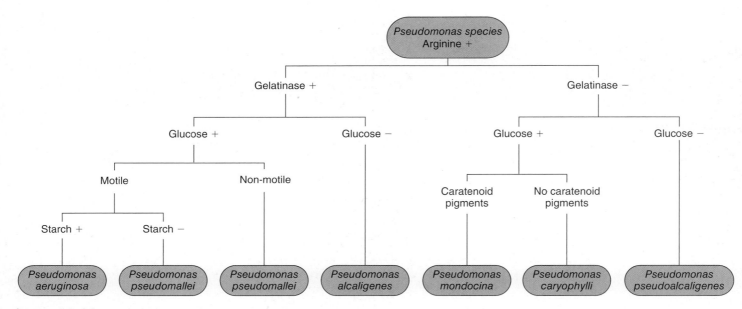

Figure 39.21 Separation outline for arginine positive *Pseudomonas* species.

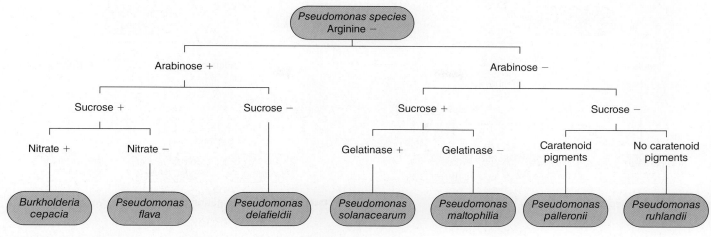

Figure 39.22 Separation outline for arginine negative *Pseudomonas* species.

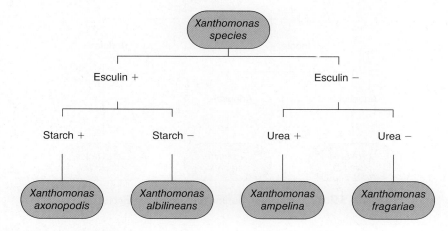

Figure 39.23 Separation outline for *Xanthomonas* species.

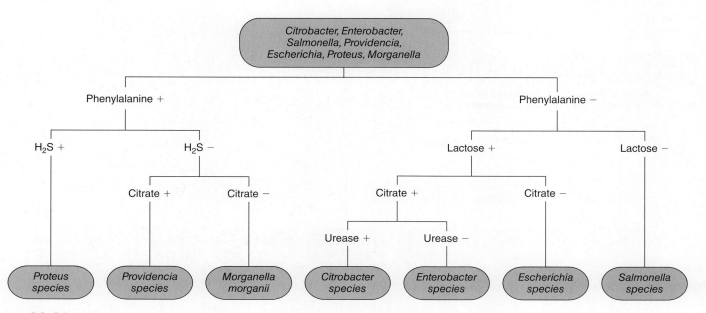

Figure 39.24 Separation outline for *Citrobacter, Enterobacter, Salmonella, Providencia, Escherichia, Proteus,* and *Morganella.*

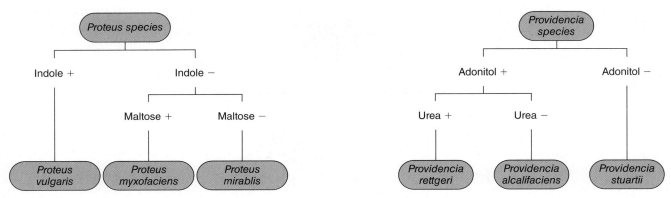

Figure 39.25 Separation outline for *Proteus* species.

Figure 39.26 Separation outline for *Providencia* species.

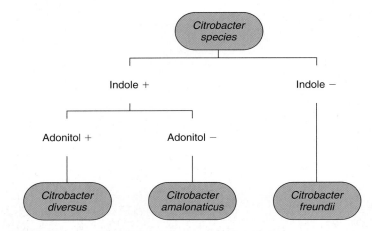

Figure 39.27 Separation outline for *Citrobacter* species.

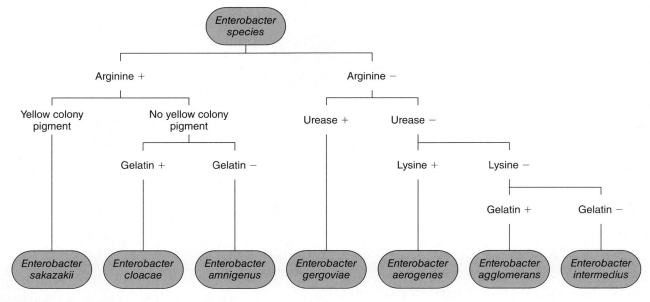

Figure 39.28 Separation outline for *Enterobacter* species.

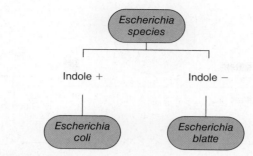

Figure 39.29 Separation outline for *Escherichia* species.

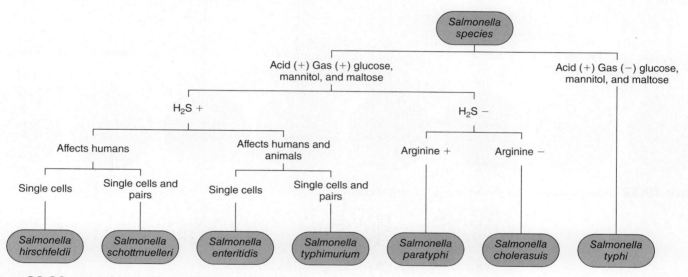

Figure 39.30 Separation outline for *Salmonella* species.*

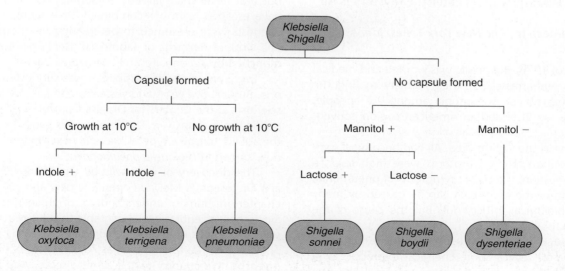

Figure 39.31 Separation outline for *Klebsiella* and *Shigella* species.

*Isolates of *Salmonella* are often named according to their serotype, a convention not commonly seen in other bacteria. The names used here correspond to the names of each species as catalogued in the American Type Culture Collection.

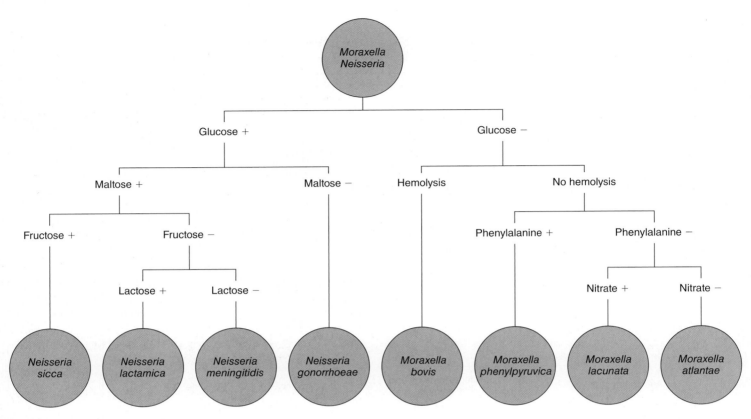

Figure 39.32 Separation outline for *Moraxella* and *Neisseria* species.

RESOLUTION OF THE CASE

Legionnaires' Disease: 5 Years Later the Mystery Is All but Gone
Harold M. Schmeck Jr., *The New York Times, Jan. 19, 1982*

In the summer of 1976, the public was worried and medical scientists were embarrassed by their inability to find the cause of an outbreak of pneumonia among 182 people, most of whom had attended an American Legion convention in Philadelphia. . . .

Twenty-nine of the victims died. All had become ill soon after the convention. Their symptoms were high fever, a cough and congestion, the classic symptoms of pneumonia, an inflammation of the lungs that can be caused by many agents, some biological, others not. But the cause of the outbreak was a total mystery.

Could it be an unusual virus? Or psittacosis, known popularly as parrot fever? Some chemical byproduct of a fluorocarbon refrigerant gas? A rare metallic poison? The swine flu epidemic for which Americans were preparing,

but that never materialized? Something bizarre that might have escaped from a recombinant DNA laboratory where scientists were presumed to be creating new forms of life? The work of terrorists or saboteurs? Contaminated pretzels munched by legionnaires in Philadelphia bars?

Every one of these possibilities was suggested by someone. But just five years ago yesterday, on Jan. 18, 1977, two scientists at the Centers for Disease Control in Atlanta made public the real answer: The disease was caused by none of the above, but by a type of bacteria never before identified, now known as *Legionella pneumophila*. . . .

The discovery of *Legionella* by Dr. Charles C. Shepard and Dr. Joseph E. McDade opened a new and still-unfolding chapter in human understanding of infectious diseases. Since then, scientists have been able to solve retrospectively several previous mysterious epidemics caused by *Legionella*. They have attributed as many as 70,000 cases of pneumonia a year in this country to the same disease. And they have learned to diagnose *Legionella* cases in time to treat them

(continued)

RESOLUTION OF THE CASE *(concluded)*

properly, and to identify, so far, no fewer than seven species of this disease-causing bacterium.

By their discovery, Dr. Shepard and Dr. McDade reminded scientists forcibly that there are still "new" and unexpected causes of infectious disease to be found on the earth, that these may dwell in unexpected places and work in strange and novel ways. Their finding opened the doors for a broad range of studies that are continuing to this day.

In short order, the original discovery also solved several longbaffling mysteries, including that of an epidemic at St. Elizabeth's Hospital in Washington in 1965, a mysterious disease outbreak in a building in Pontiac, Mich., in 1969 and another that marred an Odd Fellows convention in Philadelphia in 1974.

As the investigation continued, it became clear that the bacteria were neither particularly rare nor anything like new. The pneumonia they cause has been encountered all over the world. It makes up a substantial portion of pneumonia for which a cause has been difficult to define. . . .

The current estimate at the Centers for Disease Control [puts] the total number of people infected in the United States at 50,000 to 70,000 a year, according to Dr. Albert Balows, head of the bacteriology division at the centers.

The infections do not seem to be transmissible directly from person to person. Instead, the bacteria lurk in the water of air conditioning systems, cooling towers and sometimes even in shower heads. Their human victims inhale bacteria in the form of aerosols—fine mists and water droplets in air.

It is this link to air conditioning and other sources of standing water that probably accounts for the fact that many of the known outbreaks of the infection have occurred in hospitals, other medical institutions, or individual buildings.

The ability to find the causes in the recent cases has been important in a practical way, too, because infections with *Legionella pneumophila* usually respond well to erythromycin, an antibiotic that might not be used if the cause of the infection were unknown or misunderstood. . . .

The discovery of *Legionella pneumophila* at the Centers for Disease Control in January 1977, and all the related discoveries that have followed that achievement, show that the century-old science of medical bacteriology can still produce important surprises that are valuable to public health and to the practice of medicine.

REFERENCE

Bergey's Manual Trust. http://www.bergeys.org/outlines.html.

NOTES

Photographic Atlas for Laboratory Applications in Microbiology

Barry Chess
Pasadena City College

Connect
Learn
Succeed™

Contents

BACTERIAL COLONIAL MORPHOLOGY   A-31

BACTERIAL MICROSCOPIC MORPHOLOGY   A-37

FUNGI   A-45

PROTISTS   A-57

HELMINTHS   A-69

HEMATOLOGY AND SEROLOGY   A-79

Preface

A picture is worth a thousand words. And significantly more than a thousand when the words are filiform, xanthophyceae, and echinulate. As a student in the microbiology classroom, you are constantly being asked to evaluate things that you've never before seen, using a vocabulary that is both brand new and extraordinarily specific in most instances. Hardly seems fair. Think for just a moment about the number of *familiar* objects that are difficult to describe with words alone: a sunset, the knot in a shoelace, a Chihuahua with floppy ears. Now think about a photograph of each of these items.

A picture is worth a thousand words.

That is where this atlas comes in. Over the next 70 or so pages you'll find more than 300 color photos that **show** you what a capsule stain should look like, exactly how large a red blood cell is, and how to tell the eggs of a pinworm from those of a roundworm. Because that is what microbiologists—and microbiology students—do. What you won't find is a long (or long-winded) discussion of the biology of a schistosome or even a full explanation of why a bacterial cell may be Gram positive as opposed to Gram negative. That discussion is left for your lab manual or textbook, which presumably has a thousand words for every picture.

This atlas is organized from the point of view of a scientist trying to identify a specimen in the laboratory. By illustrating the answers to very basic questions—"What does a mold look like?" "Is my endospore stain positive?" "Is that an oocyst?"—you'll be able to evaluate your work with the eye of someone who has seen these things before, someone who knows the look of a Chihuahua.

You may also notice that the photos of single organisms are spread throughout the manual. The reason for this again goes back to the manner in which you would examine a specimen in the lab. You wouldn't, in fact couldn't, examine a bacterial specimen both macroscopically and microscopically at the same time, so you won't find those images next to one another. What you will find in every magnified image, however, is an indication of the magnification used to acquire it. In most cases, knowing what a specimen looks like when viewed at a magnification of 1000x is generally more useful than knowing that it is 7 μm in length. Always keep magnification in mind as you examine your specimen; the difference in appearance between a pea and a watermelon is mostly a matter of size.

The atlas is divided into eight major sections as follows:

- Staining techniques
- Cultural and biochemical tests
- Bacterial colonial morphology
- Bacterial microscopic morphology
- Fungi (both macroscopic and microscopic images)
- Protists
- Helminths
- Hematology and serology

Within each section, of course, microbes are further organized. For example, protozoans are classified by motility: amoeba, flagellates, ciliates, and apicomplexans. When you come across a flagellated protozoan in your studies, you will know right where to go.

One last piece of advice, before I hit my thousand words, when using this atlas: feel free to browse. Note the similarities, and the differences, between different groups of organisms. Get to know the range of tests that can be used to identify a microbe; understand the manner in which a cyst differs from an endospore and differs from an egg. Recognize the look of an acid-fast stain and know when it is useful. Take your time in the microbiology lab to interpret those things you see—it will only help in the end.

Enough of the words. On to the pictures.

Staining Techniques

At the top of the list of techniques used to identify a bacterium sit a variety of staining techniques. They range from simple and negative stains, which provide the morphology and arrangement of your specimen, to more complex differential stains that furnish some information concerning the specimen's biology.

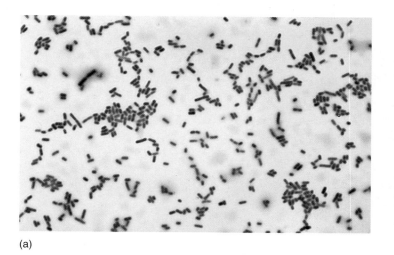

(a)

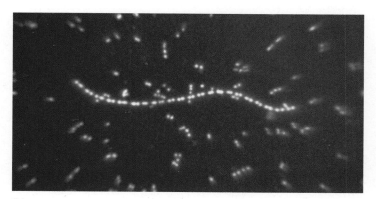

(b)

Figure 1.1 **Positive versus negative staining.** (a) In positive staining, the bacterial cell, which has an overall negative charge, is stained directly using dyes that are positively charged, revealing dark cells against a clear background (1000x). (b) In negative staining, a negatively charged dye, which will be repelled by bacterial cells, is used to stain the background, resulting in clear cells against a darkened field (1000x).

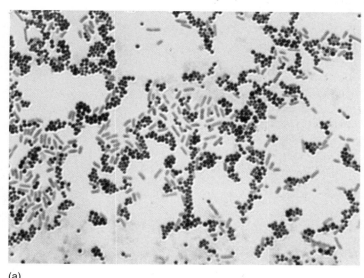

(a)

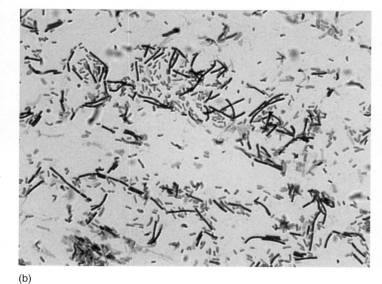

(b)

Figure 1.2 **Gram Staining.** (a) *Staphylococcus aureus* is a Gram-positive cocci while *Escherichia coli* is a Gram-negative rod (1000x). (b) *Bacillus megaterium* is a large Gram-positive rod while *Escherichia coli* is a smaller, Gram-negative rod (1000x).

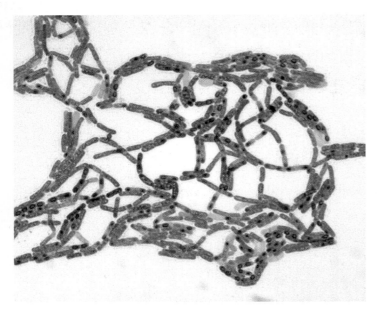

Figure 1.3 **Endospore stain of *Bacillus* sp. (1000x).** In this stain, malachite green is forced into the endospore using heat, while safranin is used to stain the vegetative portion of the cell. As nutrients are depleted from the bacterial culture, more and more cells exit the vegetative cycle and form endospores. Visible in this micrograph are pink vegetative cells, a few green endospores, and several cells in the process of sporulation, in which a green endospore is nestled within a pink vegetative sporangium.

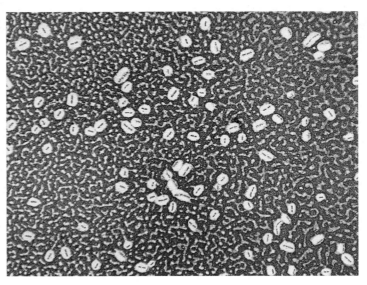

Figure 1.5 **Capsule stain of an unidentified bacillus (1200x).** The cell itself is stained with a basic dye like crystal violet while the background is obscured with an acidic stain like nigrosin. The clear space surrounding each cell is the bacterial capsule, which does not stain with either dye.

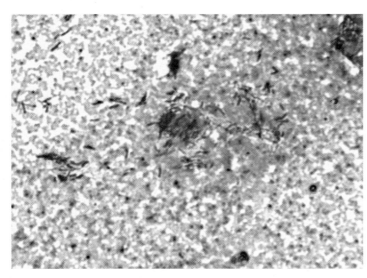

Figure 1.4 **Acid-fast Stain (1000x).** Acid-fast stain of *Mycobacterium smegmatis* (red, acid-fast rods) and *Staphylococcus aureus* (blue, non-acid-fast cocci).

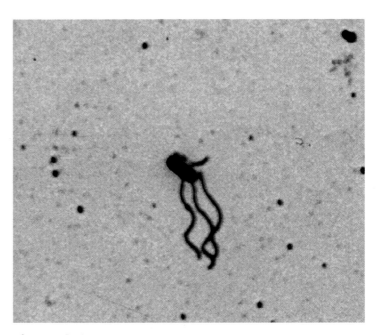

Figure 1.6 **Flagellar stain (320x).** Flagellar staining allows the visualization of bacterial flagella, which are normally too small to be seen using a light microscope and traditional staining techniques. In this case, *Clostridium tertium* was stained using a Leifson flagella stain.

Cultural and Biochemical Tests

Cultural and biochemical tests, of which there are a near infinite variety, provide information about the biology of an organism. Does it need oxygen to grow? Can it ferment lactose? Does it produce pigment at one temperature but not another? All are potential details that may prove useful as you narrow down the list of possible species. Two facts are important to keep in mind about all of these tests. First, for a test to be useful, some fraction of potential organisms must test positive and some negative. For example, testing the ability of humans to grow without oxygen would be fruitless as all human beings are obligate aerobes (who die when oxygen is lacking). By contrast, bacteria exist that grow in the presence of oxygen, the absence of oxygen, and all conditions in between. The second point to keep in mind is that a negative test result is not worse than a positive result. If we test a person for the presence of brown eyes and the result is negative, it does not mean that the individual's vision is less acute than a person whose eyes are blue, just that the pigment to darken the eyes is missing. As you move through this section, you'll notice the first few figures display isolation techniques. This is of paramount importance; in the absence of a pure culture, any work to identify a bacterium will prove fruitless.

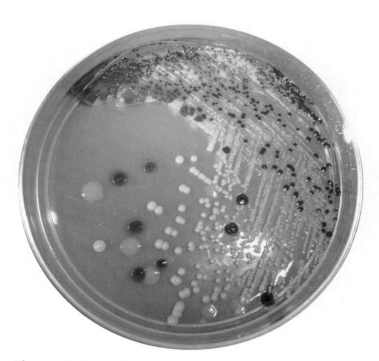

Figure 2.1 **A well-executed streak-plate after incubation.** Isolated colonies include *Serratia marcescens* (red), *Micrococcus luteus* (yellow), and *Escherichia coli* (off-white).

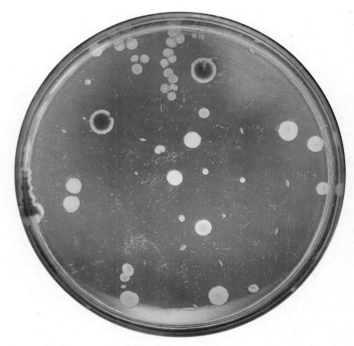

Figure 2.2 **Loop dilution of mixed bacterial culture.** A loop dilution is used to separate multiple species of bacterial by preparing serial dilutions of the sample in a liquefied agar medium. The dilutions are then poured into Petri dishes and allowed to solidify. After incubation, colonies of each bacterial species are found randomly spread across the top, as well as within, the media.

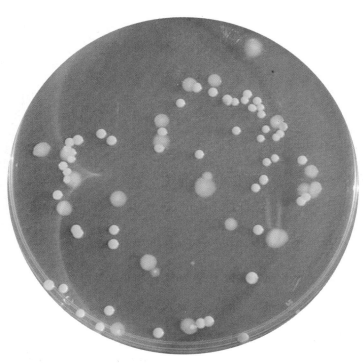

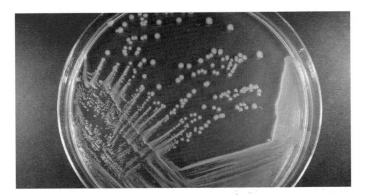

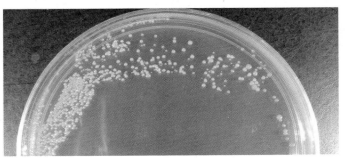

Figure 2.3 Spread plate. A spread plate may be used to separate multiple species of bacteria when very few cells are expected to be present in the sample. No attempt is made to reduce the number of cells, as is done in a streak plate or loop dilution.

Figure 2.4 Appearance of bacteria on various media. Colony color is often influenced by the type of media upon which a bacterial isolate is growing. Here, *Escherichia coli* is seen growing on (from top to bottom) nutrient agar, MacConkey agar, and EMB agar.

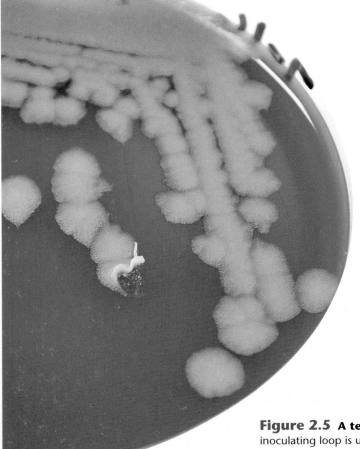

Figure 2.5 A tenacity test performed on *Bacillus anthracis*. In this test, an inoculating loop is used to lift a portion of a suspected *B. anthracis* colony from a plate. If the colony remains upright, the test is positive and is indicative of *B. anthracis*.

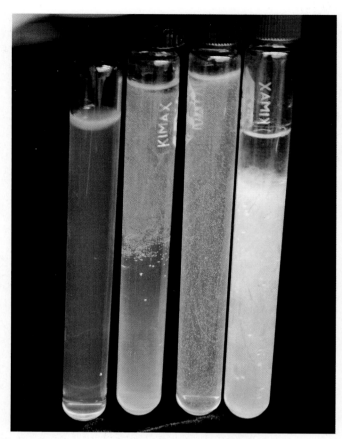

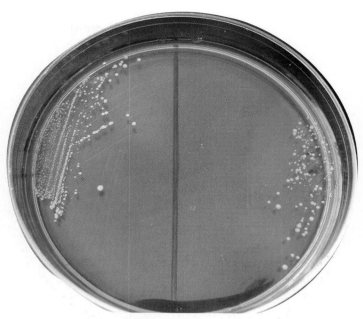

Figure 2.7 Mannitol salt agar. Mannitol salt agar is used to isolate *Staphylococcus* species from mixed cultures as well as identify those species able to ferment mannitol. *Staphylococcus aureus* (right) is able to both grow on the high salt medium and ferment mannitol. The production of acid from the fermentation lowers the pH and causes the medium to turn yellow. *Staphylococcus epidermidis* (left) can grow in the presence of high salt but does not ferment mannitol, as evidenced by the lack of any yellow coloration in the media.

Figure 2.6 Assessing oxygen requirements using fluid thioglycollate medium. Organisms display different oxygen requirements in fluid thioglycollate medium. The tube is prepared so that a top-to-bottom oxygen gradient exists within the tube, with high concentrations of O_2 at the top and no O_2 at the bottom. After incubation, the position of growth within the tube indicates the oxygen needs of the bacteria. From left to right: aerobic (*Pseudomonas aeruginosa*), facultative anaerobe (*Staphylococcus aureus*), facultative anaerobe (*Escherichia coli*), and obligate anaerobe (*Clostridium butyricum*).

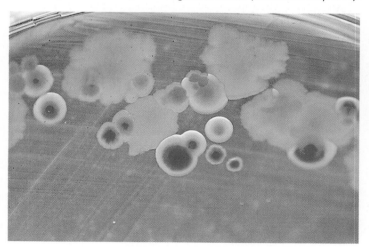

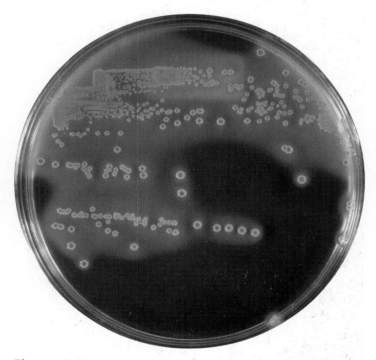

Figure 2.8 MacConkey agar. MacConkey agar is selective for Gram-negative bacteria and differential with regard to lactose fermentation. Lactose-fermenting species produce colonies with pink to red centers as a result of acid production while non-lactose fermenting species manufacture no acid and hence produce white or translucent colonies.

Figure 2.9 Desoxycholate agar. Desoxycholate agar contains lactose and the pH indicator neutral red. Bacteria that ferment lactose will produce acid that, in conjunction with neutral red, will lead to pink- to red-colored colonies. Here, *Escherichia coli*, a lactose fermenter, is shown on the medium.

Figure 2.10 Endo agar. All lactose-fermenting species produce pink to red colonies on endo agar while some, such as *E. coli,* produce colonies with a characteristic metallic sheen. In this plate, the grid pattern is used as an aid in counting the colonies.

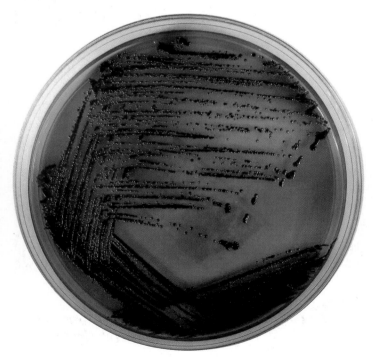

Figure 2.11 Eosin methylene blue agar with isolated colonies of *E. coli*. Eosin methylene blue (EMB) agar is a medium with both selective and differential properties. When *Escherichia coli* is grown on EMB agar, the colonies exhibit a distinct, metallic green color.

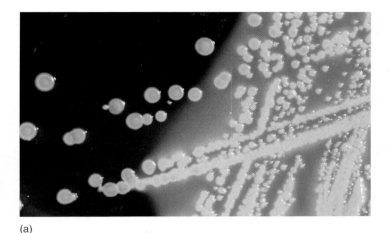

(a)

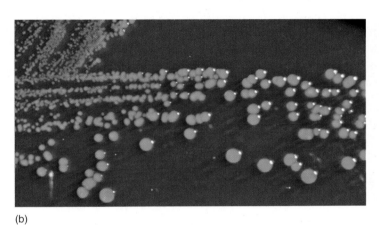

(b)

(c)

Figure 2.12 Hektoen enteric agar. (a) Bacteria such as *Escherichia coli,* which ferment any of the three carbohydrates in Hektoen enteric agar, will appear yellow to pink in color. (b) *Shigella boydii* bacteria, which do not ferment sugars in the media, produce blue-green colonies. (c) *Salmonella typhimurium* does not ferment any of the sugars in Hektoen enteric agar, leading to blue-green colonies. The black centers in the colonies are the result of hydrogen sulfide production, which can be used to distinguish *Salmonella* from *Shigella.*

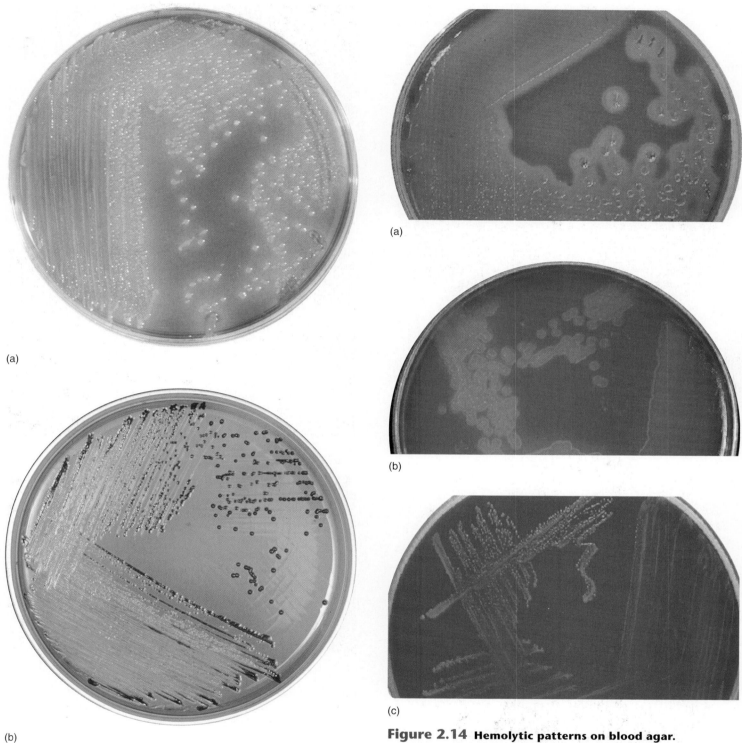

(a)

(b)

(a)

(b)

(c)

Figure 2.13   Xylose lysine desoxycholate (XLD) agar.
(a) Fermentation of xylose by *Serratia marcescens* produces yellow colonies. (b) *Salmonella* sp. produce red colonies because the decarboxylation of lysine neutralizes the acids produced from the fermentation of xylose. The black centers of the colonies are due to production of hydrogen sulfide, also characteristic of *Salmonella*.

Figure 2.14   Hemolytic patterns on blood agar.
(a) *Streptococcus pyogenes,* illustrating β-hemolysis. (b) *Streptococcus pneumoniae,* illustrating α-hemolysis. (c) *Streptococcus epidermidis,* displaying γ (no) hemolysis.

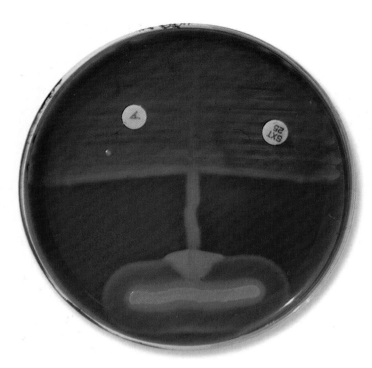

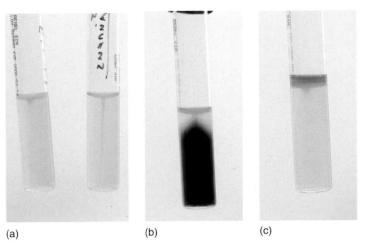

Figure 2.15 Positive CAMP reaction. The arrowhead-shaped zone of hemolysis is a result of the interaction between hemolysins produced by *Streptococcus agalactiae* (top) and *Staphylococcus aureus* (bottom). If the top organism was not *S. agalactiae,* a negative reaction would result in which hemolysis would be present around the growth of both bacteria but the arrowhead would be missing, as no synergistic interaction between the hemolysins would occur. The two discs in the photograph contain antibiotics that are also used in the classification of streptococci.

Figure 2.16 Interpretation of SIM (sulfur, indole, motility) media reactions. (a) Motility is detected as a cloudiness radiating out from a central inoculation line as shown in the tube on the left. The right-hand tube contains a nonmotile organism. (b) Reduction of sulfur is seen as a black precipitate. (c) The production of indole, from the breakdown of the amino acid tryptophan, can only be evaluated after the addition of Kovac's reagent. A deep red color at the top of the media indicates a positive reaction (seen here), while the original yellow color of the Kovac's reagent indicates a negative result.

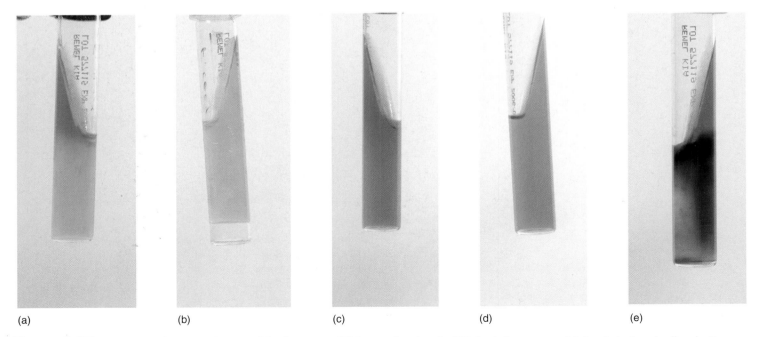

Figure 2.17 Fermentation reactions and hydrogen sulfide production in Kligler's iron agar. (a) A red slant and yellow butt indicate the fermentation of glucose. (b) A yellow slant and butt indicate that both glucose and lactose are fermented. In this case, gas was also produced, which lifted the media from the bottom of the tube. (c) No fermentation. The slight reddening of the slant is a result of aerobic peptone catabolization, resulting in alkaline end products. (d) An uninoculated control. (e) A black precipitate indicates the production of hydrogen sulfide, in addition to any fermentative reactions.

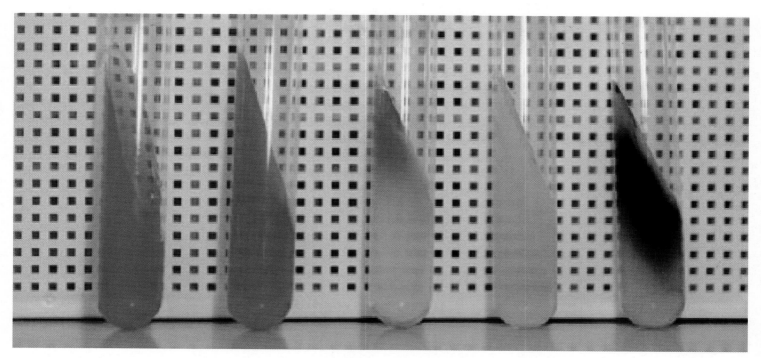

Figure 2.18 Fermentation reactions and hydrogen sulfide production in triple sugar iron agar. (a) An uninoculated control. (b) Growth without acid production. (c) Glucose fermentation with acid production. (d) Glucose along with lactose and/or sucrose fermentation. (e) A black precipitate indicates the production of hydrogen sulfide, in addition to any fermentative reactions.

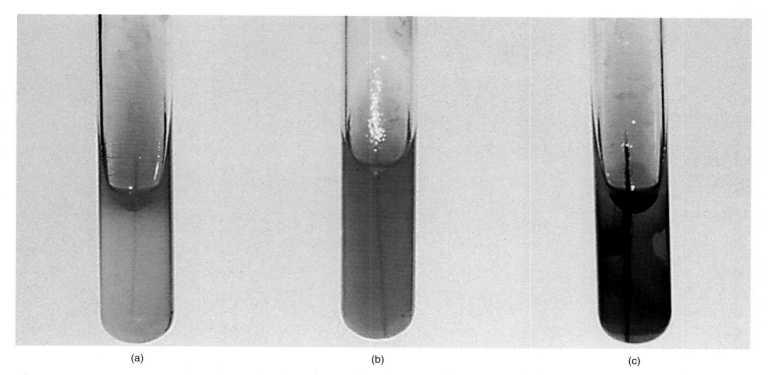

(a)   (b)   (c)

Figure 2.19 Interpretation of lysine iron agar reactions. (a) *Shigella* sp.: lysine deaminase positive (red slant), lysine decarboxylase negative (lack of purple color in the media), H_2S negative (lack of a black precipitate). (b) *Escherichia coli*: lysine deaminase negative (purple slant), lysine decarboxylase positive (purple butt), and H_2S negative (lack of black precipitate). (c) *Salmonella* sp.: lysine decarboxylase positive (purple butt), and H_2S positive (note the black precipitate surrounding the stab line).

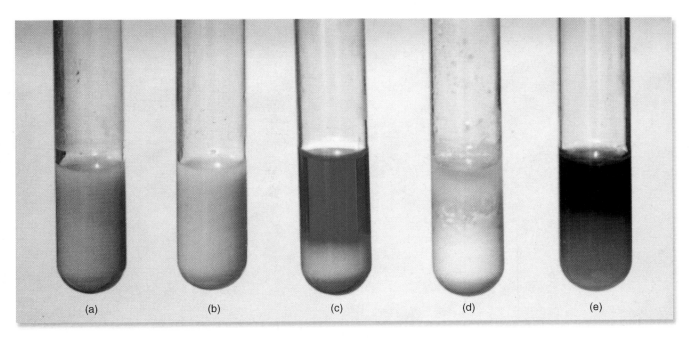

Figure 2.20 Litmus milk reactions. (a) Alkalinization. (b) Acidification due to fermentation of lactose. (c) Upper translucent portion is due to peptonization, and solid white portion in bottom is due to coagulation and litmus reduction; overall red color is interpreted as acidification. (d) Coagulation and litmus reduction in lower half, and peptonization (translucence) and acidification (pink) in top portion. (e) Peptonization is present in bottom of tube, and soluble pigments from *Pseudomonas* have obscured the litmus reaction.

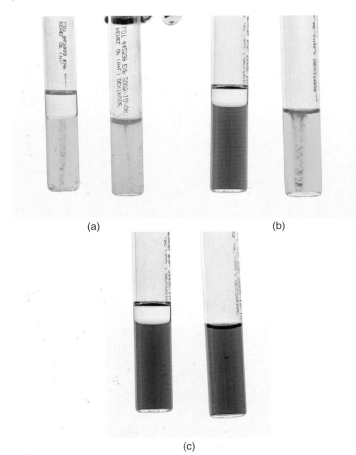

(a)   (b)

(c)

Figure 2.21 Oxidation-fermentation (O-F) media reactions.
(a) Oxidation or fermentation, or fermentation only. (b) Oxidation.
(c) No glucose metabolism.

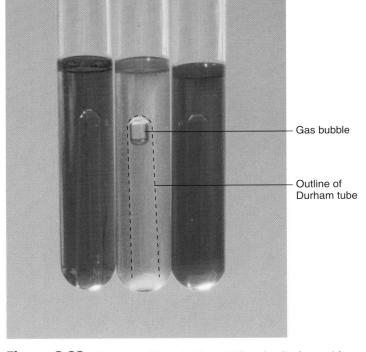

Gas bubble

Outline of Durham tube

Figure 2.22 Phenol red broth. The middle tube displays acid and gas production. The left-hand tube is an uninoculated control, and the right-hand tube shows growth (note turbidity near the bottom of the tube) but no fermentation.

Figure 2.23 Interpretation of purple broth. Fermentation of the carbohydrate within the media results in the formation of a yellow color, indicating the production of acid. Production of acid may also be accompanied by the production of gas, which is seen as a bubble within the Durham tube. In a negative reaction, the media will remain purple, and no gas bubble will form.

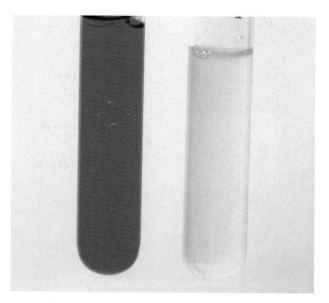

Figure 2.25 Voges Proskauer (VP) test. The production of 2,3-butanediol from the fermentation of glucose is the basis for the VP test. After the addition of Barritt's reagent A and B, the production of a red color is considered a positive reaction. The tube on the left (*Enterobacter aerogenes*) is positive while the tube on the right (*Escherichia coli*) is negative.

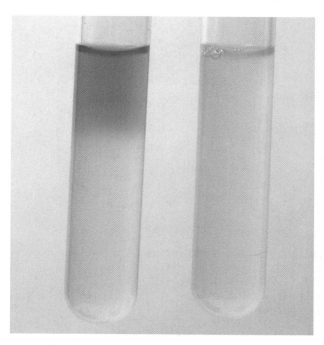

Figure 2.24 Methyl red (MR) test. Mixed acid fermentation, carried out by some species of bacteria, leads to the production of several different acids, lowering the pH of the media and causing a pink color to be seen after the addition of methyl red indicator. The tube on the left (*Escherichia coli*) is positive while the tube on the right (*Enterobacter aerogenous*) is negative.

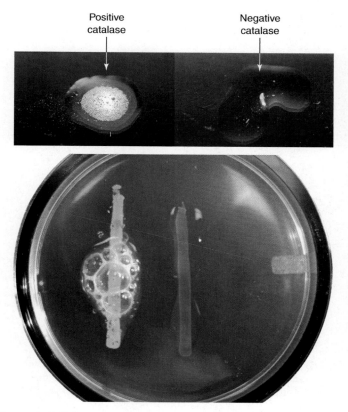

Figure 2.26 Interpretation of a catalase test. The production of bubbles of oxygen gas after the addition of hydrogen peroxide to bacteria is considered a positive result. A lack of bubbles indicates a negative result. (a) A slide catalase test. (b) Catalase test performed directly on a plate of media. In both cases *Staphylococcus aureus* (positive) is on the left, while *Streptococcus mutans* (negative) is on the right.

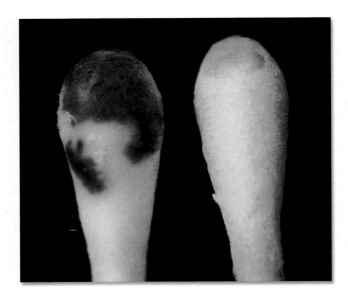

Figure 2.27 Interpretation of oxidase reaction. The left-hand swab shows a purple color due to the production of oxidase. The right-hand swab displays the yellow color typical of a negative reaction.

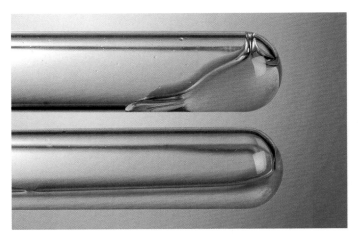

Figure 2.29 Coagulase test. A positive coagulase reaction is indicated by any degree of coagulation, or thickening, of blood plasma. Even tubes in which the media has thickened without becoming completely solid should be interpreted as a positive reaction. The top tube illustrates a negative result while the bottom displays a positive result. The coagulase test is used to distinguish *Staphylococcus aureus* (positive) from other species of *Staphylococcus* (negative).

(a)   (b)   (c)   (d)

Figure 2.28 Nitrate reduction test. Nitrate broth is used to evaluate whether or not a bacterium can reduce nitrate. (a) Initially after incubation, a turbid culture is seen, but no red color is visible. (b) If the culture reduces nitrate to nitrite, a red color will be seen immediately after the addition of nitrate reagents A and B. (c) If no red color is seen, powdered zinc (nitrate reagent C) is added to the tube and allowed to incubate for 5 min. No red color at this point means that the bacterium completely reduces nitrate, often to nitrogen gas. (d) If a red color is seen after the addition of zinc, it indicates that the bacterium does not reduce nitrate.

Figure 2.30 Citrate utilization. The utilization of citrate is indicated by the media changing color from green (negative) to blue (positive); growth may be visible, which also indicates a positive result. In this photo, the tube on the left is positive while the tube on the right is negative.

Figure 2.31 Malonate test. Malonate broth differentiates organisms that can use malonate as their sole source of carbon from those that cannot. A positive result—utilization of malonate—is indicated by a change in the color of the media from green to blue. An uninoculated tube appears on the left, *Enterobacter aerogenes* (positive) in the center, and *Escherichia coli* (negative) on the right.

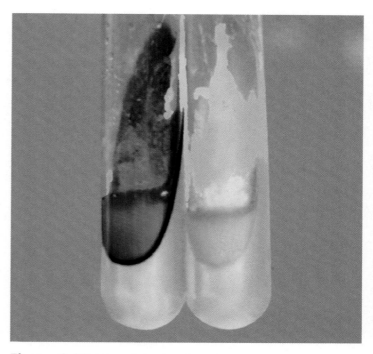

Figure 2.33 Phenylalanine deaminase test. Used to test for the presence of the enzyme phenylalanine deaminase in a bacterial species. After incubation, several drops of ferric chloride reagent are added to the slant, which is allowed to sit at room temperature for 5 min. In a positive reaction, ferric chloride reacts with phenylpyruvic acid to produce a deep green color. *Proteus mirablis* (left) gives a positive result while *Escherichia coli* (right) is negative.

Figure 2.32 Decarboxylation media. Decarboxylation media contains a specific amino acid (commonly lysine, ornithine, or arginine) as well as a pH indicator that turns the media purple if decarboxylation of the amino acid has taken place. A purple color, as seen in the right-hand tube, indicates a positive reaction while a yellow/orange color indicates that decarboxylation did not occur and is interpreted as a negative reaction.

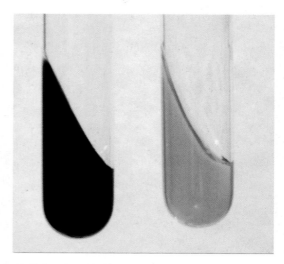

Figure 2.34 Bile Esculin Hydrolysis. Used to test the ability of a bacterial isolate to hydrolyze esculin in the presence of bile, the tube on the left exhibits bile esculin hydrolysis (a positive reaction), while the tube on the right displays a negative reaction.

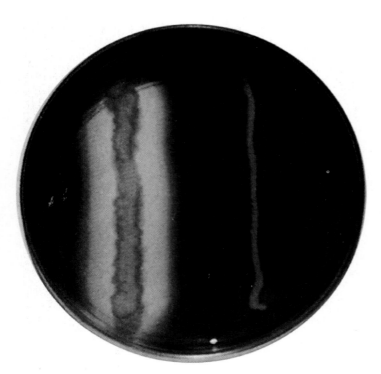

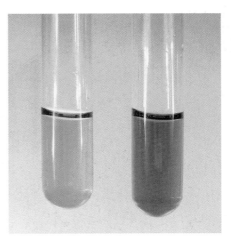

Figure 2.37 Evaluation of urea broth. Hydrolysis of urea—and the resultant production of ammonia—will turn the media pink (a positive result, right). If the organism being tested does not produce the enzyme urease, urea will not be hydrolyzed and the media will remain peach colored (a negative result, left).

Figure 2.35 Starch hydrolysis. The clear zone surrounding the streak on the left is indicative of starch hydrolysis by the enzyme amylase. The absence of a clear halo surrounding the streak on the right represents a negative reaction.

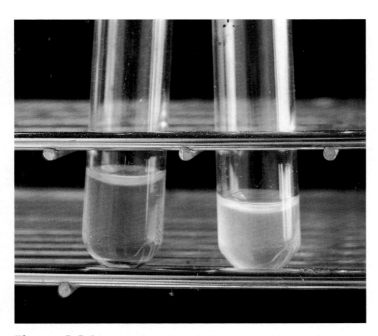

Figure 2.36 Interpretation of ONPG test. Used to determine if an organism produces the enzyme β-galactosidase. A yellow color throughout the test solution is interpreted as ONPG positive while a clear solution represents a negative result.

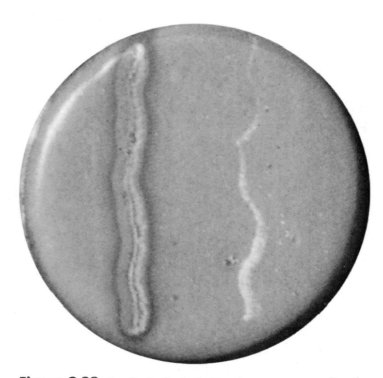

Figure 2.38 Casein hydrolysis. The clear zone surrounding the streak on the left is indicative of casein hydrolysis. The absence of a clear halo surrounding the streak on the right represents a negative reaction.

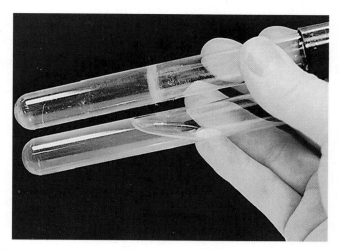

Figure 2.39 Nutrient gelatin hydrolysis. Media in the lower tube has liquefied as a result of the hydrolysis of gelatin by gelatinase. The upper tube displays a negative reaction.

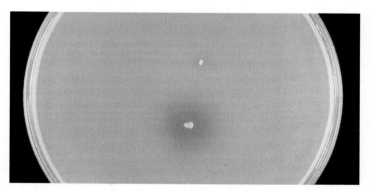

Figure 2.40 Interpretation of DNase media results. After incubation, the surface of the media is flooded with 1N HCl. A clear halo around the bacterial streak indicates the presence of DNase in the bacterium, a positive result. No halo is interpreted as a negative result.

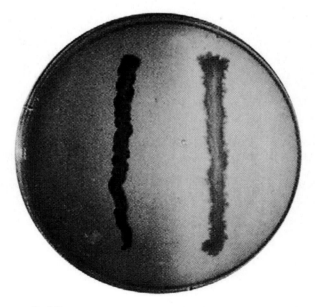

Figure 2.41 Lipid hydrolysis on spirit blue agar. Organism on the left is lipase positive, as indicated by the blue pigment. The organism on the right is lipase negative.

Figure 2.42 Antimicrobic sensitivity testing for identification. In this case, the antimicrobic novobiocin was used to aid in the identification of staphylococcal testing. *Staphylococcus aureus* (right) is sensitive to novobiocin while *Staphylococcus saprophyticus* (left) is resistant. For an organism to be considered sensitive to an antimicrobic, the zone of inhibition must be larger than a specific size, which differs for each antimicrobic used and, in some cases, for the bacteria being tested.

Figure 2.43 PYR DrySlide results. Used to aid in the identification of group A streptococci and group D enterococci, the PYR test identifies bacteria capable of enzymatically hydrolyzing L-pyrrolidonyl-β-naphthylamide. The upper left quadrant displays the pink color characteristic of a positive reaction while the lower right quadrant displays only the yellow color of the developing reagent, a negative reaction.

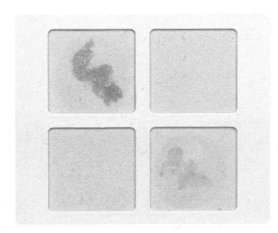

Figure 2.44 Nitrocefin DrySlide for the determination of β-lactamase production. Organisms that synthesize β-lactamase will produce a pink color (upper left) when allowed to react with nitrocefin, a type of cephalosporin. Nitrocefin remains yellow (lower right) in a negative reaction.

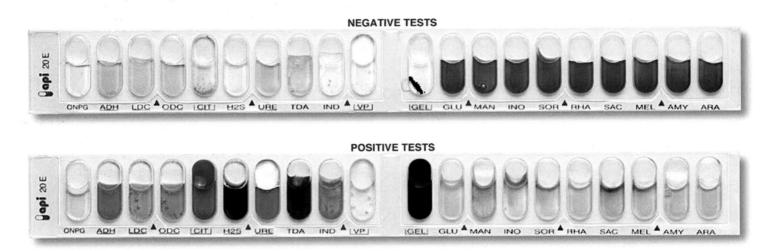

NEGATIVE TESTS

POSITIVE TESTS

Figure 2.45 API 20E system. The API 20 E system is a miniaturized collection of biochemical tests useful in the identification of Gram-negative bacteria. The top photo shows a negative reaction for each of the tests while the bottom photo displays a positive reaction.

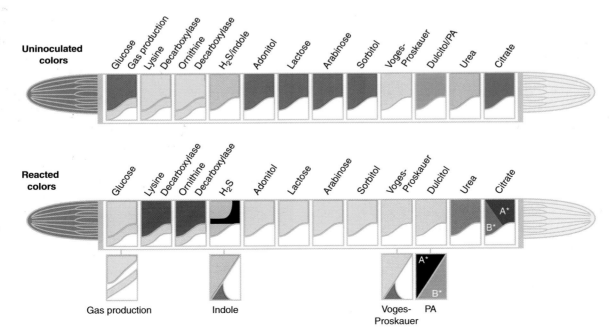

Figure 2.46 Enterotube II system. The Enterotube II system is a miniaturized collection of 12 biochemical tests useful in the identification of Gram-negative bacteria. The top illustration shows a negative reaction for each of the tests while the bottom illustration displays a positive reaction.

Figure 2.47 BCYE agar with isolated colonies of _Legionella pneumophila_. BCYE (Buffered Charcoal Yeast Extract) is used for the isolation and presumptive identification of _Legionella_ species. _Legionella pneumophila_ produces mucoid colonies which may range from colorless to blue-gray and which fluoresce yellow-green under long-wave UV light.

Figure 2.48 BG Sulfa agar. BG (Brilliant Green) Sulfa agar is a highly selective medium used for the isolation of _Salmonella_. Brilliant green and sodium pyridine inhibit Gram-positive bacteria and most Gram-negative bacteria other than _Salmonella_ species. _Salmonella sp._ are typically seen as pink to red opaque colonies surrounded by a deep red medium. Fermentation of lactose and/or glucose—an indication that the isolate is not _Salmonella_—will produce a yellow-green zone around the colony, due to the inclusion of the pH indicator phenol red. Left, uninoculated medium; right, _Salmonella typhimurium_.

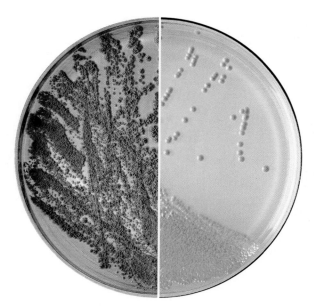

Figure 2.49 BiGGY agar. BiGGY (Bismuth Sulfite Glucose Glycine Yeast) agar is a medium with both selective and differential properties used in the isolation and identification of _Candida_ species. Bismuth sulfite suppresses bacterial growth while other ingredients allow the differentiation of various _Candida_ species. Additionally, _Candida tropicalis_ produces brown/black colonies with a metallic sheen while _Candida albicans_ produces similarly colored colonies without a metallic sheen. Left, _Candida krusei_; right _Candida albicans_.

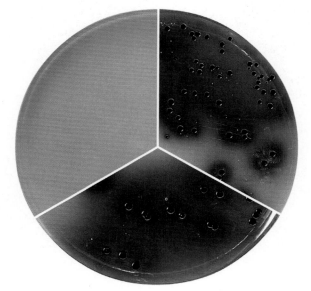

Figure 2.50 Bismuth Sulfite agar. Bismuth Sulfite agar is a highly selective medium used for isolating _Salmonella_ species, particularly _Salmonella typhi_, from food and clinical specimens. The medium contains bismuth sulfite and brilliant green to inhibit Gram-positive bacteria and coliforms. _Salmonella_ isolates precipitate iron when grown on the medium, giving the colonies a distinct brown to black color with a metallic sheen. Clockwise from upper left; uninoculated media, _Salmonella typhi_, _Salmonella typhimurium_.

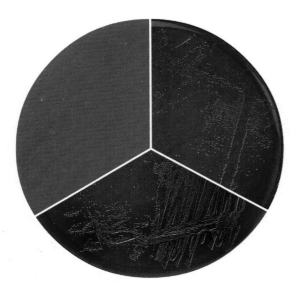

Figure 2.51 Bordet Gengou agar. A medium used to detect and isolate *Bordetella pertussis* from clinical specimens. Clockwise from upper left; uninoculated media, *Bordetella pertussis*, *Bordetella parapertussis*.

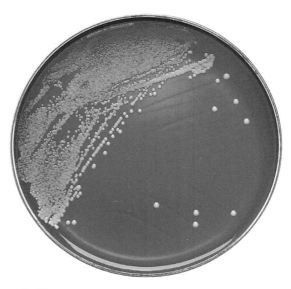

Figure 2.52 Brain Heart CC agar with isolated colonies of *Candida albicans*. Brain Heart CC agar is a selective medium used for used for the isolation of pathogenic fungi from specimens heavily contaminated with bacteria and saprophytic fungi. The inclusion of chloramphenicol inhibits the growth of bacteria while cyclohexamide inhibits most saprophytic molds.

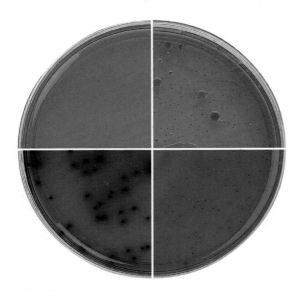

Figure 2.54 Brilliant Green Bile broth. Brilliant Green Bile broth is used for the detection of coliforms in food, dairy products, water and wastewater. Bile and brilliant green inhibit Gram-positive bacteria and most Gram-negative bacteria while coliform bacteria can grow in the presence of both inhibitors. Production of gas—as detected by a gas bubble—indicates fermentation of lactose. Left, uninoculated media; right *Escherichia coli*.

Figure 2.53 Brilliant Green Bile agar. A selective and differential media used to isolate, differentiate and enumerate coliform bacteria. Bile and brilliant green inhibit Gram-positive bacteria and most Gram-negative bacteria. Species which ferment lactose—like *E. coli*—produce deep red colonies (*E. coli* also precipitates bile, leading to colonies with dark centers) while *Salmonella* and other non-lactose fermenting species produce colorless to faint pink colonies. Clockwise from upper left; uninoculated media, *Enterobacter aerogenes*, *Salmonella enteritidis*, *Escherichia coli*.

Figure 2.55 CLED agar. CLED (Cystine-Lactose-Electrolyte –Deficient) agar is used for the isolation, enumeration, and presumptive identification of microorganisms from urine. Organisms which ferment lactose will change the color of the media from blue to yellow as a result of the inclusion of the pH indicator bromthymol blue, while a lack of electrolytes in the media reduces the swarming of *Proteus*. The appearance of commonly found urinary pathogens is as follows: *Escherichia coli*, opaque yellow colonies, with a slightly deeper yellow center; *Klebsiella*, yellow to whitish-blue colonies, extremely mucoid; *Proteus*, translucent blue colonies; *Pseudomonas aeruginosa*, green colonies with a matted surface and rough periphery; *Enterococci*, small yellow colonies about 0.5 mm in diameter; *Staphylococcus aureus*, deep yellow colonies, uniform in color, Coagulase negative staphylococci, pale yellow colonies, more opaque than *E. faecalis*. Clockwise from upper left; *Staphylococcus aureus*, *Proteus vulgaris*, *Escherichia coli*, *Enterococcus faecalis*.

Figure 2.56 Campylobacter agar with 10% sheep blood. Used to isolate and cultivate *Campylobacter jejuni* from human fecal specimen. Two types of colonies are generally seen with *C. jejuni*. The first is small, raised, grayish-brown, smooth and glistening, with an entire translucent edge, while the second is flat, mucoid, translucent, grayish, and has an irregular edge. Left, uninoculated medium; Right *Campylobacter jejuni*.

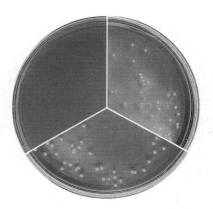

Figure 2.57 Candida BCG agar. Candida Bromcresol Green (BCG) agar is a medium with both selective and differential properties used for the detection and isolation of *Candida* in clinical specimens. *Candida* species produce convex to cone-shaped colonies surrounded by yellow media. Neomycin is added to the media to inhibit the growth of bacteria. Clockwise from upper left; uninoculated media, *Candida albicans*, *Candida tropicalis*.

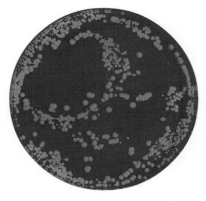

Figure 2.58 Candida Isolation agar with isolated colonies of *C. albicans*. A rich medium which supports the growth of many bacteria, yeasts and molds. Candida Isolation agar contains aniline blue, which is metabolized by *C. albicans* to produce a fluorescent end product, causing *C. albicans* to fluoresce when exposed to long-wave UV light.

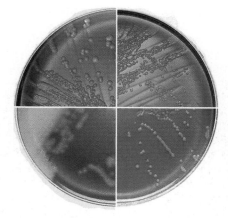

Figure 2.59 Chocolate II agar. Chocolate II agar is a rich media used for the isolation and cultivation of fastidious microorganisms, especially *Neisseria* and *Haemophilus* species, from clinical specimens. Clockwise from upper left; *Neisseria gonorrhoeae*, *Neisseria meningitidis*, *Haemophilus influenzae*, *Streptococcus pneumonia*.

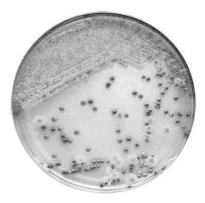

Figure 2.60 CHROMagar Candida. A medium selective for yeast and differential for the species *Candida albicans*, *Candida tropicalis* and *Candida krusei*, CHROMagar Candida allows for the detection of mixed yeast cultures in specimens. *C. albicans* colonies appear light to medium green, *C. tropicalis* colonies appear dark blue to metallic blue, and *C. krusei* colonies appear light rose with a whitish border. Other yeasts may develop either their natural color (cream) or appear light to dark mauve.

Figure 2.61 CHROMagar MRSA with isolated colonies of methicillin resistant *Staphylococcus aureus*. A medium which allows direct detection of methicillin resistant *Staphylococcus aureus* (MRSA). MRSA will grow in the presence of the antimicrobic cefoxitin and produce mauve colored colonies. Other bacteria may grow on the media but will produce colorless, white, blue, or blue-green colonies. Swab specimens from the anterior nares of patients or healthcare workers can be plated directly on the media to screen for MRSA colonization.

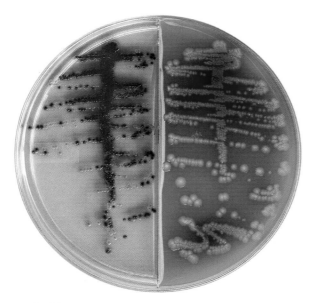

Figure 2.62 CHROMagar Orientation/TSA II agar with isolated colonies of *Escherichia coli* (pink) and *Enterococcus sp.* (blue). CHROMagar Orientation medium is a nonselective, differential medium for isolating and differentiating urinary tract pathogens. The appearance of commonly found urinary pathogens is as follows: *Escherichia coli*, dark rose to pink, transparent colonies; *Klebsiella-Enterobacter-Serratia* group, medium blue to dark blue colonies; *Proteus-Moganella-Providencia* group, pale to beige colonies surrounded by brown halos; *Enterococcus*, blue-green small colonies; *Streptococcus agalactiae*, blue green to light blue, pinpoint to small colonies; *Staphylococcus saprophyticus*, light pink to rose, small opaque colonies; other bacterial species and yeasts, natural (cream) pigmentation. The right side of the plate contains trypticase soy agar supplemented with 5% sheep's blood, a rich growth medium, without any selective or differential properties.

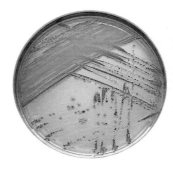

Figure 2.63 CHROMagar O157 with isolated colonies of Enterobacter cloacae (blue) and Escherichia coli O157 (pink). CHROMagar O157 is a selective medium used for the identification of *E. coli* O157, which produces pink colonies while other strains of *E. coli* produce blue colonies. Species of bacteria other than *E. coli* are either inhibited or produce blue colonies.

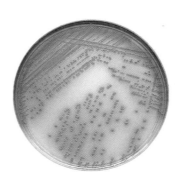

Figure 2.64 CHROMagar Salmonella with isolated colonies of *Salmonella typhimurium*. CHROMagar Salmonella is a selective medium used for the isolation of *Salmonella* species. Antimicrobics are added to the media to inhibit the growth of yeasts as well as most bacteria. *Salmonella* species appear light pink to pink while other bacteria appear colorless, blue-green, or are growth inhibited.

Figure 2.65 CHROMagar Staph aureus with isolated colonies of *Staphylococcus aureus*. CHROMagar Staph aureus is used for the isolation and identification of *Staphylococcus aureus*. Selective agents suppress the growth of Gram-negative organisms and partially suppress the growth of yeasts. *Staphylococcus aureus* will produce mauve to orange-mauve colored colonies while most Gram-negative organisms, if not inhibited, will produce white, blue, green or blue-green colonies.

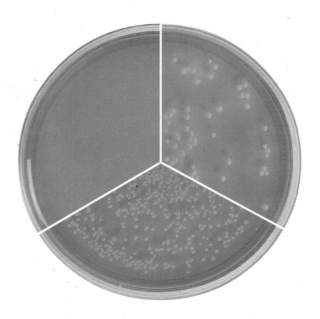

Figure 2.66 DNAse Test agar with Methyl Green. Used to detect deoxyribonuclease activity, the media contains deoxyribonucleic acid (DNA), along with the dye methyl green. Intact DNA binds to the dye giving the medium its green color. DNase activity is indicated by a clear halo surrounding a bacterial colony while the absence of a clear zone can be interpreted as a lack of DNase activity. Clockwise from upper left, uninoculated media, *Staphylococcus aureus* (DNase +), *Staphylococcus epidermidis* (DNase −).

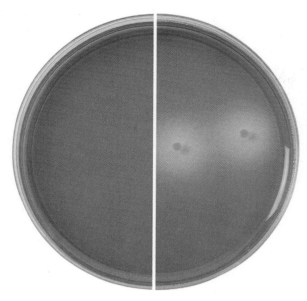

Figure 2.67 Dextrose Tryptone agar. Used to isolate bacteria responsible for "flat sour" spoilage of canned foods such as *Bacillus coagulans* and *Bacillus stearothermophilus*, both of which produce acid through the fermentation of dextrose. The media contains bromcresol purple which turns yellow as the pH of the media drops. Left, uninoculated media; right *Bacillus coagulans* (positive for dextrose fermentation).

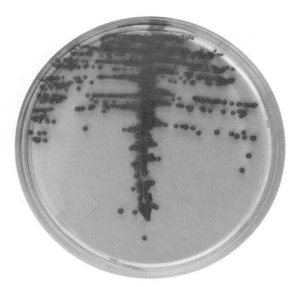

Figure 2.69 CHROMagar Orientation agar with isolated colonies of *Enterococcus sp.* CHROMagar Orientation medium is a nonselective, differential medium for isolating and differentiating urinary tract pathogens. The appearance of commonly found urinary pathogens is as follows: *Escherichia coli*, dark rose to pink, transparent colonies; *Klebsiella-Enterobacter-Serratia* group, medium blue to dark blue colonies; *Proteus-Moganella-Providencia* group, pale to beige colonies surrounded by brown halos; *Enterococcus*, blue-green small colonies; *Streptococcus agalactiae*, blue green to light blue, pinpoint to small colonies; *Staphylococcus saprophyticus*, light pink to rose, small opaque colonies; other bacterial species and yeasts, natural (cream) pigmentation.

Figure 2.68 EC Medium with MUG. Used to detect *Escherichia coli* in food, water, and milk. Bile salts are included in the media to inhibit the growth of Gram-positive bacteria. *E. coli* hydrolyzes MUG (4-methylumbelliferyl-β-D-glucuronide) to produce a fluorescent product that can be detected with long-wave UV light. *E. coli* typically both fluoresces and produces gas (as seen by a gas bubble trapped within the inverted Durham tube). A positive reaction for both fluorescence and gas production is seen here.

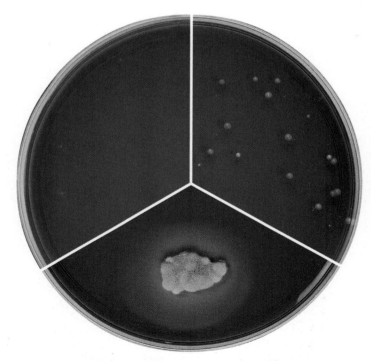

Figure 2.70 Littman Oxgall Agar. A selective medium used to isolate and cultivate fungi, Littman Oxgall Agar contains crystal violet and streptomycin to inhibit the growth of bacteria, and bile to restrict the spreading of fungus colonies, facilitating the isolation of pure cultures. Clockwise from upper left; uninoculated media, *Saccharomyces cerevisiae*, *Trichophyton mentagrophytes*.

Figure 2.71 MIO Medium. MIO (Motility-Indole-Ornithine) medium is used to aid in identification of members of the *Enterobacteriaceae* based on motility, indole production, and ornithine decarboxylase activity. Motility is indicated by a clouding of the media away from the central stab line. Ornithine decarboxylase activity is indicated by a purple to yellow-purple color; a yellow color should be interpreted as a negative reaction. After reading the first two results, 3 or 4 drops of Kovac's reagent—enough to cover the surface of the media—are added to each tube. The appearance of a red ring indicates the production of indole. From left; uninoculated media, *Enterobacter aerogenes* (motility +, indole −, ornithine +), *Escherichia coli* (all reactions positive).

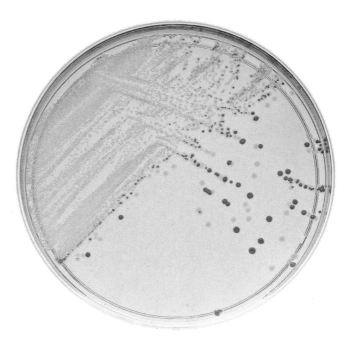

Figure 2.73 MacConkey II agar with Sorbitol. *Escherichia coli* (pink) and *Escherichia coli* O157:H7 (colorless). A selective and differential media, MacConkey II agar with sorbitol is used to culture and detect *E. coli* O157:H7. Bile salts and crystal violet in the medium inhibit the growth of Gram-positive organisms while sorbitol serves as the only fermentable sugar. Most strains of *E. coli* ferment sorbitol and produce pink colonies while *E. coli* O157:H7 does not ferment the sugar and consequently produces colorless colonies.

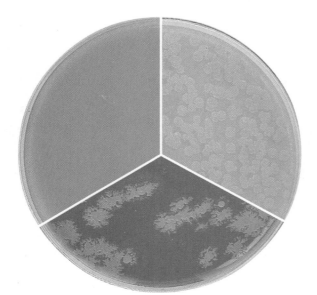

Figure 2.72 MYP agar. MYP (Mannitol-Egg Yolk-Polymyxin) agar is used to isolate and enumerate *Bacillus cereus*, a common cause of foodborne illness. Bacteria that produce lecithinase, which hydrolyzes lecithin in the media, form colonies surrounded by a zone of white precipitate. Bacteria that ferment mannitol produce acid products that result in yellow colonies on a yellow medium. *Bacillus cereus* is characteristically lecithinase positive and mannitol negative while other species differ with respect to these characteristics. Clockwise from upper left; uninoculated media, *Bacillus subtilis*, *Bacillus cereus*.

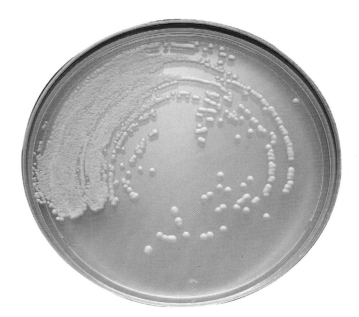

Figure 2.74 Mycosel agar with isolated colonies of *Candida albicans*. Used to isolate pathogenic fungi, Mycosel agar is a highly selective agar containing cyclohexamide and chloramphenicol to inhibit the growth of bacteria and other fungi. The plates are typically filled deep to prevent drying of the media during the 4 to 6 week incubation period commonly needed to culture fungi.

Figure 2.77 Presence-Absence broth. Used for detecting coliforms in treated water, Presence-Absence broth contains sodium laurel sulfate as a selective agent to inhibit most organisms besides coliforms, lactose as a fermentable carbohydrate, and bromcresol purple as a pH indicator. A yellow color indicates the fermentation of lactose, while gas production may be observed as a foaming of the medium when the bottle is gently shaken. Left; positive reaction (lactose fermentation and gas production); right, a negative reaction.

Figure 2.75 PC agar with isolated colonies of *Burkholderia cepacia*. A selective medium used for the isolation and detection of *Burkholderia cepacia*. The medium contains several selective agents which inhibit the growth of most organisms besides *B. cepacia*. On PC agar *B. cepacia* gives rise to grayish-white colonies surrounded by a pink-red zone in the surrounding medium.

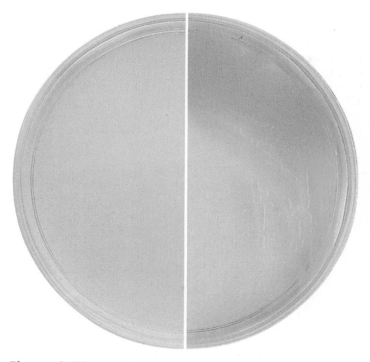

Figure 2.78 Pseudomonas Isolation agar. A selective and differential medium used for detecting *Pseudomonas aeruginosa*. Irgasan is an antimicrobic that selectively inhibits bacteria other than *Psuedomonas* species. *P. aeuruginosa* produces blue-green colonies with a the pigment diffusing into the media. Left; uninoculated media, right; *Pseudomonas aeruginosa*.

Figure 2.76 Phenol Red Agar. Used to differentiate pure cultures of bacteria based on carbohydrate fermentation, the medium is prepared with a single fermentable carbohydrate (lactose, mannitol etc.) and contains phenol red as a pH indicator. Fermentation of the carbohydrate is indicated by a change in the color of the media from red to yellow. Gas formation is indicated by the collection of gas in the bottom of the culture tube or by splitting of the agar. From left; uninoculated media, positive reaction with acid and gas, negative fermentation reaction with positive growth.

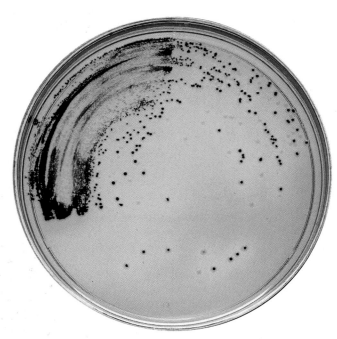

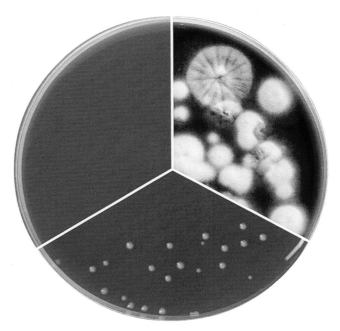

Figure 2.79 SS agar with isolated colonies of *Shigella flexneri* (clear) and *Salmonella typhimurium* (black centers). *Salmonella Shigella* (SS) agar is a moderately selective agar used primarily for the isolation of bacteria in the genus *Salmonella*. The inclusion of bile salts, brilliant green, sodium citrate and ferric citrate inhibit the growth of Gram-positive bacteria. Lactose, along with the pH indicator neutral red result in lactose fermenting bacteria producing pink colonies while lactose nonfermenters (including *Salmonella* and *Shigella*) produce colorless colonies. The production of hydrogen sulfide by *Salmonella* species results in these colonies developing a black center, while colonies of *Shigella* remain clear throughout.

Figure 2.80 Sabouraud Brain Heart Infusion agar. Used to isolate pathogenic fungi, Sabouraud Brain Heart Infusion agar is a selective medium containing cyclohexamide, chloramphenicol, and gentamicin to inhibit the growth of bacteria and saprophytic fungi. Clockwise from upper left; uninoculated media, *Aspergillus niger*, *Saccharomyces cerevisiae*.

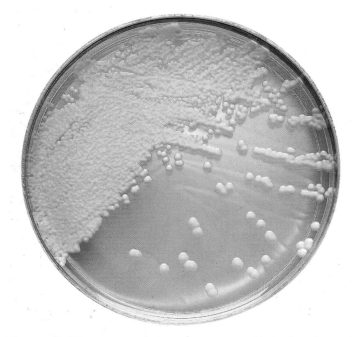

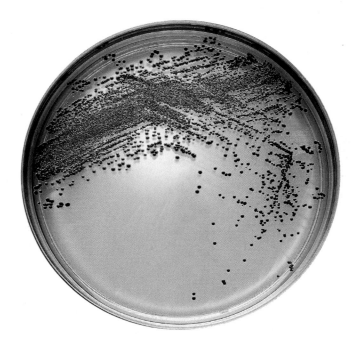

Figure 2.81 Sabouraud Dextrose agar with isolated colonies of *Candida albicans*. A general purpose medium for the cultivation of fungi, especially dermatophytes, the low pH (approximately 5.6) of the media favors the growth of fungi and slightly inhibits the growth of most bacteria.

Figure 2.82 Serum Tellurite agar with isolated colonies of *Corynebacterium diphtheriae*. A selective and differential medium used to isolate and presumptively identify *Corynebacterium* species. The medium contains potassium tellurite, which is inhibitory toward most microorganisms. *Corynebacterium* species however are resistant to tellurite, producing gray to black colonies in its presence.

Bacterial Colonial Morphology

Just as a bee hive looks different than a school of fish, the manner in which bacterial cells congregate into macroscopic aggregates can be telling. As you peruse this section, note the colonial characteristics of different bacteria. Colonies are commonly described as having a specific configuration (view from above) and margin (edge). Additionally, the elevation, or vertical growth of a colony—most easily judged by viewing the plate from the edge—is often a useful characteristic to note.

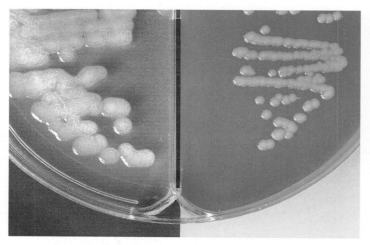

Figure 3.1 *Bacillus anthracis* **colonies.** On blood agar (right), colonies have a ground glass appearance characteristics of the species. When grown on bicarbonate agar in a CO_2 environment (left), *B. anthracis* forms capsules and the colonies exhibit a mucoid appearance.

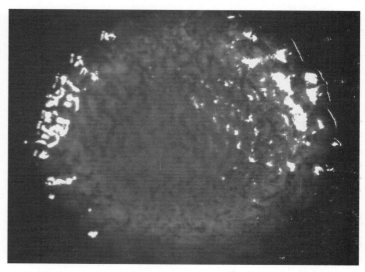

Figure 3.3 Close-up of a single colony of *Bacillus subtilus* growing on blood agar. Note the heavy capsule and ground glass appearance.

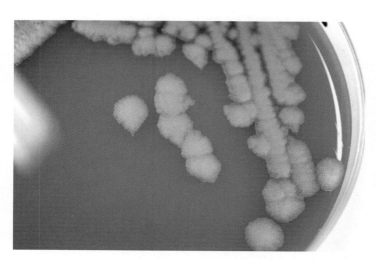

Figure 3.2 **Colonies of *Bacillus anthracis* growing on blood agar.** The typical appearance of *B. anthracis* includes rough-edged colonies with a ground glass, nonpigmented texture. The small outcroppings seen on some of colonies (particularly the colony seen in the center of the plate) are called "comma projections" and are characteristic of *B. anthracis*.

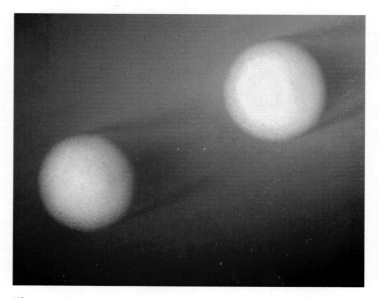

Figure 3.4 **Colonies of *Clostridium botulinum* Type A grown on blood agar.** Botulinum toxin type A is a nerve toxin produced by *C. botulinum* that causes the paralytic disease botulism and is also used cosmetically, under the trade name BOTOX, to reduce the appearance of wrinkles.

Figure 3.5 Primary nasopharyngeal culture on blood agar.
Incubation of the culture in an enhanced carbon dioxide atmosphere
allowed for the growth of *Streptococcus pneumoniae* colonies, which
can be recognized by their depressed centers. Colonies not displaying
these depressed centers are not *S. pneumoniae*.

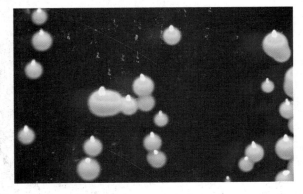

Figure 3.6 *Sporosarcinae ureae.* *Sporosarcinae ureae* colonies
growing on nutrient agar.

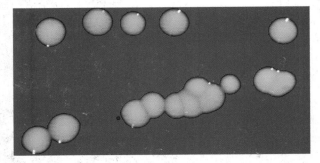

Figure 3.7 *Mycobacterium cosmeticum* colonies.

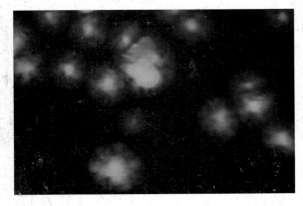

Figure 3.8 *Mycobacterium smegmatis* colonies growing on
nutrient agar.

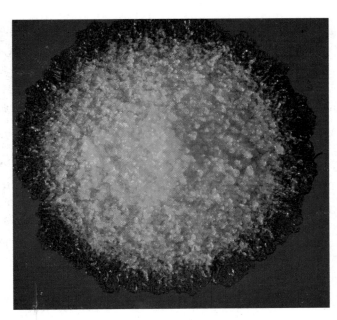

Figure 3.9 Colony of *Mycobacterium marinum*. The yellow
color seen here appears only when the bacterium is exposed to light.
When grown in the dark, the colonies are nonpigmented.

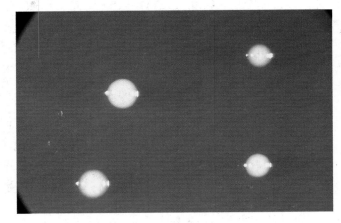

Figure 3.10 *Propionibacterium acnes* **bacterial colonies
growing on blood agar.** This bacterium is commonly found on the
skin, where it feeds on secretions from the sebaceous glands,
contributing to the development of acne.

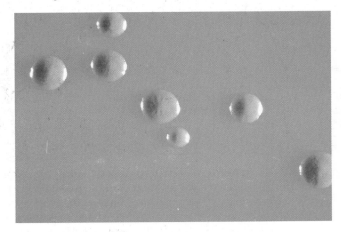

Figure 3.11 *Corynebacterium diphtheria* **colonies. Many**
biotypes of *C. diphtheria* exist, with the most severe disease being
associated with the gravis biotype, shown here.

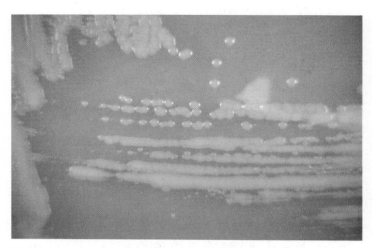

Figure 3.12 *Pseudomonas sp.* **colonies.** *Pseudomonas* sp. are often recognized by the deep green pigment they produce.

Figure 3.14 *Alcaligenes faecalis* **colonies.** *Alcaligenes faecalis* colonies growing on blood agar.

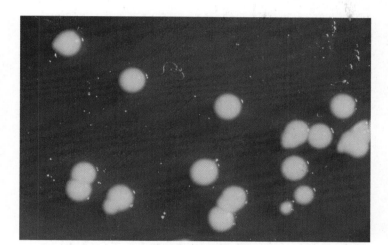

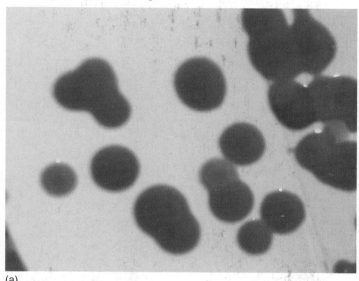

(a)

Figure 3.13 *Klebsiella pneumoniae* **growing on blood agar (top) and MacConkey agar (bottom).** The mucoid appearance of the colonies, especially apparent in the bottom photo, is indicative of an encapsulated bacterium. *K. pneumoniae* is a common cause of hospital-acquired infections involving the urinary and pulmonary systems.

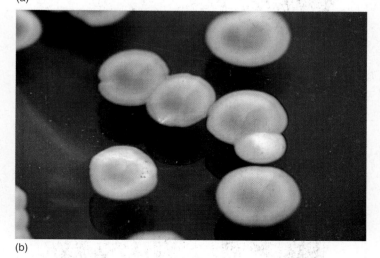

(b)

Figure 3.15 *Serratia marcescens* **colonies.** *S. marcesens* is a thermally dimorphic bacterial species. (a) At low incubation temperatures (<35°C), red colonies are produced while at temperatures above 38°C, white colonies are seen. (b) At 37°C, cells in the center of the growing colony are cool enough to produce red pigment, leading to white colonies with pink centers.

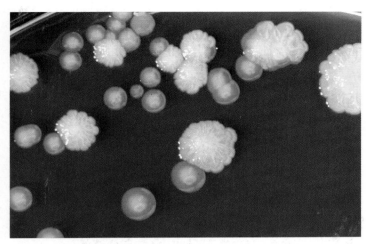

Figure 3.16 *Enterobacter cloacae* **colonies.** In an unusual display, this pure culture produces both smooth- and rough-appearing colonies.

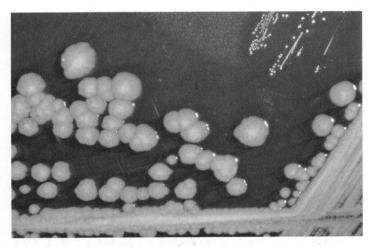

Figure 3.17 **Colonies of** *Proteus mirablis* **bacteria grown of XLD medium.** Approximately one-fourth of the population carries *Proteus* as part of their normal intestinal microbiota.

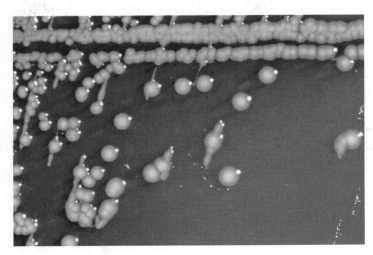

Figure 3.18 *Salmonella choleraesuis.* Colonies of *Salmonella choleraesuis* growing on blood agar. Colonies generally range from 2 to 4 mm in diameter.

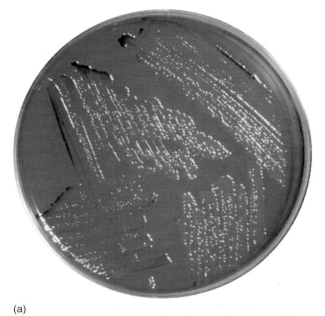

(a)

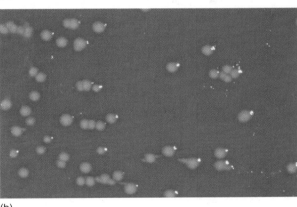

(b)

Figure 3.19 *Citrobacter freundii.* (a) *Citrobacter freundii* growing on XLD agar. The black color seen on the left side of the plate indicates the deposition of hydrogen sulfide, a characteristic of the bacterium. (b) Colonies of *C. freundii* growing on blood agar display a smooth, convex appearance with a moist, shiny surface and entire edge.

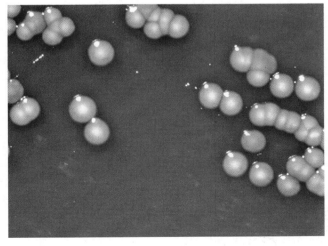

Figure 3.20 *Providencia alcalifaciens.* Colonies of *Providencia alcalifaciens* growing on blood agar.

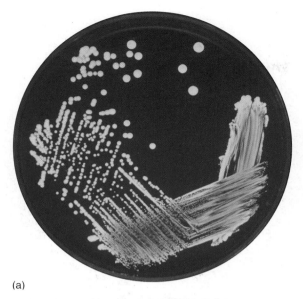

(a)

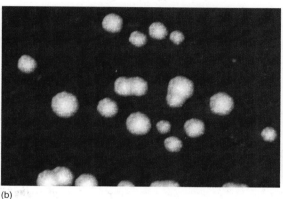

(b)

Figure 3.21 *Legionella pneumophila.* (a) *Legionella pneumophila*, illuminated with ultraviolet light. (b) Colonies of *L. pneumophila* growing on Feeley-Gorman agar appear round, off-white, and have a "cut glass" appearance.

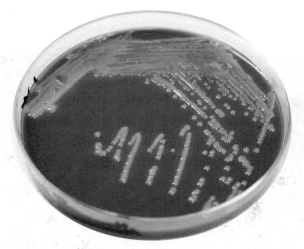

Figure 3.22 *Francisella tularensis* **colonies on chocolate agar plate.** Infection with this bacterium causes tularemia, a disease typically restricted to small mammals such as rodents or rabbits. Because infection can proceed through several routes, including ingestion of contaminated water or inhalation of aerosols, and tularemia is generally lethal if not treated, *F. tularensis* is considered a potential agent of bioterrorism.

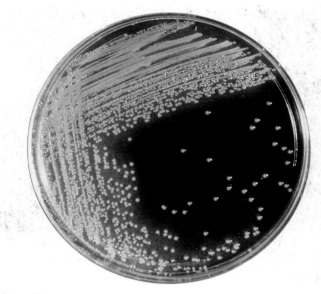

Figure 3.23 *Vibrio cholerae.* Streak plate of *Vibrio cholerae,* the causative agent of cholera.

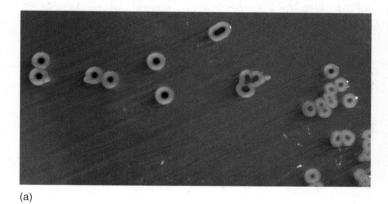

(a)

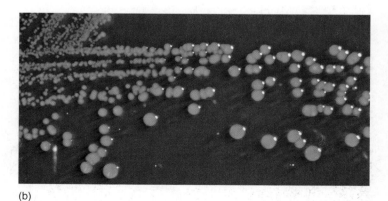

(b)

Figure 3.24 *Salmonella* **and** *Shigella* **on Hektoen enteric agar.** (a) Colonies of *Salmonella* serotype Typhimurium growing on Hektoen enteric agar, a medium commonly used to differentiate *Salmonella* from *Shigella*. The black centers of the colonies are due to production of hydrogen sulfide, which is indicative of *Salmonella*. (b) The lack of a dark center within colonies of *Shigella boydii* growing on HE agar indicates an absence of hydrogen sulfide production and points toward *Shigella*.

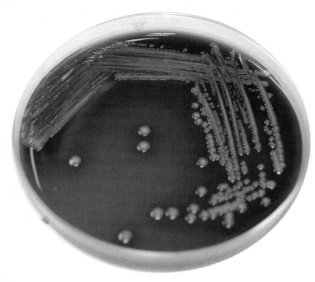

Figure 3.25 *Yersinia pestis.* Colonies of *Yersinia pestis* growing on blood agar. Infection with *Y. pestis*, usually by way of being bitten by an infected flea, leads to plague.

(a)

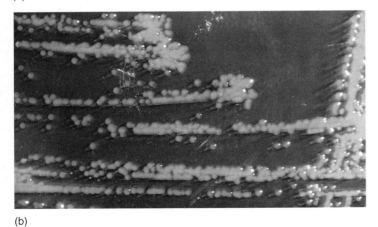

(b)

Figure 3.26 *Yersinia enterocolitica.* (a) Colonies of *Yersinia enterocolitica* growing on blood agar. Infection with *Y. enterocolitica* can result in fever, abdominal pain, and bloody diarrhea lasting 1 to 3 weeks. The disease primarily affects children. (b) The same bacteria growing on Hektoen enteric agar appears blue-green in color. The lack of a dark center in the colonies indicates that hydrogen sulfide is not produced by the bacterium.

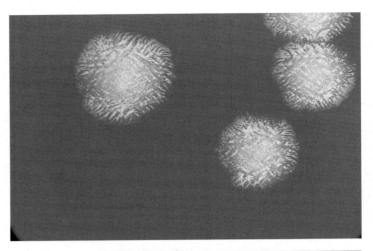

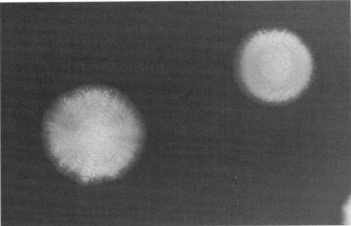

Figure 3.27 *Fusobacterium* and *Bacteroides.* Colonies of *Fusobacterium* sp. (top) and *Bacteroides fragilis* (bottom) growing on blood agar. Both bacteria are obligate anaerobes and common members of the gastrointestinal microbiota.

Bacterial Microscopic Morphology

As a rule, bacterial cells don't differ dramatically from species to species, at least when compared to their multicellular counterparts in other domains. Nevertheless, there are a variety of staining techniques that provide a great deal of information about bacteria on a microscopic level. In this section, bacteria are classified first and foremost as being either Gram positive or Gram negative. Within each cate- gory, bacteria are subdivided according to the three common bacterial shapes: bacilli, cocci, or spirals. You will also see examples of differential stains that demonstrate the existence of endospores, capsules, and flagella. These structures are as close as bacteria come to possessing cellular organelles, and their visualization in the laboratory can be quite helpful in identifying unknown cultures.

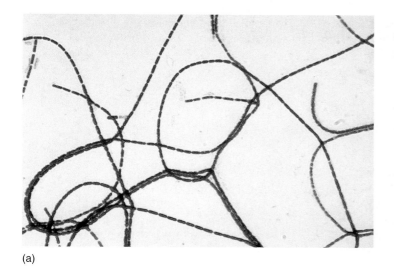

(a)

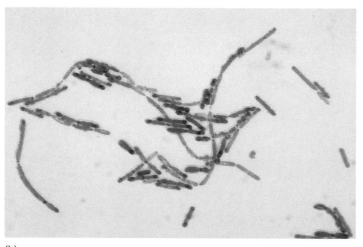

(b)

Figure 4.1 *Bacillus* **Sp., Gram stain.** (a) Gram stain of *Bacillus anthracis* (1000x). The causative agent of anthrax in humans and animals, *B. anthracis* often forms long chains in culture. (b) Gram stain of *Bacillus cereus* subspecies *mycoides* (1000x). This species of *Bacillus* commonly causes food poisoning. Although classified as Gram-positive, many species of *Bacillus* will begin to stain Gram-negative, as seen here, when the culture ages.

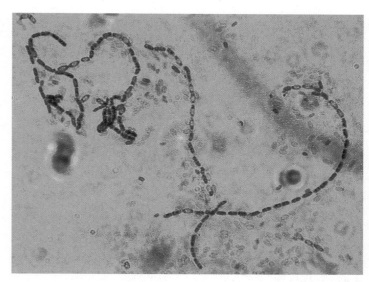

Figure 4.2 **Simple stain of** *Bacillus subtilis* **(1000x).** Note the unstained areas within many of the cells. These are developing endospores that resist normal staining techniques and consequently appear clear after staining.

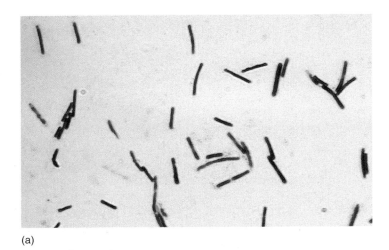

(a)

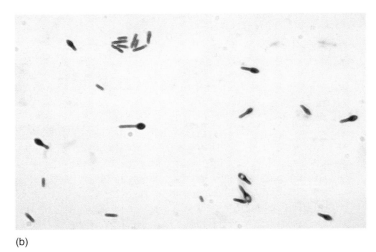

(b)

Figure 4.3 Gram Stain of *Clostridium* sp. (a) Gram stain of *Clostridium perfringens* (1000x). *C. perfringens* endospores are found in the soil where they can, if introduced into a wound, cause gas gangrene. They also have the propensity to contaminate animal flesh and vegetables, leading to food intoxication. (b) Gram stain of *Clostridium botulinum* type A (1000x). The botulin toxin produced by this bacterium is the cause of botulism, an illness that can cause death by paralyzing the muscles of the diaphragm. The location of the developing endospores—the swelling seen at the end of some cells—is characteristic of *Clostridium*.

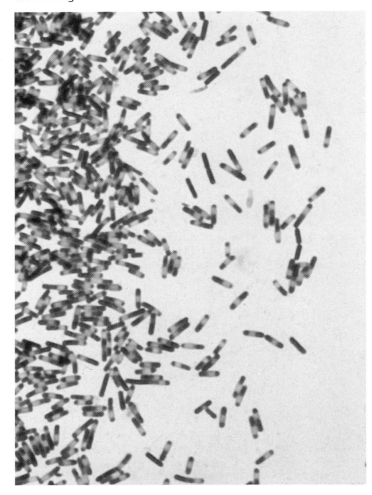

Figure 4.4 Gram stain of *Clostridium difficile* (1000x). *C. difficile* is a part of the gastrointestinal flora of most persons but can also cause *Clostridium difficile* disease, which is marked by watery diarrhea, nausea, and abdominal pain. Note the numerous developing endospores, which can be seen as light staining areas in many of the cells.

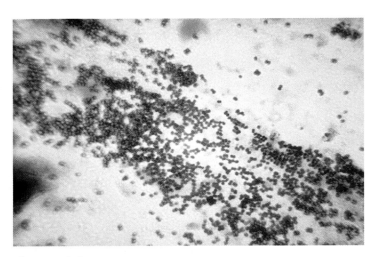

Figure 4.5 Gram stain of *Staphylococcus aureus* (1000x). *S. aureus* cells are typically found in grapelike clusters. *S. aureus* can cause skin and bone infections, food intoxications, and a number of other diseases.

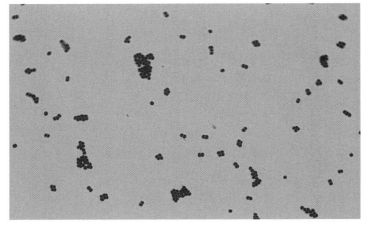

Figure 4.6 Gram stain of *Staphylococcus saprophyticus* (1000x).

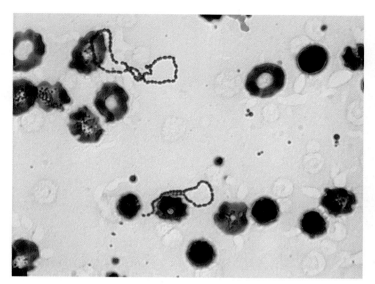

Figure 4.7 **Gram stain of** *Streptococcus viridins* **(1000x) growing in blood culture.** Part of the normal microbiota, *S. viridins* routinely enters the bloodstream during dental procedures, presenting a risk of bacterial endocarditis to persons with certain types of cardiac aberrations.

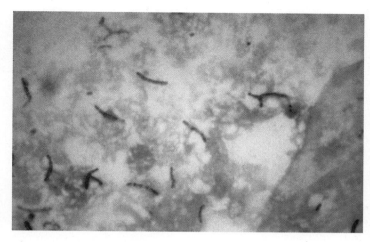

Figure 4.9 *Mycobacterium tuberculosis,* **acid-fast stain (1000x).** When stained using an acid-fast procedure, the bacterium stains red while non-acid fast species stain blue. *M. tuberculosis* is the causative agent of tuberculosis, a disease affecting nearly one-third of the people on Earth.

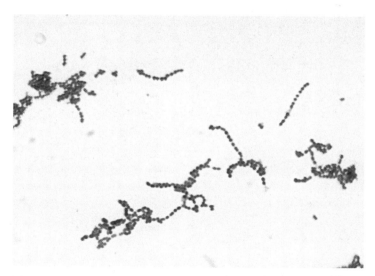

Figure 4.8 **Gram stain of** *Streptococcus* **sp. (1000x).** Gram staining reveals long chains of Gram-positive cocci. These chains develop only in liquid cultures and increase in length as the culture ages.

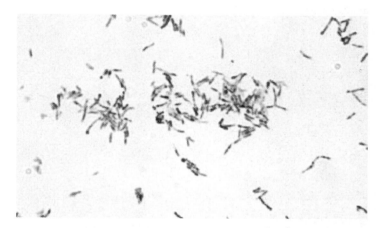

Figure 4.10 *Corynebacterium diphtheria,* **simple stain (1200x).** *C. diphtheria* is the causative agent of diphtheria. The irregular staining seen here is characteristic of *Corynebacterium.*

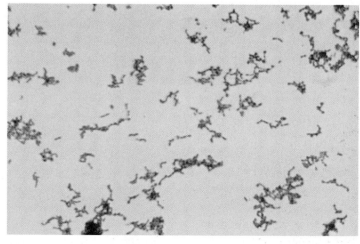

Figure 4.11 **Gram stain of** *Propionibacterium acnes* **(1000x).** This Gram-positive rod is the causative agent of acne vulgaris.

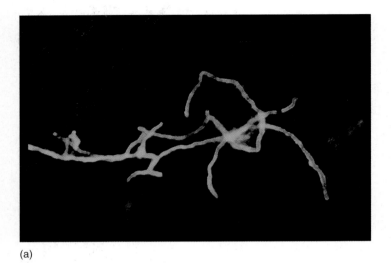

(a)

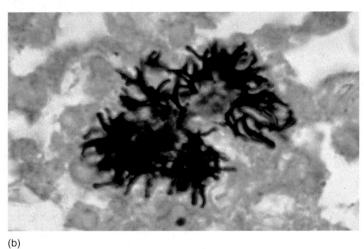

(b)

Figure 4.12 *Actinomyces species* **(1000x).** (a) *Actinomyces* are Gram-positive rods related to *Mycobacterium*. In this fluorescent micrograph, antibodies specific to *A. israelii* were allowed to bind to the specimen prior to examination. When viewed with a fluorescent microscope and illuminated with ultraviolet light, the specimen fluoresces. (b) Silver stain of a brain abscess displaying *Actinomyces naeslundii* bacteria (1000x).

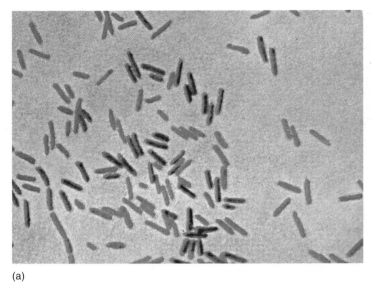

(a)

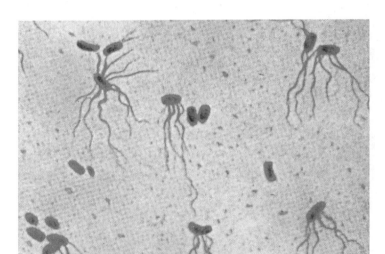

(b)

Figure 4.14 **Two views of** *Salmonella typhi,* **the bacterium responsible for typhoid fever.** (a) A Gram stain shows *S. typhi* to be a Gram-negative rod (1000x). (b) A flagellar stain reveals the flagella that provide *S. typhi* with its motility.

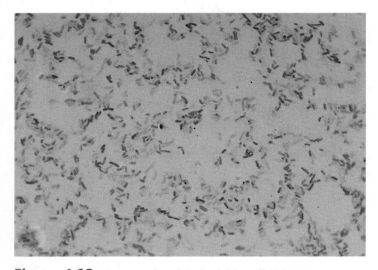

Figure 4.13 **Gram stain of** *Escherichia coli* **(1000x).**

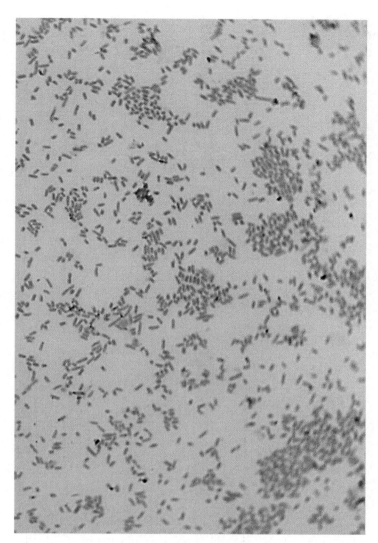

Figure 4.15 **Gram stain of _Proteus mirablis_ (1000x).**

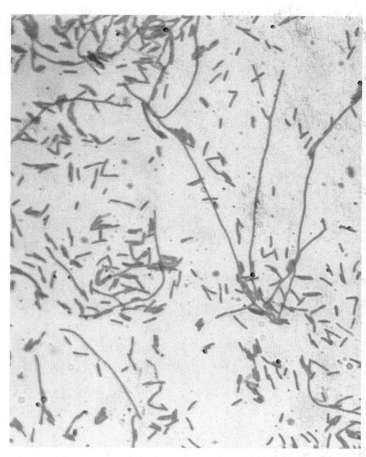

Figure 4.17 **Gram stain of _Legionella pneumophila_ (1000x).** Gram staining reveals solitary cells and long chains of _L. pneumophila_, a Gram-negative bacillus responsible for legionellosis, a pneumonia beginning 2 to 10 days after exposure to the bacterium. _L. pneumophila_ is also the cause of Pontiac fever, a less serious respiratory illness occurring a few hours to two days after exposure.

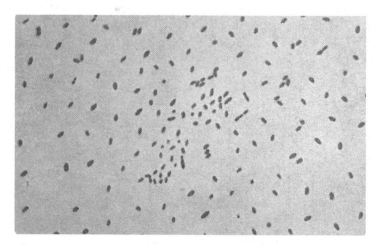

Figure 4.16 **Gram stain of _Bordetella pertussis_ (1000x).** The cause of pertussis, or whooping cough, _B. pertussis_ is a very small, encapsulated, Gram-negative coccobacillus. Pertussis is highly contagious and is most common in children and those who have not completed the primary vaccination series.

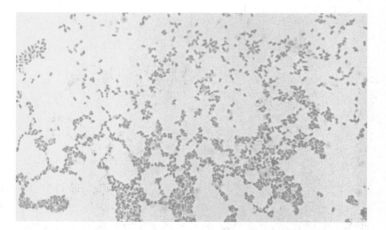

Figure 4.18 **Gram stain of _Brucella melitensis_ (1000x).** _Brucella melitensis_, a Gram-negative coccobacillus that is the cause of brucellosis (undulant fever), a zoonotic disease. The cells characteristically stain poorly and often appear as fine sand.

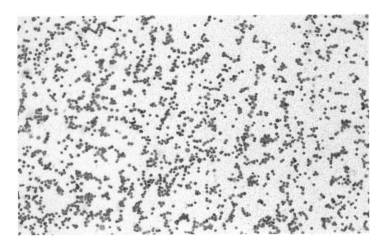

Figure 4.19 **Gram stain of *Francisella tularensis* (1000x),** a tiny (0.2μm–0.7μm) Gram-negative coccobacillus that is the cause of tularemia. The cells characteristically stain poorly and usually appear singly.

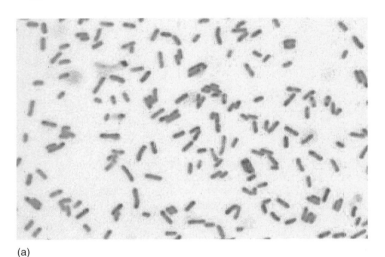

(a)

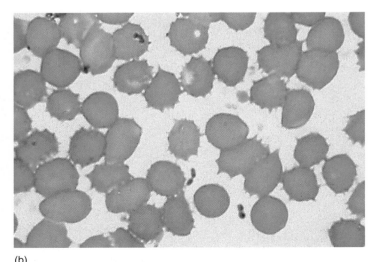

(b)

Figure 4.20 ***Yersinia pestis.*** (a) Gram stain of *Yersinia pestis* (1000x), a small Gram-negative bacillus that is the causative agent of plague. Bipolar staining occasionally occurs with *Y. pestis*, leaving the ends of the cells stained darker than middle portion. (b) A blood smear containing *Y. pestis* bacteria (1000x).

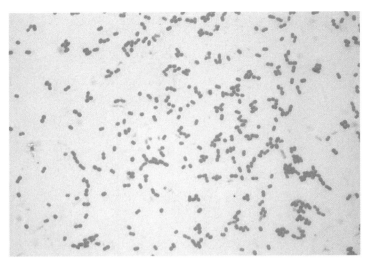

Figure 4.21 **Gram stain of *Acinetobacter calcoaceticus* (1000x).** *Acinetobacter* is a small, Gram-negative cell, often found in pairs, and varying in morphology from coccus to short rod. Another species in this genus, *A. baumannii*, is an emergent agent of disease in nosocomial settings, especially military hospitals.

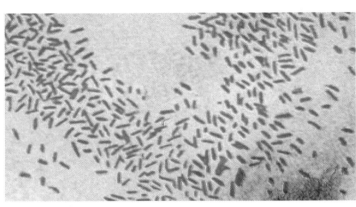

(a)

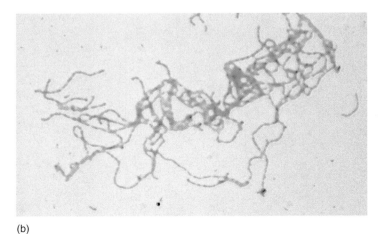

(b)

Figure 4.22 ***Haemophilus spp.*** (a) Gram stain of *Haemophilus influenza* (1000x). Mistakenly thought to be the cause of influenza after it was isolated from patients suffering from the flu, *H. influenza* was later determined to be a causative agent of bacterial meningitis. (b) *Haemophilus ducreyi* (1000x). *H. ducreyi* is a small, Gram-negative, encapsulated rod that causes chancroid, a sexually transmitted disease prevalent in tropical areas.

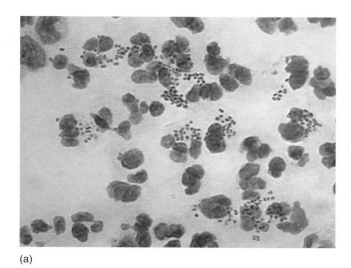

(a)

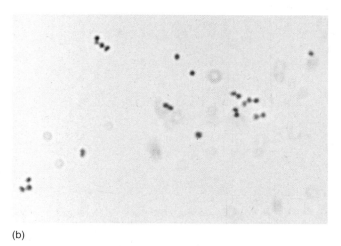

(b)

Figure 4.23 *Neisseria spp.* (a) Gram stain of urethral exudates revealing numerous *Neisseria gonorrhoeae*, the cause of gonorrhea (1000x). *N. gonorrhoeae* is a Gram-negative coccus that, rather than being perfectly round, is flattened on one side. The cells are usually found as diplococci, with their flat sides touching. (b) Gram stain of *Neisseria meningitidis* (1150x), the bacterium responsible for meningococcal meningitis.

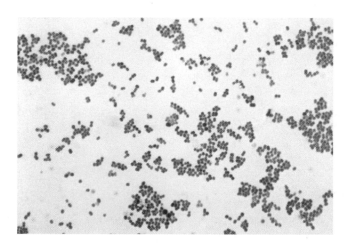

Figure 4.24 **Gram stain of *Neisseria subflava* (1000x).** Though generally nonpathogenic, this Gram-negative bacterium has been responsible for some cases of endocarditis and meningitis.

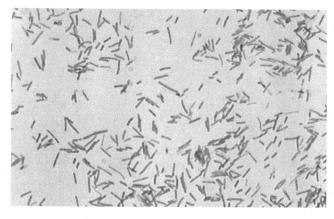

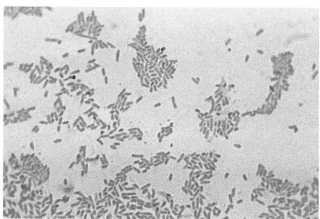

Figure 4.25 **Microbiota of the oral cavity and gastrointestinal tract (Gram stain 1000x).** Both *Bacteroides fragilis* (top) and *Fusobacterium novum* (bottom) are found in the oral cavity and gastrointestinal tract. Although generally regarded as commensals, both Gram-negative organisms are linked to the development of periodontal disease.

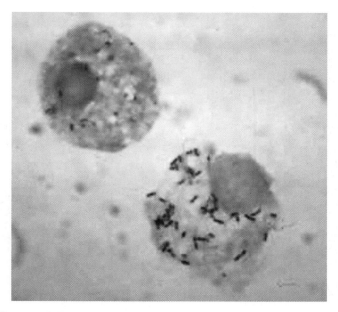

Figure 4.26 *Rickettsia rickettsia* **(1000x).** The cause of Rocky Mountain Spotted Fever, these small, parasitic bacteria are growing within the cells of an infected tick.

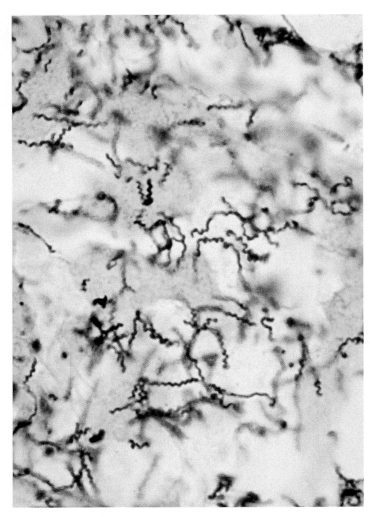

Figure 4.27 *Treponema pallidum* **(1000x).** Silver stain of *Treponema pallidum*, the spirochete responsible for syphilis, a sexually transmitted disease.

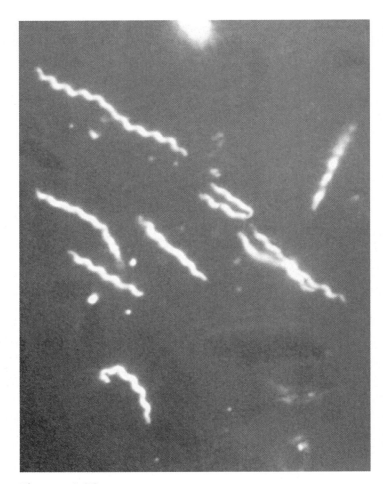

Figure 4.29 *Borrelia burgdorferi* **(400x).** Darkfield view of *Borrelia burgdorferi*, the spirochete responsible for causing Lyme disease. Spirochetes usually stain poorly, making darkfield microscopy the best technique for examining them.

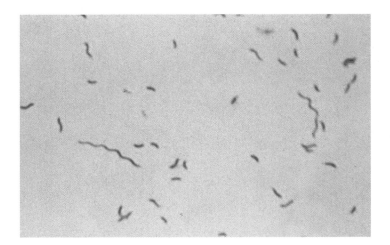

Figure 4.28 **Gram stain of** *Campylobacter fetus* **(1000x).** *Campylobacter* is a Gram-negative spiral rod. At least three species of *Campylobacter* are known to cause disease in humans.

Fungi

There are a number of characteristics that are helpful in the identification of an unknown fungal specimen. Initially, fungi are examined with regard to their macroscopic colonial morphology, and both the topside and underside of the colony hold characteristic clues. Next, a microscopic examination reveals information about a fungus's hyphal structure, sporangium and spores. Often this information is enough to allow its identification. This section begins with images of the four fungal genera that are primary pathogens, *Histoplasma*, *Blastomyces*, *Coccidioides*, and *Paracoccidioides*. After that, less virulent pathogenic fungi are presented, and finally, opportunistic pathogens are presented. In addition, a single plant pathogen is shown, which, although it has no medical importance to humans, it is of tremendous economic significance.

(a)   (b)

Figure 5.1 **Macroscopic appearance of *Histoplasma capsulatum*.** (a) Slant cultures of *H. capsulatum* growing on Sabourard's agar. The colonies are creamy and moist looking. Found growing in soils containing high concentrations of bird and bat guano in the southern and eastern United States, as well as portions of South America and Africa. Inhalation of conidia can lead to histoplasmosis—also known as Ohio Valley fever—a disease typically confined to the lungs, but which can be lethal if other organs are affected. (b) A slant culture containing both *H. capsulatum* (bottom) and *Cryptococcus neoformans* (top). *C. neoformans* is an opportunistic pathogen, with infections generally restricted to AIDS patients and others who are immunosuppressed.

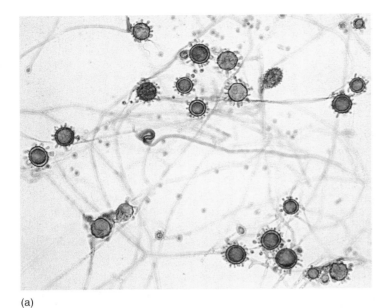

(a)

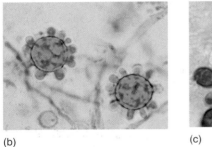

(b)

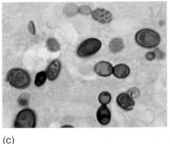

(c)

Figure 5.2 Microscopic morphology of *Histoplasma capsulatum*. Found growing in soils containing high concentrations of bird and bat guano in the southern and eastern United States, as well as portions of South America and Africa. Inhalation of conidia can lead to histoplasmosis—also known as Ohio Valley fever—a disease typically confined to the lungs, but which can be lethal if other organs are affected. (a) Macroconidia of *H. capsulatum* (400x) are large and covered with fingerlike projections while microconidia are smaller, round, and possess a smooth surface. (b) Close up of macroconidia (1000x). Also visible are the septate hyphae characteristic of *H. capsulatum*. (c) At 37°C, *H. capsulatum* converts to a yeast form. In this micrograph, several cells can be seen undergoing replication by budding.

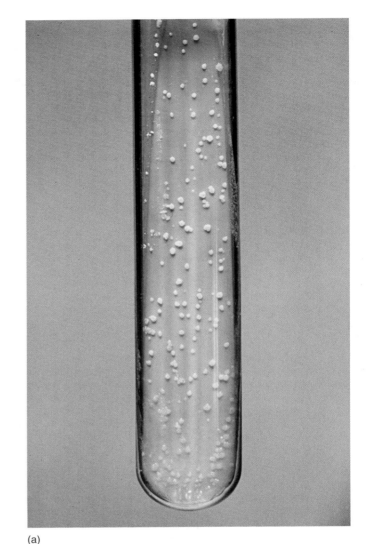

(a)

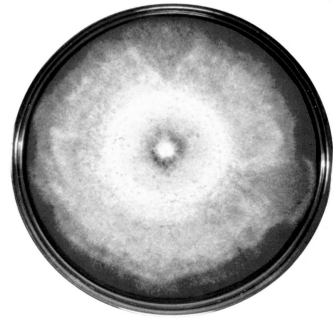

(b)

Figure 5.3 *Blastomyces dermatitidis* growing on Lowenstein-Jensen medium. (a) *B. dermatitidis* is a highly infectious fungus restricted primarily to parts of the south-central, south-eastern and midwestern United States, where it is found in moist soil enriched with decomposing organic debris. Inhalation of *B. dermatitidis* conidia can give rise to North American blastomycosis—also known as Gilchrist disease and Chicago disease—which is marked by fever, chills, productive cough, muscle and joint pain, along with pleuritic chest pain. Laboratory personnel must be aware of the highly infectious nature of *Blastomyces* and confine work with the organism to an isolation cabinet. (b) Mycelial phase of *B. dermatitidis*.

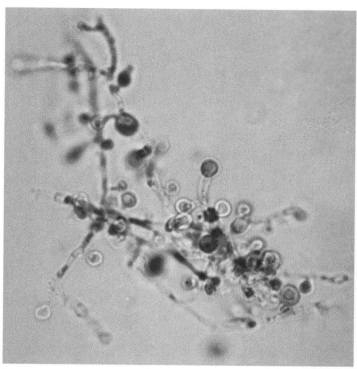

(a)

(b)

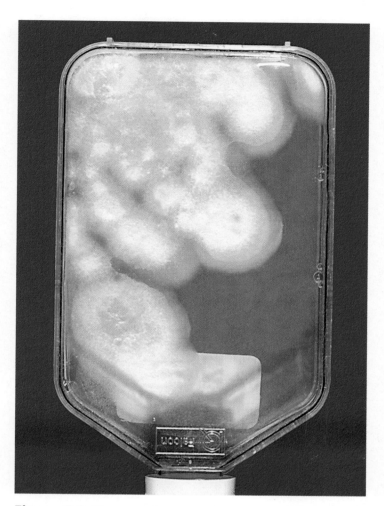

Figure 5.5 Macroscopic appearance of *Coccidioides immitis*. Sabouraud's dextrose agar culture of *C. immitis*, a highly infectious fungus endemic to parts of the southwestern United States. Inhalation of *C. immitis* arthrospores can give rise to coccidioidomycosis, or valley fever. In the laboratory, *C. immitis* grows rapidly, producing colonies that are initially moist, membranous, and grayish, progress to white and cottony, and eventually turn tan to brown in color.

Figure 5.4 *Blastomyces* (600x). (a) *Blastomyces* produce conidia at a 90-degree angle to hyphae, often resembling lollipops. (b) A budding, yeast-phase cell of *Blastomyces*. The species is known for the development of thick-walled yeast cells.

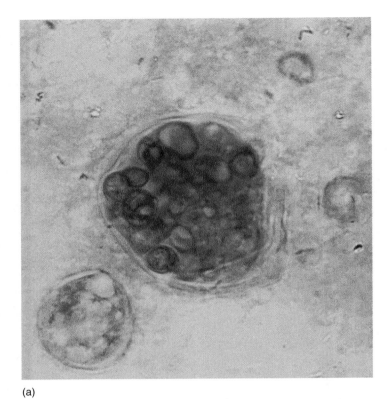

(a)

Figure 5.7 Slant culture of *Paracoccidioides brasilensis* during its yeast phase. Found in the rich soils of central and South America, *Paracoccidioides* is one of the four genera of true fungal pathogens. Inhalation of conidia can lead to paracoccidioidomycosis, which ranges from asymptomatic to chronic infection. Yeast colonies are described as white, heaped, wrinkled or folded and appear only after incubation at 37°C for 10–20 days.

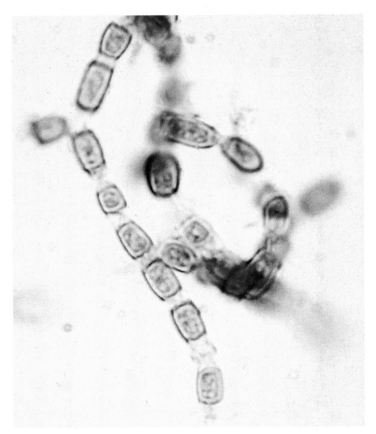

(b)

Figure 5.6 *Coccidioides immitis* (1000x). (a) Chlamydospore of *Coccidioides immitis*. (b) Arthroconidia of *C. immitis*.

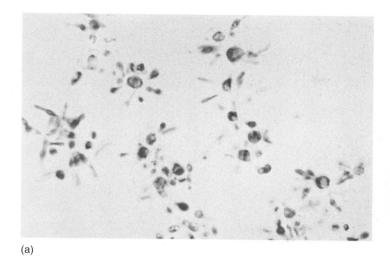

(a)

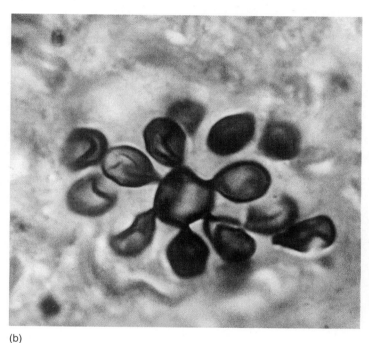

(b)

Figure 5.8 Two views of the yeast phase of *Paracoccidioides brasilensis.* (a) At 37°C, the fungus produces multiple buds that surround the mother yeast cells, giving the appearance of a steering wheel (1000x). (b) At higher magnification (2000x), the narrow neck that attach each daughter cell to the mother cell are clearly visible.

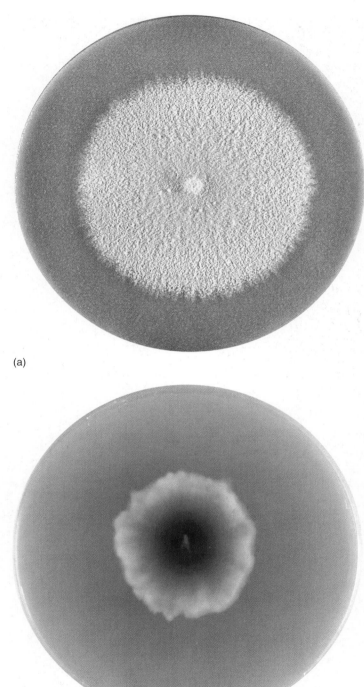

(a)

(b)

Figure 5.9 (a) Frontal view of a *Microsporum persicolor* colony. A member of the phylum Ascomycota, the genus *Microsporum* contains seventeen species, all of which are the causative agents of dermatophytoses, or superficial skin, hair, and, rarely, nail infections (although *M. persicolor* does not infect hair). The colonial morphology is glabrous, downy, wooly or powdery. From the front, colony color is white, with other species of the genus being beige, yellow, or light reddish-brown. (b) A reverse view of *Microsporum persicolor* shows the bottom of the colony to be yellow to reddish-brown in color.

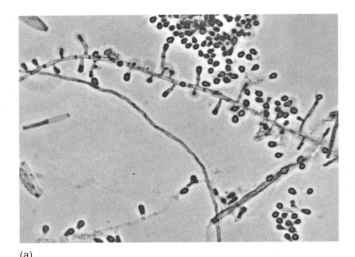

(a)

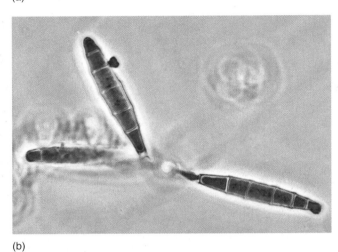

(b)

Figure 5.10 Morphology of *Microsporum persicolor* (500x).
(a) A Member of the phylum ascomycota, *M. persicolor* is one of the causative agents of dermatophytosis, or superficial skin, hair and, rarely, nail infections. Septate hyphae, along with numerous small, oval, unicellular microconidia are visible. (b) Macroconidia are composed of multiple microconidia, separated by cell walls.

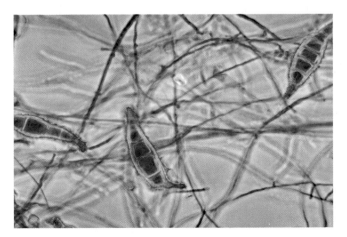

Figure 5.11 Macroconidia of *Microsporum distortum* (1000x).
M. distortum produces septate hyphae, along with irregularly shaped macroconidia. The organism occasionally causes tinea capitas in humans but is far more commonly seen as a pathogen of dogs and cats.

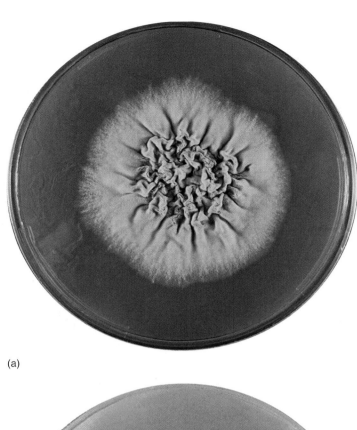

(a)

(b)

Figure 5.12 *Trichophyton rubrum*. (a) Frontal view of a *Trichophyton rubrum* colony. A member of the phylum Ascomycota, the genus *Trichophyton* contains several species, all of which are the causative agents of dermatophytoses, or superficial skin, hair, and nail infections. The colonial morphology is waxy and glabrous (flat to cottony) and white to beige in color. (b) A reverse view of the same fungus shows the colony to be yellow to reddish-brown in color.

Figure 5.13 **Frontal view of a *Trichophyton tonsurans* colony.** A member of the phylum Ascomycota, the genus *Trichophyton* contains several species, all of which are the causative agents of dermatophytoses, or superficial skin, hair and nail infections. The colonial morphology has a smooth, velvety (glabrous) appearance, with a raised center, and yellowish-beige coloration.

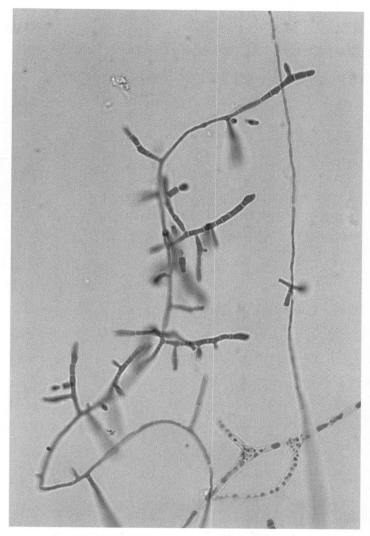

Figure 5.14 **Microconidia of *Trichophyton mariatii* (40x).** A member of the phylum Ascomycota, the genus *Trichophyton* contains several species, all of which are the causative agents of dermatophytoses, or superficial skin, hair and nail infections. Microscopic examination reveals septate hyphae and numerous pyriform (pear-shaped) microconidia, which are the predominant type of conidia produced by *Trichophyton*.

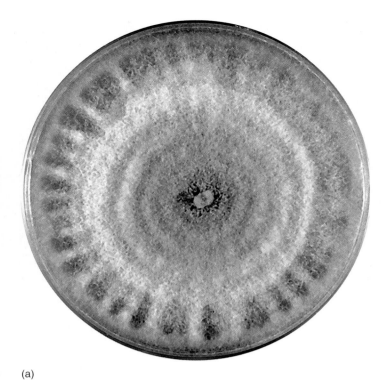

(a)

(b)

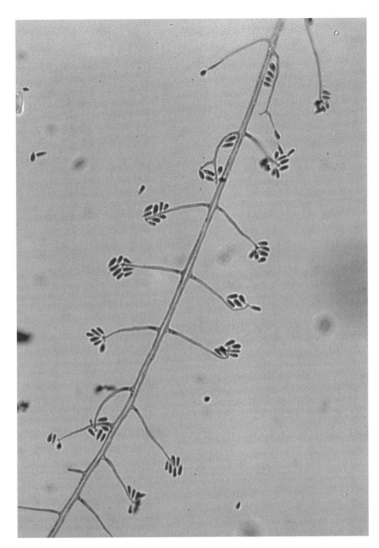

Figure 5.16 *Sporothrix schenckii* **(500x).** A member of the phylum Ascomycota, *S. schenckii* is the causative agent of sportrichosis, or rose handlers disease. The fungus typically displays septate hyphae (although no septa are visible in this micrograph) and conidiophores that arise at right angles to the vegetative hyphae. Conidia are brown, oval, and typically arranged in rosette-like clusters at the tips of the conidiophores, although freshly isolated strains may display conidia attached directly to the sides of the hyphae.

Figure 5.15 (a) Frontal view of a *Corynespora cassiicola* colony. A member of the phylum Ascomycota, *C. cassiicola* is a plant pathogen, causing damage—known as leaf spot—to cucumber, cantaloupe, pumpkins, and other members of the genus *Cucurbita*. The colonial morphology has a cottony, gray to brown color. (b) Slide culture of *Corynespora cassiicola* (400x).

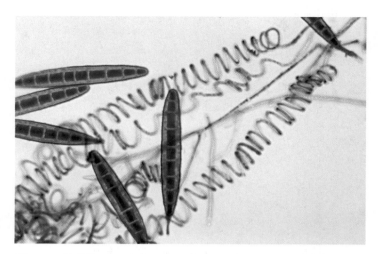

Figure 5.17 Macroconidia of *Arthroderma grubyi* (400x). Often found inhabiting dung, *Arthroderma* species are dermatophytes that may live off the skin, hair, or feathers of animals.

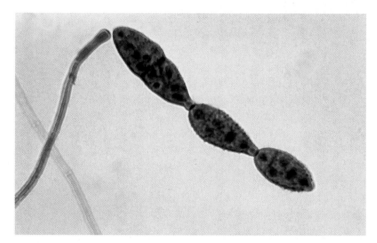

Figure 5.18 *Alternaria* sp. conidia (1000x). A member of the phylum Ascomycota, *Alternaria* sp. are opportunistic pathogens capable of causing phaeohyphomycosis, sinusitis, ulcerated cutaneous infections, and keratitis. Microscopically, *Alternaria* have brown, septate hyphae and bear large conidia with both transverse and longitudinal septa. The conidia may be borne either singly, or in chains, as seen here.

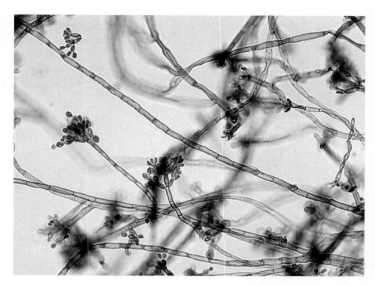

Figure 5.19 *Cladosporium sphaerospernum* (400x). Found in rotten organic material and frequently isolated as a contaminant on foods, *Cladosporium* sp. are causative agents of skin lesions, keratitis, sinusitis, and pulmonary infections. Conidia of *Cladosporium* sp. are elliptical in shape and pale to dark brown in color. They are borne at the end of conidiophores in clustered groups that easily break apart.

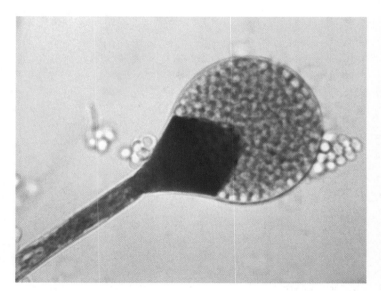

Figure 5.20 Mature sporangium of *Mucor* sp. (1000x). A ubiquitous member of the phylum Zygomycota, *Mucor* sp. can cause infections known as zygomycoses in amphibians, cattle, swine, and humans. *Mucor* displays nonseptate hyphae and large (300 μm) sporangia filled with sporangiospores. Upon rupture of the sporangiuma, the spores are broadcast over a relatively wide area.

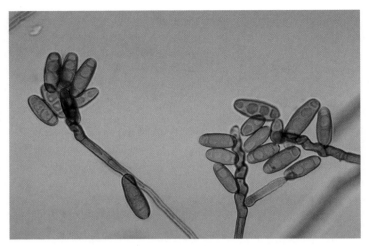

Figure 5.21 Conidiophores and conidia of *Curvularia harveyi* (1000x). Infection with *Curvularia* species cause a variety of infections, including mycetomas, wound infections, keratitis, allergic sinusitis, cerebral abscess, pneumonia, endocarditis, and disseminated infections, even in persons with fully functioning immune systems. The fungus displays brown hyphae and conidiophores, which are bent at the points where the conidia originate. The conidia themselves are brown and straight to pyriform in shape.

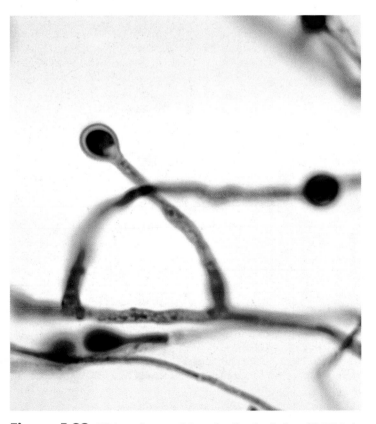

Figure 5.22 Slide culture of *Pseudoallescheria boydii* (800x). Found in soil, sewage, contaminated water, and the manure of farm animals, *P. boydii.* is an emerging opportunistic pathogen and can cause various infections in humans. Microscopic examination reveals septate hyphae and oval conidia.

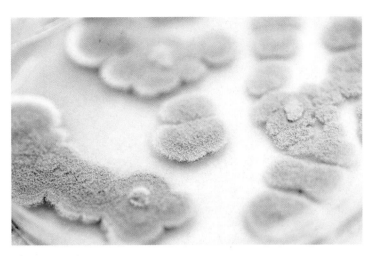

Figure 5.23 *Penicillium* colonies growing in a Petri dish.

Figure 5.24 *Penicillium marneffei* (800x). A member of the phylum Ascomycota, *P. marneffei* is endemic to Southeast Asia (Thailand, Taiwan, and India). Microscopic examination reveals brush-like clusters of phialides appear at the tip of conidiophores. The phialides in turn support unbranched chains of round conidia. Inhalation of spores can lead to penicilliosis in immunocompromised individuals; the infection is often fatal.

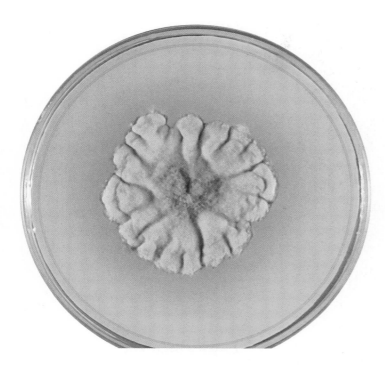

Figure 5.25 Frontal view of a *Madurella grisea* colony. A member of the phylum Ascomycota, *M. grisea* is found in soil, particularly in tropical and subtropical areas of Africa, India, and South America. The fungus enters the body via trauma, often to the foot, and causes slow progressing infections known as mycetomas that are characterized by large black masses of hyphae.

Figure 5.27 Asexual conidia of *Botrytis* sp. (400x). The genus *Botrytis* falls within the phylum Ascomycota and is strictly a plant pathogen, affecting strawberries, grapes, and tomatoes. One species commonly found on grapes, *B. cinerea*, may cause "winegrower's lung," a rare allergic reaction.

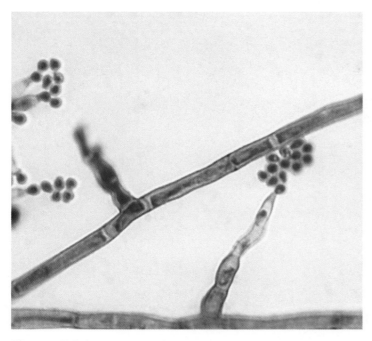

Figure 5.26 *Madurella mycetomatis* fungus (800x). A member of the phylum Ascomycota, *M. mycetomatis* is found in soil, particularly in tropical and subtropical areas of Africa, India, and South America. The fungus enters the body via trauma, often to the foot, and causes slow progressing infections known as mycetomas that are characterized by large black masses of hyphae. The fungus has septate hypha and produces oval conidia at the tips of phialides, as seen in the micrograph.

Figure 5.28 Conidiophores and conidia of *Fusarium verticillioides* (400x). Typically a pathogen of corn, *F. verticillioides* can also cause fatal toxicoses of horses and swine as well as opportunistic mycoses in immunosuppressed humans. The fungus is a member of the phylum Ascomycota and displays septate hypha.

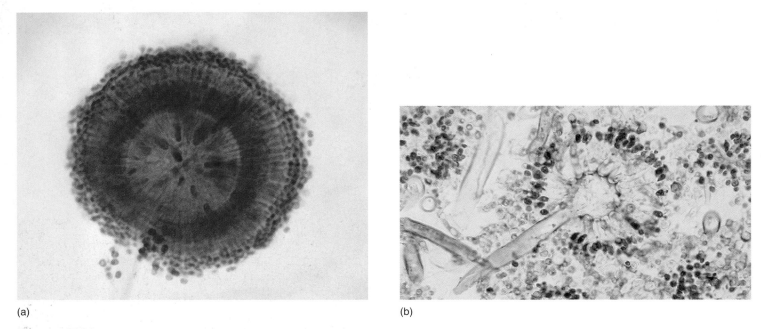

(a)   (b)

Figure 5.29 Two views of the conidial head of *Aspergillus niger* (500x). *Aspergillus* is an opportunistic fungus that typically infects individuals whose immune systems have been compromised. A member of the phylum Ascomycota, *Aspergillus* displays septate hypha. The conidiophores are long (up to 3 mm) and terminate in darkly colored, rough, spherical conidia 30–74 μm in diameter.

Protists

Traditionally, the protists have been defined as a diverse group of primarily single-celled eukaryotic organisms. Those that were heterotrophic were classified as protozoans while photosynthetic protists were deemed algae. This view has changed over the past several years as molecular biology techniques have allowed us to look at an organism's underlying genetics and not simply at its outward appearance. Currently, major shifts in taxonomy are underway that will change the way we look at these organisms, but for now, especially within the context of the laboratory, outward appearances remain important. With this in mind, this section follows a traditional taxonomy, with heterotrophic protists classified as protozoa and autotrophic protists classified as algae. Within the protozoa, organisms are subdivided into four groups according to their means of motility: amoeba (pseudopods), flagellates (flagella), ciliates (cilia), and apicomplexans (nonmotile). The algae are grouped according to their primary photosynthetic pigments, the type of cell covering they possess, and the form in which they store the energy created during photosynthesis. While this organizational structure will certainly change in the near future, with far less emphasis placed on the differences between protozoa and algae, for now the system remains well suited to the microbiology lab.

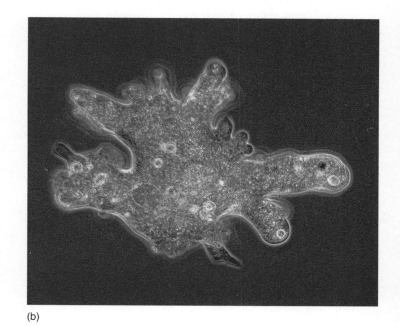

(b)

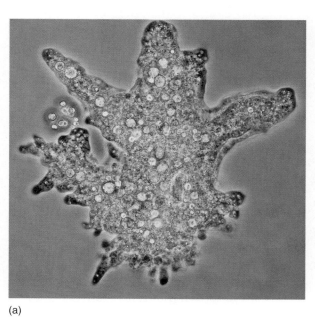

(a)

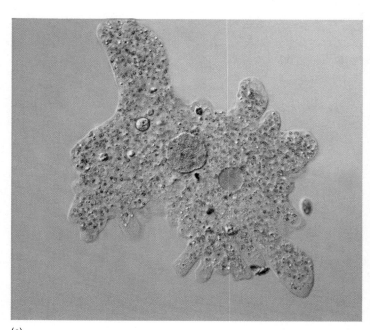

(c)

Figure 6.1 **Three views of *Amoeba*.** (a) Phase contrast (160x). (b) Darkfield (160x). (c) Differential interference contrast (200x).

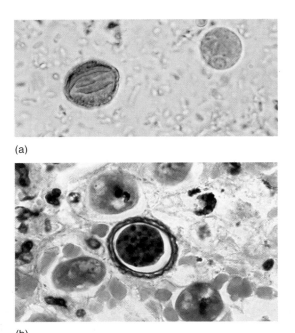

(a)

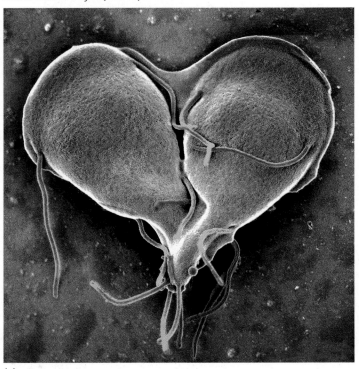

(b)

Figure 6.2 *Acanthamoeba.* (a) *Acanthamoeba* cyst (1000x). *Acanthamoeba* sp. are free-living amebae that may be found in lakes, swimming pools, tap water, and heating and air conditioning units. In individuals with compromised immune systems, infection can lead to granulomatous amebic encephalitis (GAE). Diagnosis of infection can be made from microscopic examination of biopsy specimens, which reveal trophozoites and cysts. (b) Brain tissue containing an *Acanthamoeba* cyst (1200x).

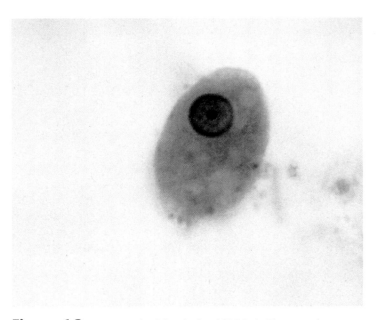

Figure 6.3 *Entamoeba histolytica* **(1200x).** The amoeba *E. histolytica* causes amebiasis, the course of which may range from asymptomatic to, in rare cases, fatal. Ingestion of cysts in feces-contaminated food or water is the most common means of infection. After passing through the stomach, excystation occurs in the small intestine and the motile trophozoite moves quickly to the large intestine. As the parasite reproduces, both cysts and trophozoites are released in the stool, with the presence of either in a stool sample being indicative of infection.

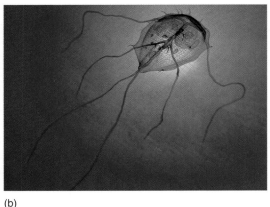

(b)

(c)

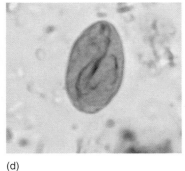

(d)

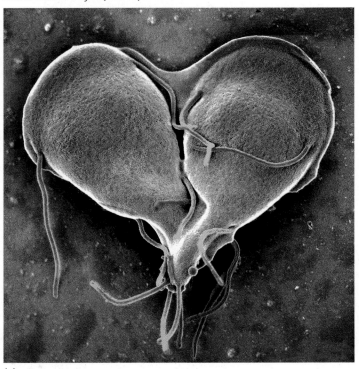

(a)

Figure 6.4 *Giardia intestinalis.* (a) This colorized scanning electron micrograph recorded a *G. intestinalis* trophozoite as it was concluding cell division, just prior to becoming two separate organisms (10,000x). Infection with *Giardia*, typically through the ingestion of water contaminated with cysts, leads to giardiasis, a diarrheal disease. Trophozoites in the large intestine give rise to cysts are shed in the feces. (b) In this micrograph (5000x), the flagella that provide the *Giardia* trophozoite with its motility are clearly visible. (c) The elongated morphology of the *Giardia* trophozoite can be easily seen in this trichrome-stained fecal specimen (1000x). (d) The thick-walled cyst of *G. intestinalis* (1000x), here stained with iodine, allows the parasite to survive for several months outside the body of the host.

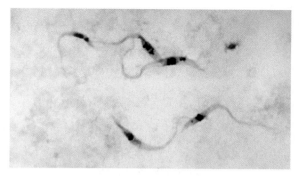

Figure 6.5 *Trypanosoma cruzi* **(1260x).** *T. cruzi* causes Chagas disease, a zoonotic disease that can be transmitted to humans by blood- sucking triatomine bugs. The disease, which is fatal in some instances, is found in the Americas from the southern United States to southern Argentina, with the highest concentration found in Mexico and Central and South America. Visualization of trypanosomes in fresh blood or fixed stained smears is diagnostic of an infection.

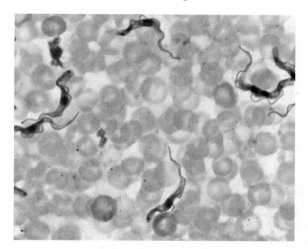

Figure 6.6 *Trypanosoma brucei* **parasites within a blood smear (1000x).** The causative agent of African sleeping sickness, *T. brucei* enters humans through the bite of an infected tsetse fly (genus *Glossina*). In turn, the tsetse fly becomes infected when taking a blood meal from an infected mammalian host. Two subspecies of the organism exist: *T. b. gambienese* causes West African sleeping sickness while *T. b. rhodesiense* causes East African sleeping sickness. Although the course of each disease is distinct, the organisms responsible are morphologically identical. Visualization of trypanosomes in the blood, bone marrow, lymph nodes, or cerebrospinal fluid indicates an infection.

Figure 6.7 *Leishmania.*
(a) Promastigote stage of *Leishmania* sp (1000x). The protozoan genera *Leishmania* includes 30 species, 21 of which cause disease in humans and other mammals. The disease results when an infected female sandfly bites a host and transmits the protozoan into the bloodstream. The parasite is engulfed by a mononuclear phagocytic blood cell where it continues to multiply.
(b) *L. donovani* in a histiocyte, a type of phagocytic bone marrow cell (1000x).

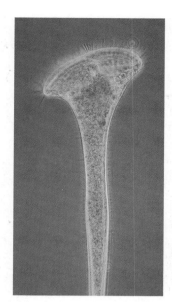

Figure 6.8 Phase contrast photomicrograph of *Stentor* (250x), a large ciliate that is commonly found in freshwater habitats. It uses its cilia to create a vortex that draws in algae, bacteria, and other protozoa, upon which it feeds.

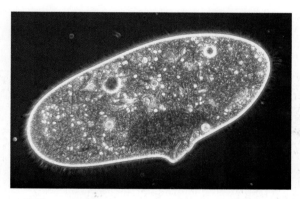

Figure 6.9 **Darkfield view of** *Paramecium* **sp. (400x).**

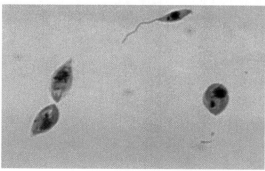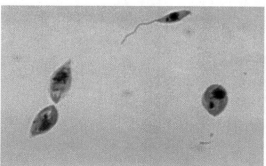

(a)

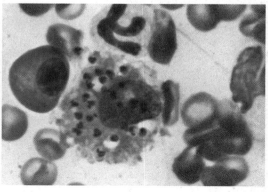

(b)

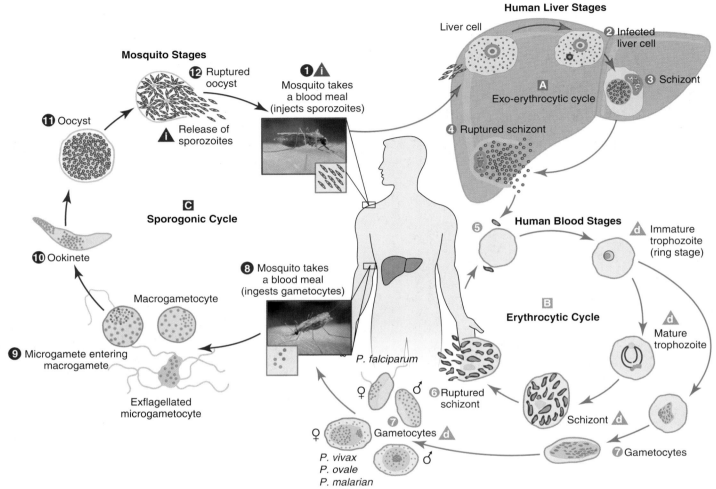

= Infective Stage

= Diagnostic Stage

Figure 6.10 Life cycle of malaria. Malaria is a complex disease that is responsible for about 1,000,000 deaths each year worldwide. The disease may be caused by at least four species of *Plasmodium* (*P. falciparum, P. vivax, P. ovale, P. malariae,* and perhaps *P. knowlesi*), involves two hosts, and has a complex life cycle composed of three distinct components. An understanding of the life cycle is necessary when diagnosing the disease. During a blood meal, a malaria-infected female *Anopheles* mosquito inoculates sporozoites into the human host ❶. Sporozoites infect liver cells ❷ and mature into schizonts ❸, which rupture and release merozoites ❹. After this initial replication in the liver (exoerythrocytic cycle Ⓐ), the parasites undergo asexual multiplication in the erythrocytes (erythrocytic cycle Ⓑ). Merozoites infect red blood cells ❺. The ring-stage trophozoites mature into schizonts, which rupture releasing merozoites ❻. Some parasites differentiate into sexual erythrocytic stages, producing male and female gametocytes. The gametocytes, male (microgametocytes) and female (macrogametocytes), are ingested by an *Anopheles* mosquito during a blood meal ❽. The parasites' multiplication in the mosquito is known as the sporogonic cycle Ⓒ. While in the mosquito's stomach, the microgametes penetrate the macrogametes, generating zygotes ❾. The zygotes in turn become motile and elongated ookinetes ❿, which invade the midgut wall of the mosquito where they develop into oocysts ⓫. The oocysts grow, rupture, and release sporozoites ⓬, which make their way to the mosquito's salivary glands. Inoculation of the sporozoites into a new human host perpetuates the malaria life cycle ❶.

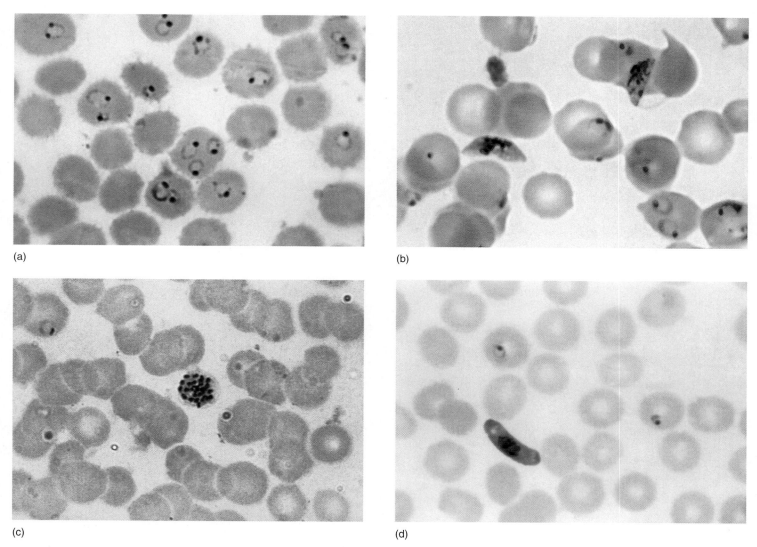

(a)   (b)

(c)   (d)

Figure 6.11 Stages of the *Plasmodium falciparum* erythocytic cycle. (a) Ring-form trophozoites (1150x). This stage represents the asexual, immature stage of *P. falciparum*. Note the central red blood cell, which contains three trophozoites. (b) Ring forms as well as crescent-shaped gametocytes are seen in this Giemsa stained blood smear (1200x). (c) Development of gametocytes leads to the production of a schizont (1000x) filled with as many as 24 merozoites. (d) Continued development of gametocytes leads to the production of microgametocytes (male) and megagametocytes (female). Here the sausage-shaped macrogametocyte is visible (1150x).

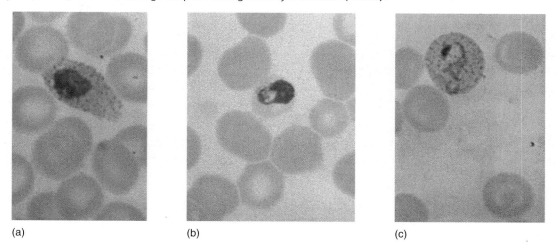

(a)   (b)   (c)

Figure 6.12 Comparison of trophozoite stage of three *Plasmodium* species (1150x). As the trophozoite matures, its morphology changes, transitioning from ring form to mature trophozoite. The morphology of the mature trophozoite stage allows species identification of *Plasmodium*. (a) *Plasmodium vivax*. (b) *Plasmodium malariae*. (c) *Plasmodium ovale*.

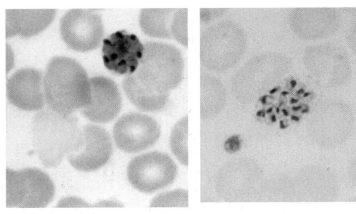

Figure 6.13 Immature (left) and mature (right)
***Plasmodium vivax* schizonts (1200x).** As the schizont matures, the
red blood cell in which it resides becomes 1.5 to 2 times larger than
an uninfected cell. The schizont may completely fill the cell, and 12 to
24 merozoites will be visible (16 are visible in the cell on the right).

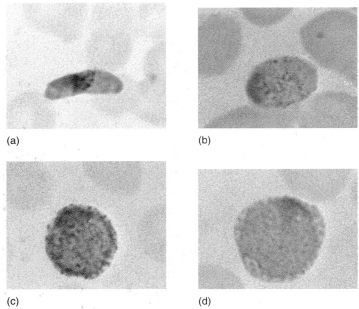

(a)

(b)

(c)

(d)

**Figure 6.14 Comparison of the macrogametocyte stage
of four *Plasmodium* species (1200x).** Male microgametocytes
and female macrogametocytes are ingested by an *Anopheles* mosquito
during a blood meal. While in the mosquito's stomach, the
macrogametocyte is penetrated by the microgametocyte, generating
zygotes. (a) *P. falciparum*. (b) *P. Malariae*. (c) *P. ovale*. (d) *P. vivax*.

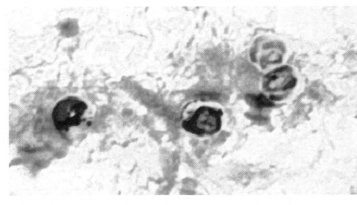

Figure 6.15 *Cryptosporidium parvum* oocysts (1000x). The
cells have been stained with a modified acid-fast technique, which
gives them the bright red coloration seen here. Ingestion of oocysts
leads to cryptosporidiosis, which may range from asymptomatic to
potentially life threatening, especially in patients who are
immunocompromised.

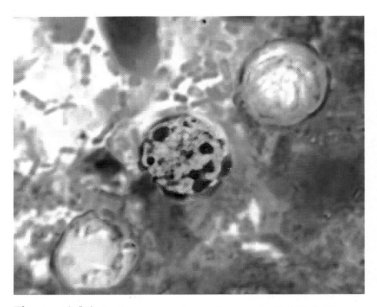

Figure 6.16 *Cyclospora caytanensis* oocysts (1000x). The
four *C. caytanensis* oocysts seen in this fecal sample indicate the
variable staining properties of the species. Infection with the parasite
leads to cyclosporiasis, a gastrointestinal disease that may range from
asymptomatic to debilitating. Although most common in tropical and
subtropical areas, at least 11 foodborne outbreaks of cyclosporiasis,
affecting approximately 3600 persons, have been documented in the
United States and Canada since 1990.

Figure 6.17 *Caulerpa taxifolia*, **a green algae (Chlorophyta) that is used in marine aquaria because of its hardiness.** When released into coastal waters, it quickly dominates, crowding out native species of algae. For this reason, many states, including California, have banned its use.

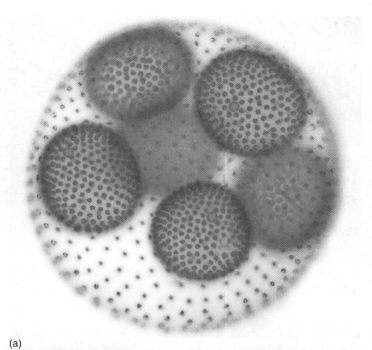

(a)

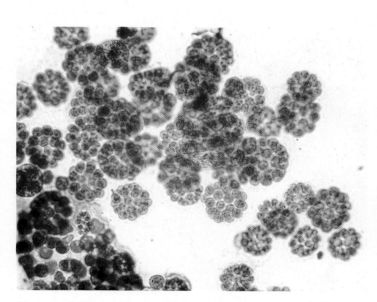

Figure 6.18 *Eudorina* **(100x).** Colonies of *Eudorina*, a colonial green algae found, most often, in colonies of 32 or 64 cells.

(b)

Figure 6.19 **Volvox.** (a) Phase contrast view (400x) of *Volvox*, a freshwater colonial algae. The colony consists of thousands of flagellated chlamydomonas-like cells embedded within an extracellular matrix. Each cell is connected to the nearest one with a cytoplasmic thread. The cells are motile and move with a common purpose. In this image, six daughter colonies can be seen within the larger mature colony. (b) Darkfield image of *Volvox* (100x).

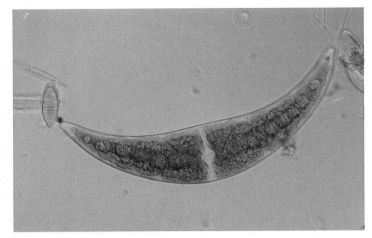

(a)

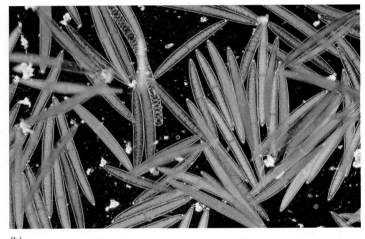

(b)

Figure 6.20 *Closterium* species are green alga (chlorophyta) typically appearing as elongated cylindrical cells, each of which contains two symmetrical semicells. Each semicell contains a single chloroplast. (a) Brightfield view (40x). (b) Darkfield view (10x).

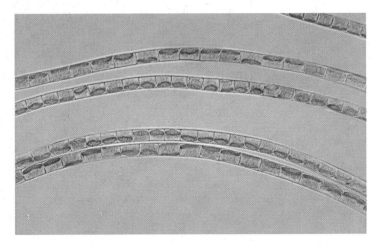

Figure 6.21 *Hormidium*, a freshwater filamentous algae (400x).

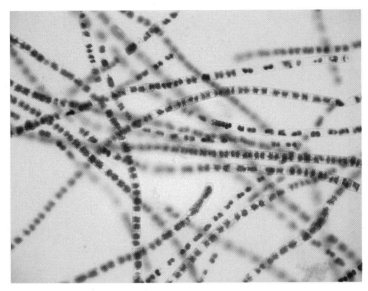

Figure 6.22 *Zygnema* **species (100x).** *Zygnema* species are composed of unbranched filaments with two stellate (star-shaped) chloroplasts in each cell.

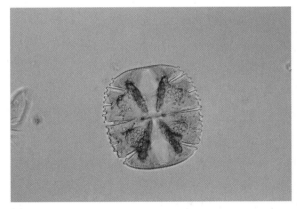

(a)

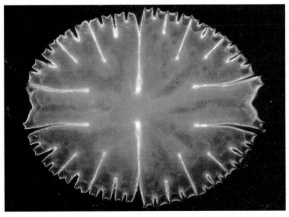

(b)

Figure 6.23 (a) Brightfield view of *Micrasterias truncata* (100x) a member of the Chlorophyta (green algae). All members of the genus *Micrasterias* display bilateral symmetry, with two mirror image semicells joined by a narrow band containing the nucleus of the organism. Each semicell contains a single large chloroplast. (b) Darkfield view of *Micrasterias* (400x).

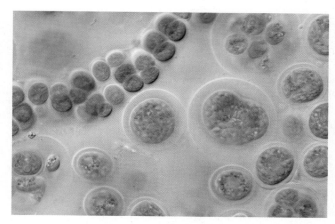

Figure 6.24 *Stigonema* and *Gloeocystis* (1000x). Differential inference contrast image of the filamentous cyanobacterium *Stigonema* viewed along with the green algae *Gloeocystis*. The photosynthetic pigment phycocyanin is responsible for the blue-green color seen in the cyanobacterium.

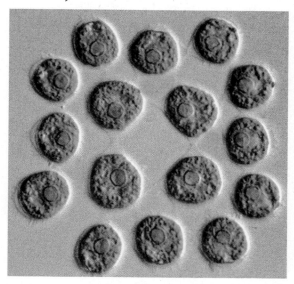

Figure 6.25 Differential interference contrast micrograph of *Gonium* (400x), a colonial freshwater algae. A member of the Chlorophyta (green algae), *Gonium* colonies have 4, 8, or 16 cells arranged in a plane and connected by a gelatinous matrix. In a 16-cell colony, such as the one seen here, four cells are in the center and three cells are on each of the four sides.

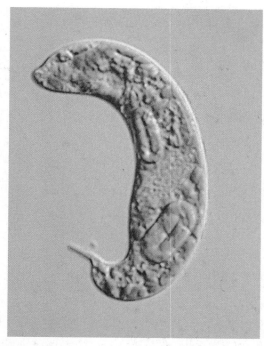

Figure 6.26 *Euglena* (640x). *Euglena* has characteristics of both plants and animals, leading to an ever-controversial taxonomic status. *Euglena* can both ingest food, like animals, and photosynthesize, like plants. *Euglena* are able to move through aquatic environments using a large flagellum for locomotion. Each cell contains a stigma, or eyespot, that detects light, allowing the *Euglena* to adjust its position to enhance photosynthesis. The motility of *Euglena* also allows for hunting when feeding as a heterotroph. Members of the genus lack cell walls, but instead have a pellicle consisting of protein bands that spiral down the length of the cell, just below the plasma membrane.

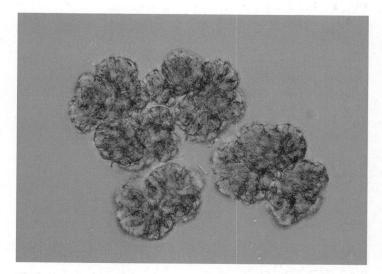

Figure 6.27 *Dictyosphaerium* (900x), a colonial green algae.

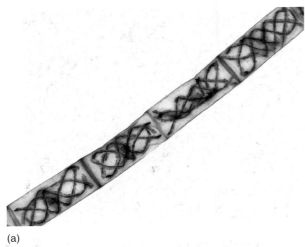
(a)

(b)

Figure 6.28 Two different views of _Spirogyra_, a green algae (190x). Chloroplasts of _Spirogyra_ are wound into a spiral within the cell, giving it a unique appearance. The algae is common in warm, clean water, where it forms filamentous masses, often forming large filamentous mats with two other genera of algae _Mougeotia_ and _Zygnema_.

(a)

(b)

Figure 6.29 Phaeophyceae. (a) _Nereocystis luetkeana_, a brown algae (Phaeophyceae) that can grow to 75 m in length, is found along rocky shores of the Pacific Ocean, from Monterey to the Aleutian Islands. Members of this group possess relatively high amounts of fucoxanthin, a photosynthetic pigment that gives the algae a brownish hue. (b) _Macrocystis pyrifera_, or California giant kelp, is also a member of the Phaeophyceae. The small air bladders near the beginning of each blade provide buoyancy, lifting the kelp to the surface of the water and allowing it to gather more light for photosynthesis.

Figure 6.30 Marine algae. A mixture of marine green algae (Chlorophyta) and brown algae (Phaeophyta) growing in a rocky tide pool.

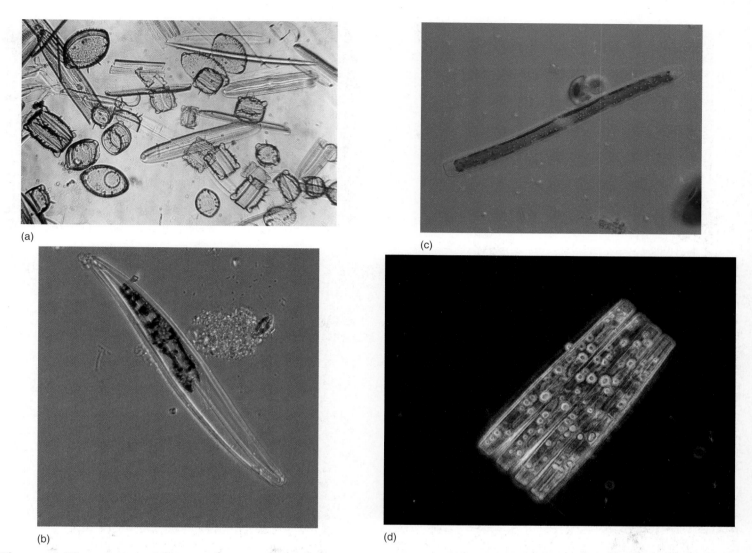

(a)

(c)

(b)

(d)

Figure 6.31 Diatoms. Diatoms found within the Chromalveolata are a tremendously large group composed of more than 200 genera and 100,000 different species. They tend to be easily recognized as their rigid silica cell walls provide them with distinctive shapes. (a) Diatom frustules (100x). (b) *Gyrosigma* (100x). (c) *Nitzschia* (200x). (d) *Fragillaria* (400x).

(a)

(b)

(c)

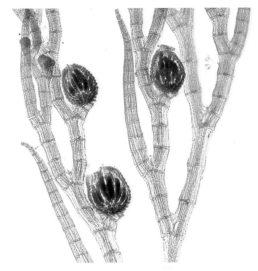

Figure 6.33 *Polysiphonia*. *Polysiphonia* is a red, filamentous, marine alga, reaching a length of about 30 cm. The haploid gametophyte supports cuplike carposporangia, each of which contains several carposporophytes.

Figure 6.32 Rhodophyta. Red algae (Rhodophyta) contain high amounts of a red photosynthetic pigment called phycoerythrin, which allows them to photosynthesize deep below the surface the water. Energy produced from photosynthesis is eventually stored in the form of starch. (a) *Corralline vancouverienis*, found along the Pacific Coast where it contributes to the formation of limestone reefs. (b) *Rhodoglossum affine*, found in the lower intertidal zone of the Pacific Coast. (c) *Chondrus crispus* displays the orange caratenoid pigments and the blue phycocyanin pigments often seen in members of the Rhodophyta.

Helminths

The study of worms that parasitize humans offers an entirely different experience from that of studying prokaryotic or even smaller eukaryotic parasites. Helminths are comparatively large, multicellular animals with tissues and organ systems. Their biology is in most cases far closer to the hosts they parasitize than to other microbes, a fact that makes effective treatment of helminthic infections difficult. Epidemiologists estimate that about 3.8 billion helminth infections are occurring worldwide at any given time, with most confined to Earth's tropical regions. In the laboratory, helminths are easily distributed among three categories through their physical appearance: nematodes (roundworms), cestodes (flatworms, or tapeworms), and trematodes (flukes). Although the biology of the organisms in each category is complex, with complicated life cycles, multiple hosts, and various levels of pathogenicity, diagnosis of infection generally relies on the visualization of the parasites. Oftentimes in fact, a diagnosis of helminthic infection is made simply by observing eggs in the feces, with a skilled parasitologist able to identify the organism responsible for an infection without ever actually laying eyes on it. This fact helps to explain the importance of helminth egg identification as a diagnostic tool.

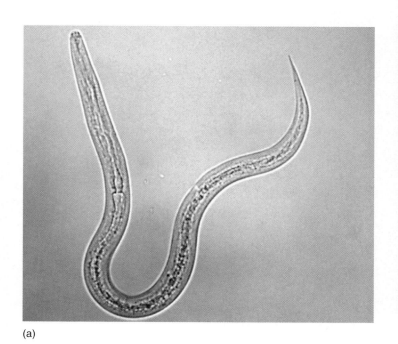

(a)

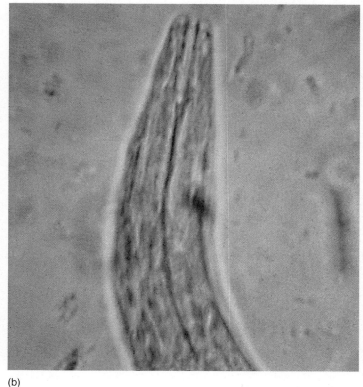

(b)

Figure 7.1 Hookworm. (a) Hookworm filariform larvae (100x). Two species of hookworm, *Ancylostoma duodenale* and *Necator americanus*, infect humans. As part of their life cycle, both species pass through an infectious filariform stage and a noninfectious rhabditiform stage. (b) Mouthparts of hookworm rhabditiform stage (1000x). Microscopic identification of eggs in the stool is the most common method for diagnosing hookworm infection.

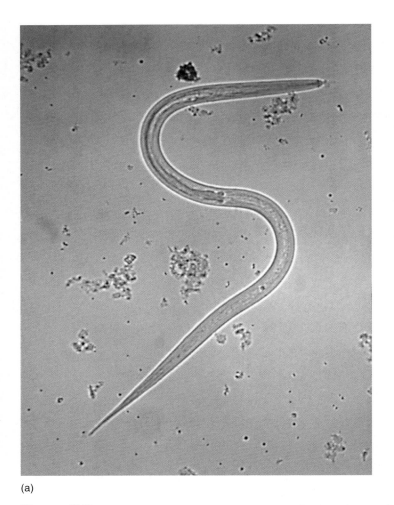

(a)

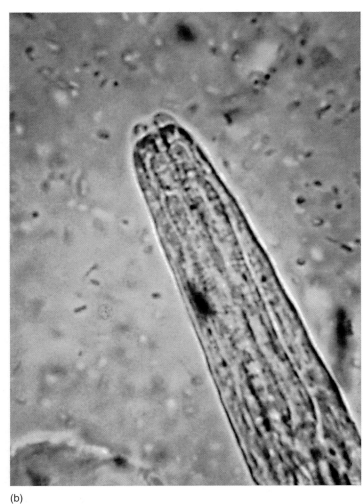

(b)

Figure 7.2 *Strongyloides.* (a) *Strongyloides* (100x) filariform larvae. The *Strongyloides* nematode has a complex life cycle that alternates between free-living (rhabditiform) and parasitic (filariform) cycles. (b) Mouthparts of *Strongyloides* rhabditiform larva (1000x). The presence of rhabditiform larvae in the stool is diagnostic of an infection.

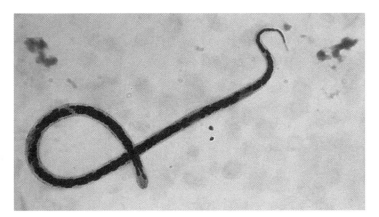

Figure 7.3 *Mansonella ozzardi* **(630x).** A filarial worm that is endemic to the Americas, with infections commonly seen in Central and South America. Midges and blackflies act as vectors, infecting the host, when the vector bites in search of a blood meal.

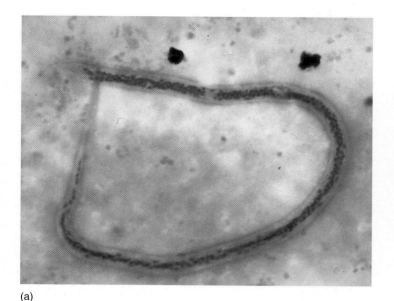

(a)

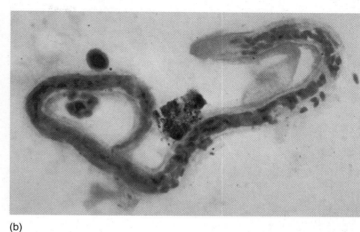

(b)

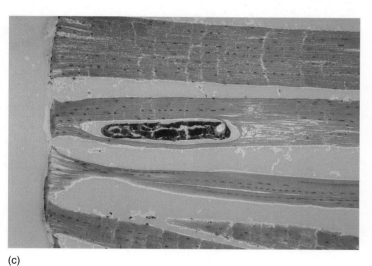

(c)

Figure 7.4 **Filarial worms.** Filariasis is caused by the infiltration of small roundworms (nematodes) into the lymphatic and subcutaneous tissue. Microscopic identification of microfilariae is the most common diagnostic procedure. While eight species infect humans, most morbidity due to filariasis can be traced to just three species. (a) *Wuchereria bancrofti* (460x) is the most common filarial parasite in humans and is found in tropical areas throughout the world. (b) *Brugia malayi* (630x) causes lymphatic filariasis and is limited to Asia. (c) *Onchocerca volvulus* (460x), seen here developing in its host, the black fly (*Simulium ochraceum*), causes onchocerciasis, or river blindness. It is found primarily in Africa, with a small area of endemicity in Latin America and the Middle East.

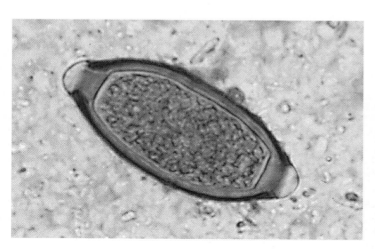

Figure 7.5 **Egg of *Trichuris trichiura*, the causative agent of trichuriasis, or human whipworm infection.** The eggs are recognized by their elongated shape, thick shell, and polar plugs. After ingestion, eggs hatch in the small intestine and release larvae, which mature into adults within the ascending colon, where they live for about a year. Sixty to seventy days after infection, the female will begin to oviposit, shedding between 3000 and 20,000 eggs per day into the feces. Whipworm is the third most common roundworm infection of humans, with an estimated 800,000 persons infected worldwide, including in the United States.

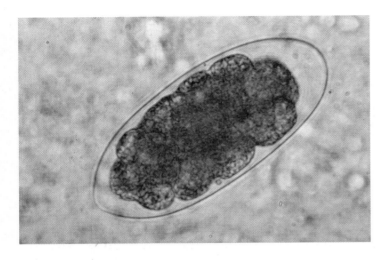

Figure 7.6 **Egg of *Trichostrongylus* sp. (1000x).** Ingestion of eggs of this parasitic nematode may result in trichostrongylosis. Microscopic identification of eggs in the feces is evidence of an infection. While primarily a disease of animals, several species of *Trichostrongylus* can cause human infection, including *T orientalis*, *T. colubriformis*, and *T. axei*.

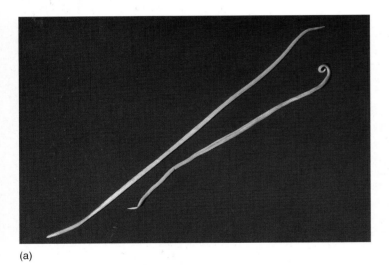

(a)

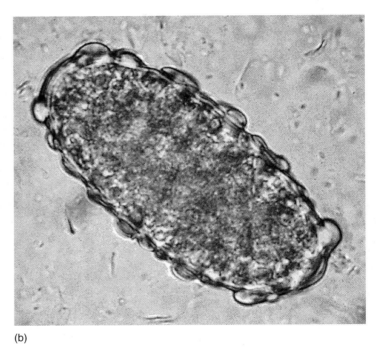

(b)

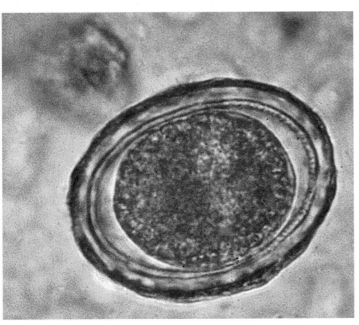

(c)

Figure 7.7 ***Ascaris lumbricoides.*** (a) These nematodes (roundworms) live in the small intestine where females—the larger of the two sexes— can grow to 30 cm in length. A female may produce 200,000 eggs per day, which pass from the body along with feces, but only fertilized eggs are infectious. (b) An unfertilized egg (400x) is elongated, covered with a thick, often rough layer, and larger than a fertilized egg. (c) A fertilized egg (400x) is rounded, thick shelled, and smaller than its unfertilized counterpart.

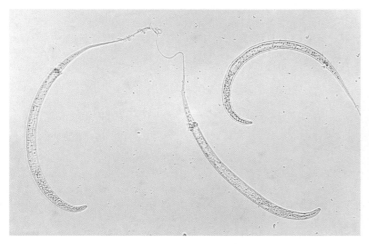

Figure 7.8 *Dranunculus medinensis* **(125x), also known as Guinea worm.** Humans become infected by drinking unfiltered water containing small crustaceans infected with *D. medinensis*. Male and female worms emerge from the crustacean and mate within the human host. After the death of the male worm, the female migrates toward the skin. About one year after infection, the female induces the formation of a painful blister on the lower extremity. When the lesion comes in contact with water (often done by the host to relieve the pain of the lesion), the female worm emerges from the body and releases larvae. Dracunculiasis has been drastically reduced as a result of intensive eradication efforts and is now restricted to rural, isolated areas in a small number of African countries.

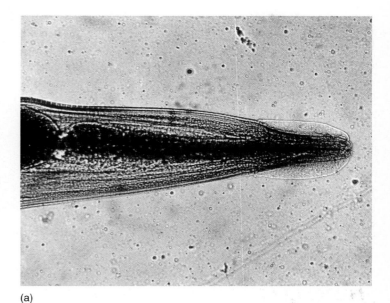

(a)

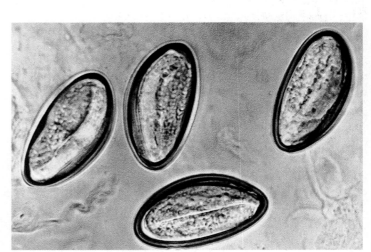

(b)

Figure 7.10 *Enterobius vermicularus.* (a) Head region of *Enterobius vermicularis* (100x). Enterobiasis, or pinworm infection, is the most common helminthic infection in the United States and is especially prevalent in young children. (b) *Enterobius* eggs. Gravid females migrate out of the anus of an infected individual and deposit eggs on the perianal folds. Hands that have scratched the perianal area may transfer the eggs to the mouth (self-infection) while contaminated clothes, bedding, or carpeting can spread eggs to others (400x).

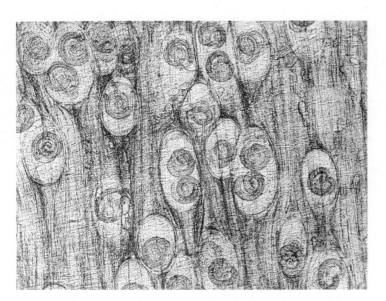

Figure 7.9 *Trichinella spiralis* **cysts embedded in muscle tissue in a case of trichinellosis (100x).** Trichinellosis is acquired by ingesting meat, commonly bear or pig, containing encysted larvae (cysts). Exposure to the harsh environment of the stomach releases the larvae, which mature into adult worms in the small intestine. After approximately a week, female worms release larvae that migrate to striated muscle, where they encyst, potentially remaining viable for years.

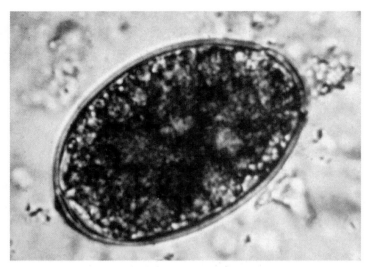

Figure 7.11   Egg of *Diphyllobothrium latum* (1000x). Also known as the fish tapeworm or broad tapeworm, *D. latum* is the largest human tapeworm, routinely growing to 10 m in length. Several intermediate hosts are involved in the life cycle, including freshwater crustaceans and minnows, but humans tend to be infected by larger freshwater fish such as trout or perch that may be eaten raw or undercooked. Mature worms reside in the small intestine where they can release more than 1,000,000 eggs per worm per day. Eggs are passed in the feces and infect the crustaceans, continuing the cycle.

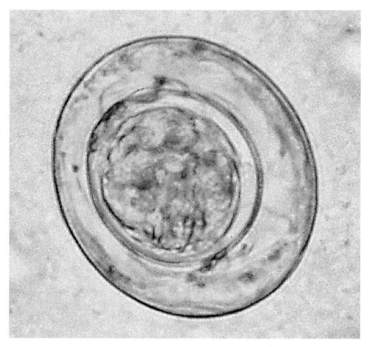

Figure 7.12   *Hymenolepsis nana* egg (1000x). Ingestion of *H. nana* eggs in contaminated food and water leads to hymenolepiasis, or dwarf tapeworm infection. A species measure 15mm to 40 mm in length. *Hymenolepsis nana* is the most common of all cestode infections and is encountered worldwide.

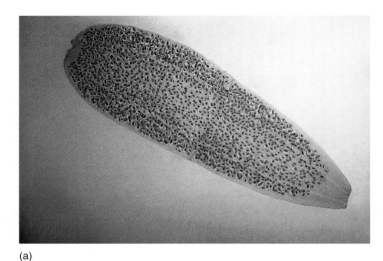

(a)

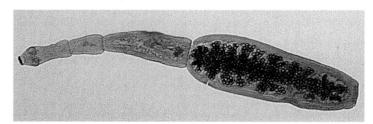

(b)

Figure 7.13   *Dipylidium caninum*. (a) *Dipylidium caninum* proglottid (8x). Dogs are the principal host of this tapeworm although humans, mostly young children, may also become infected if they ingest an infected flea. Adult worms measure up to 60 cm in length and reside within the small intestine of the host. Eventually, gravid proglottids (i.e., those full of eggs) are released in the feces.
(b) *D. caninum* egg packet. After exiting the host within the feces, the proglottid disintegrates, releasing the egg packet (1000x). Egg packets are ingested by larval stages of cat or dog fleas to continue the cycle.

Figure 7.14   *Echinococcus granulosus* (45x). *Echinococcus granulosus*, the cestode responsible for echinococcosis—also known as hydatidosis, or hydatid disease—will grow to 3–6 mm long as an adult, with a scolex and three proglottids. Dogs and other canids are the only definitive hosts, with humans becoming infected only after ingesting food or water contaminated with dog feces.

Figure 7.15 **Scolex of the tapeworm *Taenia pisiformis* (13x).**
Dogs, cats, foxes, bobcats, coyotes, and wolves are the definitive host for this tapeworm while rabbits and hares are the most common intermediate hosts.

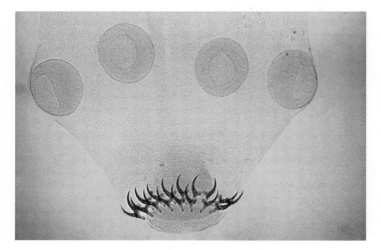

Figure 7.16 **Scolex of *Taenia solium*.** Scolex of the pork tapeworm *Taenia solium*, with its characteristic two rows of hooks and four suckers, all of which allow the tapeworm to attach to the intestine and grow to a length of 7 m. The beef tapeworm *Taenia saginata* typically grows to 5 m or less but otherwise appears very similar to *T. solium*. Humans are the only definitive hosts for either species.

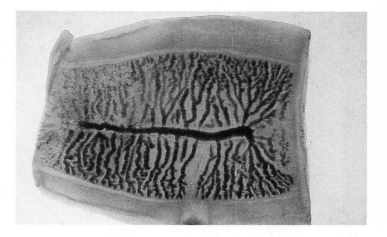

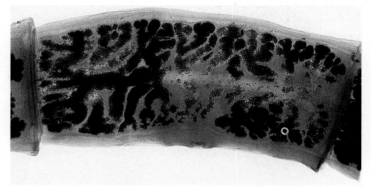

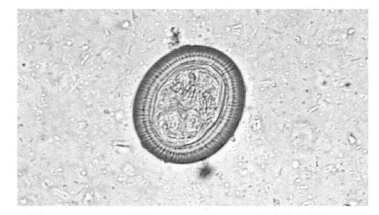

Figure 7.17 **Gravid proglottis of *Taenia saginata* and *Taenia solium*.** The presence of proglottids or eggs in the feces is diagnostic of infection, and the morphology of the proglottid is specific for each species of cestode. (a) *T. saginata* (8x) has 12–30 lateral uterine branches (the dark staining regions) on each side of the uterus. Adults consist of a scolex along with 1000–2000 proglottids, each containing about 100,000 eggs. (b) *T. solium* (8x) has about 1000 proglottids, each containing about 50,000 eggs. For both species, an average of 6 proglottids per day are passed through the stool. (c) A single *Taenia* sp. egg (400x). Eggs from all *Taenia* species are morphologically indistinguishable from each other and can survive for months in the environment. Pigs (*T. solium*) or cows (*T. saginata*) become infected by ingesting vegetation contaminated with eggs or gravid proglottids. Humans eventually become infected by eating raw or undercooked meat. Both species are found throughout the world, with *T. solium* being more prevalent in poorer communities where humans live in close contact with pigs and are more likely to eat undercooked pork. Because the dietary restrictions of Islam forbid the eating of pork, *T. solium* infection is very rare in Muslim countries.

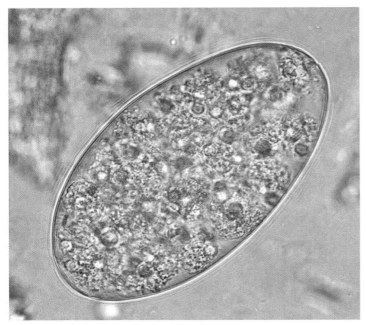

(a)

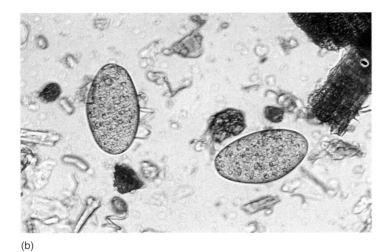

(b)

Figure 7.18 *Fasciolopis buski.* (a) *Fasciolopsis buski* trematode egg (500x). *F. buski* is the largest intestinal fluke found in humans and is most prevalent in Asia and the Indian subcontinent. Humans and pigs serve as mammalian hosts and become infected when they ingest freshwater plants harboring parasitic cysts. Maturation into adult flukes—measuring up to 75 mm long—occurs in the duodenum of the mammalian host, where adults live for about a year before dying. Microscopic identification of eggs, or rarely adult flukes, in the feces is the most common means of diagnosis. (b) *Fasciola buski* eggs in a stool sample (125x).

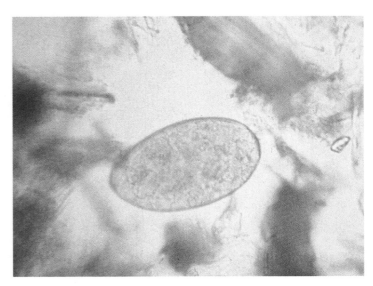

Figure 7.19 *Fasciola hepatica* (125x). The egg of the trematode *Fasciola hepatica*, or sheep liver fluke, is morphologically identical to that of *Fasciola buski*. Humans become infected with *F. hepatica* by ingesting metacercariae-containing freshwater plants, especially watercress. Development of the adult fluke from the infective metacercariae takes place over 3–4 months in the liver, where the adult flukes reside within the biliary ducts. Human infections with *F. hepatica* are found in areas where sheep and cattle are raised, and where humans consume raw watercress, including Europe, the Middle East, and Asia.

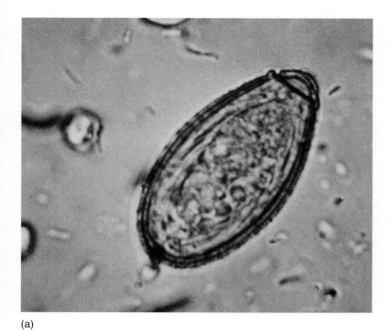

(a)

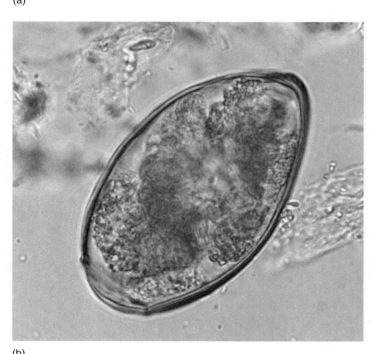

(b)

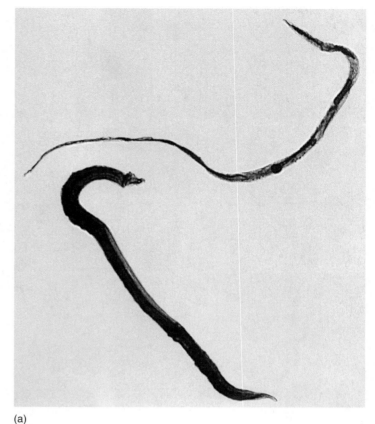

(a)

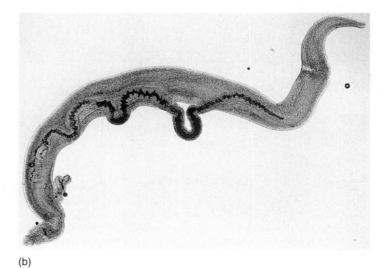

(b)

Figure 7.20 Trematode eggs. (a) The egg of the trematode
Clonorchis sinensis (400x), or Chinese liver fluke, is generally 27–35 μm
in length, 11–20 μm in width and contains an easily visible operculum
at the smaller end of the egg. *C. sinensis* is endemic to parts of Asia
including Korea, China, Taiwan, and Vietnam. It has also been seen in
nonendemic areas—including the United States—where it is invariably
linked to recent immigrants or the ingestion of imported, undercooked,
or pickled freshwater fish containing metacercariae of the parasite.
(b) Egg of the trematode *Paragonimus westermani* (128x), or Oriental
lung fluke, is several times larger than *Clonorchis* and appears
asymmetrical, with one side slightly flattened. The operculum is visible
on the large end while the opposite end of the egg is thickened.
Human infection with *P. westermani* occurs by eating inadequately
cooked or pickled crab or crayfish that harbor metacercariae of the
parasite, which is found throughout southeast Asia and Japan.

Figure 7.21 *Schistosoma mansoni.* (a) *Schistosoma mansoni*
trematodes (80x). The female (top) is much smaller than the male
(bottom). (b) During mating, the female is cradled within the
gynecophoral canal of the male (100x).

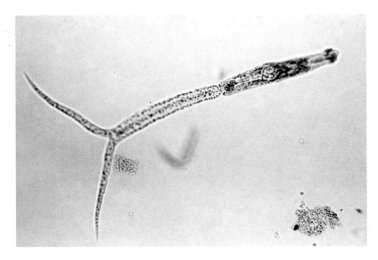

Figure 7.22 *Schistosoma* **cercaria (150x).** The cercaria is the infective, free-living stage of the *Schistosoma* parasite. This stage is produced in the snail, which is the intermediate host of *Schistosoma*. After exiting the snail, the cercaria penetrates the skin of a human host, shedding its forked tail as it does so.

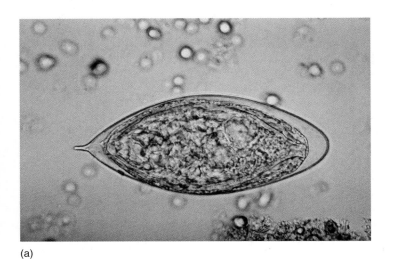

(a)

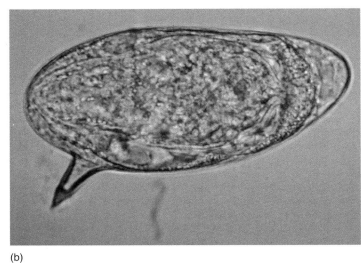

(b)

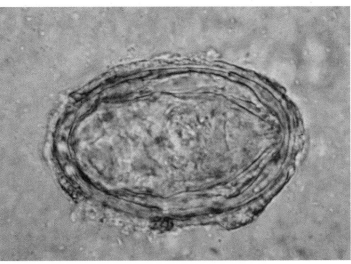

(c)

Figure 7.23 *Schistosoma* **eggs.** Eggs from trematodes responsible for schistosomiasis. Three species are responsible for the vast majority of human infection. The presence of eggs in the feces or urine, depending on the species, is indicative of an infection as well as aids in the identification of the species responsible. (a) *Schistosoma haematobium* (500x). The egg is recognized by its terminal spine. (b) *Schistosoma mansomi* (500x) possesses a lateral spine. (c) *Schistosoma japonicum* (500x) lacks a distinct spine.

Hematology and Serology

The study of blood, as a means of evaluating the health of the body, is a familiar activity for anyone who has ever provided a blood sample during a doctor's visit. In terms of microbiology, however, blood is generally studied with one of three goals in mind: 1) blood typing prior to a transfusion, 2) diagnosing infection with bloodborne pathogens, or 3) using the antibodies within the blood to identify specific antigens. This section presents a very small slice of this complex field, concentrating first on the major cellular components of blood, moving on to blood typing, and finally to serological reactions.

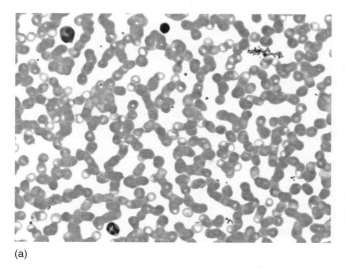
(a)

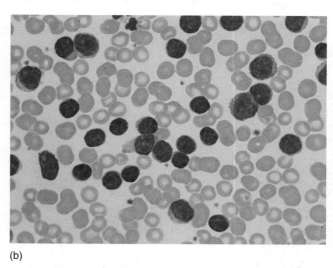
(b)

Figure 8.1 Peripheral blood smears. (a) A peripheral blood smear from a healthy person reveals several thousand doughnut-shaped red blood cells (erythrocytes) and only a few of the larger, darker staining leukocytes—or white blood cells—a typical ratio in a healthy person (400x). Also visible are many blue/gray platelets, which are involved in blood clotting and are easily distinguished by their small size. (b) In leukemia, the number of leukocytes increases dramatically in comparison to erythrocytes. Changes in the appearance of the leukocytes themselves, such as an increase in the amount of the cell occupied by the nucleus, help to identify the condition seen here as chronic lymphocytic leukemia (400x).

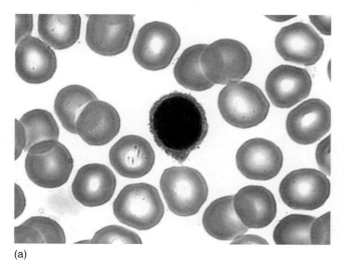
(a)

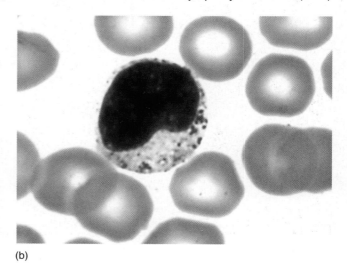
(b)

Figure 8.2 Agranulocytic leukocytes. (a) Lymphocytes (1600x) are the second most common white blood cell (WBC), comprising 20%–35% of total WBCs. Lymphocytes appear as small spherical cells with a round, darkly staining nucleus surrounded by a thin layer of cytoplasm. Lymphocytes are central to the specific immune reactions within the body. (b) Monocytes (410x) are the largest WBCs and display a large, often indented, nucleus. The principle activity of monocytes is phagocytosis. Eventually these cells leave the bloodstream and mature to become macrophages and dendritic cells within the tissues.

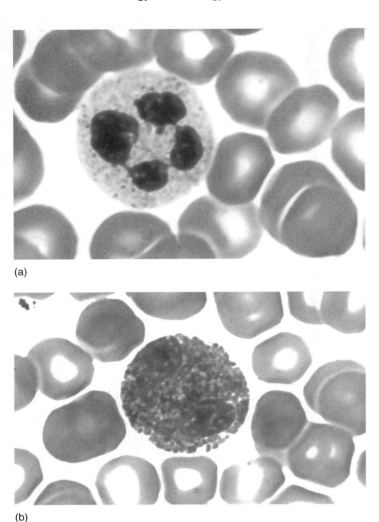

(a)

(b)

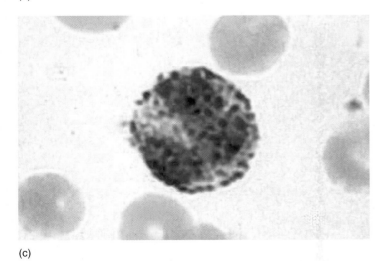

(c)

Figure 8.3 Granulocytic leukocytes. (a) Neutrophils are the most prevalent of leukocytes, comprising 50%–90% of all white blood cells. They function in general phagocytosis and can be recognized by their multilobed nucleus and small, purple, cytoplasmic granules (1600x). (b) Eosinophils make up 1%–3% of circulating white blood cells and display a bilobed nucleus along with large orange granules. Eosinophils generally increase in number during infection with eukaryotic pathogens. (c) Basophils (320x) are the rarest of white blood cells, making up only 0.5% of circulating leukocytes. Basophils are involved in the allergic response.

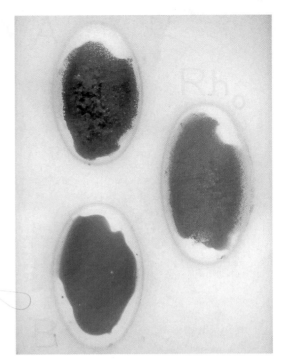

Figure 8.4 Blood Typing. Each well of the blood-typing slide contains a drop of the blood to be tested along with a single type of antibody: anti-A, anti-B, or anti-Rh. If the red blood cells express a cell surface antigen recognized by the antibodies in the well, agglutination of the cells occurs. Here, because agglutination has occured in the "A" well (containing anti-A antibody) on the upper left and the "Rh" well (containing anti-Rh antibody) seen on the right-hand side of the slide, the blood can be designated as A+.

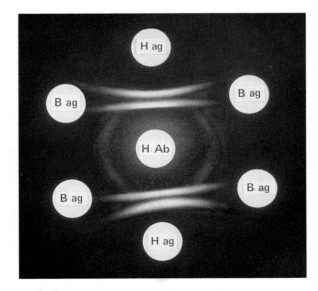

Figure 8.5 Double immunodiffusion (Ouchterlony) test in agar gel. The central well in the agar contains antibodies to *Histoplasma capsulatum*. The two wells at the top and bottom contain *Histoplasma* extract while the four wells to the left and right contain extract from *Blastomyces dermatitidis*. The bands appearing above and below the center well are caused by a precipitation reaction occurring between the *Histoplasma* antigens and the antibodies to *Histoplasma*. The fact that no bands appear between the center well and any of the *Blastomyces* antigens indicates no antigen/antibody binding.

Colony Morphology

PRINCIPLES AND APPLICATIONS

When inoculated onto a solid media, bacterial cells typically grow in tight associations known as **colonies,** which are simply the result of a single bacterial cell dividing to produce two cells, two cells dividing to produce four, four producing eight, and so on. After just 20 divisions, a single cell will produce a colony of over one million cells, easily visible to the naked eye. Careful evaluation of the physical characteristics of a bacterial colony can help in its identification and allow you to distinguish between multiple bacterial species growing on an agar plate. Shape, color, texture, and size are important properties that differ from one bacterial species to another, and all may aid you in identification of an unknown specimen. In some cases, morphology is used to include or exclude bacterial colonies from further study. For example, a sputum specimen will contain bacteria that represent the normal flora of the throat as well as species that may be responsible for a lower respiratory infection; colony morphology can often be used to differentiate between the two, speeding laboratory work immensely. Some bacteria produce colonies so large that details can be seen with the naked eye, in most cases, however, a small degree of magnification will prove quite helpful in differentiating details (Figure 40.1).

Figure 40.1 Colony morphology shows tremendous variation between different bacterial species.

Shape, which is often one of the most obvious difference between colonies of different species, is described with reference to the entire colony (**configuration**), colony edge (**margin**), and colony height (**elevation**). Examples of these characteristics can be seen in Figure 40.2. Colony configuration and margin are usually best examined using transmitted light, while elevation is best seen from the side using reflected light.

Texture also varies from species to species, with colonies being described as **moist, mucoid,** or **dry.** This characteristic is a function of the adherence of the individual bacterial cells to one another and can be easily seen when a colony is lifted from the plate using a needle. Dry colonies will break apart and have difficulty adhering to the needle while moist colonies can be pulled apart more easily, with growth adhering to both the plate and the needle. Mucoid colonies may leave a string of extracellular material a centimeter or more in length stretching from the needle to the colony. As this mucoid string may be several thousand times longer than the individual cells themselves, it says much about the structure and relationship of individual cells in the colony. Color is also a consideration but is often influenced by the type of media upon which a culture is growing, the temperature of incubation, or the age of the culture (Figure 40.3). Lastly, colony color may be combined with the descriptors **shiny, dull, opaque,** or **translucent** to provide a more accurate depiction. Images seen in this exercise provide examples of many of the characteristics you may encounter.

MATERIALS

Each person should obtain or have access to:
Streak plate cultures of your unknown organism
Metric rulers
Colony counter and/or dissecting microscope

PROCEDURE

1. Evaluate the growth of your unknown bacteria using the terms seen in this exercise. Measure the diameter of a few representative colonies, and record all the information on the data sheet (Appendix E) for your unknown.

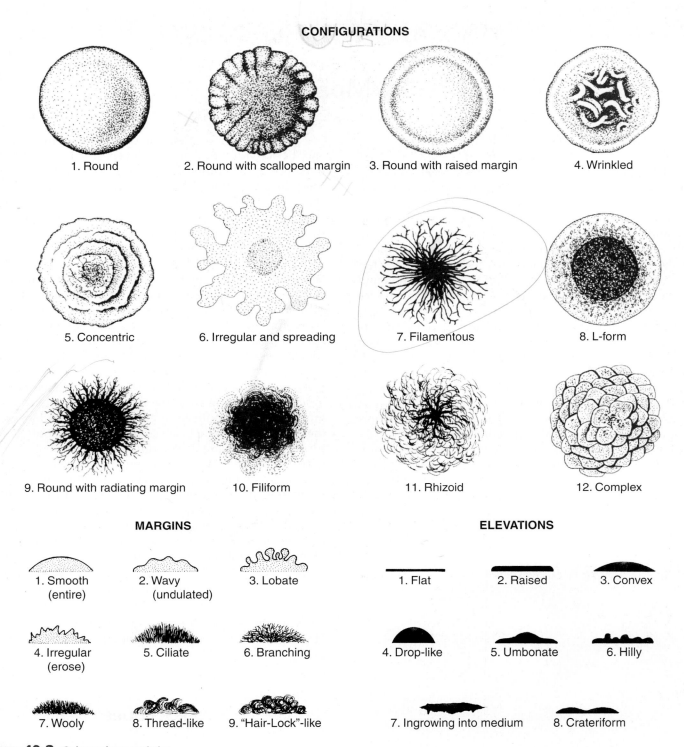

Figure 40.2 Colony characteristics.

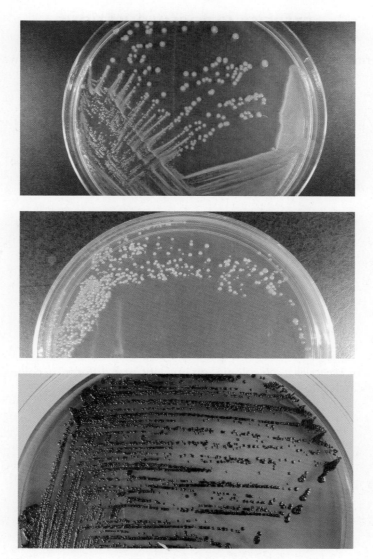

Figure 40.3 Colony color is often influenced by the type of media upon which a bacterial isolate is growing. Here, *Escherichia coli* is seen growing on (from top to bottom) nutrient agar, MacConkey agar, and EMB agar.

- Round
- Smooth
- Flat
- light yellow

- Round
- Smooth
- Flat
- light yellow

EXERCISE 40 REVIEW QUESTIONS

1. Distinguish between the terms **configuration, margin,** and **elevation.** Explain the best way to view each of these characteristics.

2. How would the appearance of *Escherichia coli* colonies on nutrient agar differ from colonies of the same species on MacConkey agar (a little research into MacConkey agar will be necessary)?

REFERENCES

Reddy, C.A., Beveridge, T.J., Breznak, J.A., Marzluf, G.A., Schmidt, T.M., and Snyder, L.R. (eds.). 2007. *Methods for General and Molecular Microbiology, 3rd ed.,* chap. 15. Washington D.C.: ASM Press.

Wiley, J., Sherwood, L., and Woolverton, C. 2011. *Prescott, Harley & Klein's Microbiology, 8th ed.,* chap. 6. New York: McGraw-Hill.

Growth in Solid and Liquid Media

PRINCIPLES AND APPLICATIONS

Although cultures are often streaked for isolation on an agar plate, they tend to be more commonly held on agar slants or in broths. Bacteria growing in either type of media display specific growth patterns, and while it would be impossible to conclusively identify an organism just from these characteristics, they can be helpful in ruling in or out certain species.

Growth on an agar slant may be evaluated in several ways. First, the amount of growth may be described as **none, slight, moderate,** or **abundant.** The quantity of growth generally depends on both the type of media used and the incubation temperature; some organisms will grow well under a variety of conditions while others require enriched media or a special incubation environment.

Colored pigments may be produced by some bacteria under certain conditions. *Serratia marcescens* for instance produces a deep red pigment at 25°C but not at 38°C. Other bacteria, such as *Pseudomonas fluorescens* may produce diffusible pigments (Figure 41.1), which cause the media itself (as opposed to the bacteria) to become colored. Most bacteria though do not produce pigments, instead producing white to off-white colonies.

Opacity is related both to the degree of growth as well as any pigments produced. Bacteria may be described as **transparent, translucent,** or **opaque.** Holding your slant up

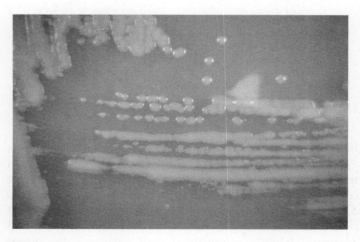

Figure 41.1 *Pseudomonas sp.* are often recognized by the deep green pigment they produce.

to a light source will allow you to determine if any pigment is present as well as the degree of opacity.

The manner in which a bacterial species grows on the surface of an agar slant or within a broth can also be characterized. In the case of a slant, growth may be termed **filiform, echinulate, beaded, effuse, arborescent,** or **rhizoid,** all of which are illustrated in Figure 41.2. Within a broth, bacteria may be visible on the surface, below the surface, and at the bottom of the tube. When describing surface growth, growth

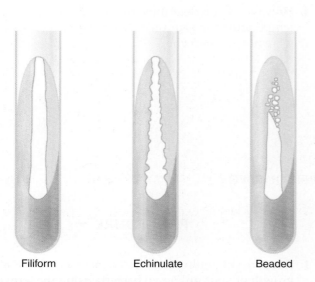
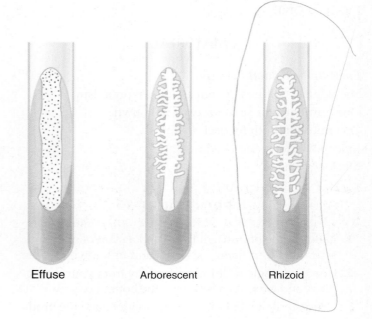

| Filiform | Echinulate | Beaded | Effuse | Arborescent | Rhizoid |

Figure 41.2 Bacterial growth characteristics on agar slants.

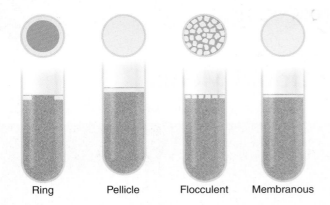

Figure 41.3 Surface growth characteristics in liquid media.

that completely covers the surface of the broth may be described as **membranous** if thin and as a **pellicle** if much thicker. If bacteria adhere to the sides of the tube but do not completely cover the surface of the media, it is described as a **ring** while floating masses of bacteria are termed **flocculent** (Figure 41.3). Below the surface, the broth may be described as **turbid** (cloudy), **granular** (small particles), **flocculent** (large particles), or **flaky** (large flakes). Sediment at the bottom of the tube may also be characterized, which generally requires agitating the tube to suspend the bacteria that have settled. Once this is done, the sediment may be described as **granular, flocculent,** or **flaky,** depending on the size of the particles. The broth may be described as **viscid** if a thick mass of bacterial cells and extracellular matrix appear when the tube is shaken or stirred with a sterile loop. Finally, just as with growth on a slant, the abundance of growth in a broth may be categorized using the terms **slight, moderate,** and **abundant.**

PERIOD ONE

MATERIALS

Each student should obtain:
Streak plate cultures of your unknown organism
One nutrient agar slant for each unknown
One nutrient broth for each unknown
Inoculating loop and needle
Marking pen

PROCEDURE

1. Label a tube of nutrient broth and a slant of nutrient agar with your name, lab time, and unknown number.
2. Choose a single well-isolated colony from your streak plate, and circle its position on the bottom of your plate.
3. Use a sterile inoculating needle to inoculate the broth. After sterilizing the needle, use the same colony to

inoculate the slant with a fishtail streak. Be sure to scrupulously follow aseptic technique as you complete your inoculations.
4. Incubate the broth and slant at the optimum temperature for your unknown (your instructor can provide this information) for 24–48 h.

QUESTIONS—PERIOD ONE

1. What do the terms *arbor* and *rhizoid* mean? Distinguish between arborescent growth and rhizoidal growth.

2. Distinguish between a diffusible pigment and a nondiffusible pigment.

3. Can a bacterial culture be described as both membranous and turbid? Explain.

4. How could you determine the optimum incubation temperature for your unknown bacterial culture if it were not given to you?

PERIOD TWO

PROCEDURE

1. Retrieve your cultures from the incubator. Evaluate the growth of your unknown bacteria using the terms seen in this exercise, and record all the information for your unknown on the data sheet found in Appendix E.

EXERCISE 41 REVIEW QUESTIONS

1. Why is longer-term storage of bacterial cultures done using agar slants rather than agar plates?

2. What characteristic of a bacterial culture would be most likely to change if the culture were grown at a less than optimal temperature?

REFERENCES

Reddy, C.A., Beveridge, T.J., Breznak, J.A., Marzluf, G.A., Schmidt, T.M., and Snyder, L.R. (eds.). 2007. *Methods for General and Molecular Microbiology, 3rd ed.,* chap. 15. Washington D.C.: ASM Press.

Wiley, J., Sherwood, L., and Woolverton, C. 2011. *Prescott, Harley & Klein's Microbiology, 8th ed.,* chap. 6. New York: McGraw-Hill.

NOTES

Growth in Solid and Liquid Media

335

Jamie

Simple Stain

PRINCIPLES AND APPLICATIONS

Bacterial cells tend to lack optical contrast, essentially meaning that the cells themselves are very nearly transparent. As you can imagine, microscopic differentiation of nearly clear cells from a nearly clear background is difficult. To aid in this endeavor, bacterial cells are commonly stained to increase the contrast with their surroundings. One of the most straightforward applications of dye to increase contrast is the **simple stain.** In this technique, a dye with a positively charged **chromophore** (the colored portion of the dye molecule) is used to stain a bacterial **smear** on a slide. The vast majority of bacterial cells carry a large number of negatively charged proteins on their surface, and these proteins are attracted to the positive charge of the dye (Figure 42.1).

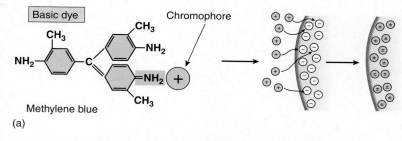

(a)

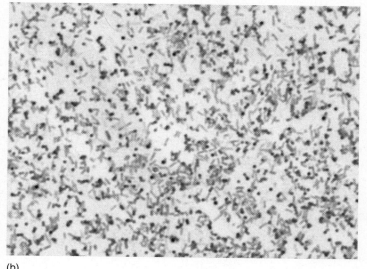

(b)

Figure 42.1 (a) In simple staining, a positively charged dye such as methylene blue is allowed to react with the negatively charged proteins on the surface of the bacterial cell. Electrostatic attraction results in the cells retaining the color of the dye, and all types of cells stain approximately equally. (b) A simple stain of *Bacillus subtilis* and *Staphylococcus* stained with methylene blue.

The end result is that nearly all bacteria stain readily with positively charged, or basic, dyes. Examples of basic dyes commonly used in the microbiology lab include methylene blue, crystal violet, safranin, and malachite green. In all cases, it is the charge of the dye, not the color, which is responsible for the final result.

Simple staining is useful for determining the morphological characteristics of a cell, such as size, shape, and arrangement, but little else. Because the attraction between the dye and the cell is based simply on the overall charge of the cell and not on a specific biological property, all cells will stain to approximately the same extent, regardless of any important biological differences between them.

Prior to most staining techniques, a bacterial smear is prepared and **heat-fixed.** The purpose of the smear is to apply an adequate number of cells to the slide while heat fixation kills the cells (making them safe to work with) and binds them to the slide so that they do not come off during the staining process. When using a solid culture to prepare a smear, it is important to remember that too many organisms will tend to stain unevenly, and the arrangement of bacterial cells will be difficult to determine. Broth cultures are the best choice for viewing bacterial arrangements, but care must be taken to use enough bacteria in the preparation of a smear. The process of making a smear is illustrated in Figure 42.2 along with a review of aseptic handling of cultures in Figure 42.3.

In this exercise, smears will be prepared from both solid and liquid cultures of bacteria. Cells of an avirulent strain of *Corynebacterium diphtheriae*, growing in a solid culture, will be used to illustrate three common characteristics of the genus. The first is **pleomorphism,** or a difference in the size and shape of cells within a single species. The second is the **palisades arrangement,** a side-by-side arrangement of cells that is characteristic of the genus (Figure 42.4). The third distinguishing trait is **metachromatic granules,** small granules of phosphate that appear red when cells are stained with methylene blue. With careful observation, you should be able to see all of these properties on your slide. For the second slide, a liquid culture of *Escherichia coli* will be used to prepare a smear for staining and microscopic observation.

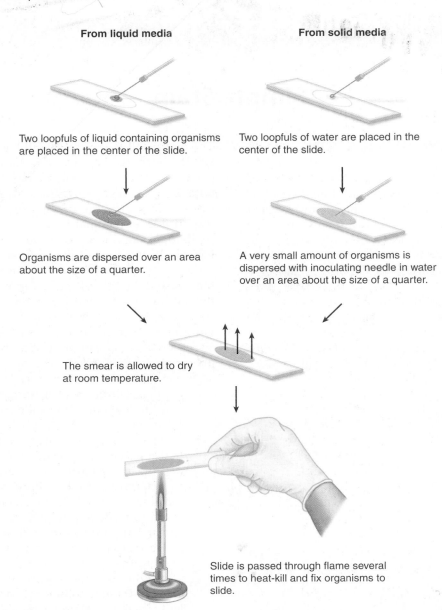

From liquid media

Two loopfuls of liquid containing organisms are placed in the center of the slide.

Organisms are dispersed over an area about the size of a quarter.

From solid media

Two loopfuls of water are placed in the center of the slide.

A very small amount of organisms is dispersed with inoculating needle in water over an area about the size of a quarter.

The smear is allowed to dry at room temperature.

Slide is passed through flame several times to heat-kill and fix organisms to slide.

Figure 42.2 Preparing a heat-fixed smear from solid or liquid media.

Figure 42.3 Aseptic procedure for organism removal.

(1) Shake the culture tube from side to side to suspend organisms. Do not moisten cap on tube.

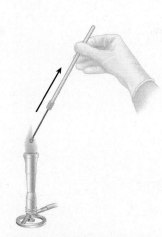

(2) Heat the loop and wire to red-hot. Flame the handle slightly, also.

(3) Remove the cap and flame the neck of the tube. Do not place the cap down on the table.

(4) After allowing the loop to cool for at least 5 sec, remove a loopful of organisms. Avoid touching the side of the tube.

(5) Flame the mouth of the culture tube again.

(6) Return the cap to the tube, and place the tube in a test tube rack.

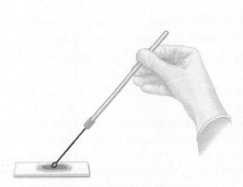

(7) Place the loopful of organisms in the center of the slide.

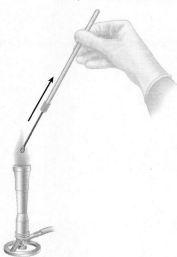

(8) Flame the loop again before removing another loopful from the culture or setting the inoculating loop aside.

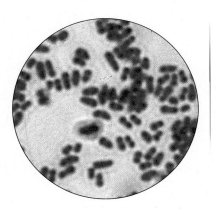

Figure 42.4 Simple stain of *Corynebacterium diphtheriae* with methylene blue. Note the variation in size between cells (pleomorphism) and the parallel arrangement of cells (palisades).

PRE-LAB QUESTIONS

1. Methylene blue, crystal violet, and safranin are all different colors. What characteristic do they share that is important for simple staining?

2. What is the importance of limiting the number of cells used to prepare a smear?

3. Name two reasons that cells are heat-fixed prior to staining.

MATERIALS

Each student should obtain:

Cultures of avirulent *C. diphtheriae* grown on Loeffler's medium

Broth culture of *E. coli*

Two glass slides

Methylene blue stain

Wash bottle

Bibulous paper

Inoculating loop and needle

Marking pen

PROCEDURE

This procedure is also illustrated in Figure 42.5.

1. Label a slide in the upper left-hand corner with the name of the organism to be stained.
2. Prepare a heat-fixed smear of *C. diphtheriae* and *E. coli* following the procedure outlined in Figures 42.2 and 42.3.
3. Flood each heat-fixed smear with methylene blue, and allow the stain to penetrate for 1 min.
4. Briefly (2–3 sec) wash excess stain from the slide with water.
5. Gently blot the slide with bibulous paper to remove excess water.
6. Locate the stained specimen using the low power objective of your microscope; keep the specimen centered and in focus as you move to high power and then to oil immersion. Examine the stained specimen under oil immersion. Record your observations in the results section of this exercise.

 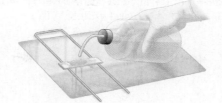 

(1) A bacterial smear is stained with methylene blue for 1 min.

(2) Stain is briefly washed off slide with water.

(3) Water drops are carefully blotted off slide with bibulous paper.

Figure 42.5 Procedure for simple staining.

RESULTS

Draw images from your slides that show the three noteworthy characteristics of *C. diphtheriae*.

EXERCISE 42 REVIEW QUESTIONS

1. *Staphylococcus aureus* is a coccus-shaped bacterium that grows in large irregular bunches (like grapes). How would *S. aureus* differ visually from *C. diphtheriae* and *E. coli* when examined using a simple stain?

2. A small number of bacterial species have a preponderance of *positive charges* on the outside of their cells. How would these cells stain with a dye such as methylene blue or crystal violet? Why?

REFERENCES

CDC. *Diphtheria.* http://www.cdc.gov/ncidod/dbmd/diseaseinfo/diptheria_t.htm.

Reddy, C.A., Beveridge, T.J., Breznak, J.A., Marzluf, G.A., Schmidt, T.M., and Snyder, L.R. (eds.). 2007. *Methods for General and Molecular Microbiology, 3rd ed.*, chap. 2. Washington D.C.: ASM Press.

Wiley, J., Sherwood, L., and Woolverton, C. 2011. *Prescott, Harley & Klein's Microbiology, 8th ed.*, chap. 2. New York: McGraw-Hill.

Negative Stain

PRINCIPLES AND APPLICATIONS

In most cases, increasing the contrast between bacterial cells and the background for microscopic examination is accomplished by staining the cells with a positively charged, or basic, dye so the cells appear far darker than the background. In the case of negative staining, however, a negatively charged, or acidic, dye is used to stain the background while leaving the bacterial cells unchanged. When examined, the cells are seen as light objects against a dark field (Figure 43.1). Key to this type of staining are negatively charged stains such as nigrosin and india ink.

The procedure for negative staining involves mixing cells with the acidic dye and spreading them over the surface of a slide in a very thin layer. The negative-charged chromophore of the acidic dye allows the dye to bind to the glass of the slide—which carries a slight positive charge—while being repelled by the negatively charged bacterial cells. Negative staining provides a rapid means of determining morphological features such as shape and arrangement. Furthermore, because heat fixation—which causes cell shrinkage—is not involved, negative staining is often the best choice for determining the size of a cell. This relatively gentle process is also used for observing spirochaetes, which are quite thin and do not stain well with basic stains.

PRE-LAB QUESTIONS

1. *The dyes normally used for negative staining, nigrosin and india ink, are negatively charged black dyes. How would a bacterial cell look if it were stained with eosin, a negatively charged* red *dye?*

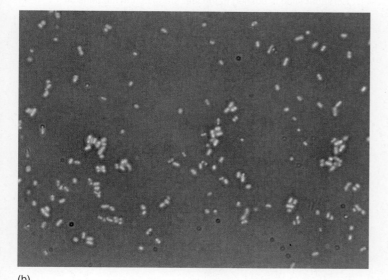

Acidic dye

nigrosine

(a)

(b)

2. *What dye could be used to perform a negative stain of a rare bacterial species that is covered with positively charged proteins?*

Figure 43.1 (a) In negative staining, a negatively charged dye such as nigrosin is used to stain bacterial cells. The dye is attracted to the glass slide and repelled by the negatively charged proteins on the surface of the bacterial cell, resulting in clear cells against a dark background. (b) A negative stain of bacterial cells.

MATERIALS

Each student should obtain:

Slant cultures of *Bacillus megaterium* and *Staphylococcus aureus*

Nigrosin or india ink

Inoculating loop and needle

Marking pen

Two glass slides

PROCEDURE

This procedure is illustrated in Figure 43.2.

1. Label a slide in the upper left-hand corner with the name of the organism to be stained.

2. Use a dropper to place a small (less than 1/8 in.) drop of nigrosin or india ink near one end of a clean microscope slide.

3. Use an inoculating needle to disperse a small amount of each organism into the dye, remembering to flame your needle completely between organisms.

4. Place a spreader slide at an angle in front of the drop, and slowly move it backward until the slide contacts the bacteria/dye suspension.

5. Push the spreader along the slide, dragging the suspension across the bottom slide.

6. Allow the slide to air dry.

7. Examine the stained specimen under low power, high power, and oil immersion. Record your observations in the Results section of this exercise.

RESULTS

Draw images from your slides that show two microscopic fields.

Organisms are dispersed into a small drop of nigrosin or india ink. Drop should not exceed 1/8" diameter and should be near the end of the slide.

Once spreader slide contacts the drop on the bottom slide, the suspension will spread out along the spreading edge as shown.

Spreader slide is moved toward drop of suspension until it contacts the drop causing the liquid to be spread along its spreading edge.

Spreader slide is pushed to the left, dragging the suspension over the bottom slide.

Once the preparation has completely air-dried, it may be examined under oil immersion. No heat should be used to hasten drying.

Figure 43.2 Procedure for negative staining.

EXERCISE 43 REVIEW QUESTIONS

1. How can you differentiate between *S. aureus* and *B. megaterium* in a negative stain?

2. What could happen if too much nigrosin is used in the preparation of a negative stain?

REFERENCES

Reddy, C.A., Beveridge, T.J., Breznak, J.A., Marzluf, G.A., Schmidt, T.M., and Snyder, L.R. (eds.). 2007. *Methods for General and Molecular Microbiology, 3rd ed.*, chap. 2. Washington D.C.: ASM Press.

Wiley, J., Sherwood, L., and Woolverton, C. 2011. *Prescott, Harley & Klein's Microbiology, 8th ed.*, chap. 2. New York: McGraw-Hill.

NOTES

Capsule Stain

PRINCIPLES AND APPLICATIONS

Bacterial cells of some species are protected against harsh environmental conditions by a complex layer of sugars and proteins called a **capsule.** The capsule adheres tightly to the cell and aids the cell by helping it to resist dehydration, adhere to invasive devices such as catheters, and initiate the formation of biofilms. With few exceptions, when an encapsulated bacteria loses the ability to form capsules, it also loses much of its pathogenicity.

Because not all species of bacteria form capsules, determining that an unknown bacterial species produces a capsule can be helpful in its identification. Ironically, the process of capsular staining ends up staining everything except the capsule. Capsular staining is a two-step process that can be thought of as a combination of negative staining followed by simple staining. In the first step, the acidic dyes nigrosin or india ink are used to obliterate the background. Because these dyes are negatively charged, they will adhere to the glass of the slide but be repelled by the negatively charged bacterial cells, leaving the cells and capsules colorless against a dark background. The smear is allowed to air dry, but heat fixation is kept to a minimum as excessive heat can shrink the cells, resulting in a halo around the cells that could be mistaken for a capsule. Conversely, excess heat can also destroy the capsule, resulting in a false-negative identification of a capsule-containing cell. In the second step, the basic dye crystal violet is used to stain the bacterial cell. Because the capsule will not stain with either dye, it is visible as a clear halo between the cell and the background. In a properly prepared capsule stain, the background will be gray while the cells themselves are purple, surrounded by a clear halo, which represents the capsule (Figure 44.1). Nonencapsulated cells will still appear as purple cells against a gray background, but the halo will be absent.

PRE-LAB QUESTIONS

1. *What would an encapsulated cell look like if the capsule-staining process was stopped early, just prior to staining with crystal violet?*

2. *Draw the manner in which an encapsulated cell would differ from a nonencapsulated cell in a capsule stain.*

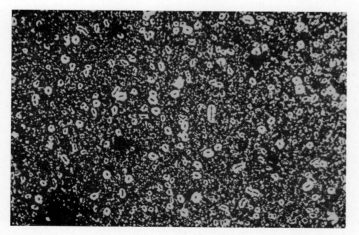

Figure 44.1 Capsule stain of *Klebsiella pneumoniae.* Note the dark background, purple cells, and clear capsule.

MATERIALS

Each student should obtain:

Skim milk agar cultures of *Klebsiella pneumoniae*
Nigrosin or india ink
Crystal violet
Two glass slides
Inoculating needle
Marking pen

PROCEDURE

This procedure is illustrated in Figure 44.2.

1. Label a slide in the upper left-hand corner with the name of the organism to be stained.
2. Use a dropper to place a small (less than 1/8 in.) drop of nigrosin or india ink near one end of a clean microscope slide. Add a small drop of sheep serum to the slide next to, but not touching, the nigrosin.
3. Use an inoculating needle to disperse a small amount of *K. pneumoniae* into the dye. Agitate the needle to break up any clumps of organism.

4. Place a spreader slide at an angle in front of the drop, and slowly move it backward until the slide contacts the bacteria/dye suspension.
5. Push the spreader along the slide, dragging the suspension across the bottom slide.
6. Allow the slide to air dry.
7. Flood the slide with crystal violet, and stain for 1 min.
8. Gently wash the slide with water to remove excess crystal violet. Because the smear has not been thoroughly heat-fixed, if the washing is too vigorous, the bacteria will be washed off the slide.
9. Blot dry with bibulous paper.
10. Locate the stained specimen under low power. Increase the magnification until you can examine the stained specimen under oil immersion. Record your observations in the Results section of this exercise.

(1) A needle is used to mix a small amount of organism into a drop of india ink/sheep serum.

(2) The ink suspension of bacteria is spread over slide and air-dried.

(3) The slide is *gently* heat-dried to fix the organisms to the slide.

(4) Smear is stained with crystal violet for 1 min.

(5) Crystal violet is *gently* washed off with water.

(6) Slide is blotted dry with bibulous paper and examined with oil immersion objective.

Figure 44.2 Procedure for capsule staining.

RESULTS

Draw images from your slides that show two microscopic fields.

EXERCISE 44 REVIEW QUESTIONS

1. Imagine that your lab ran out of nigrosin and crystal violet, and the only dyes available were eosin (a red, acidic dye) and malachite green (a green, basic dye). How could you substitute these dyes for nigrosin and crystal violet in a capsule stain, and what would be the appearance of your slide when you examined it?

2. Encapsulated bacteria easily adhere to smooth-walled, medical devices such as catheters, shunts, and artificial joints. Furthermore, the capsule interferes with phagocytosis, enhancing the virulence of these organisms. Explain how the structure and consistency of the capsule contributes to both of these problems.

REFERENCE

Wiley, J., Sherwood, L., and Woolverton, C. 2011. *Prescott, Harley & Klein's Microbiology, 8th ed.*, chap. 2. New York: McGraw-Hill.

Gram Stain

PRINCIPLES AND APPLICATIONS

Without a doubt, the most important stain in the microbiology lab is the **Gram stain.** Developed by Christian Gram in 1884, this stain is routinely the first stain used in the processing of any laboratory sample, and the Gram reaction is generally the very first piece of information given when discussing an organism. The Gram stain even dictates initial medical treatment in many cases, with Gram-positive organisms being treated with different antibiotics than Gram-negative organisms. The Gram stain is an example of a **differential** stain; a type of stain in which cells of different species may take on dissimilar appearances after the staining process based on underlying biological differences. To be more specific, after Gram staining, a Gram-positive cell will appear purple and a Gram-negative cell will appear pink, and in both cases, the color change is based on the structure of the cell wall (Figure 45.1).

Differential staining, and Gram staining is no exception, relies on a **primary stain, mordant, decolorizer,** and **counterstain.** The primary stain in a Gram stain is crystal violet, which stains both Gram-negative and Gram-positive cells a deep purple color. The second step involves the addition of Grams' iodine, which acts as a mordant, causing the crystal violet to form large insoluble complexes within the cell wall of Gram-positive cells. At this point in the Gram-staining process, both Gram-positive and Gram-negative cells appear purple, but that is about to change. The decolorization step uses ethyl alcohol, or a combination of ethyl alcohol and acetone, to differentially remove the crystal violet from cells based on the thickness of their cell wall. The thick peptidoglycan layer found in Gram-positive cells decolorizes slowly but evenly, remaining purple until the last of the crystal violet is washed away. Gram-negative cells, by contrast, have a peptidoglycan layer roughly one-tenth as thick as a Gram-positive cell, but they also possess an outer membrane composed of lipids. Gram-negative cells destain rapidly as the ethyl alcohol permeabilizes the outer membrane and the dye is removed from the thin peptidoglycan layer of the cell wall very quickly. This leaves Gram-negative cells (and only Gram-negative cells) colorless at this point. Finally, safranin, the counterstain, is used to stain Gram-negative cells pink. The color change accompanying each step is illustrated in Figure 45.2.

The Gram stain is a reasonably simple process, but its importance in the laboratory merits a few final thoughts. First, best results are obtained with young cultures (less than 24-h old). After this, some Gram-positive cultures may act Gram negative or Gram variable, leading to incorrect results. (It is important to note that Gram-negative cultures never change

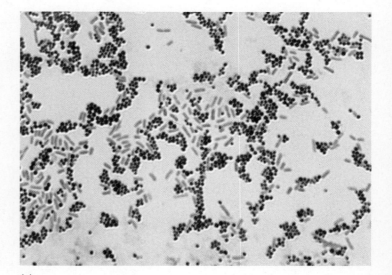

(a)

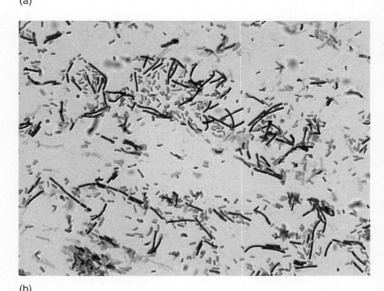

(b)

Figure 45.1 Gram staining. (a) *Staphylococcus aureus* is a Gram-positive cocci while *Escherichia coli* is a Gram-negative rod. (b) *Bacillus megaterium* is a large Gram-positive rod while *Escherichia coli* is a smaller Gram-negative rod.

as they age, always remaining Gram negative.) Secondly, the decolorization step is the most important step of the entire process. Too much alcohol will leach the crystal violet from even Gram-positive cells while too little will not remove all of the purple color from Gram-negative ones. Lastly, thin smears are essential for good results. Thick smears almost always result in inadequate staining or decolorization as well as making cellular arrangements difficult or impossible to see.

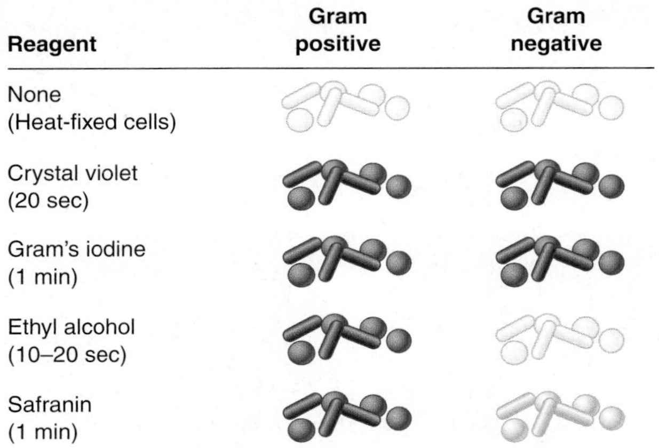

Reagent	Gram positive	Gram negative
None (Heat-fixed cells)		
Crystal violet (20 sec)		
Gram's iodine (1 min)		
Ethyl alcohol (10–20 sec)		
Safranin (1 min)		

Figure 45.2 Color changes that occur in each step of the Gram-staining process.

PRE-LAB QUESTIONS

1. List below each of the reagents used in a Gram stain along with the amount of time typically used and its purpose in the staining process.

2. What would happen if you inadvertently attempted to decolorize a Gram stain using water instead of ethyl alcohol?

3. What would happen if the step involving Gram's iodine was inadvertently left out of the Gram-staining procedure?

4. What is the most critical step in a Gram stain, and how can inaccuracy in this step affect your results?

5. Why is it important to use fresh cultures for Gram staining?

MATERIALS

Each student should obtain:

Fresh slant culture of *Bacillus megaterium*

Fresh broth cultures of:

　Staphylococcus aureus

　Escherichia coli

　Moraxella (Branhamella) catarrhalis

Gram-staining kit (crystal violet, Gram's iodine, ethyl
　alcohol, safranin)

Wash bottle

Bibulous paper

Six glass slides

Inoculating loop and needle

Marking pen

　　You will be responsible for several sets of slides as outlined below. It is acceptable to make all of the smears ahead of time, but stain and examine only a single slide at a time to ensure that your technique is leading to acceptable results.

- For slides with only a single bacterial specimen, begin with a heat-fixed smear as instructed in Exercise 42.
- For slides containing two specimens, prepare them as follows: On the left one-third of the slide, prepare a smear of the first bacterium. On the right one-third of the slide, prepare a smear of the second bacterium. In the middle one-third of the slide, prepare a smear containing both bacterium mixed together, being careful to flame adequately to prevent contamination of the stock cultures. If you've stained your mixed bacterial smears correctly, you should see organisms that differ both in shape and Gram reaction, as seen in Figure 45.3. If you do not see these results, inform your instructor who will be able to determine what went wrong by examining all three smears on your slide.

Single bacterial smears:

　S. aureus

　E. coli

　B. megaterium

　M. catarrhalis

Mixed bacterial smears:

　E. coli and *S. aureus*

　B. megaterium and *M. catarrhalis*

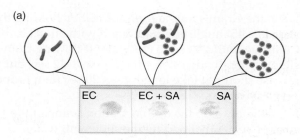

(a)

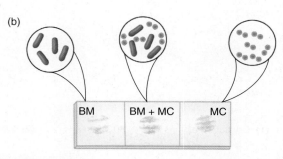

(b)

Figure 45.3 Expected results for mixed organism Gram stains. (a) *Escherichia coli* cells should appear as pink rods while *Staphylococcus aureus* cells appear as purple cocci. If purple rods are present, underdecolorization has occurred due to not enough alcohol being used in the decolorization process while pink cocci indicate that too much alcohol was used, resulting in overdecolorization of the specimen. (b) *Bacillus megaterium* is Gram positive and should appear as large purple rods while *Moraxella (Branhamella) catarrhalis* is a Gram-negative cocci and should appear pink. The presence of pink rods indicates overdecolorization (less alcohol should be used in the future) while the presence of purple cocci is indicative of underdecolorization and more alcohol should be used on future attempts.

PROCEDURE

This procedure is illustrated in Figure 45.4. When Gram staining, complete the process from beginning to end with no gaps, leaving no "extra" time between steps. If your timing is consistent from slide to slide, your results will also be much more consistent.

1. Label a slide in the upper left-hand corner with the name of the organism or organisms to be stained and prepare a heat-fixed smear.
2. Cover the smear with crystal violet and allow the stain to penetrate for 30 sec.
3. Briefly wash the slide with water to remove excess stain. Drain the slide of excess water.
4. Cover the smear with Gram's iodine for 1 min.

5. Wash the Gram's iodine from the cell by holding the slide at a 45° angle and flooding it with ethyl alcohol. Continue until you see a clear runoff flowing from the slide. Generally this step should take 8–12 sec, but in no case should it take longer than 20 sec for a properly prepared smear.

6. Immediately stop the decolorization action of the alcohol by gently flooding the slide with water.

7. Cover the smear with safranin for 1 min.

8. Rinse the slide gently to remove excess safranin.

9. Blot dry with bibulous paper.

10. Examine the stained specimens under oil immersion. Record your observations in the Results section of this exercise.

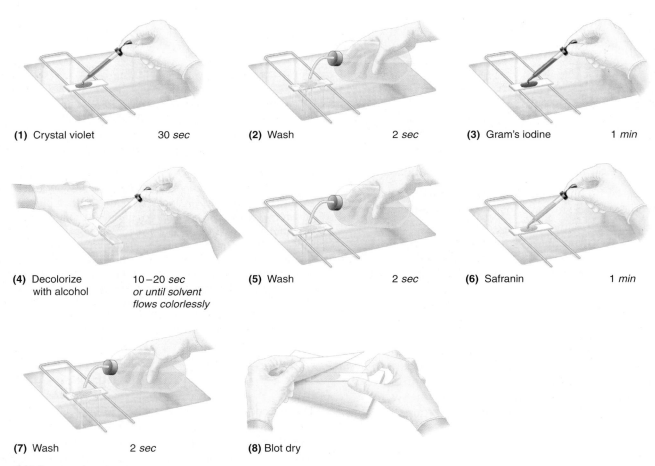

(1) Crystal violet   30 *sec*

(2) Wash   2 *sec*

(3) Gram's iodine   1 *min*

(4) Decolorize with alcohol   10–20 *sec or until solvent flows colorlessly*

(5) Wash   2 *sec*

(6) Safranin   1 *min*

(7) Wash   2 *sec*

(8) Blot dry

Figure 45.4 Procedure for Gram staining.

RESULTS

Draw images from your each of your slides, being sure to use proper colors and bacterial morphologies.

S. aureus   E. coli   E. coli plus S. aureus

B. megaterium   M. catarrhalis   B. megaterium plus M. catarrhalis

EXERCISE 45 REVIEW QUESTIONS

1. Predict the color of each of the organisms at the indicated step in a properly done Gram stain.

   S. aureus after the mordant

   E. coli after the primary stain

   E. coli after the counterstain

   B. megaterium after decolorization

Moraxella (Branhamella) catarrhalis after decolorization

E. coli after the counterstain

2. One wash step in the Gram stain occurs after the crystal violet has been used to stain the cell while a second wash occurs after the decolorizing step. What is the difference in importance between these two washes?

REFERENCES

Reddy, C.A., Beveridge, T.J., Breznak, J.A., Marzluf, G.A., Schmidt, T.M., and Snyder, L.R. (eds.). 2007. *Methods for General and Molecular Microbiology, 3rd ed.,* chap. 2. Washington D.C.: ASM Press.

Wiley, J., Sherwood, L., and Woolverton, C. 2011. *Prescott, Harley & Klein's Microbiology, 8th ed.,* chap. 2. New York: McGraw-Hill.

Endospore Stain

PRINCIPLES AND APPLICATIONS

Most bacteria have little recourse when environmental conditions deteriorate, simply dying when the conditions become too hot, too cold, or too lacking in nutrients for life to continue. However, species within the genera *Bacillus* and *Clostridium* (as well as a few other, less medically relevant genera) have the ability to form **endospores;** protective structures that allow the cell to survive harsh conditions. The process of endospore formation takes approximately 6–12 h and is initiated by a depletion of nutrients in the environment. During the first part of this period, the vegetative cell is termed a **sporangium** and the endospore forms within; when the spore is fully formed, the sporangium will be sloughed off. The metabolically inactive endospore is resistant to heat, radiation, and chemicals that would kill a vegetative cell, and the spore may remain viable for thousands of years. When conditions once again become favorable, the endospore proceeds through a process of germination, producing a new vegetative cell and reentering a cycle of vegetative growth (Figure 46.1). Note that although endospores are often spoken of as "spores," they should not be confused with the reproductive spores found in fungi. Bacterial endospores are protective structures only with no reproductive capabilities.

The ability of endospores to withstand most chemicals not only makes them difficult to kill but also makes them difficult to stain using standard techniques. When endospore-forming cells are present in a specimen that has been Gram stained, the occasional formation of endospores in some of the cells may be seen as a clear area within the Gram-positive cell. Endospore staining, a differential stain that can help distinguish between endospore formers and nonformers, uses heat as a mordant to drive the primary stain, **malachite green,** into both the endospore and any vegetative cells. Decolorization with water removes the primary dye from vegetative cells and the sporangium of spore formers but not from the endospore. Finally, counterstaining with **safranin** imparts a pink color to the vegetative cells, ensuring that they are visible during microscopic examination (Figure 46.2). The size, shape, and location of the endospore are important in the identification of an unknown bacterial species, with endospores being either larger or smaller than the vegetative cell, spherical or elliptical in shape, and located in the center of the cell, at the extreme end of the cell (terminal), or somewhere in between (subterminal).

Two procedures for endospore staining are provided here. The first is the Schaeffer-Fulton method, a traditional technique that uses steam from boiling water to heat the primary stain while the second is a newer method that employs a microwave oven as a source of heat and is consequently quite a bit faster. Use the method indicated by your instructor.

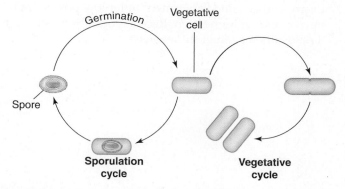

Figure 46.1 When nutrients are in adequate supply, endospore-forming organisms remain in a vegetative cycle, multiplying as rapidly as conditions allow. If nutrients become depleted, the vegetative cell will become a sporangium, and an endospore will be produced within. Eventually, the sporangium is sloughed off, leaving only the metabolically inert and highly resistant endospore. The endospore will remain viable for thousands of years until water and a specific chemical stimulus induce germination of the endospore and a resumption of the vegetative cycle.

Reagent	Endospore former	Non-Endospore former
None (heat-fixed smear)		
Malachite green and heat		
Water		
Safranin		

Figure 46.2 Color changes that occur at each step of the endospore-staining process.

PRE-LAB QUESTIONS

1. *List each of the reagents used in an endospore stain along with its purpose in the staining process.*

2. *What is the mordant in an endospore stain?*

3. *What would you observe if you performed an endospore stain on a culture of* Bacillus megaterium *but neglected to apply any heat during the staining process?*

MATERIALS

Each student or group should obtain:

48–72 h nutrient agar culture of *Bacillus megaterium,*
   B. subtilis, or *B. cereus*

Electric hot plate and small (25 ml) beaker (Schaeffer-Fulton method only)

Microwave oven and empty Petri dish (microwave method only)

Endospore-staining kit (5% malachite green, safranin)

Water bottle

Glass slides

Bibulous paper

Inoculating loop and needle

Marking pen

PROCEDURE

Schaeffer-Fulton Method

This procedure is illustrated in Figure 46.3.

1. Label a slide in the upper left-hand corner for identification, and prepare a heat-fixed smear of the appropriate culture.

2. Saturate the smear with malachite green, and steam the slide over boiling water for 5 min. Do not allow the stain to dry on the smear. Add additional stain if required to keep the smear wet during the heating process.

3. Allow the slide to cool slightly so that it will not crack. Rinse with water for 30 sec.

4. Counterstain with safranin for 30 sec.

5. Rinse briefly to remove excess safranin.

6. Blot dry with bibulous paper.

7. Examine the stained specimen under low power, high power, and oil immersion. Record your observations in the Results section of this exercise.

(1) Cover the heat-fixed smear with malachite green. Steam over boiling water for 5 min. Add additional stain if smear begins to dry.

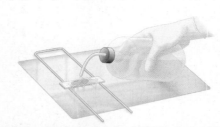

(2) After the slide has cooled sufficiently, rinse with water for 30 sec.

(3) Counterstain with safranin for about 30 sec.

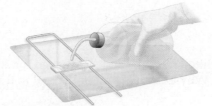

(4) Rinse briefly with water to remove safranin.

(5) Blot dry with bibulous paper and examine slide under oil immersion.

Figure 46.3 Procedure for the Schaeffer-Fulton method of endospore staining.

Microwave Method

1. Label a slide in the upper left-hand corner for identification and prepare a heat-fixed smear of the appropriate culture.
2. Place a cut-to-fit piece of paper towel (two layers of paper) into the bottom of an empty Petri dish. Saturate the toweling with water.
3. Place the heat-fixed slide on top of the paper towel, and flood the slide with malachite green.
4. Place the slide in the microwave, and heat for 30 sec at full power.
5. Remove the Petri dish from the microwave, and allow the slide to cool slightly so that it will not crack. Rinse with water for 30 sec.
6. Counterstain with safranin for 30 sec.
7. Rinse briefly to remove excess safranin.
8. Blot dry with bibulous paper.
9. Examine the stained specimen under low power, high power, and oil immersion. Record your observations in the Results section of this exercise.

RESULTS

Draw images from your slide, being sure to use proper colors and bacterial morphologies.

EXERCISE 46 REVIEW QUESTIONS

1. Why are older cultures typically used for endospore staining?

2. After the specimen is stained with malachite green, the slide is washed for 30 sec. In most other staining techniques, washes are only a few seconds in length. Why is such a long rinse used in this case?

REFERENCES

Reddy, C.A., Beveridge, T.J., Breznak, J.A., Marzluf, G.A., Schmidt, T.M., and Snyder, L.R. (eds.). 2007. *Methods for General and Molecular Microbiology, 3rd ed.,* chap. 2. Washington D.C.: ASM Press.

Wiley, J., Sherwood, L., and Woolverton, C. 2011. *Prescott, Harley & Klein's Microbiology, 8th ed.,* chap. 2. New York: McGraw-Hill.

Acid-Fast Stain

PRINCIPLES AND APPLICATIONS

Cells of certain bacterial genera contain a waxy substance in their cell walls known as mycolic acid, a complex lipid that inhibits the passage of aqueous dyes through the cell wall. Most methods of staining these cells rely on carbolfuschin, a dark red dye containing 5% phenol as the primary stain. The dye is lipid soluble, and its penetration of the cell is further enhanced by heating, which acts as a mordant, driving the stain into the cell. Once in the cell wall, the carbolfuschin is equally hard to remove. A solution of hydrochloric acid in ethyl alcohol (acid-alcohol) is used to decolorize the cells. Those cells that retain the red color of the carbolfuschin under these harsh conditions are called **acid fast** while non-acid-fast cells are clear after this step. Methylene blue is then used to counterstain the smear, staining non-acid-fast cells blue while the acid-fast cells remain red (Figure 47.1).

The acid-fast stain is primarily used to detect members of the genus *Mycobacterium* (Figure 47.2), such as the pathogens *M. tuberculosis*—the cause of tuberculosis (TB)—and *M. leprae*—the cause of leprosy. Other bacterial species, including members of the genus *Nocardia* and even protozoan parasites (*Cryptosporidium* and *Isospora*) show at least some degree of acid-fast behavior. Because so few organisms are acid fast, the stain is generally only used when infection by an acid-fast organism is suspected. Acid-fast staining of sputum samples often directs the treatment, allowing the physician to administer drugs such as isoniazid and rifampin, which are specific for TB infection, long before the bacteria could be cultured in the laboratory.

Three procedures for acid-fast staining are provided here. The first is the **Ziehl-Neelsen** method that uses steam

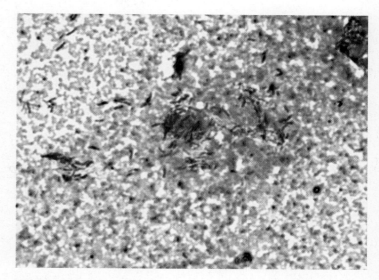

Figure 47.2 Acid-fast stain of *Mycobacterium smegmatis* (red, acid-fast rods) and *Staphylococcus aureus* (blue, non-acid-fast cocci).

from boiling water to heat the carbolfuschin while the second is a newer method that employs a microwave oven as a source of heat to force the stain into the cells. Not only is this second method quicker, it produces fewer fumes from the heating of phenol than does the Ziehl-Neelsen method. These fumes can be quite irritating to the eyes and mucous membranes. The third procedure is the Kinyoun method, which relies on a more concentrated carbolfuschin stain rather than heat to force the stain into the acid-fast cell wall. Your instructor will inform you of the method routinely used in your laboratory.

PRE-LAB QUESTIONS

1. *List below each of the reagents used in an acid-fast stain along with its purpose in the staining process.*

Reagent	Acid fast	Non-Acid fast
None (heat-fixed smear)		
Carbolfuschin and heat		
Acid-alcohol		
Methylene blue		

Figure 47.1 Color changes that occur at each step of the acid-fast-staining process.

2. *What is the mordant in an acid-fast stain?*

3. *What would you observe if you performed an acid-fast stain on a culture of* Mycobacterium smegmatis *but neglected to apply any heat during the staining process?*

MATERIALS

Each student should obtain:

Slant culture of *Mycobacterium smegmatis*

Nutrient broth culture of *Staphylococcus aureus*

Electric hot plate and small (25 ml) beaker (Ziehl-Neelsen method only)

Microwave oven and empty Petri dish (microwave method only)

One of the following:

  Ziehl-Neelsen acid-fast-staining kit (Ziehl's carbolfuschin, acid alcohol, methylene blue)

  Kinyoun acid-fast-staining kit (Kinyoun carbolfuschin, acid-alcohol, brilliant green stain)

Sheep serum (Kinyoun method only)

Glass slides

Inoculating loop and needle

Marking pen

PROCEDURE

Ziehl-Neelsen Method

This procedure is illustrated in Figure 47.3.

1. Label a slide in the upper left-hand corner for identification.
2. Prepare a mixed smear by transferring two loopfuls of *S. aureus* to a slide and adding to it a very small amount of *M. smegmatis*. The acid-fast bacteria tend to form clumps that will have to broken apart with the inoculating needle. Air dry and heat-fix the slide.
3. Saturate the smear with carbolfuschin, and steam the slide over boiling water for 5 min. Add additional stain if required to keep the smear from drying out.
4. Allow the slide to cool slightly so that it will not crack. Rinse with acid-alcohol for 8–12 sec.
5. Rinse briefly with water to stop the decolorizing effect of the acid-alcohol.
6. Counterstain with methylene blue for 30 sec.
7. Rinse briefly to remove excess methylene blue.
8. Blot dry with bibulous paper.
9. Examine the stained specimen under low power, high power, and oil immersion. Record your observations in the Results section of this exercise.

Microwave Method

1. Label a slide in the upper left-hand corner for identification.
2. Prepare a mixed smear by transferring two loopfuls of *S. aureus* to a slide and adding to it a very small amount of *M. smegmatis*. The acid-fast bacteria tend to form clumps that will have to broken apart with the inoculating needle. Air dry and heat-fix the slide.
3. Place a cut-to-fit piece of paper towel (two layers of paper) into the bottom of an empty Petri dish. Saturate the toweling with water.
4. Place the heat-fixed slide on top of the paper towel, and flood the slide with carbolfuschin.
5. Place the slide in the microwave, and heat for 30 sec at full power.
6. Remove the Petri dish from the microwave, and allow the slide to cool slightly so that it will not crack. Rinse with acid-alcohol for 8–12 sec.
7. Counterstain with methylene blue for 30 sec.
8. Rinse briefly to remove excess methylene blue.
9. Blot dry with bibulous paper.
10. Examine the stained specimen under low power, high power, and oil immersion. Record your observations in the Results section of this exercise.

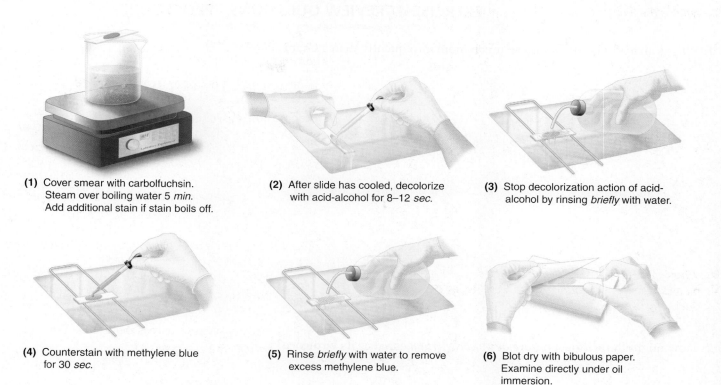

(1) Cover smear with carbolfuchsin. Steam over boiling water 5 *min*. Add additional stain if stain boils off.

(2) After slide has cooled, decolorize with acid-alcohol for 8–12 *sec*.

(3) Stop decolorization action of acid-alcohol by rinsing *briefly* with water.

(4) Counterstain with methylene blue for 30 *sec*.

(5) Rinse *briefly* with water to remove excess methylene blue.

(6) Blot dry with bibulous paper. Examine directly under oil immersion.

Figure 47.3   Ziehl-Neelsen acid-fast-staining procedure.

Kinyoun Method

1. Label a slide in the upper left-hand corner for identification.
2. Add a single drop of sheep serum to a clean slide, and add to it two loopfuls of *S. aureus* and a small amount of *M. smegmatis*. The acid-fast bacteria tend to form clumps that will have to broken apart with the inoculating needle. Air dry and heat-fix the slide.
3. Flood the smear with Kinyoun carbolfuschin stain for 5 min. Perform this step in a chemical fume hood or otherwise ensure adequate ventilation.
4. Rinse gently with water for 2–3 sec.
5. Destain the slide with acid-alcohol for 2–3 min until the runoff from the slide is clear. Follow with a gentle rinse with water to remove the acid-alcohol.
6. Counterstain with methylene blue for 1–3 min.
7. Rinse briefly to remove excess methylene blue.
8. Blot dry with bibulous paper.
9. Examine the stained specimen under low power, high power, and oil immersion. Record your observations in the Results section of this exercise.

RESULTS

Draw images from your slide, being sure to use proper colors and bacterial morphologies.

EXERCISE 47 REVIEW QUESTIONS

1. Why is an acid-fast stain used so much more infrequently than a Gram stain?

2. How might the acid-fast nature of *Mycobacterium* enhance its virulence?

REFERENCES

CDC. Division of Tuberculosis Elimination (DTBE). http://www.cdc.gov/tb/.

Forbes, B.A., Sahm, D.F., and Weissfeld, A.S. 2007. *Bailey and Scott's Diagnostic Microbiology, 12th ed.*, chap. 45. St. Louis: Mosby.

Reddy, C.A., Beveridge, T.J., Breznak, J.A., Marzluf, G.A., Schmidt, T.M., and Snyder, L.R. (eds.). 2007. *Methods for General and Molecular Microbiology, 3rd ed.*, chap. 2. Washington D.C.: ASM Press.

Wiley, J., Sherwood, L., and Woolverton, C. 2011. *Prescott, Harley & Klein's Microbiology, 8th ed.*, chap. 2. New York: McGraw-Hill.

Motility Methods: Wet Mount and Hanging Drop

PRINCIPLES AND APPLICATIONS

Motile bacteria display **chemotaxis,** a complex movement of the cell toward nutrients or away from harmful substances. Roughly speaking, about half of all bacteria are motile, with that total including most spirilla, about half of the bacilli, and a few cocci. While motility offers a bacterial cell many advantages, in the laboratory, it is often used as an aid in the identification of an isolate. The major motility structures found in bacteria are flagella, which may be viewed only after specialized staining processes have made them visible. While these techniques are required if we are to characterize the bacterial flagella themselves, motility may be more simply determined by directly viewing a small amount of a liquid culture at high magnification.

Two methods of directly determining motility exist. The first is the **wet mount,** a procedure in which a small amount of a liquid culture is placed on a microscope slide and covered with a cover glass. The slide is then examined, and any rapidly swimming cells noted. Although quick and straightforward, wet mounts suffer from several problems, including a tendency to dry out, current movements that may mimic motility, and the potential for contamination of instruments and self with pathogenic microbes. A second method, the **hanging drop** technique, eliminates most problems with drying and water currents and reduces the chances of contamination. In this method, a cover glass is prepared with a small dab of petroleum jelly (Vaseline) on each corner and a single loopful of a liquid culture in the center. A special slide with a depression in the middle is then placed over the cover glass with the concavity in the slide centered over the drop. The petroleum jelly causes the cover glass to adhere to the slide as it is inverted. In this way, the culture hangs from the cover-slip, reducing both water currents and evaporation. Examination of the slide can then reveal if motile cells are present.

Unfortunately, just because a cell moves, it is not necessarily motile. True bacterial motility must be differentiated from **Brownian motion** or movement due to water currents. Brownian motion is due to the bombardment of cells by surrounding molecules, primarily water. This impact causes the cell to jiggle or shake but without displaying any true directional movement, which is required if a cell is to be deemed motile. Water currents, caused by evaporation of the media or through pressure exerted on the cover slip by the oil immersion lens, will make all cells move in the same direction. This is generally easy to differentiate from true motility.

In this lab, you will compare the motility of two organisms using wet mounts, hanging drop slides, or both, as directed by your instructor. When using either technique, speed is of the essence; time and evaporation of the media will reduce bacterial motility while condensation forming on a hanging drop slide can reduce the clarity of the image.

PRE-LAB QUESTIONS

1. *Describe the type or types of movement you would expect to see for:*

A motile bacterium

A nonmotile bacterium

A motile bacterium that had been autoclaved to kill all the cells within the culture

2. *Do all the cells in a culture need to be moving for the culture to be described as motile? Why or why not?*

3. *Unstained cells have very little contrast. What single adjustment could you make to your microscope that would make the cells easier to see while at the same time reducing evaporation of your sample?*

MATERIALS

Each student should obtain:

24–48 h broth cultures of:

  Micrococus luteus

  Proteus vulgaris

Microscope cover glasses

Inoculating loop

Microscope slides (wet mount only)

Depression slide (hanging drop technique only)

Petroleum jelly and a few toothpicks (hanging drop technique only)

Marking pen

PROCEDURE

Wet Mount

1. Label a slide in the upper left-hand corner for identification.
2. Transfer two or three loopfuls of culture to be studied on the center of the slide.
3. Gently lower a cover glass onto the culture, using your loop to prevent "splashing" of the drop and the introduction of air bubbles.
4. Adjust the voltage control and diaphragm on your microscope to reduce light to a minimum. Low light intensity favors a clearer view of the specimen and slows evaporation of the culture.
5. Examine your specimens under low and high power. The use of the oil immersion lens is usually impossible when examining a wet mount. Look for true motility that is marked by movement of a cell in a single direction that is several times the length of the bacterium. If even only a few cells are moving, the culture should be considered motile. Ignore Brownian motion as well as movements due to water currents, which are generally marked by everything on the slide moving in the same direction.
6. Because the slide and cover glass contain live bacteria, be sure to dispose of them in a hard-sided biohazard container when you have finished your examination.
7. Record your observations in the Results section of this exercise.

Hanging Drop

This procedure is illustrated in Figure 48.1.

1. Apply a dab of petroleum jelly to each corner of a cover glass.
2. Apply two loopfuls of a liquid bacterial culture to the center of the cover glass.
3. Orient a hanging drop slide so that the depression is facing downward, and place it carefully on top of the cover glass.

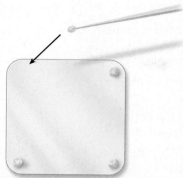

(1) A small amount of Vaseline is placed near each corner of the cover glass with a toothpick.

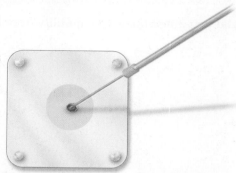

(2) Two loopfuls of organisms are placed on the cover glass.

(3) Depression slide is pressed against Vaseline on cover glass and quickly inverted.

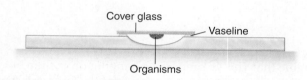

(4) The completed preparation can be examined under oil immersion.

Figure 48.1 Procedure for the preparation of a hanging drop slide.

4. Invert the slide so that the cover glass is now on top of the slide.

5. Adjust the voltage control and diaphragm on your microscope to reduce light to a bare minimum. Low light intensity favors a clearer view of the specimen and slows condensation on the slide.

6. Examine the slide using the low power objective. Because bacteria will be drawn to the edge of the drop by surface tension, focus on the edge of the drop.

7. Switching to high power, look for true motility that is marked by movement in a single direction that is several times the length of the bacterium. If even a few cells are moving, the culture should be considered motile. Ignore Brownian motion as well as movements due to water currents, which are generally marked by everything on the slide moving in the same direction. If more detail is needed, add a drop of oil to the coverslip and switch to oil immersion before continuing your observation.

8. Depression slides are used repeatedly, but you should make no attempt to disinfect them yourself. Place the slide and cover glass into a container of disinfectant, which will kill the bacteria, allowing the slide to be cleaned at a later time.

9. Record your observations in the Results section of this exercise.

RESULTS

Using images and written descriptions, depict the images from your slides. Which of the two bacteria tested were motile?

EXERCISE 48 REVIEW QUESTIONS

1. How did you differentiate true motility from Brownian motion and water currents?

2. Imagine that you've prepared a hanging drop slide but were called out of the room for an hour before being able to look at it. When you return, examination of the slide shows motility. Would you be satisfied with this result? What if the slide showed no motility when you returned?

REFERENCES

Reddy, C.A., Beveridge, T.J., Breznak, J.A., Marzluf, G.A., Schmidt, T.M., and Snyder, L.R. (eds.). 2007. *Methods for General and Molecular Microbiology, 3rd ed.,* chap. 2. Washington D.C.: ASM Press.

Wiley, J., Sherwood, L., and Woolverton, C. 2011. *Prescott, Harley & Klein's Microbiology, 8th ed.,* chap. 3. New York: McGraw-Hill.

Flagella Stain

PRINCIPLES AND APPLICATIONS

Bacterial motility is usually identified in one of two ways. In the first, a liquid culture is observed under the microscope, and motility is evaluated directly. In the second, cells are used to inoculate a semisolid agar that is firm enough to restrict nonmotile cells to the inoculation site but fluid enough to allow motile cells freedom of movement. After a period of incubation, motile cultures may be easily distinguished from nonmotile cultures.

While both methods reveal whether or not a particular cell is motile, neither provides any information about the **flagellum** itself, which is the organelle responsible for motility in the vast majority of species. The reasons for this seemingly incomplete examination are twofold. First, for purposes of identification, simply knowing whether or not a bacterium is motile is almost always sufficient, and this information is easily obtained. Second, bacterial flagellum are so slender, between 14–28 nm in diameter, that they fall below the level of resolution of the light microscope and so must be thickened to many times their original diameter if they are to be seen. Flagella staining has been designed to do exactly this. Leifson staining solution, used in this exercise, contains tannic acid, pararosaniline acetate, and pararosaniline hydrochloride, which bind tightly to flagella, increasing their thickness and allowing them to be seen in a light microscope. Unfortunately, dealing with such delicate cellular structures is not a trivial matter, with even gentle transferring of cells and the drying of a culture on a slide being enough to break the flagella or dislodge them from the cell. Consequently, this type of staining does not have a spectacular success rate.

The information obtained from flagella staining can be useful, and the images that are seen are dramatic (Figure 49.1). Bacterial flagella appear in several different arrangements, which can be helpful in some identification schemes. If a bacterial cell displays only a single flagellum, it is said to be **monotrichous** and polar; several flagella emanating from the same end of the cell is described as a **lophotrichous** arrangement; small groups of flagella at each end are an **amphitrichous** arrangement; and flagella exiting the cell from many points is a **peritrichous** arrangement (Figure 49.2).

In this lab, you will use Leifson staining to stain and observe three bacterial species with different flagellar arrangements. This staining process works by coating the flagella with stain until they are thick enough to visualize with a light microscope (generally greater than 0.2 μm in diameter). Both bacterial cells and flagella will stain red.

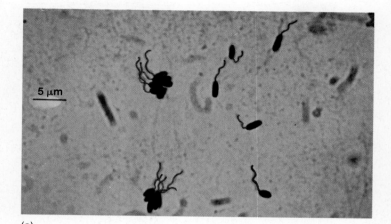

(a)

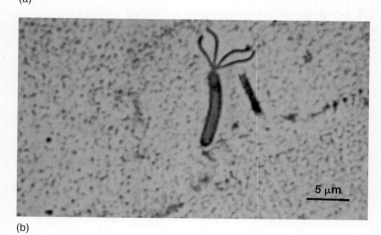

(b)

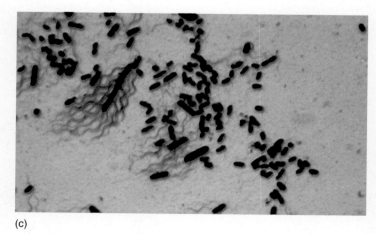

(c)

Figure 49.1 Flagellar arrangements. (a) *Pseudomonas*, displaying a single polar flagella on each cell (a monotrichous polar arrangement). (b) *Spirillum*. Several flagella on each cell is referred to as a lophotrichous arrangement. (c) *Proteus vulgaris* displays a peritrichous arrangement in which flagella emanate from all around the bacterial cell.

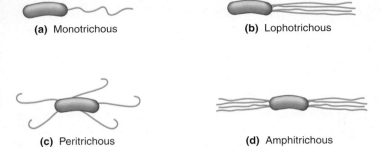

(a) Monotrichous   (b) Lophotrichous

(c) Peritrichous   (d) Amphitrichous

Figure 49.2 Flagellar arrangements.

3. *What do you think could be the result of overly rough handling of a culture during preparation of a flagella stain?*

PRE-LAB QUESTIONS

1. *Define the following terms:*

Motility

Flagellum

2. *What is meant by* resolution of the light microscope, *and how does this relate to the viewing of bacterial flagella?*

MATERIALS

Each student should obtain:
24–48 h broth cultures of:
Spirillum serpens
Proteus vulgaris
Pseudomonas aeruginosa
Inoculating loop
Microscope slides
Leifson staining solution
Wax pencil
Marking pen

PROCEDURE

1. Label a slide in the upper left-hand corner for identification. Your slide should be absolutely clean for best results.
2. Prepare a bacterial smear in the center of the slide. Allow the slide to air dry and do not heat-fix.
3. Draw a circle around the smear using the wax pencil.
4. Flood the slide with dye, using the wax pencil line to restrict the dye to the area inside the circle.
5. Watch the slide for 7–15 min. When a golden film appears on the dye surface and a precipitate appears throughout the film, rinse the stain gently with water.
6. Allow the slide to air dry. Do not blot.
7. Examine the stained specimen under oil immersion. Record your observations in the Results section of this exercise.

RESULTS

Draw images of each of your specimens. Indicate the morphological arrangement of the flagella for each specimen.

S. serpens

P. aeruginosa

P. vulgaris

EXERCISE 49 REVIEW QUESTIONS

1. Why can't a flagella stain be combined with a hanging drop slide to allow flagella to be seen as they move?

2. A typical flagella is about 20 nm. About how many times must the diameter of the flagella be increased to make it visible in a light microscope?

3. If, when examining a flagella stain of a pure culture, you saw some cells that had an amphitrichous arrangement while others displayed a monotrichous arrangement, how would you categorize the flagellar morphology of the culture? What probably happened to the culture to cause this result?

REFERENCES

Reddy, C.A., Beveridge, T.J., Breznak, J.A., Marzluf, G.A., Schmidt, T.M., and Snyder, L.R. (eds.). 2007. *Methods for General and Molecular Microbiology, 3rd ed.,* chap. 2. Washington D.C.: ASM Press.

Wiley, J., Sherwood, L., and Woolverton, C. 2011. *Prescott, Harley & Klein's Microbiology, 8th ed.,* chap. 3. New York: McGraw-Hill.

Streak-Plate Isolation

PRINCIPLES AND APPLICATIONS

Properly studying a single species of bacteria in the laboratory requires that it be isolated from many other bacteria in the population. In a successful separation, one cell in a mixed population of bacteria will be separated from all others and immobilized atop or within a solid growth media. As this separated cell continues to reproduce over many generations, it will give rise to a single **colony** containing millions of cells, all of which are derived from a single cell (Figure 50.1). Each colony can now be considered a **pure culture** and can be used for further study of the bacterium, usually after **subculturing** the colony to its own tube of media for convenience.

Several methods exist to separate bacterial cells but chief among them is the **streak-plate method.** This method is the most economical in terms of time and materials, requiring just a few minutes and only a single plate of media. Its main drawback is that a certain degree of skill is required, which takes time to fully develop. All streak-plates are essentially dilutions of bacteria using a solid media. For this dilution to be effective, it is absolutely essential to realize that bacteria are added only once to the plate. After streaking the first sector, a sterile loop is used to reduce the number of cells, with no more bacteria being added to the plate. The streak-plate method of isolation is outlined in Figure 50.2. Four different streaking techniques are shown, with personal preference being the greatest reason to use one technique over another. Each of the techniques shown here will result in isolated colonies; use the one recommended by your instructor.

PRE-LAB QUESTIONS

1. *Define the following terms:*

Colony

Pure culture

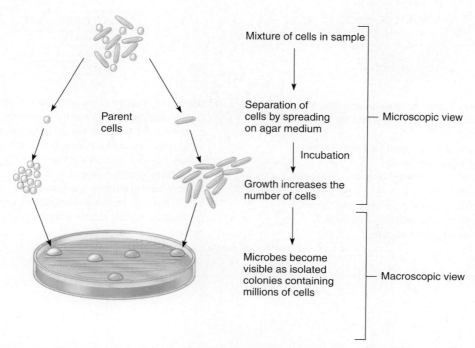

Figure 50.1 Development of isolated colonies. By streaking the cells across the surface of an agar plate, they are separated over a large area, resulting in isolated cells. After incubation, each isolated cell will give rise to a colony, a visible mound of bacterial cells, all of which are descendents of the original isolated cell.

Quadrant streak
(Method A)

(1) Streak one loopful of organisms over area 1 near edge of the plate. Apply the loop lightly. Don't gouge into the medium.
(2) Flame the loop, cool 5 sec, and make five or six streaks from area 1 through area 2. Momentarily touching the loop to a sterile area of the medium before streaking ensures a cool loop.
(3) Flame the loop again, cool it, and make six or seven streaks from area 2 through area 3.
(4) Flame the loop again, and make as many streaks as possible from area 3 through area 4, using up the remainder of the plate surface.
(5) Flame the loop before putting it aside.

Quadrant streak
(Method B)

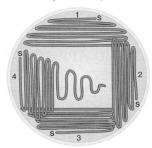

(1) Streak one loopful of organisms back and forth over area 1, starting at point designated by "s." Apply loop lightly. Don't gouge into the medium.
(2) Flame the loop, cool 5 sec, and touch the medium in a sterile area momentarily to ensure coolness.
(3) Rotate dish 90° while keeping the dish closed. Streak area 2 with several back and forth strokes, hitting the original streak a few times.
(4) Flame the loop again. Rotate the dish and streak area 3 several times, hitting the area several times.
(5) Flame the loop, cool it, and rotate the dish 90° again. Streak area 4, contacting area 3 several times and drag out the culture as illustrated.
(6) Flame the loop before putting it aside.

Radiant streak

(1) Spread a loopful of organisms in small area near the edge of the plate in area 1. Apply the loop lightly. Don't gouge into the medium.
(2) Flame the loop and allow it to cool for 5 sec. Touching a sterile area will ensure coolness.
(3) **From the edge** of area 1, make 7 or 8 straight streaks to the opposite side of the plate.
(4) Flame the loop again, cool it sufficiently, and cross streak over the last streaks, **starting near area 1.**
(5) Flame the loop before putting it aside.

Continuous streak

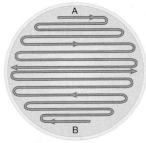

(1) Starting at the edge of the plate (area A) with a loopful of organisms, spread the organisms in a single continuous movement to the center of the plate. Use light pressure and avoid gouging the medium.
(2) Rotate the plate 180° so that the uninoculated portion of the plate is away from you.
(3) Without flaming the loop again, and using the same face of the loop, continue streaking the other half of the plate by starting at area B and working toward the center.
(4) Flame the loop before putting it aside.

Figure 50.2   Four different streaking techniques.

2. *What is the difference between a pure culture and a mixed culture?*

One nutrient agar pour
One empty Petri dish
Hot plate
Inoculating loop
Marking pen
Thermometer

PROCEDURE

This procedure is illustrated in Figures 50.3 and 50.4.

1. Label the bottom of your Petri dish with your name and lab time. Label around the periphery of the plate so your view of the bacterial colonies will be as complete as possible after the incubation.

2. Place the agar tube into a beaker containing enough water to cover the agar in the tube. Bring the water in the beaker to a boil, and allow 5 min for the agar to liquefy. Remove the beaker from the hot plate, and allow the agar to cool to 50°C. Adding fresh water to the beaker will allow it to cool faster.

3. Remove the cap and flame the neck of the tube. Carefully pour the liquefied agar into the bottom of the Petri plate, while holding the top of the plate with your other hand. Swirl the plate gently if needed to ensure that the media covers the entire bottom of the plate. Allow 10 min for the plate to fully solidify.

PERIOD ONE

MATERIALS

Each group should obtain:
A mixed broth culture containing:
    Escherichia coli
    Serratia marcescens
    Micrococcus luteus

(1) Liquefy a nutrient agar pour by boiling for 5 min.

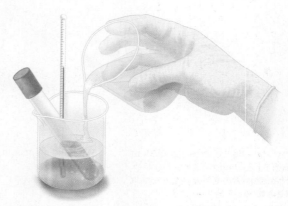

(2) Cool down the nutrient agar pour to 50°C by pouring off some of the hot water and adding cold water to the beaker. Hold at 50°C for 5 min.

(3) Remove the cap from the tube, and flame the open end of the tube.

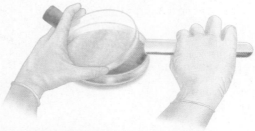

(4) Pour the contents of the tube into the bottom of the Petri plate, and allow it to solidify.

Figure 50.3 Pouring an agar plate for streaking.

(1) Shake the culture tube from side to side to suspend organisms. Do not moisten cap on tube.

(2) Heat loop and wire to red-hot. Flame the handle slightly also.

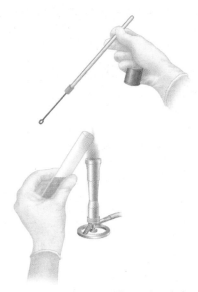

(3) Remove the cap and flame the neck of the tube. Do not place the cap down on the table.

(4) After allowing the loop to cool for at least 5 sec, remove a loopful of organisms. Avoid touching the side of the tube.

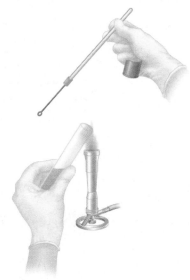

(5) Flame the mouth of the culture tube again.

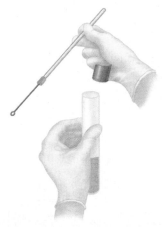

(6) Return the cap to the tube, and place the tube in a test tube rack.

(7) Streak the plate, holding it as shown. Do not gouge into the medium with the loop.

(8) Flame the loop before placing it down.

Figure 50.4   Inoculation of an agar plate.

4. Carefully agitate the tube containing the mixed culture until the bacteria is suspended in the media.

5. Flame the loop to red hot, and allow it to cool. Remove the cap from the culture tube and, while holding the cap with the pinky finger of your dominant hand, flame the neck of the tube.

6. Remove a single loopful of broth from the tube.

7. Flame the neck of the tube and replace the cap; return the tube to the test tube rack.

8. Streak the organism onto the plate using the streaking technique selected from Figure 50.2. Hold the Petri dish cover over the plate to guard against airborne contamination as you work. Use as little pressure as possible to avoid gouging the medium.

9. Flame the loop before placing it aside.

10. Incubate the plate in an inverted position at 30°C for 24–48 h. Inverting the plates prevents condensation (which may have accumulated on the lid of the plate) from dropping onto the agar and causing the organisms to spread over the agar surface, ruining the entire isolation procedure.

QUESTIONS—PERIOD ONE

1. What is the importance of generating isolated bacterial colonies?

2. Draw the results you would expect to see in a well-done streak-plate.

3. What is a subculture?

PERIOD TWO

RESULTS

Retrieve your plate from the incubator. To make differentiation easier, each bacterium used in this exercise exhibits a unique color: *E. coli* produces off-white growth, *M. luteus* produces yellow growth, and *S. marcescens* is red. Rarely, if ever, will you encounter outside the laboratory a mixed culture with such vibrant colors, but here the colors will help you to more fully evaluate the success of your separation. A successful outcome will always be marked by many well-isolated colonies (Figure 50.5). If your plate looks beautiful, all the better, but as long as several well-isolated colonies are present, the plate can be judged a success.

Record below the appearance of your streak-plate. Critically analyze the plate with regard to the number of well-isolated colonies. Note any actions that could have improved the success of your plate.

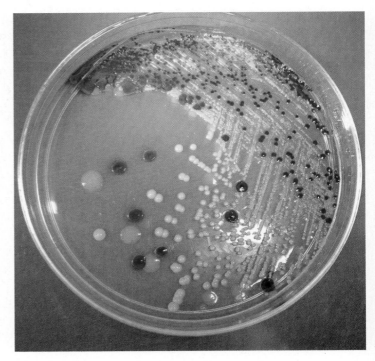

Figure 50.5   A well-executed streak-plate after incubation. Isolated colonies include *Serratia marcescens* (red), *Micrococcus luteus* (yellow), and *Escherichia coli* (off-white).

EXERCISE 50 REVIEW QUESTIONS

1. Why are cuts in the agar made during the streaking process so detrimental to the goal of getting isolated colonies?

2. In this exercise, the bacterial sample to be isolated was taken from a broth. What problem do you think is commonly encountered when isolating cells from a solid culture, and how would you deal with this problem?

3. What would be the negative effects of a drop of condensed water falling onto your streak-plate during incubation?

4. Presumably any isolated colony represents a pure culture. What could happen that would prevent an isolated colony from being a pure culture? What could you do to verify that a specific colony on your plate was pure?

REFERENCES

Cowan, M.K. 2012. *Microbiology: A Systems Approach, 3rd ed.,* chap. 3. New York: McGraw-Hill.

Reddy, C.A., Beveridge, T.J., Breznak, J.A., Marzluf, G.A., Schmidt, T.M., and Snyder, L.R. (eds.). 2007. *Methods for General and Molecular Microbiology, 3rd ed.,* chap. 11. Washington D.C.: ASM Press.

Talaro, K.P., and Chess, B. 2012. *Foundations in Microbiology, 8th ed.,* chap. 3. New York: McGraw-Hill.

Wiley, J., Sherwood, L., and Woolverton, C. 2011. *Prescott, Harley & Klein's Microbiology, 8th ed.,* chap. 6. New York: McGraw-Hill.

Loop Dilution

PRINCIPLES AND APPLICATIONS

Properly studying a single species of bacteria in the laboratory requires that it be isolated from other bacteria in the population. In a successful separation, one cell in a mixed population of bacteria will be separated from all the others and immobilized atop or within a solid growth media. As this separated cell continues to reproduce over many generations, it will give rise to a single **colony** containing millions of cells, all of which are derived from a single cell. Each colony can now be considered a **pure culture** and can be used for further study of the bacterium, usually after **subculturing** the colony to its own tube of media for convenience.

One method that almost always produces satisfactory results is loop dilution. In this method, a small sample of a bacterial culture is diluted into a tube of liquid agar, and the cells are dispersed throughout the agar. A single loopful from this tube is diluted into a second tube, and a loopful from the second tube is then diluted into a third tube. Each tube of agar is then poured into a Petri dish and allowed to solidify. Following incubation, isolated colonies are present throughout the media, each having arisen from a single isolated cell (Figure 51.1). While this method requires less practice than a streak-plate to achieve adequate results, it is much more expensive in terms of time and media used.

Loop dilutions depend on the unique qualities of agar as a solidifying agent. When heated, agar liquefies at 100°C and solidifies at 42°C. By holding the agar at 50°C, it can be inoculated in the liquid state but will be cool enough to not kill the bacteria. Once poured, the temperature of the agar will quickly drop below 42°C, and the agar will solidify, allowing discrete colonies to form from individual cells.

PRE-LAB QUESTIONS

1. *Define the following terms:*

   Colony

   Pure culture

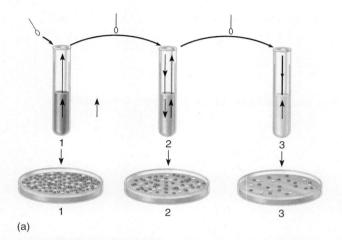

(a)

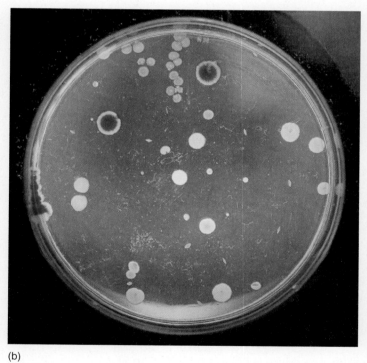

(b)

Figure 51.1 Loop dilution. (a) An inoculating loop is used to transfer a small amount of a mixed culture to a tube of melted agar, and the cells are distributed throughout the agar. In a similar manner, a sample from the first tube is diluted into a second tube and from the second tube into a third tube. The loop is flamed after each inoculation. The contents of each tube is then poured into an empty Petri dish, allowed to solidify and incubate for 48 h. (b) One of the three plates will contain well-isolated colonies.

2. *What is the temperature of boiling water? If a thermometer sitting in boiling water registers less than this temperature, what is wrong?*

3. *Why must media be held at 50°C prior to inoculation?*

PERIOD ONE

MATERIALS

Each group should obtain:

A mixed broth culture containing:

   Escherichia coli

   Serratia marcescens

   Micrococcus luteus

Three nutrient agar pours

Three empty Petri dishes

Hot plate

Inoculating loop

Marking pen

Thermometer

PROCEDURE

This procedure is illustrated in Figure 51.2.

1. Label the bottom of your Petri dishes with your name and lab time. Also label each dish "I," "II," or "III." Place the label around the periphery so your view of the plate will be as complete as possible after incubation.

2. Label the agar tubes "I," "II," or "III," and place them into a beaker containing enough water to cover the agar in the tubes. Bring the water in the beaker to a boil, and allow 5 min for the agar to liquefy. Remove the beaker from the hot plate, and allow the agar to cool to 50°C. Adding fresh water to the beaker will allow it to cool faster. Leave the melted agar tubes in the beaker of warm water when not actively working with them as this will hold them at 50°C.

3. Carefully agitate the tube containing the mixed culture until the bacteria are suspended in the media.

4. Flame the loop to red hot, and allow it to cool. Remove the cap from the culture tube and, while holding the cap with the pinky finger of your dominant hand, flame the neck of the tube.

5. Remove a single loopful of broth from the tube.

6. Flame the neck of the tube, and replace the cap. Return the culture to the test tube rack.

7. Remove the cap from tube I, and flame the neck of the tube. Inoculate the media by submerging the loop into the agar and gently swirling it to dislodge and mix the bacterial cells. After removing the loop, flame the neck of the tube and replace the cap. Place the tube back in the water-filled beaker.

8. Flame the loop, and allow it to cool for 5–10 sec. Remove the cap from tube I and flame the neck of the tube. Carefully remove a single loopful of inoculated agar. Flame the tube, replace the cap, and return the tube to the beaker.

9. Remove the cap from tube II, and flame the neck of the tube. Inoculate the media by submerging the loop into the agar and swirling it to dislodge and mix the bacterial cells. After removing the loop, flame the neck of the tube and replace the cap. Return the tube to the water-filled beaker.

10. Flame the loop, and allow it to cool for 5–10 sec. Remove the cap from tube II, and flame the neck of the tube. Carefully remove a single loopful of inoculated agar. Flame the tube, replace the cap, and return the tube to the beaker.

11. Remove the cap from tube III, and flame the neck of the tube. Inoculate the media by submerging the loop into the agar and swirling it to dislodge and mix the bacterial cells. After removing the loop, flame the neck of the tube, and replace the cap. Return the tube to the water-filled beaker. Flame the loop before placing it aside.

12. Remove the cap from the tube I, and flame the neck of the tube. Carefully pour the liquefied agar into the bottom of plate number I. Gently swirl the plate until the agar has completely covered the base of the plate. Flame the neck of the tube, and replace the cap. Repeat this process for tubes II and III.

13. After the media has completely solidified, incubate the plates in an inverted position at 30°C for 24–48 h.

(1) Liquefy three nutrient agar pours, cool to 50°C, and let stand for 10 min.

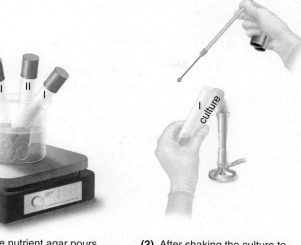

(2) After shaking the culture to disperse the organisms, flame the loop and necks of the tubes.

(3) Transfer one loopful of the culture to tube I.

(4) Flame the loop and the necks of both tubes.

(5) Replace the caps on the tubes, and return culture to the test tube rack.

(6) Disperse the organisms in tube I by shaking the tube or rolling it between the palms.

(7) Flame the loop and allow it to cool. Transfer one loopful from tube I to tube II. Return tube I to the water bath.

(8) Flame the loop and allow it to cool. After shaking tube II and transferring one loopful to tube III, flame the necks of each tube.

(9) Pour the inoculated pours into their respective Petri plates.

Figure 51.2 Procedure for a loop dilution.

QUESTIONS—PERIOD ONE

1. Why is it important to generate isolated bacterial colonies?

2. Using a drawing, predict the density of colonies on each of the three plates in a typical loop dilution series. Which plates would you expect to contain isolated colonies?

3. What is a subculture?

PERIOD TWO

RESULTS

Retrieve your plates from the incubator. To make differentiation easier, each bacterium used in this exercise exhibits a unique color: *E. coli* produces off-white growth, *M. luteus* produces yellow growth, and *S. marcescens* is red. Rarely, if ever, will you encounter outside the laboratory a mixed culture with such vibrant colors, but here the colors will help you to more fully evaluate the success of your separation. A loop dilution will produce a series of three plates in which the first plate has thousand of colonies, none of which are isolated, while the second and third plates have dramatically fewer colonies, many of which are well isolated. A successful loop dilution will always have at least one plate with many well-isolated colonies (see Figure 51.1).

Record below the appearance of your loop dilution plates. Critically analyze the plates with regard to the number of well-isolated colonies. Note any actions that could have improved the success of your plate.

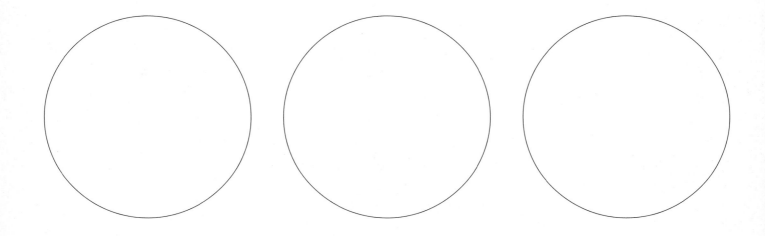

EXERCISE 51 REVIEW QUESTIONS

1. Why is the temperature of the agar at the time of the inoculation so critical to the success of a loop dilution? What would you expect if a tube was inoculated at 45°C? What about 65°C?

2. What would be the negative effects of a drop of condensed water falling onto your plate during incubation?

3. Presumably any isolated colony represents a pure culture. What could happen that would prevent an isolated colony from being a pure culture? What could you do to verify that a specific colony on your plate was pure?

4. Suppose that after incubation, your first (least diluted) plate had thousands of colonies, but your second plate had no colonies at all. How many colonies would you expect on your third plate?

5. With regard to the previous question, why do you think the first plate contained thousands of colonies while the second contained none at all?

REFERENCES

Cowan, M.K. 2012. *Microbiology: A Systems Approach, 3rd ed.*, chap. 3. New York: McGraw-Hill.

Reddy, C.A., Beveridge, T.J., Breznak, J.A., Marzluf, G.A., Schmidt, T.M., and Snyder, L.R. (eds.). 2007. *Methods for General and Molecular Microbiology, 3rd ed.*, chap. 11. Washington D.C.: ASM Press.

Talaro, K.P., and Chess, B. 2012. *Foundations in Microbiology, 8th ed.*, chap. 3. New York: McGraw-Hill.

Wiley, J., Sherwood, L., and Woolverton, C. 2011. *Prescott, Harley & Klein's Microbiology, 8th ed.*, chap. 6. New York: McGraw-Hill.

Spread-Plate

PRINCIPLES AND APPLICATIONS

Properly studying a single species of bacteria in the laboratory requires that it be isolated from other bacteria in the population. In a successful separation, one cell in a mixed population of bacteria will be separated from all others and immobilized atop or within a solid growth media. As this separated cell continues to reproduce over many generations, it will give rise to a single **colony** containing millions of cells, all of which are derived from a single cell. Each colony can now be considered a **pure culture** and can be used for further study of the bacterium, usually after **subculturing** the colony to its own tube of media for convenience.

In most cases, the enormous number of cells in a bacterial culture requires that the culture be greatly thinned, utilizing either a streak-plate or a loop dilution to separate the cells from one another. However, when the concentration of cells in a culture is small, or a highly selective media prevents all but a small number of cells in the culture from growing, less effort is needed to separate them from one another. A spread-plate is a method of isolating just such a culture. In this method, a small volume of a bacterial culture is used to inoculate the surface of an agar plate, and a sterile glass rod, bent into the shape of an L and often referred to as a hockey stick, is used to spread the cells over the surface of the plate. In this way, the relatively few cells in the culture will be physically separated from one another and will grow into isolated colonies.

PRE-LAB QUESTIONS

1. *Define the following terms:*

Colony

Pure culture

Selective media

2. *Only about 300 colonies can be adequately isolated on a standard Petri dish. Given this fact, what is the greatest number of cells that can be used to inoculate a plate of nutrient agar using a spread-plate technique?*

3. *If 5,000,000 cells are used to inoculate a plate using the spread-plate method but only 231 colonies grow, what can you say about the media that was used?*

PERIOD ONE

MATERIALS

Each group should obtain:

A previously diluted mixed broth culture containing:

   Escherichia coli

   Serratia marcescens

   Micrococcus luteus

One nutrient agar plate

Approximately 50 ml of ethyl or isopropyl alcohol in a 250-ml beaker

Bent glass spreading rod

One sterile dropper or pipette

Sterile water

Petri plate turntable (optional)

Marking pen

PROCEDURE

This procedure is illustrated in Figure 52.1.

1. Place the spreader rod in the beaker, and add enough ethanol to the beaker to cover the lower portion of the rod.

2. Arrange the items on your bench so that the Bunsen burner is between the alcohol and the Petri dish. This reduces the chance of accidentally setting the alcohol aflame.

3. Mark the bottom of your Petri dishes with your name and lab time. Place the label around the periphery so your view of the plate will be as complete as possible after incubation.

4. Carefully agitate the tube containing the diluted mixed culture until the bacteria is suspended in the media.

5. Place a single drop of sterile water in the center of the plate. Flame your loop, and retrieve a single loopful of the dilute culture. Add this to the drop of water on the plate. Flame the loop prior to placing it aside.

6. Holding the upper end of the spreader rod, remove it from the alcohol and pass it through the flame of the Bunsen burner, allowing the alcohol to ignite. Be sure to keep your hand above the spreader so that the flaming alcohol doesn't run onto your hand. Wait until the alcohol has completely burned off before continuing.

7. Place the lower portion of the spreader flat against the agar plate. Rotate the plate using your thumb and middle finger while moving the spreader back and forth. If a turntable is used, gently spin the turntable while moving the spreader back and forth. In both cases, finish by rotating the plate one complete revolution while holding the spreader against the edge of the plate.

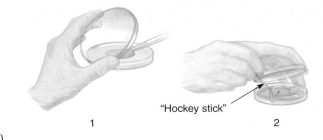

"Hockey stick"

1   2

(a)

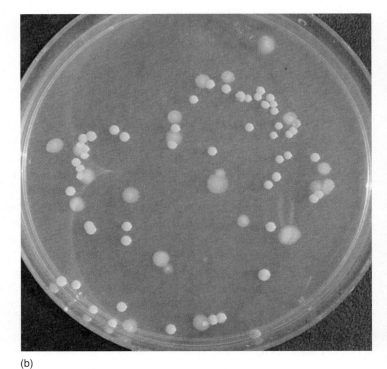

(b)

Figure 52.1   (a) Procedure for a spread-plate. (b) Typical results of a spread-plate used to separate two species of bacteria from a mixed culture.

8. Allow the spreader to cool before returning it to the alcohol. It is possible for a hot spreader to ignite the alcohol. There is no need to flame it again.

9. Allow the plate to sit upright for 5 min, and then incubate it in an inverted position at 25°C for 24–48 h.

QUESTIONS—PERIOD ONE

1. What is the importance of generating isolated bacterial colonies?

2. What would happen if the spread-plate method was used to isolate colonies from a culture with a high concentration of bacterial cells?

3. What is a subculture?

RESULTS

Retrieve your plate from the incubator. To make differentiation easier, each bacterium used in this exercise exhibits a unique color: *E. coli* produces off-white growth, *M. luteus* produces yellow growth, and *S. marcescens* is red. Rarely, if ever, will you encounter outside the laboratory a mixed culture with such vibrant colors, but here the colors will help you to more fully evaluate the success of your separation. A spread-plate produces random colonies across the surface of the plate, and a successful plate will have many well-isolated colonies (see Figure 52.1).

Record below the appearance of your spread-plate. Critically analyze the plate with regard to the number of well-isolated colonies. Note any actions that could have improved the success of your plate.

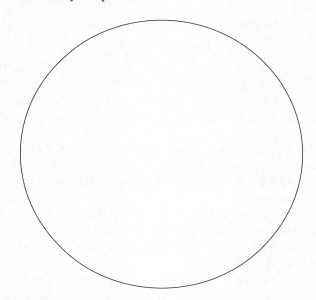

EXERCISE 52 REVIEW QUESTIONS

1. For each of the media seen in the accompanying table:

   a. Fill in the first column by indicating what cells are selected for (i.e., what type of cells will survive on the media). Referring to the exercises associated with each type of media will help you in this task.

   b. Complete the second column by deciding whether, for each type of media, a spread-plate could be an appropriate isolation technique for a sample with heavy bacterial growth.

	Selective/nonselective	Appropriate for spread-plate technique (Y/N)
Nutrient agar		
MacConkey agar		
Mannitol salt agar		
Trypticase soy agar		

2. Under what circumstance would a selective media *not* be helpful when performing a spread plate.

3. Would a differential media be of use when using a spread plate to isolate a pure culture from a sample with heavy bacterial growth? Why or why not?

REFERENCES

Cowan, M.K. 2012. *Microbiology: A Systems Approach, 3rd ed.*, chap. 3. New York: McGraw-Hill.

Reddy, C.A., Beveridge, T.J., Breznak, J.A., Marzluf, G.A., Schmidt, T.M., and Snyder, L.R. (eds.). 2007. *Methods for General and Molecular Microbiology, 3rd ed.*, chap. 11. Washington D.C.: ASM Press.

Talaro, K.P., and Chess, B. 2012. *Foundations in Microbiology, 8th ed.*, chap. 3. New York: McGraw-Hill.

Wiley, J., Sherwood, L., and Woolverton, C. 2008. *Prescott, Harley & Klein's Microbiology, 7th ed.*, chap. 6. New York: McGraw-Hill.

Fluid Thioglycollate Medium

A general purpose, liquid medium used to determine the oxygen requirements of bacterial specimens. Fluid thioglycollate is also commonly used to grow microaerophiles and anaerobes.

PRINCIPLES AND APPLICATIONS

One key piece of information often sought in the microbiology lab is the oxygen requirements, or tolerance, of an organism. Knowing whether an organism is an **aerobe, microaerophile, facultative anaerobe,** or **obligate anaerobe** is important information that can aid in the identification of an unknown organism; additionally, some organisms will only grow in an atmosphere partially or completely devoid of oxygen. Fluid thioglycollate medium is useful in both these situations. This media, a thick broth because of the addition of a small amount of agar, has dissolved oxygen expelled during autoclaving. During cooling, oxygen diffuses into the media, moving from top to bottom, producing an oxygen gradient within the top 1–2 cm or so of the tube (Figure 53.1); below this level the media provides an anaerobic environment. The indicator **resazurin** shows the location of oxygen in the tube, turning pink where oxygen is present. The medium is inoculated with a vertical stab from top to bottom, ensuring that organisms are initially present throughout the media. After incubation, the position of growth within the media indicates the oxygen requirements of the bacterium (Figure 53.2).

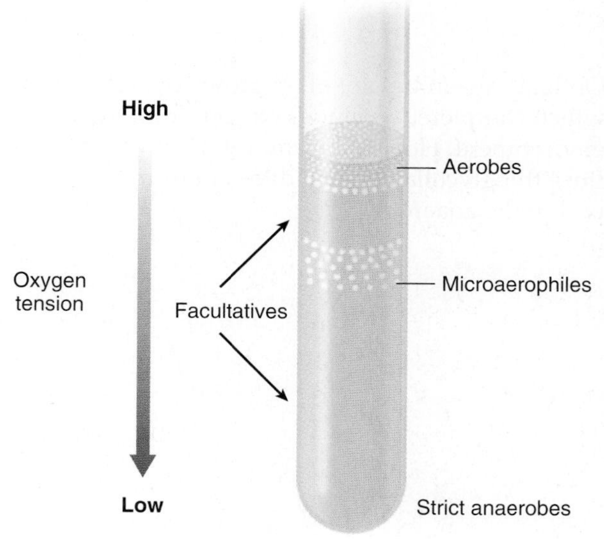

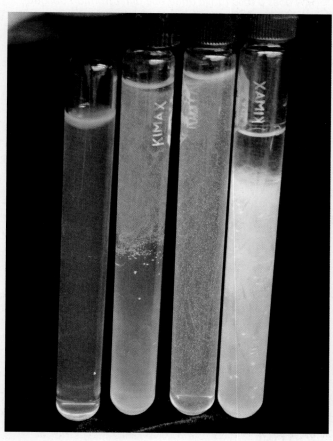

Figure 53.2 Organisms displaying different oxygen requirements in fluid thioglycollate medium. The tube is prepared so that a top-to-bottom oxygen gradient exists within the tube, with high concentrations of O_2 at the top and no O_2 at the bottom. After incubation, the position of growth within the tube indicates the oxygen needs of the bacteria. From left to right: aerobe (*Pseudomonas aeruginosa*), facultative anaerobe (*Staphylococcus aureus*), facultative anaerobe (*Escherichia coli*), and obligate anaerobe (*Clostridium butyricum*).

High

Oxygen tension

Low

Aerobes

Microaerophiles

Facultatives

Strict anaerobes

Figure 53.1 Oxygen gradient within a tube of fluid thioglycollate medium.

PRE-LAB QUESTIONS

1. *Differentiate between the terms* aerobe *and* microaerophile.

2. *How does a facultative anaerobe differ from an obligate anaerobe?*

3. *What does fastidious mean, and how does it relate to this exercise?*

PERIOD ONE

MATERIALS

Each student should obtain:
Three tubes of fluid thioglycollate media
Broth cultures of:
    Pseudomonas aeruginosa
    Staphylococcus aureus
    Clostridium sporogenes
Inoculating needle
Marking pen

PROCEDURE

1. Examine each of the fluid thioglycollate tubes for the present of a pink color in the upper 1 cm of the tube. This represents the oxygen-containing region of the media; if this color extends further than 2 cm toward the bottom of the tube, the tube should be boiled for 10 min to drive off the excess oxygen and restore the gradient.
2. Label each tube with the name of the appropriate organism: *P. aeruginosa, S. aureus,* or *C. sporogenes.*
3. Use an inoculating needle to inoculate each tube with the appropriate organism. Handle and inoculate the tubes gently so as not to disturb the oxygen gradient.
4. Incubate the tubes at 37°C for 24–48 h.

QUESTIONS—PERIOD ONE

1. Predict where in the tube you would see growth of *Clostridium sporogenes*? Why is this so?

2. Obligate anaerobes are often grown in an anaerobe jar, which completely excludes oxygen from the environment. How is the environment within a tube of fluid thioglycollate media different from that found within the anaerobe jar?

3. Sketch the appearance of growth in the accompanying fluid thioglycollate tubes.

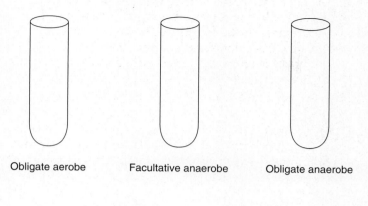

Obligate aerobe   Facultative anaerobe   Obligate anaerobe

PERIOD TWO

Retrieve your tubes from the incubator, being careful not to disturb the growth within the tube. Sketch the appearance of each of your fluid thioglycollate tubes, and indicate the oxygen preference (aerobe, facultative anaerobe, etc.) seen for each organism.

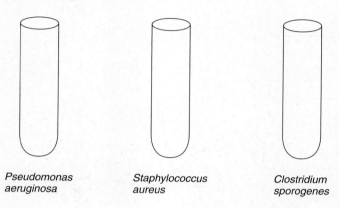

Pseudomonas aeruginosa   *Staphylococcus aureus*   *Clostridium sporogenes*

EXERCISE 53 REVIEW QUESTIONS

1. Fluid thioglycollate tubes need to be handled gently (i.e., not shaken or mixed).

  a. What will happen to the tube if it is shaken?

  b. What would be the effects of shaking the tube thoroughly prior to inoculating it?

  c. What would be the effects of shaking the tube thoroughly after incubation?

2. Prior to inoculating a tube of fluid thioglycollate media, you notice that the pink color in the media runs from the surface almost all the way to the bottom of the tube. What should you do and why?

3. The enzymes catalase and superoxide dismutase are present in some bacteria to detoxify the toxic by-products of aerobic metabolism. If a bacterial isolate lacked both of these enzymes, where would you expect it to grow in a tube of fluid thioglycollate media?

REFERENCES

Cowan, M.K. 2012. *Microbiology: A Systems Approach, 3rd ed.,* chap. 7. New York: McGraw-Hill.

Reddy, C.A., Beveridge, T.J., Breznak, J.A., Marzluf, G.A., Schmidt, T.M., and Snyder, L.R. (eds.). 2007. *Methods for General and Molecular Microbiology, 3rd ed.,* chap. 15. Washington D.C.: ASM Press.

Talaro, K.P., and Chess, B. 2012. *Foundations in Microbiology, 8th ed.,* chap. 7. New York: McGraw-Hill.

Wiley, J., Sherwood, L., and Woolverton, C. 2011. *Prescott, Harley & Klein's Microbiology, 8th ed.,* chap. 7. New York: McGraw-Hill.

Mannitol Salt Agar

A solid media with both selective and differential properties, mannitol salt agar is used for the isolation of staphylococci from a mixed culture of bacteria. The media contains the carbohydrate mannitol as a carbon source, 7.5% NaCl as a selective agent, and phenol red to detect the fermentation of mannitol to acidic end products.

PRINCIPLES AND APPLICATIONS

The genus *Staphylococcus* is populated with a few dozen species, several of which are potential human pathogens. All staphylococcal species share a number of characteristics, one of which is an ability to grow at sodium chloride concentrations as high as 15%. Mannitol salt agar contains 7.5% sodium chloride, which inhibits the growth of most other bacterial species. This selective aspect of mannitol salt agar allows staphylococcal species to be isolated from a mixed culture.

The differential aspect of mannitol salt agar is based on the fact that it contains mannitol along with the pH indicator **phenol red.** If a bacterial species is able to ferment mannitol, acid will be produced, lowering the pH and causing the phenol red indicator to turn from red to yellow. Thus, if the media around the bacterial growth turns yellow after incubation, this indicates that the organism growing on the plate ferments mannitol. If the media around the growth remains red, the organism growing on the plate does not ferment mannitol (Figure 54.1).

While all the species of *Staphylococcus* can grow in the high salt concentration provided by mannitol salt agar, not all of them are able to ferment mannitol. The most important pathogen of the group, *S. aureus*, ferments mannitol while most of the other, mostly nonpathogenic, species do not. This information is summarized in Table 54.1.

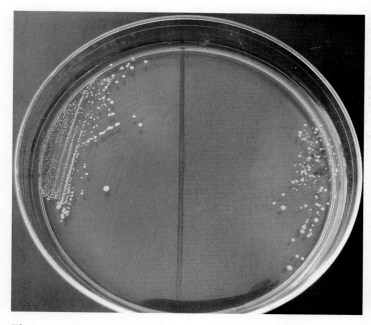

Figure 54.1 Mannitol salt agar is used to isolate *Staphylococcus* species from mixed cultures as well as identify those species able to ferment mannitol. *Staphylococcus aureus* (right) is able to both grow on the high salt medium and ferment mannitol. The production of acid from the fermentation lowers the pH and causes the medium to turn yellow. *Staphylococcus epidermidis* (left) can grow in the presence of high salt but does not ferment mannitol, as evidenced by the lack of any yellow coloration in the media.

TABLE 54.1			
Growth	**Fermentation of mannitol**	**Interpretation**	**Presumptive identification**
Little to none	N/A	Inhibited by NaCl	Not *Staphylococcus*
Good growth	Negative (media remains red)	Not inhibited by NaCl, no acid produced	*Staphylococcus species*
Good growth	Positive (media turns yellow)	Not inhibited by NaCl, acid produced from mannitol	*Staphylococcus aureus*

PRE-LAB QUESTIONS

1. *Differentiate between the terms* selective *and* differential *as they apply to mannitol salt agar.*

2. *Differentiate between the terms* facultative halophile *and* obligate halophile.

PERIOD ONE

MATERIALS

Each student should obtain:
One plate of mannitol salt agar
Broth cultures of:
   Staphylococcus aureus
   Escherichia coli
   Staphylococcus epidermidis
Inoculating loop
Marking pen

PROCEDURE

1. Use a marker to divide a mannitol salt agar plate into thirds. Label the plate with your name and lab time, and label each section of the plate with the name of the appropriate organism.
2. Using aseptic technique, use a loop to make a single small streak of each organism in the appropriate sector.
3. Incubate the plate at 37°C for 24–48 h.

QUESTIONS—PERIOD ONE

1. How would you differentiate:

   Staphylococcus species from non-*Staphylococcus* species?

   S. aureus from *S. epidermidis?*

PERIOD TWO

Retrieve your plate from the incubator, and examine the growth in each section of your plate. Sketch the appearance of each organism on the plate, and indicate the salt tolerance and ability to ferment mannitol for each organism.

EXERCISE 54 REVIEW QUESTIONS

1. What would be the effect of removing the sodium chloride from mannitol salt agar plates?

2. What would be the effect of removing the mannitol from mannitol salt agar plates?

3. There are a few genera of bacteria besides *Staphylococcus* that can grow on mannitol salt agar. What other (simple) laboratory test could you do that would help to confirm that the isolate on your mannitol salt agar plate was *Staphylococcus*?

REFERENCES

Cowan, M.K. 2012. *Microbiology: A Systems Approach, 3rd ed.,* chap. 3. New York: McGraw-Hill.

Reddy, C.A., Beveridge, T.J., Breznak, J.A., Marzluf, G.A., Schmidt, T.M., and Snyder, L.R. (eds.). 2007. *Methods for General and Molecular Microbiology, 3rd ed.,* chap. 11. Washington D.C.: ASM Press.

Talaro, K.P., and Chess, B. 2012. *Foundations in Microbiology, 8th ed.,* chap. 3. New York: McGraw-Hill.

Wiley, J., Sherwood, L., and Woolverton, C. 2011. *Prescott, Harley & Klein's Microbiology, 8th ed.,* chap. 6. New York: McGraw-Hill.

MacConkey Agar

A media with both differential and selective properties, MacConkey agar contains bile salts and crystal violet to select for Gram-negative bacteria as well as the carbohydrate lactose and the pH indicator neutral red to differentiate between lactose fermenters and lactose nonfermenters.

PRINCIPLES AND APPLICATIONS

MacConkey agar is often used as the first step in the isolation of members of the Enterobacteriaceae, providing two important pieces of information: Gram reaction and ability to ferment lactose. The selective aspect of the media is a result of the inclusion of bile salts and crystal violet, which prevents the growth of Gram-positive bacteria, allowing most Gram-negative species to grow. The differential characteristic of the media results from the inclusion of lactose and neutral red, a pH indicator that is colorless above a pH of 6.8 and red below. Fermentation of the lactose produces lactic acid and, consequently, a lowering of the pH of the media. The lower pH causes colonies of lactose-fermenting species to exhibit a pink or red color while nonfermenting species appear white or translucent (Figure 55.1).

MacConkey agar has a number of variations that allow the differentiation of normal microbial flora from potential pathogens. For example, MacConkey sorbitol allows *Escherichia coli* O157:H7, a pathogen, to be differentiated from most other strains of the bacterium that are found in the feces of healthy people. The medium works by replacing the lactose normally present in the media with sorbitol. Because *E. coli* O157:H7 cannot ferment sorbitol, it produces white colonies while most other strains, which can ferment the sugar, produce pink colonies. Several other varieties of MacConkey agar exist, with each offering slightly different differential and selective properties. The careful selection of such a media can provide a great deal of information and shorten the time between culturing and identification.

PRE-LAB QUESTION

1. Differentiate between the terms selective and differential as they apply to MacConkey agar.

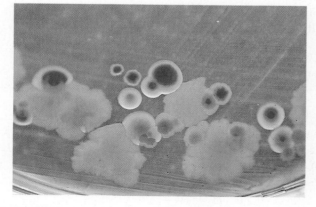

Figure 55.1 MacConkey agar is selective for Gram-negative bacteria as well as being differential with regard to lactose fermentation. Lactose-fermenting species produce colonies with pink to red centers as a result of acid production while non-lactose-fermenting species manufacture no acid and hence produce white or translucent colonies.

PERIOD ONE

MATERIALS

Each student should obtain:
One plate of MacConkey agar
Broth cultures of:
  Morganella morganii
  Escherichia coli
  Enterococcus faecalis
Inoculating loop
Marking pen

PROCEDURE

1. Use a marker to divide a MacConkey agar plate into thirds. Label the plate with your name and lab time, and label each section of the plate with the name of the appropriate organism.
2. Using a loop, make a single small streak of each culture onto the appropriate section of the plate.
3. Incubate the plate at 37°C for 24 h. If reading the plate cannot be accomplished after 24 h, the plates should be removed from the incubator and refrigerated until the lab period.

QUESTIONS—PERIOD ONE

1. A mixed culture containing *Escherichia coli, Providencia stuartii*, and *Staphylococcus aureus* is streaked onto MacConkey agar. *E. coli* is a Gram-negative lactose fermenter, *P. stuartii* is a Gram-negative nonfermenter, and *S. aureus* is a Gram-positive fermenter. Explain the appearance of each species after incubation of the plate.

2. How would the appearance of each of the three organisms in the previous question change if trypticase soy agar was used instead of MacConkey agar?

PERIOD TWO

Retrieve your plate from the incubator and examine both the growth and ability to ferment lactose for each of the three organisms on the plate. Sketch the appearance of your plate.

Complete the table for each of the organisms in your culture based on their reaction on MacConkey agar.

Species	Growth on MacConkey agar(+/−)	Gram reaction	Color	Fermentation of lactose
E. coli				
M. morganii				
E. faecalis				

EXERCISE 55 REVIEW QUESTIONS

1. Phenol red is another indicator commonly used to detect the presence of acid. This indicator is red above pH 7 and yellow below. What would be the appearance of the same three organisms as in the previous questions—*E. coli, P. stuartii,* and *S. aureus* —if the neutral red in MacConkey agar was replaced with phenol red?

2. A variation of MacConkey agar exists that allows for the growth of both Gram-negative and Gram-positive bacterial species. What ingredients are purposefully left out of this formulation of the media?

REFERENCES

Cowan, M.K. 2012. *Microbiology: A Systems Approach, 3rd ed.,* chap. 3. New York: McGraw-Hill.

Reddy, C.A., Beveridge, T.J., Breznak, J.A., Marzluf, G.A., Schmidt, T.M., and Snyder, L.R. (eds.). 2007. *Methods for General and Molecular Microbiology, 3rd ed.,* chap. 11. Washington D.C.: ASM Press.

Talaro, K.P., and Chess, B. 2012. *Foundations in Microbiology, 8th ed.,* chap. 3. New York: McGraw-Hill.

Wiley, J., Sherwood, L., and Woolverton, C. 2011. *Prescott, Harley & Klein's Microbiology, 8th ed.,* chap. 6. New York: McGraw-Hill.

NOTES

Desoxycholate Agar

Desoxycholate agar is a slightly selective medium, using sodium desoxycholate, ferric citrate, and sodium citrate to inhibit the growth of Gram-positive bacteria. The medium is also differential, containing the pH indicator neutral red to differentiate between lactose fermenters and lactose nonfermenters.

PRINCIPLES AND APPLICATIONS

Desoxycholate agar is commonly used for the isolation and differentiation of Gram-negative enteric bacilli, such as members of the Enterobacteriaceae. The selective agents sodium desoxycholate, ferric citrate, and sodium citrate inhibit the growth of Gram-positive bacteria while the inclusion of lactose and neutral red provide a means of differentiating between lactose fermenters and lactose non-fermenters. Lactose-fermenting bacteria will produce red colonies as the acid they produce from fermentation of the lactose reacts with the indicator neutral red. Lactose nonfermenters do not produce acid and consequently produce clear colonies on desoxycholate agar. Desoxycholate agar is commonly used to screen for species of *Salmonella* and *Shigella* in clinical specimens. The majority of species making up the normal intestinal microbiota are coliforms that ferment lactose and therefore produce red colonies (Figure 56.1) while *Salmonella* and *Shigella* are lactose nonfermenters, producing clear colonies.

PRE-LAB QUESTIONS

1. *In desoxycholate agar, list the ingredients that cause the medium to be:*

Selective

Differential

2. *What is a coliform? How would you recognize a coliform on desoxycholate agar?*

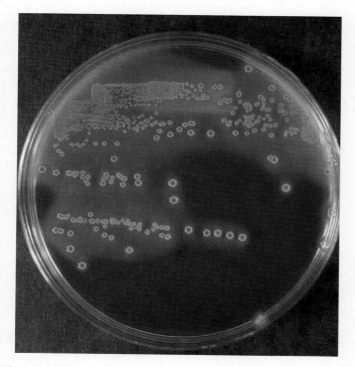

Figure 56.1 Desoxycholate agar contains lactose and the pH indicator neutral red. Bacteria that ferment lactose will produce acid that, in conjunction with neutral red, will lead to pink- to red-colored colonies. Here, *Escherichia coli*, a lactose fermenter, is shown on the medium.

PERIOD ONE

MATERIALS

Each student should obtain:
One plate of desoxycholate agar
Broth cultures of:
   Shigella flexneri
   Escherichia coli
   Enterococcus faecalis
Inoculating loop
Marking pen

PROCEDURE

1. Use a marker to divide a desoxycholate agar plate into thirds. Label the plate with your name and lab time, and label each section of the plate with the name of the appropriate organism.
2. Using a loop, make a single small streak of each culture onto the appropriate section of the plate.
3. Incubate the plate at 37°C for 24–48 h.

QUESTIONS—PERIOD ONE

1. A mixed culture containing *Escherichia coli*, *Salmonella choleraesuis*, and *Staphylococcus aureus* is streaked onto desoxycholate agar. *E. coli* is a Gram-negative lactose fermenter, *S. choleraesuis* is a Gram-negative nonfermenter, and *S. aureus* is a Gram-positive fermenter. Explain the appearance of each species after incubation of the plate.

2. How would the appearance of each of the three organisms in the previous question change if trypticase soy agar was used instead of desoxycholate agar?

PERIOD TWO
Retrieve your plate from the incubator, and examine both the growth and ability to ferment lactose for each of the three organisms on the plate. Sketch the appearance of your plate.

EXERCISE 56 REVIEW QUESTIONS

1. If lactose and neutral red were left out of desoxycholate agar, which bacteria in this exercise (*S. flexneri, E. coli,* and/or *E. faecalis*) could no longer be as easily differentiated from one another?

2. Phenol red is another indicator commonly used to detect the presence of acid. This indicator is red above pH 7 and yellow below. What would be the appearance of the same three organisms as in the previous questions—*E. coli, S. choleraesuis,* and *S. aureus* —if the neutral red in desoxycholate agar was replaced with phenol red?

REFERENCES

Cowan, M.K. 2012. *Microbiology: A Systems Approach, 3rd ed.,* chap. 3. New York: McGraw-Hill.

Talaro, K.P., and Chess, B. 2012. *Foundations in Microbiology, 7th ed.,* chap. 3. New York: McGraw-Hill.

Wiley, J., Sherwood, L., and Woolverton, C. 2011. *Prescott, Harley & Klein's Microbiology, 8th ed.,* chap. 6. New York: McGraw-Hill.

NOTES

Endo Agar

Endo agar is a slightly selective medium, using a combination of sodium sulfite and basic fuschin to inhibit the growth of Gram-positive bacteria. The medium is also differential, with lactose-fermenting bacteria producing pink to red colonies while lactose nonfermenters produce clear colonies.

PRINCIPLES AND APPLICATIONS

Used for the detection of bacteria in potable water, wastewater, dairy products, and foods, endo agar uses sodium sulfite and basic fuschin to suppress the growth of Gram-positive organisms. Basic fuschin also serves as a pH indicator, turning red as acid accumulates from the fermentation of lactose. **Coliforms** ferment lactose, and because of this, they produce pink to red colonies on the medium. Some strong lactose fermenters, such as *Escherichia coli*, produce colonies with a metallic sheen (Figure 57.1). Organisms that do not ferment lactose produce clear colonies. Historically an important media, in the last several years, endo agar has been replaced by more contemporary formulations for the examination of water, food, and milk.

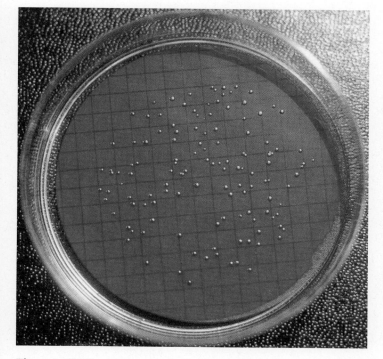

Figure 57.1 All lactose-fermenting species produce pink to red colonies on endo agar while some, such as *E. coli*, produce colonies with a characteristic metallic sheen. In this plate, the grid pattern is used as an aid in counting the colonies.

PRE-LAB QUESTIONS

1. *In endo agar, list the ingredients that cause the medium to be:*

Selective

Differential

2. *What is a coliform? How would you recognize a coliform on endo agar?*

PERIOD ONE

MATERIALS

Each student should obtain:
One plate of endo agar
Broth cultures of:
  Shigella flexneri
  Escherichia coli
  Enterococcus faecalis
Inoculating loop
Laboratory marker

PROCEDURE

1. Use a marker to divide an endo agar plate into thirds. Label the plate with your name and lab time, and label each section of the plate with the name of the appropriate organism.
2. Using a loop, make a single small streak of each culture onto the appropriate section of the plate.
3. Incubate the plate at 37°C for 24–48 h.

QUESTIONS—PERIOD ONE

1. A mixed culture containing *E. coli, Salmonella choleraesuis,* and *Staphylococcus aureus* is streaked onto endo agar. *E. coli* is a Gram-negative lactose fermenter, *S. choleraesuis* is a Gram-negative nonfermenter, and *S. aureus* is a Gram-positive fermenter. Explain the appearance of each species after incubation of the plate.

2. How would the appearance of each of the three organisms in the previous question change if trypticase soy agar was used instead of endo agar?

PERIOD TWO

Retrieve your plate from the incubator, and examine both the growth and ability to ferment lactose for each of the three organisms on the plate. Sketch the appearance of your plate.

EXERCISE 57 REVIEW QUESTIONS

1. If lactose was left out of endo agar, which bacteria from the earlier questions (*E. coli, S. choleraesuis,* and/or *S. aureus*) could no longer be as easily differentiated from one another?

2. Endo agar has been supplanted by newer types of media. List two of those types of media, and indicate how coliforms and noncoliforms are identified on each type.

REFERENCES

Cowan, M.K. 2012. *Microbiology: A Systems Approach, 3rd ed.,* chap. 24. New York: McGraw-Hill.

Talaro, K.P., and Chess, B. 2012. *Foundations in Microbiology, 8th ed.,* chap. 27. New York: McGraw-Hill.

Wiley, J., Sherwood, L., and Woolverton, C. 2011. *Prescott, Harley & Klein's Microbiology, 8th ed.,* chap. 6. New York: McGraw-Hill.

NOTES

Eosin Methylene Blue Agar

Eosin methylene blue agar is a slightly selective medium for the isolation of Gram-negative enteric bacteria. The dyes eosin and methylene blue are used to inhibit the growth of Gram-positive bacteria. The medium is also differential, with lactose-fermenting bacteria producing blue-black colonies and non-lactose-fermenting bacteria producing colonies that are colorless to amber.

PRINCIPLES AND APPLICATIONS

Used for the detection of enteric bacteria in dairy products and foods, eosin methylene blue (EMB) agar uses two dyes, eosin and methylene blue, to suppress the growth of Gram-positive organisms. **Coliforms** and other bacteria that ferment lactose produce acid from the fermentation. This acid causes the precipitation of methylene blue eosinate, which is absorbed by the cells, producing blue-black colonies on the medium. Colonies of some strong lactose fermenters, such as *Escherichia coli*, have a characteristic, often green, metallic sheen (Figure 58.1). Organisms that do not ferment lactose, such as *Salmonella* and *Shigella*, produce colorless or lightly pigmented colonies. The dyes in EMB agar do not completely inhibit the growth of all Gram-positive microorganisms;

fecal enterococci and staphylococci will grow on the media, usually producing pinpoint colonies. EMB is recommended by the American Public Health Association for the microbiological examination of food and dairy products.

PRE-LAB QUESTIONS

1. *In EMB agar, list the ingredients that cause the medium to be:*

Selective

Differential

2. *What is a coliform? How would you recognize a coliform on EMB agar?*

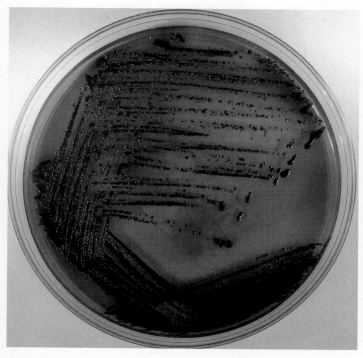

Figure 58.1   *Escherichia coli* growing on EMB agar.

PERIOD ONE

MATERIALS

Each student should obtain:
One plate of EMB agar
Broth cultures of:
   Shigella flexneri
   Escherichia coli
   Enterococcus faecalis
Inoculating loop
Marking pen

PROCEDURE

1. Use a marker to divide an endo agar plate into thirds. Label the plate with your name and lab time, and label each section of the plate with the name of the appropriate organism.
2. Using a loop, make a single small streak of each culture onto the appropriate section of the plate.
3. Incubate the plate at 37°C for 24–48 h.

QUESTIONS—PERIOD ONE

1. A mixed culture containing *E. coli, Salmonella choleraesuis,* and *Staphylococcus aureus* is streaked onto EMB agar. *E. coli* is a Gram-negative lactose fermenter, *S. choleraesuis* is a Gram-negative nonfermenter, and *S. aureus* is a Gram-positive fermenter. Explain the appearance of each species after incubation of the plate.

2. How would the appearance of each of the three organisms in the previous question change if trypticase soy agar was used instead of EMB agar?

PERIOD TWO
Retrieve your plate from the incubator, and examine both the growth and ability to ferment lactose for each of the three organisms on the plate. Sketch the appearance of your plate.

EXERCISE 58 REVIEW QUESTION

1. If lactose was left out of EMB agar, which bacteria from the earlier questions—*E. coli, S. choleraesuis,* and *S. aureus*—could no longer be as easily differentiated from one another?

REFERENCE

Wiley, J., Sherwood, L., and Woolverton, C. 2011. *Prescott, Harley & Klein's Microbiology, 8th ed.,* chap. 6. New York: McGraw-Hill.

NOTES

Hektoen Enteric Agar

Hektoen enteric agar is a selective medium used to isolate Gram-negative enteric bacteria, especially Salmonella *and* Shigella, *from complex specimens. Bile salts are used to inhibit the growth of Gram-positive bacteria, three carbohydrates are included to distinguish between groups of enteric bacteria, and the production of hydrogen sulfide is used to distinguish* Salmonella *from* Shigella.

PRINCIPLES AND APPLICATIONS

Hektoen enteric agar is used to isolate and culture several types of Gram-negative bacteria from a variety of clinical and nonclinical specimens. The selective aspect of the media is due to the inclusion of bile salts, which inhibit the growth of Gram-positive organisms. The carbohydrates lactose, sucrose, and salicin are included in the media to differentiate fermentative from nonfermentative enteric bacteria. Fermentation of these sugars by organisms such as *Escherichia* will result in the production of acid, which in turn lowers the pH of the media surrounding the colony. The dyes acid fuschin and bromthymol blue are included in the media as pH indicators and impart a yellow to pink color to organisms that ferment any of the three sugars. Organisms that do not ferment any of the three sugars (*Salmonella* and *Shigella*) produce blue-green colonies. Ferric ammonium citrate and sodium thiosulfate in the media allow for the detection of hydrogen sulfide (H_2S) production as noted by a black precipitate; consequently, H_2S-producing colonies such as *Salmonella* produce a black precipitate in the center of their colonies (Figure 59.1).

PRE-LAB QUESTIONS

1. In Hektoen enteric agar, list the ingredients that cause the medium to be:

   Selective

   Differential

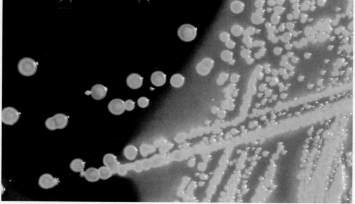
(a)

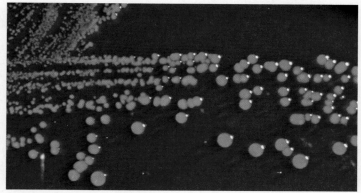
(b)

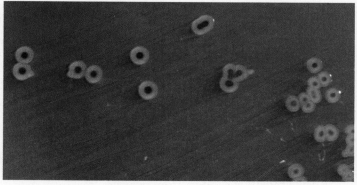
(c)

Figure 59.1 (a) Bacteria such as *Escherichia coli,* which ferment any of the three carbohydrates in Hektoen enteric agar, will appear yellow to pink in color. (b) *Shigella boydii* bacteria, which do not ferment sugars in the media, produce blue-green colonies. (c) *Salmonella typhimurium* does not ferment any of the sugars in Hektoen enteric agar, leading to blue-green colonies. The black centers in the colonies are the result of hydrogen sulfide production, which can be used to distinguish *Salmonella* from *Shigella.*

2. *In what two ways is Hektoen enteric agar a differential medium?*

3. *What is a coliform? How would a coliform appear on Hektoen enteric agar?*

PERIOD ONE

MATERIALS

Each student should obtain:
One plate of Hektoen enteric agar
Broth cultures of:
   Shigella flexneri
   Escherichia coli
   Salmonella typhimurium
Inoculating loop
Marking pen

PROCEDURE

1. Use a marker to divide a Hektoen enteric agar plate into thirds. Label the plate with your name and lab time, and label each section of the plate with the name of the appropriate organism.
2. Using a loop, make a single small streak of each culture onto the appropriate section of the plate.
3. Incubate the plate at 37°C for 24–48 h.

QUESTIONS—PERIOD ONE

1. A mixed culture containing *E. coli, Salmonella choleraesuis,* and *Enterococcus faecalis* is streaked onto Hektoen enteric agar. *E. coli* is a Gram-negative lactose fermenter that produces hydrogen sulfide, *S. choleraesuis* is a Gram-negative nonfermenter that produces hydrogen sulfide, and *E. faecalis* is a Gram-positive, lactose fermenter. Explain the appearance of each species after incubation of the plate.

2. How would the appearance of each of the three organisms in the previous question change if eosin methylene blue agar was used instead of Hektoen enteric agar? What if trypticase soy agar was used?

PERIOD TWO

Retrieve your plate from the incubator, and examine the growth, ability to ferment lactose, and ability to produce hydrogen sulfide for each of the three organisms on the plate. Sketch the appearance of your plate.

EXERCISE 59 REVIEW QUESTIONS

1. If bromthymol blue and acid fuschin were left out of the Hektoen enteric agar, which bacteria, of the three used in this exercise, could no longer be as easily differentiated from the others?

2. If ferric ammonium citrate and sodium thiosulfate were left out of the Hektoen enteric agar, which bacteria, of the three used in this exercise, could no longer be as easily differentiated from the others?

REFERENCES

Cowan, M.K. 2012. *Microbiology: A Systems Approach, 3rd ed.,* chap. 3. New York: McGraw-Hill.

Talaro, K.P., and Chess, B. 2012. *Foundations in Microbiology, 8th ed.,* chap. 3. New York: McGraw-Hill.

Wiley, J., Sherwood, L., and Woolverton, C. 2011. *Prescott, Harley & Klein's Microbiology, 8th ed.,* chap. 35. New York: McGraw-Hill.

Xylose Lysine Desoxycholate Agar

Xylose lysine desoxycholate agar is a selective medium used to isolate Gram-negative enteric bacteria, especially Shigella, from complex specimens. Sodium desoxycholate is used to inhibit the growth of Gram-positive bacteria. Fermentation of xylose, decarboxylation of lysine, and production of H₂S are all used to differentiate Shigella from other enteric species.

PRINCIPLES AND APPLICATIONS

Xylose lysine desoxycholate (XLD) agar is used to isolate *Shigella* and *Salmonella* from complex specimens containing a variety of enteric bacteria, and the medium is recommended for the testing of foods, dairy products, and water. The selective aspect of the media is due to the inclusion of sodium desoxycholate, which inhibits the growth of Gram-positive organisms.

The carbohydrate **xylose** along with the pH indicator phenol red are used to differentiate *Shigella* from other enteric bacteria. Phenol red is included in the media so that the production of acid may be recognized, appearing red under basic conditions and yellow when acidic. While almost all enteric bacteria ferment xylose, producing acid and turning the media yellow, *Shigella* species are nonfermenters and the media retains a pink to red color.

Salmonella is distinguished from other enteric bacteria through the inclusion of the amino acid *lysine* in the media. Like most enteric bacteria, *Salmonella* ferments xylose, producing acid and turning the colonies and surrounding media yellow (due to the presence of phenol red). However, *Salmonella* also **decarboxylates** the amino acid lysine, producing alkaline end products, which neutralizes any acid produced. In this way, both *Salmonella* and *Shigella* species produce red colonies on XLD agar.

The final differential aspect of the media involves discriminating between *Salmonella* and *Shigella* by using a hydrogen sulfide (H₂S) indicator system. Sodium thiosulfate and ferric ammonium citrate are included to indicate the production of hydrogen sulfide: in the presence of H₂S production and these reagents, the result is a blackening of the colonies. Hydrogen sulfide is produced by *Salmonella* but not by *Shigella; Salmonella* colonies, therefore, have a black center while *Shigella* colonies are red throughout (Figure 60.1).

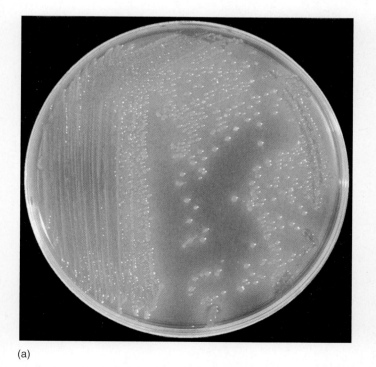

(a)

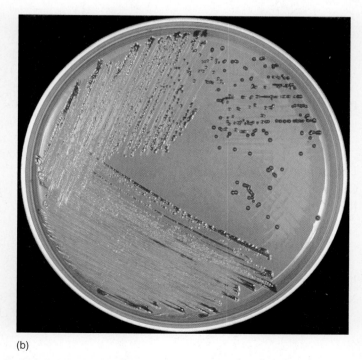

(b)

Figure 60.1 (a) *Serratia marcescens* plated on xylose lysine desoxycholate agar. Like most enteric bacteria, *S. marcescens* ferments lactose, giving colonies and the surrounding media a yellow color. *Shigella* species do not ferment xylose and produce red colonies if they are present in the culture. (b) *Salmonella sp.* plated on xylose lysine desoxycholate agar. *Salmonella* ferments xylose, but the decarboxylation of the lysine neutralizes any acids produced, resulting in red colonies. Manufacture of hydrogen sulfide in *Salmonella* produces colonies with a black center.

PRE-LAB QUESTIONS

1. *In XLD agar, list the ingredients that cause the medium to be:*

Selective

Differential

2. *In what two ways is XLD agar a differential medium?*

PERIOD ONE

MATERIALS

Each student should obtain:

One plate of XLD agar

Broth cultures of:

   Shigella flexneri

   Escherichia coli

   Salmonella typhimurium

Inoculating loop

Marking pen

PROCEDURE

1. Use a marker to divide a XLD agar plate into thirds. Label the plate with your name and lab time, and label each section of the plate with the name of the appropriate organism.
2. Using a loop, make a single small streak of each culture onto the appropriate section of the plate.
3. Incubate the plate at 37°C for 24–48 h.

QUESTIONS—PERIOD ONE

1. A mixed culture containing *Escherichia coli*, *Salmonella choleraesuis*, and *Enterococcus faecalis* is streaked onto XLD agar. *E. coli* is a Gram-negative xylose fermenter, *S. choleraesuis* is a Gram-negative nonfermenter and hydrogen sulfide producer, and *E. faecalis* is a Gram-positive lactose fermenter. Explain the appearance of each species after incubation of the plate.

2. How would the appearance of *Salmonella* change if lysine were inadvertently left out of XLD agar? What if xylose were left out?

PERIOD TWO

Retrieve your plate from the incubator, and examine the growth, ability to ferment xylose, and ability to produce hydrogen sulfide for each of the three organisms on the plate. Sketch the appearance of your plate.

EXERCISE 60 REVIEW QUESTIONS

1. If sodium thiosulfate and ferric ammonium citrate were left out of XLD agar, which bacteria *(E.coli, S. choleraesuis,* and/or *E. Faecalis)* could no longer be as easily differentiated from the others?

2. XL agar, a medium similar to XLD agar, is also widely used for many microbiological applications. How do you think the composition of XL agar differs from XLD agar? How would the appearance of organisms on XL agar differ from those on XLD agar?

Talaro, K.P., and Chess, B. 2012. *Foundations in Microbiology, 8th ed.,* chap. 3. New York: McGraw-Hill.

61

Blood Agar

Blood agar is commonly used for cultivating fastidious organisms as well as for observing any hemolytic reactions such as those seen with certain species of Streptococcus.

PRINCIPLES AND APPLICATIONS

Blood agar provides a rich environment for growth, which is particularly helpful for the cultivation of fastidious microbes. This media is also commonly used to display hemolytic characteristics of cultivated bacteria and provide presumptive identification of *Streptococcus agalactiae* through the CAMP reaction, a synergistic effect that occurs when hemolysins produced by *Streptococcus agalactiae* react with hemolysins produced by a specific strain of *Staphylococcus aureus.*

Many species of bacteria produce hemolysins, toxins that have the ability to **hemolyze** (destroy) red blood cells, releasing the hemoglobin within the cells. The type of toxin possessed by a bacterial species, and therefore the type of hemolysis produced, can be used to partially identify many bacterial isolates. The three major types of hemolysis displayed by bacterial cells are **α (alpha), β (beta),** and **γ (gamma),** each of which is displayed in Figure 61.1.

β-Hemolysis is the complete destruction of red blood cells, leading to a clear zone around each bacterial colony. The appearance of β-hemolysis is a worrisome sign in the clinical lab as it is indicative of *S. pyogenes*, the organism responsible for "strep throat," scarlet fever, glomerulonephritis, and rheumatic fever. The toxins responsible for β-hemolysis in *S. pyogenes* are streptolysin O (SLO) and streptolysisn S (SLS). While both toxins cause beta hemolysis, SLO is oxygen labile and displays maximum hemolysis under anaerobic conditions; SLS also shows its greatest effects under anaerobic conditions, but the toxin itself is oxygen stable. The isolation of β-hemolytic streptococci from healthy individuals is rare.

The most common type of hemolysis seen in the laboratory is α-hemolysis, a partial clearing of the blood around colonies combined with a greening of the media due to oxidation of the released hemoglobin. Streptococcal species that are part of the normal flora of the nasopharynx (*S. mitis* and *S. salivarius*) as well as pathogens (*S. pneumoniae*) display this type of hemolysis. γ-Hemolysis is ironically the description used to depict a lack of hemolysis. Like α-hemolysis, γ-hemolysis is a common characteristic of the flora of the nasopharynx.

(a)

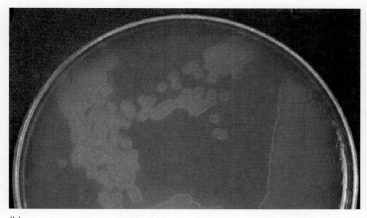

(b)

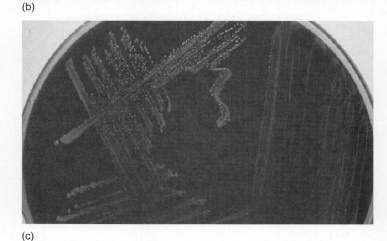

(c)

Figure 61.1 Hemolytic patterns on blood agar. (a) *Streptococcus pyogenes*, illustrating β-hemolysis. (b) *Streptococcus pneumoniae*, illustrating α-hemolysis. (c) *Staphylococcus epidermidis*, displaying γ (no) hemolysis.

PRE-LAB QUESTION

1. *Could blood agar be classified as a selective or differential medium? Explain.*

PERIOD ONE

MATERIALS

Each student should obtain:
One plate of blood agar
Sterile cotton swab
Tongue depressor
Inoculating loop
Marking pen

PROCEDURE

1. Label the plate with your name and lab time.

2. Using the tongue depressor to hold down the tongue, use the swab to sample your lab partner's throat. Sample the back of the throat along with any white patches on or near the tonsils. Avoid touching the cheeks or tongue.

3. Roll the swab over an area approximately equal to one-fifth of the surface of the blood agar plate. Be sure that the entire surface of the swab contacts the agar.

4. Exchange plates with your lab partner so that you are each manipulating your own bacteria.

5. Using a loop, streak out the bacteria using a standard four quadrant streak. After streaking the final quadrant of your plate but **prior to flaming the loop,** stab the loop several times into the agar.

6. Incubate the plate at 37°C for 24 h.

7. Be sure to dispose of the swab and tongue depressor in a hard-sided biohazard container.

QUESTIONS—PERIOD ONE

1. If multiple colonies are seen on your blood agar plate after the incubation and all of them exhibit α-hemolysis, are they all members of the same species? Why or why not?

2. Why do you think you stabbed the blood agar plate with your loop? (Hint: What kind of environment surrounds those cells that were inoculated below the surface of the agar?)

PERIOD TWO
Retrieve your plate from the incubator. To prevent the spread of potential pathogens through the laboratory, allow your instructor to examine your plate prior to your own inspection. Examine the plate for growth as well as any hemolysis. Sketch the appearance of your plate.

EXERCISE 61 REVIEW QUESTION

1. What hemolysins would be more active in areas where the agar was stabbed? Why?

REFERENCES

Cowan, M.K. 2012. *Microbiology: A Systems Approach, 3rd ed.*, chaps. 13 and 21. New York: McGraw-Hill.

Talaro, K.P., and Chess, B. 2012. *Foundations in Microbiology, 8th ed.*, chaps. 13 and 18. New York: McGraw-Hill.

Wiley, J., Sherwood, L., and Woolverton, C. 2011. *Prescott, Harley & Klein's Microbiology, 8th ed.*, chap. 21. New York: McGraw-Hill.

NOTES

Motility Media

A semisolid media used for the detection of bacterial motility.

PRINCIPLES AND APPLICATIONS

Motility media contains just enough agar to produce a jelly-like consistency, thick enough to retard the movement of nonmotile organisms while fluid enough to allow motile cells freedom of movement. The media is inoculated with a single stab to the center of the tube and incubated. After incubation, if growth is restricted to the stab line, the bacterium can be considered nonmotile, while growth away from the stab line, and perhaps throughout the tube, is indicative of a motile species (Figure 62.1).

PRE-LAB QUESTION

1. *Why is it important to inoculate motility media with a single stab?*

PERIOD ONE

MATERIALS

Each student should obtain:
Two tubes of motility media
Fresh broth cultures of:
    Corynebacterium diphtheriae
    Proteus vulgaris
Inoculating needle
Marking pen

PROCEDURE

1. Label each tube with your name and lab time. Label one tube *Proteus* and the other *Corynebacterium.*
2. Using a needle, stab each tube with the appropriate organism approximately two-thirds of the way to the bottom of the tube, as seen in Figure 62.2.
3. Incubate the tubes at 37°C for 24–48 h.

Figure 62.1 Interpretation of motility media. The tube on the right contains a nonmotile organism. The tube on the left contains a motile organism.

QUESTIONS—PERIOD ONE

1. What results would you expect if the motility media in your lab was accidentally made with 2% agar instead of 0.4% agar?

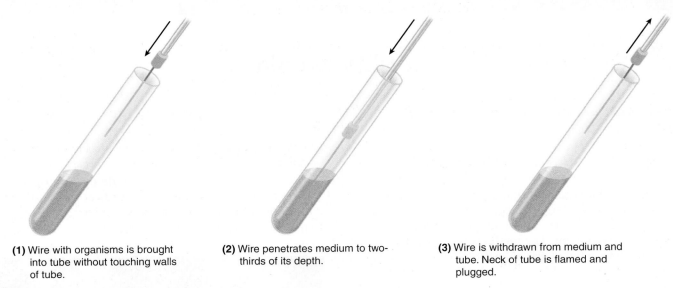

(1) Wire with organisms is brought into tube without touching walls of tube.

(2) Wire penetrates medium to two-thirds of its depth.

(3) Wire is withdrawn from medium and tube. Neck of tube is flamed and plugged.

Figure 62.2   Procedure for inoculating motility media.

2. How would you classify (i.e., motile or nonmotile) a tube in which the bacteria had spread across the surface of the media but was restricted to the stab line within the media?

PERIOD TWO
Retrieve your tubes from the incubator, and examine the tubes for motility. Sketch the appearance of each tube.

EXERCISE 62 REVIEW QUESTION

1. Provide one advantage and one disadvantage that the motility media method has relative to the hanging drop slide method for the determination of motility.

REFERENCES

Cowan, M.K. 2012. *Microbiology: A Systems Approach, 3rd ed.,* chap. 3. New York: McGraw-Hill.

Talaro, K.P., and Chess, B. 2012. *Foundations in Microbiology, 8th ed.,* chap. 3. New York: McGraw-Hill.

Wiley, J., Sherwood, L., and Woolverton, C. 2011. *Prescott, Harley & Klein's Microbiology, 8th ed.,* chap. 21. New York: McGraw-Hill.

NOTES

SIM Medium

SIM medium is a combination differential media used to differentiate bacteria based on the reduction of sulfur, production of indole from the breakdown of tryptophan, and motility.

PRINCIPLES AND APPLICATIONS

SIM (sulfur-indole-motility) medium is a semisolid media that is used to determine three bacterial activities simultaneously: the reduction of **sulfur,** production of **indole** from the breakdown of the amino acid tryptophan, and **motility.** This particular combination of tests is especially useful in the differentiation of many enteric bacilli. The media is inoculated with a single central stab to a depth approximately two-thirds of the way to the bottom of the tube. The media is then incubated at 37°C for 24–48 h.

The first of the three characteristics to be determined is motility. SIM media is a semisolid medium containing enough agar (0.4%) to retard nonmotile bacteria while allowing motile species freedom of movement. Motility is detected as turbidity radiating from the central stab line. It is important to inoculate as cleanly as possible as even slight back and forth movement of the inoculating needle may distribute bacteria throughout the tube, producing what may be mistaken for motility in a nonmotile culture. Results of the test are most easily determined when they are compared to an uninoculated tube (Figure 63.1).

Reduction of sulfur is next evaluated. Bacteria able to reduce sulfur do so through one of two different biochemical pathways, each utilizing unique enzymes. In both cases, however, hydrogen sulfide (H_2S) gas is produced. SIM media includes an iron-containing compound that will react with H_2S to produce ferrous sulfide (FeS), a black precipitate (Figure 63.2). Any blackening of the media is indicative of sulfur reduction and is considered a positive reaction while no blackening is considered a negative result and can be interpreted as a lack of sulfur reduction.

Lastly, bacteria that possess the enzyme tryptophanase catabolize the amino acid tryptophan to produce pyruvic acid, ammonia, and indole. Detection of indole is

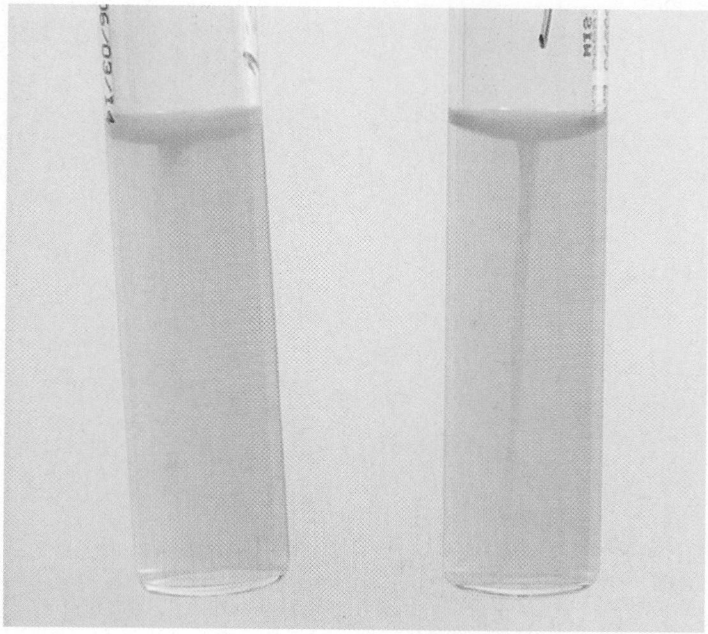

$$H_2S + FeSO_4 \longrightarrow H_2SO_4 + FeS \downarrow$$
(a)

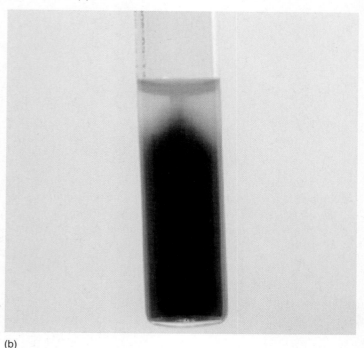
(b)

Figure 63.1 Motility (left) is seen as bacterial growth spreads from the central inoculation line, leading to cloudiness throughout the media. The growth of nonmotile bacteria (right) is restricted to the central inoculation line.

Figure 63.2 (a) Production of ferrous sulfide. Hydrogen sulfide gas produced by some bacteria reacts with ferrous sulfate in the medium to produce the black precipitate ferrous sulfide.
(b) Organisms producing a black precipitate are said to be H_2S + while those that do not produce a black precipitate are H_2S −.

TABLE 63.1	Possible reactions of SIM media		
	Sulfur reduction	Indole production (after adding Kovac's reagent)	Motility
Positive reaction	Black precipitate	Red ring	Growth radiating from stab line
Negative reaction	No black precipitate	No red ring	Growth restricted to stab line

accomplished by layering the top of the media with 3–4 drops of Kovac's reagent and noting the appearance of a deep red ring in the reagent layer. A red ring indicates an indole-positive reaction while no change indicates that the sample is indole negative (Figure 63.3). All the possible results using the SIM media are summarized in Table 63.1.

PRE-LAB QUESTIONS

1. *You receive a note from your laboratory technician that an error in media preparation resulted in the SIM media that you've already inoculated containing 1.5% agar instead of the 0.4% normally seen in the media. What result may be inaccurate and why?*

2. *A bacterial culture identified as nonmotile using a hanging drop slide is inoculated into SIM media. After incubation, a film of bacteria is seen across the surface of the media, seemingly indicating the bacterium's ability to spread. What went wrong?*

Figure 63.3 Indole production. After adding several drops of Kovac's reagent to the SIM tube, the presence of indole from tryptophan metabolism is detected as a red band at the top of the tube.

PERIOD ONE

MATERIALS

Each student should obtain:

Three tubes of SIM media
Broth cultures of:
   Shigella flexneri
   Escherichia coli
   Salmonella typhimurium
Inoculating needle
Marking pen

PROCEDURE

1. Label each SIM tube with your name, lab time, and the name of the organism that will be used to inoculate that tube.
2. Use an inoculating needle to stab each tube, with the appropriate organism, approximately two-thirds of the way to the bottom of the tube.
3. Incubate the tubes at 37°C for 24–48 h.

QUESTION—PERIOD ONE

1. Identify the organism growing in each of the following SIM tubes; use the accompanying table to make your identification.

	H$_2$S Produced	Indole	Motility
Citrobacter	+	+	+
Escherichia	–	+	+
Enterobacter	–	–	+
Klebsiella	–	–	–
Salmonella	+	–	+

a. Black precipitate throughout the media, and red ring upon the addition of Kovac's reagent

b. No black precipitate, and media appears cloudy throughout with a red ring upon the addition of Kovac's reagent

c. Black precipitate throughout the media, and no red ring upon the addition of Kovac's reagent

d. No black precipitate, bacterial growth appears as a solid line in the center of the media, and no color change upon the addition of Kovac's reagent

e. No black precipitate, media appears cloudy, and no red ring upon the addition of Kovac's reagent

PERIOD TWO

1. Retrieve your tubes from the incubator. Evaluate each organism for motility, H$_2$S production, and the presence of indole. As you evaluate your tubes, record your results in Table 63.2.
2. Motility is indicated by growth radiating from the stab line or turbidity throughout the media. If bacterial growth is restricted to the stab line, motility is absent.
3. Evaluate H$_2$S production by examining the media for the presence of a black precipitate. Blackening of the media along the stab line or throughout the tube is a positive result.
4. Add three to four drops of Kovac's reagent to the surface of the media. A red ring indicates the presence of indole and is interpreted as a positive reaction.

TABLE 63.2	SIM tube results		
	Motility	H$_2$S production	Indole
E. coli			
S. typhimurium			
S. flexneri			

EXERCISE 63 REVIEW QUESTION

1. If you were not concerned about detecting H_2S production, which ingredient(s) could you leave out of SIM media? How would a media missing this ingredient(s) have affected your identification of the organisms used in this lab?

REFERENCES

Cowan, M.K. 2012. *Microbiology: A Systems Approach, 3rd ed.,* chap. 3. New York: McGraw-Hill.

Talaro, K.P., and Chess, B. 2012. *Foundations in Microbiology, 8th ed.,* chap. 3. New York: McGraw-Hill.

Kligler's Iron Agar

Kligler's iron agar is a differential media used to evaluate fermentation of glucose and lactose as well as the production of hydrogen sulfide. The media is most often used to help identify Gram-negative rods, such as members of the Enterobacteriaceae.

PRINCIPLES AND APPLICATIONS

Of the characteristics that help to define members of the family Enterobacteriaceae in the lab, perhaps none is more informative than the ability or inability to ferment the carbohydrates lactose and glucose. Kligler's iron agar (KIA) contains a small amount of glucose and a much larger quantity of lactose as well as a pH indicator that signifies the production of acid resulting from the fermentation of either sugar.

An organism that ferments only glucose will produce acid, turning the entire medium yellow within just a few hours. After about 12 h, the small amount of glucose in the KIA media will have been exhausted. The organism will then begin to break down amino acids in the medium, producing ammonia (NH_3) and raising the pH. Toward the end of a 24-h incubation, the accumulation of NH_3 will be enough to alkalinize the slant, but not the butt, of the tube. Hence, organisms that produce a red slant and yellow butt after a 24-h incubation indicate the fermentation of glucose alone.

In the case of an organism that can ferment lactose, production of acid end products will result in a medium that is yellow in both the slant and butt. Because there is 10 times as much lactose as glucose in the medium (1.0% versus 0.1%), the sugar will not be exhausted over the course of a 24-h incubation, and both the slant and butt will remain yellow. Also realize that fermentation of lactose involves the initial breakdown of lactose to glucose (and galactose) and the subsequent fermentation of glucose to produce an acid end product; in other words, if an organism ferments lactose, it must also ferment glucose. If gas is produced by the fermentation of either carbohydrate, it can be easily detected as the medium will appear cracked or in some cases lifted entirely from the bottom of the tube.

If a bacterium reduces sulfur, either from thiosulfate (a component of the medium) or the amino acid cysteine, hydrogen sulfide (H_2S) will be produced. Iron-containing compounds in the media will react with the H_2S to form a black precipitate, most often seen in the butt of the tube. Because the reduction of sulfur will only occur in an acidic environment, the presence of a black precipitate indicates both the reduction of sulfur and fermentation. If the production of the precipitate interferes with evaluation of the butt, the color of the slant can be used to determine if glucose alone has been fermented (red slant) or if both lactose and glucose have been fermented (yellow slant). The various reactions of KIA are shown in Figure 64.1.

(a) (b) (c) (d) (e)

Figure 64.1 Fermentation reactions and hydrogen sulfide production of Kligler's iron agar. (a) Alkaline (slant)/acid (butt), (b) acid/acid with gas production, (c) alkaline/no change, (d) uninoculated, and (e) alkaline/acid with hydrogen sulfide production.

PRE-LAB QUESTIONS

1. *Complete the accompanying table by providing the color of the slant and butt and indicating how you would visually determine the production of H₂S and/or gas.*

Visual result (color of slant/butt)	Interpretation
	Glucose and lactose fermentation
	Glucose fermentation
	Reduction of sulfur (production of H₂S)
	Production of gas

2. *In what three ways is KIA a differential medium?*

PERIOD ONE

MATERIALS

Each student should obtain:

Three tubes of KIA

Solid cultures (slants or plates) of:

   Morganella morganii

   Escherichia coli

   Salmonella typhimurium

Inoculating needle

Marking pen

PROCEDURE

1. Mark each KIA tube with the name of the appropriate organism.
2. For each organism, use an inoculating needle to pick up a small amount of growth from a plate or slant. Inoculate the KIA tube by streaking the surface of the slant and then stabbing deep into the butt of the tube.
3. Incubate the tubes at 37°C for 24–48 h.

QUESTIONS—PERIOD ONE

1. Interpret each of the reactions described in the accompanying table.

	Lactose fermentation	Glucose fermentation	H₂S production	Gas production
Red slant, yellow butt				
Yellow slant, butt color obscured by black precipitate				
Red slant, red butt				
Yellow slant, yellow butt, cracks in the media				

2. If you did not have to worry about hydrogen sulfide production, which two ingredients could be left out of the formulation of the media?

PERIOD TWO

Retrieve your tubes from the incubator, and examine them for the ability to ferment lactose and glucose as well as the ability to produce hydrogen sulfide and/or gas. Sketch the appearance of your tubes.

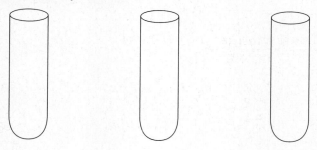

EXERCISE 64 REVIEW QUESTIONS

1. If KIA tubes were accidentally prepared with 10 times the normal amount of glucose (so that the amounts of glucose and lactose were identical), how would the results of the test be changed?

2. If phenol red was accidentally omitted from a batch of KIA, how would this affect the differential capabilities of the media?

REFERENCES

Zimbro, M.J. and Power, D.A. 2003. *Difco and BBL Manual, Manual of Microbiological Culture Media.* Sparks, Maryland: Becton, Dickinson and Co., p. 282.

Triple Sugar Iron Agar

Triple sugar iron agar is a differential media used to evaluate fermentation of sucrose, glucose, and lactose as well as the production of hydrogen sulfide and/or gas. The media is most often used to help identify Gram-negative rods, such as members of the Enterobacteriaceae.

PRINCIPLES AND APPLICATIONS

Of the characteristics that help to define members of the family Enterobacteriaceae in the lab, perhaps none is more informative than the ability or inability to ferment carbohydrates. In addition to beef extract, yeast extract, and peptone, which serve as sources of carbon and energy, triple sugar iron (TSI) agar contains a small amount of glucose along with much larger quantities of lactose and sucrose; a pH indicator is included to signal the production of acid from the fermentation of any of the three sugars.

An organism that ferments only glucose will produce acid, turning the entire medium yellow within just a few hours. After about 12 h, the small amount of glucose present will have been exhausted. The organism will then begin to break down amino acids in the medium, producing ammonia (NH_3) and raising the pH. Toward the end of a 24-h incubation, the accumulation of NH_3 will be enough to alkalinize the slant, but not the butt, of the tube. Hence, organisms that produce a red slant and yellow butt after 24-h incubation indicate the fermentation of glucose alone.

In the case of an organism that can ferment lactose or sucrose, production of acid end products will result in a medium that is yellow in both the slant and butt. Because both lactose and sucrose are included in the media at 10 times the concentration of glucose (1.0% for both lactose and sucrose versus 0.1% for glucose), the sugar will not be exhausted over the course of a 24-h incubation, and both the slant and butt will remain yellow. If gas is produced by fermentation of any of the three carbohydrates, it can be easily detected as the medium will appear cracked or in some cases lifted entirely from the bottom of the tube.

If a bacterium reduces sulfur, either from thiosulfate (a component of the medium) or the amino acid cysteine, hydrogen sulfide (H_2S) will be produced. Ferrous sulfate in the media will react with the H_2S to form a black precipitate, most often seen in the butt of the tube. Because the reduction of sulfur will only occur in an acidic environment, the presence of a black precipitate indicates both the reduction of sulfur and fermentation. If the production of the precipitate interferes with evaluation of the butt, the color of the slant can be used to determine if glucose alone has been fermented (red slant) or if glucose plus lactose and/or sucrose have been fermented (yellow slant). The various reactions of TSI agar are shown in Figure 65.1

PRE-LAB QUESTIONS

1. Complete the accomanying table by providing the color of the slant and butt and indicating how you would visually determine the production of H_2S and/or gas in TSI agar.

Visual result (Color of slant/butt)	Interpretation
	Glucose and lactose or sucrose fermentation
	Glucose fermentation
	Reduction of sulfur (production of H_2S)
	Production of gas

2. In what three ways is TSI agar a differential medium?

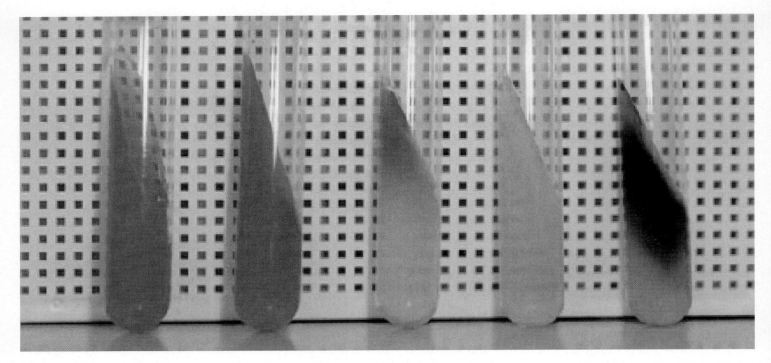

Figure 65.1   Fermentation reactions and hydrogen sulfide production of triple sugar iron agar. (a) Uninoculated, (b) growth without acid production, (c) alkaline (slant)/acid (butt), (d) acid/acid, and (e) alkaline/acid with hydrogen sulfide production.

3. *Fermentation of which two sugars cannot be differentiated using TSI agar?*

PERIOD ONE

MATERIALS

Each student should obtain:

Three tubes of TSI agar

Solid cultures (slants or plates) of:

 Shigella flexneri

 Escherichia coli

 Salmonella typhimurium

Inoculating needle

Marking pen

PROCEDURE

1. Mark each TSI tube with your name and lab time, along with the name of the appropriate organism.
2. For each organism, use an inoculating needle to pick up a small amount of growth from a plate or slant. Inoculate the TSI tube by streaking the surface of the slant and then stabbing deep into the butt of the tube.
3. Incubate the tubes at 37°C for 24 h.

QUESTIONS—PERIOD ONE

1. Interpret each of the following reactions:

	Lactose and/or sucrose fermentation	Glucose fermentation	H₂S production	Gas production
Red slant, yellow butt				
Yellow slant, butt color obscured by black precipitate				
Red slant, red butt				
Yellow slant, yellow butt, cracks in the media				

2. If you did not have to worry about H₂S production, which two ingredients could be left out of the formulation of the media?

PERIOD TWO

Retrieve your tubes from the incubator, and examine them for the ability to ferment lactose (and/or sucrose) and glucose as well as the ability to produce hydrogen sulfide and/or gas. Sketch the appearance of your tubes.

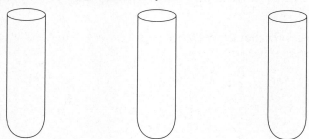

EXERCISE 65 REVIEW QUESTIONS

1. If TSI agar were accidentally prepared with 10 times the normal amount of glucose (so that the amounts of glucose, sucrose, and lactose were identical), how would the results of the test be changed?

2. If phenol red was accidentally omitted from a batch of TSI agar, how would this affect the differential capabilities of the media?

3. If a loop is used to inoculate a tube of TSI agar, splitting of the agar can occur when the loop is stabbed into the media. In what way could this mechanical splitting of the media affect your evaluation of the tube after incubation?

REFERENCES

Cowan, M.K. 2012. *Microbiology: A Systems Approach, 3rd ed.,* chap. 3. New York: McGraw-Hill.

Reddy, C.A., Beveridge, T.J., Breznak, J.A., Marzluf, G.A., Schmidt, T.M., and Snyder, L.R. (eds.). 2007. *Methods for General and Molecular Microbiology, 3rd ed.,* chap. 11. Washington D.C.: ASM Press.

Talaro, K.P., and Chess, B. 2012. *Foundations in Microbiology, 8th ed.,* chap. 3. New York: McGraw-Hill.

Wiley, J., Sherwood, L., and Woolverton, C. 2011. *Prescott, Harley & Klein's Microbiology, 8th ed.,* chap. 35. New York: McGraw-Hill.

Lysine Iron Agar

Lysine iron agar is a combination media used for the differentiation of enteric organisms based on their ability to decarboxylate or deaminate lysine as well as form hydrogen sulfide. Lysine iron agar is designed to be used along with other media, such as triple sugar iron agar, to aid in the differentiation of members of Salmonella and Shigella.

PRINCIPLES AND APPLICATIONS

Lysine iron agar (LIA) was developed in the early 1960s, primarily as a means of differentiating species of *Salmonella* and *Shigella* from one another based on each species ability to deaminate or decarboxylate the amino acid lysine as well as form hydrogen sulfide. The media contains a small amount of glucose as a fermentable carbohydrate and bromcresol purple to indicate the pH of the media. This indicator is purple above pH 6.8 and turns yellow below pH 5.2. Lysine is provided as a substrate for the enzymes lysine decarboxylase and lysine deaminase.

The media is poured as a slant with a deep butt and inoculated with a streak along the surface of the media and two stabs to the bottom of the butt. Acid production from the fermentation of glucose results in the media turning yellow within the first few hours of incubation; if the gene coding for the enzyme lysine decarboxylase is present in the isolate, it will be induced during this period as well. Decarboxylation of lysine produces alkaline end products that raise the pH of the media and turn it purple. In this way, a purple slant and butt after incubation indicate the presence of lysine decarboxylation.

Deamination of lysine, via the enzyme lysine deaminase, produces compounds that react with ferric ammonium citrate to produce a red color. Because deamination requires oxygen, it occurs to a much greater extent in the slant as opposed to the anaerobic butt of the tube, which remains yellow. A red slant with a yellow butt is indicative of lysine deamination. If neither decarboxylation nor deamination occurs, degradation of the peptone in the media will produce alkaline end products, raising the pH enough that the slant will turn purple while the butt remains yellow.

Lastly, if a bacterium reduces sulfur, either from thiosulfate (a component of the medium) or the amino acid cysteine, hydrogen sulfide (H_2S) will be produced. Ferric ammonium citrate in the media will react with the H_2S to form a black precipitate, most often seen in the butt of the tube. The reactions seen in LIA are shown in Figure 66.1

PRE-LAB QUESTIONS

1. *Define the following terms:*

   Decarboxylation

   Deamination

2. *Complete the accompanying table by providing the color of the slant and butt and indicating how you would visually determine the production of H_2S in LIA.*

Visual result (color of slant/butt)	Interpretation
	Lysine deaminase negative *Lysine decarboxylase positive*
	Lysine deaminase negative *Lysine decarboxylase negative* *Glucose fermentation*
	Lysine deaminase positive *Lysine decarboxylase negative* *Glucose fermentation*
	Reduction of sulfur (production of H_2S)

3. *In what three ways is LIA a differential medium?*

PERIOD ONE

MATERIALS

Each student should obtain:
Three tubes of LIA
Solid cultures (slants or plates) of:
    Shigella flexneri
    Proteus mirablis
    Salmonella typhimurium
Inoculating needle
Marking pen

PROCEDURE

1. Mark each LIA tube with your name and lab time, along with the name of the appropriate organism.
2. For each organism, use an inoculating needle to pick up a small amount of growth from a plate or slant. Inoculate the LIA tube by stabbing the butt twice and then streaking the surface of the slant.
3. Incubate the tubes at 37°C for 24 h.

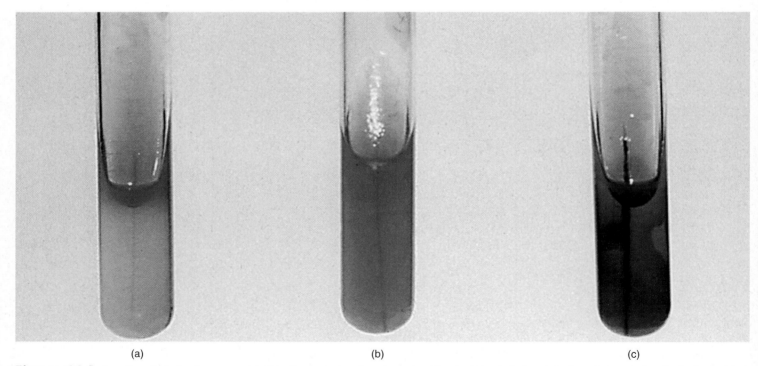

     (a)   (b)   (c)

Figure 66.1 Reactions in lysine iron agar. (a) *Shigella* sp.: lysine deaminase positive (red slant), lysine decarboxylase negative (lack of purple), and H_2S negative (lack of black precipitate). (b) *Escherichia coli:* lysine deaminase negative (weakly purple slant), lysine decarboxylase positive (weakly purple butt), and H_2S negative (lack of black precipitate). (c) *Salmonella* sp.: lysine deaminase negative (purple slant), lysine decarboxylase positive (purple butt), and H_2S positive (note black precipitate around stab line).

QUESTIONS—PERIOD ONE

1. Interpret each of the following reactions by placing a check mark into the appropriate boxes.

	Lysine deaminase	Lysine decarboxylase	Glucose fermentation	H₂S production
Red slant, yellow butt				
Purple slant, yellow butt				
Purple slant, purple butt				
Black precipitate				

2. If you did not have to worry about H₂S production, which two ingredients could be left out of the formulation of the media?

PERIOD TWO

Retrieve your tubes from the incubator, and examine them for the ability to deaminate or decarboxylate lysine as well as the ability to produce hydrogen sulfide. Sketch the appearance of your tubes.

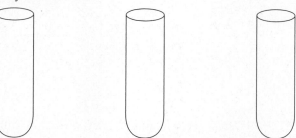

EXERCISE 66 REVIEW QUESTIONS

1. If LIA tubes were accidentally prepared with no glucose, how would the results of the test be changed?

2. If bromcresol purple was accidentally omitted from a batch of LIA, how would this affect the differential capabilities of the media?

3. Draw the amino acid lysine (your textbook should prove helpful). Circle and label the amino and carboxyl portions of the molecule.

4. What do you think the enzyme phenylalanine deaminase does?

REFERENCES

Reddy, C.A., Beveridge, T.J., Breznak, J.A., Marzluf, G.A., Schmidt, T.M., and Snyder, L.R. (eds.). 2007. *Methods for General and Molecular Microbiology, 3rd ed.,* chap. 15. Washington D.C.: ASM Press.

Wiley, J., Sherwood, L., and Woolverton, C. 2011. *Prescott, Harley & Klein's Microbiology, 8th ed.,* chap. 35. New York: McGraw-Hill.

Litmus Milk

Litmus milk is a complex differential medium used as an aid in the classification of many bacterial species. It is also commonly used for the maintenance of lactic acid bacteria.

PRINCIPLES AND APPLICATIONS

Individual species of bacteria utilize skim milk in very precise ways, with each species producing various metabolic products. In combination, these reactions can be helpful in identifying unknown bacteria. Litmus milk is a 10% solution of sterile skim milk fortified with the pH indicator azolitmin. Azolitmin is pink below pH 4.5, blue above pH 8.3, and varying shades of purple between these two points. Milk provides nutrients for growth, including the carbohydrate lactose and the protein casein. After an incubation of up to 14 days, several reactions are possible (Figure 67.1), including:

No Change　In some cases, the bacteria neither ferments the available carbohydrates nor changes the pH of the media.

Production of Acid　If lactose is fermented by the enzyme beta-galactosidase, acid is produced and the media will turn pink. In addition, casein may precipitate, forming an acid clot that may be entirely pink or consist of a pink band atop a white clot. Cracks in the clot indicate the formation of gas. Exceptional gas production that breaks apart the clot is known as **stormy fermentation** and is indicative of *Clostridium*.

Alkalinization　This change to the media will impart a blue color throughout the media or cause a blue band to form at the top of the media. This pH change is a result of proteolytic and degradative enzymes producing ammonia or amines from the breakdown of proteins.

Coagulation　This change to the casein leads to the formation of a curd, or clot. A casein clot can be distinguished from an acid clot as an acid clot will dissolve in an alkaline environment. If the bacterium being tested possesses the enzyme rennin, casein will be converted to paracasein, and a translucent, watery liquid called **whey** will appear at the top of the tube.

Peptonization　This change is evident when proteolytic bacteria digest milk proteins, leading to the formation of a translucent, often brown, liquid and the dissolution of any clots.

Reduction　This change of litmus is seen as a whitening of the media due to the action of reductase enzymes that remove oxygen from the media.

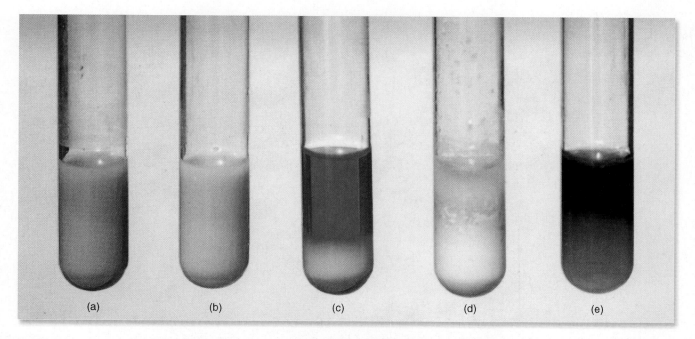

(a)　　(b)　　(c)　　(d)　　(e)

Figure 67.1　Litmus milk reactions. (a) Alkalinization. (b) Acidification due to fermentation of lactose. (c) Upper translucent portion is due to peptonization, and solid white portion in bottom is due to coagulation and litmus reduction; overall red color is interpreted as acidification. (d) Coagulation and litmus reduction in lower half, and peptonization (translucence) and acidification (pink) in top portion. (e) Peptonization is present in bottom of tube, and soluble pigments from *Pseudomonas* have obscured the litmus reaction.

PRE-LAB QUESTIONS

1. *Define the following terms:*

Fermentation

Alkalinization

Coagulation

Peptonization

Reduction

2. *How would you (quickly) tell the difference between an acid reaction and an acid clot?*

PERIOD ONE

MATERIALS

Each student should obtain:
Six tubes of litmus milk
Fresh broth cultures of:
   Alcaligenes faecalis
   Enterococcus faecium
   Klebsiella pneumoniae
   Lactococcus lactis
   Pseudomonas aeruginosa
Inoculating loop
Marking pen

PROCEDURE

1. Mark each litmus milk tube with your name and lab time, along with the name of the appropriate organism. Mark the sixth tube "control."
2. Use a loop to inoculate each tube with the appropriate organism.
3. Incubate the tubes at 37°C for 7–14 days.

QUESTION—PERIOD ONE

1. Interpret each of the reactions described in the left-hand
   side of the accompanying table by supplying the correct
   term from the list. Each term is to be used only once.

Acid reaction   Alkaline reaction   Acid clot
Curd   Gas   Stormy fermentation
Peptonization   Reduction of litmus No change

Reaction	Interpretation
No change when compared to control.	
Pink color throughout liquid media.	
Medium has remained liquid but turned blue.	
Medium is pink and solidified.	
Cracks running through solidified media.	

Medium is translucent with a brown tint.	
Solid medium has been broken apart.	
Semisolid clot surrounded by grayish liquid.	
White color in lower portion of the medium.	

PERIOD TWO
Retrieve your tubes from the incubator, and examine them for any of the reactions discussed earlier. Use the control tube for comparison. Sketch the appearance of your tubes.

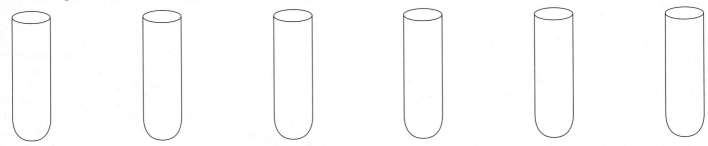

EXERCISE 67 REVIEW QUESTION

1. How would you interpret a control tube that was pink and completely solidified after incubation?

REFERENCE

Reddy, C.A., Beveridge, T.J., Breznak, J.A., Marzluf, G.A., Schmidt, T.M., and Snyder, L.R. (eds.). 2007. *Methods for General and Molecular Microbiology, 3rd ed.*, chap. 15. Washington D.C.: ASM Press.

68

Oxidation-Fermentation Test

The oxidation-fermentation test is used to determine a bacterial species' ability to oxidatively metabolize a carbohydrate as opposed to fermenting the same carbohydrate. This information is especially useful in the classification of Gram-negative bacteria.

PRINCIPLES AND APPLICATIONS

When bacterial species metabolize carbohydrates, two potential paths may be utilized, depending on the particular bacterium and whether or not oxygen is present. In **oxidative metabolism,** the carbohydrate is oxidized to pyruvate that then passes through the Krebs cycle (also referred to as the tricarboxylic acid cycle or citric acid cycle) and the electron transport pathway, producing CO_2 as an end product. In **fermentative metabolism,** carbohydrates are converted to pyruvate, but the primary end product is an acid. This difference in end products means that fermentative bacteria will acidify media to a much greater extent than will oxidative bacteria.

Oxidation-fermentation (O-F) media contains a single carbohydrate such as glucose as well as the pH indicator bromthymol blue that is yellow below pH 6.0, blue above pH 7.6, and green between the two. Acidification of the media is detected by a change in color from green to yellow. Each species to be tested is used to inoculate two tubes, one of which is then layered with mineral oil to exclude air; both tubes are then incubated. Acidification of the media within the unsealed tube (no oil) is due to oxidative metabolism of the carbohydrate while acidification of the media within the sealed tube (with oil) is a result of fermentative metabolism. If no acidification appears in either tube, then the carbohydrate is not utilized by the bacterium being studied (Figure 68.1).

Aerobic organisms have the potential to alkalinize the media through the metabolism of peptone, which can neutralize small amounts of acid produced by aerobic metabolism. To prevent this from happening, O-F media has a high concentration of carbohydrates relative to the peptone concentration. The media also contains just enough agar to provide a semisolid consistency, allowing for the determination of motility if desired.

PRE-LAB QUESTION

1. *Define the following terms:*

Fermentative metabolism

Oxidative metabolism

Carbohydrate

Anaerobic

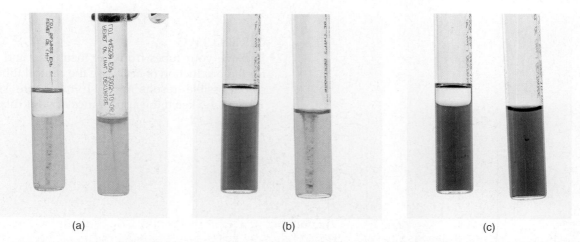

(a)   (b)   (c)

Figure 68.1 O-F glucose reactions. (a) Oxidation or fermentation, or fermentation only. (b) Oxidation. (c) No glucose metabolism.

PERIOD ONE

MATERIALS

Each student or group should obtain:

Six tubes of O-F media

Fresh agar slants of:
  Escherichia coli
  Pseudomonas aeruginosa

Sterile mineral oil

Sterile transfer pipettes

Inoculating needle

Marking pen

PROCEDURE

1. Mark each tube with your name and lab time. Label two tubes with the name of the first organism, two tubes with name of the second organism, and two tubes "control."
2. Use a needle to inoculate each tube with the appropriate organism by stabbing the media to within 1 cm of the bottom of the tube.
3. Using a sterile pipette, overlay one of each pair of tubes with 4 mm of sterile mineral oil.
4. Incubate the tubes at 37°C for 48 h.

QUESTIONS—PERIOD ONE

1. What is the purpose of the mineral oil, and why is it only added to one tube of each pair?

2. What reactions would you expect to see in your control tubes?

PERIOD TWO

Retrieve your tubes from the incubator, and examine them for the production of acid. Use the control tubes for comparison. Possible results for O-F tubes are summarized in Table 68.1. Sketch the appearance of your tubes.

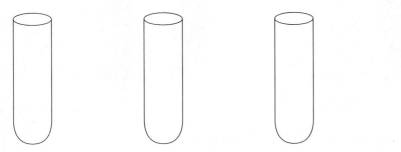

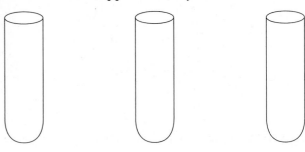

	Appearance	
Interpretation (Symbol)	**Sealed tubes**	**Unsealed tubes**
Oxidation (O)	Green or blue	Yellow (any amount)
Fermentation or oxidation plus fermentation (F or O/F)	Yellow throughout media	Yellow throughout media
Slow fermentation or oxidation plus slow fermentation (F or O/F)	Slightly yellow near surface of media	Slightly yellow near surface of media
Sugar not metabolized	Green or blue	Green or blue

TABLE 68.1 Oxidative-fermentation media results and interpretations

EXERCISE 68 REVIEW QUESTIONS

1. Enteric bacteria are facultative anaerobes, meaning they are capable of both aerobic and fermentative metabolism. What results would you expect to see in both sealed and unsealed O-F tubes inoculated with such a bacterium?

2. How would you interpret a yellow color throughout your control tubes?

3. If bromthymol blue were accidentally left out of the media, what would be the result?

REFERENCE

Reddy, C.A., Beveridge, T.J., Breznak, J.A., Marzluf, G.A., Schmidt, T.M., and Snyder, L.R. (eds.). 2007. *Methods for General and Molecular Microbiology, 3rd ed.,* chap. 15. Washington D.C.: ASM Press.

Phenol Red Broth

Phenol red broth is a differential media used to detect the fermentation of specific carbohydrates. This information can be used as an aid in the identification of many species of bacteria.

PRINCIPLES AND APPLICATIONS

Fermentation of carbohydrates by bacteria results in the production of acid and, in some cases, gas. Because bacteria differ with regard to the carbohydrates they are able to ferment, fermentation results for a series of carbohydrates can be used as an aid in the bacteria's identification.

Phenol red broth is a complete medium to which a single carbohydrate (glucose, lactose, xylose, etc.) has been added to a concentration of one-half percent The media also contains phenol red as a pH indicator, which is yellow below pH 6.8, purple-pink above pH 7.4, and red between these points. The media is initially adjusted to a pH of 7.3 and so appears red. In addition, a small test tube called a **Durham tube** is placed in the media, with the opening of the tube facing downward.

When inoculated with a bacterial species that can ferment the carbohydrate within the media, acid end products are produced, resulting in a yellow color throughout the broth. The production of gas may sometimes accompany fermentation, with at least some of the gas forming a bubble as it is trapped within the Durham tube. If the bacterium does not ferment the carbohydrate contained within the media, the media will remain red (Figure 69.1). Finally, some bacteria are able to deaminate amino acids in the media, producing ammonia (NH_3), raising the pH, and turning the media purple or pink. In overly long incubations of fast-growing organisms, the deamination of amino acids may cause a positive reaction to *revert*, becoming pink or red as alkaline products raise the pH of a positive (i.e., yellow) reaction.

PRE-LAB QUESTIONS

1. *Define the following terms:*

Ferment

Carbohydrate

2. *What class of biological molecule is responsible for the ability of different bacteria to ferment different sugars?*

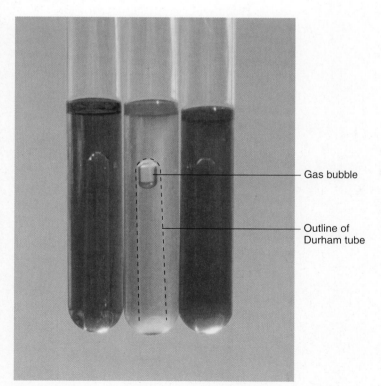

Gas bubble

Outline of Durham tube

Figure 69.1 Phenol red broth results. The middle tube displays acid and gas production. The left-hand tube is an uninoculated control, and the right-hand tube shows growth (note turbidity near the bottom of the tube) but no fermentation.

3. *Should a gas bubble be present within a Durham tube prior to it being incubated with a microorganism?*

QUESTIONS—PERIOD ONE

1. How do you recognize acid production in phenol red glucose broth? What about gas production?

PERIOD ONE

MATERIALS

Each student or group should obtain:
Five tubes of phenol red glucose broths with Durham tubes
Five tubes of phenol red lactose broths with Durham tubes
Five tubes of phenol red mannitol broths with Durham tubes
Five tubes of phenol red sucrose broths with Durham tubes
Fresh broth cultures of:
   Escherichia coli
   Pseudomonas aeruginosa
   Staphylococcus aureus
   Proteus vulgaris
Inoculating loop
Marking pen

2. How would you expect your control tubes to look?

PROCEDURE

1. Mark each tube with your name and lab time. Label one tube of each carbohydrate with the name of the appropriate organism such that each organism will be inoculated into four tubes, each with a different carbohydrate.

2. Use a loop to inoculate each tube with the appropriate organism. Do not inoculate the control tube.

3. Incubate the tubes at 37°C for 48 h.

PERIOD TWO

Retrieve your tubes from the incubator, and examine them for the production of acid and gas. Note that a positive result for gas should only be recorded if a bubble equal to at least 10% of the volume of the Durham tube is present. Tiny bubbles should be ignored. Use the control tubes for color comparison. Possible results for phenol red tubes are summarized in Table 69.1. Sketch the appearance of your tubes.

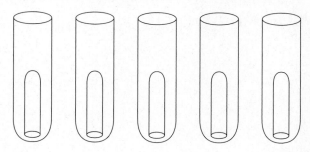
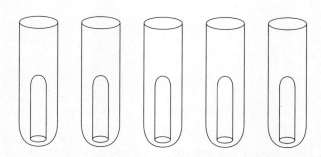

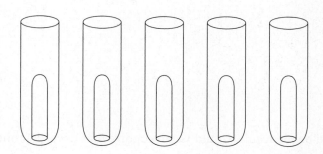
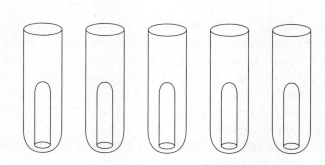

TABLE 69.1	Phenol red broth results and interpretations
Interpretation (Symbol)	**Appearance**
Fermentation with acid and gas end products (A/G)	Yellow and bubble in Durham tube
Fermentation with acid end products but no gas (A/−)	Yellow and no bubble in Durham tube
No fermentation (−/ −)	Red and no bubble in Durham tube
Degradation of peptone and alkaline end products (−/ −)	Pink and no bubble in Durham tube

EXERCISE 69 REVIEW QUESTIONS

1. How would you explain a pink or red tube with a large gas bubble within the Durham tube?

2. How would you interpret a yellow color throughout your control tubes?

3. If phenol red were accidentally left out of the media, what would be the result? Could you recognize that it was left out prior to inoculating your tubes?

REFERENCES

Cowan, M.K. 2012. *Microbiology: A Systems Approach, 3rd ed.,* chap. 3. New York: McGraw-Hill.

Reddy, C.A., Beveridge, T.J., Breznak, J.A., Marzluf, G.A., Schmidt, T.M., and Snyder, L.R. (eds.). 2007. *Methods for General and Molecular Microbiology, 3rd ed.,* chap. 15. Washington D.C.: ASM Press.

Talaro, K.P., and Chess, B. 2012. *Foundations in Microbiology, 8th ed.,* chap. 3. New York: McGraw-Hill.

Purple Broth

Purple broth is a differential media used to detect the fermentation of specific carbohydrates. This information can be used as an aid in the identification of many bacterial species, especially Gram-negative enteric bacilli.

PRINCIPLES AND APPLICATIONS

Fermentation of carbohydrates by bacteria results in the production of acid and, in some cases, gas. Because bacteria differ with regard to the carbohydrates they are able to ferment, fermentation results for a series of carbohydrates can be used as an aid in the bacteria's identification.

Purple broth is a complete medium to which a single carbohydrate (glucose, lactose, xylose, etc.) has been added to a concentration of one percent. The media also contains bromcresol purple as a pH indicator, which is yellow below pH 5.2 and purple above pH 6.8. The media is initially adjusted to a pH of 6.8 and so appears purple. In addition, a small test tube called a **Durham tube** is placed in the media, with the opening of the tube facing downward.

When inoculated with a bacterial species that can ferment the carbohydrate within the media, acid end products are produced, resulting in a yellow color throughout the broth. The production of gas may sometimes accompany fermentation, with at least some of the gas forming a bubble as it is trapped within the Durham tube. If the bacterium does not ferment the carbohydrate contained within the media, the media will remain purple. Lastly, some bacteria are able to deaminate amino acids in the media, producing ammonia (NH_3), raising the pH, and keeping the media purple despite some fermentation of the carbohydrate within the media (Figure 70.1). In fact, in all cases, the media will only turn yellow when the alkaline products produced by the degradation of peptone are neutralized by the production of acid from carbohydrate fermentation.

Figure 70.1 Interpretation of purple broth. Fermentation of the carbohydrate within the media results in the formation of a yellow color, indicating the production of acid. Production of acid may also be accompanied by the production of gas, which is seen as a bubble within the Durham tube. In a negative reaction, the media will remain purple, and no gas bubble will form.

PRE-LAB QUESTION

1. *Define the following terms:*

Ferment

Carbohydrate

PERIOD ONE

MATERIALS

Each student or group should obtain:
Three tubes of purple glucose broths with Durham tubes
Fresh broth cultures of:
   Escherichia coli
   Alcaligenes faecalis
Inoculating loop
Marking pen

PROCEDURE

1. Mark each tube with your name and lab time. Label one tube with the name of each microorganism. Label the final tube "control."
2. Use a loop to inoculate each tube with the appropriate organism. Do not inoculate the control tube.
3. Incubate the tubes for 48 h at 37°C.

QUESTIONS—PERIOD ONE

1. How do you recognize acid production in purple glucose broth? What about gas production?

2. How would you expect your control tube to look?

PERIOD TWO

Retrieve your tubes from the incubator, and examine them for the production of acid and gas. Note that a positive result for gas should only be recorded if a bubble equal to at least 10% of the volume of the Durham tube is present. Tiny bubbles should be ignored. Use the control tube for color comparison. Possible results for purple broth tubes are summarized in Table 70.1. Sketch the appearance of your tubes.

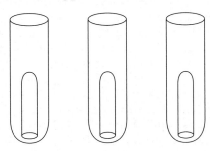

TABLE 70.1	Purple broth results and interpretations
Interpretation (Symbol)	**Appearance**
Fermentation with acid and gas end products (A/G)	Yellow and bubble in Durham tube
Fermentation with acid end products but no gas (A/−)	Yellow and no bubble in Durham tube
No fermentation and alkaline end products from degradation of peptone (−/−)	Purple, turbid broth, and no bubble in Durham tube
No fermentation, little to no degradation of peptone with minimal to no alkaline end products (−/−)	Purple, not turbid, and no bubble in Durham tube

EXERCISE 70 REVIEW QUESTIONS

1. How would you interpret a yellow color throughout your control tube?

2. If bromcresol purple were accidentally left out of the media, what would be the result? Could you recognize that it was left out prior to inoculating your tubes?

3. If an unknown organism results in yellow media and a gas bubble when tested in purple broth, what result would you expect when using phenol red broth with the same organism?

REFERENCES

Zimbro, M.J. and Power, D.A. 2003. *Difco and BBL Manual, Manual of Microbiological Culture Media.* Sparks, Maryland: Becton, Dickinson and Co., p. 465.

Methyl Red and Voges-Proskauer Tests

The methyl red and Voges-Proskauer tests are two related assays that use the same media formulation. The methyl red test determines if a bacterium is capable of a mixed acid fermentation while the Voges-Proskauer test identifies organisms that ferment sugars to produce the alcohol end products acetoin and 2,3-butanediol. The results of the tests are often helpful in the identification of Gram-negative intestinal bacteria.

PRINCIPLES AND APPLICATIONS

Determining what carbohydrates a particular bacterial species can ferment is a common technique used in the laboratory to aid in its identification, and many techniques exist for gathering this information. In most cases, the acids produced by fermentation are unstable and are quickly converted to more neutral products. A mixed acid fermentation results when glucose is fermented to produce several acids that are fairly stable, lowering the pH of the media. In addition, small amounts of carbon dioxide, ethanol, and hydrogen gas are generated. Methyl red indicator is red at pH 4.4, yellow at pH 6.2, and various shades of orange between the two endpoints. After incubation of an organism in methyl red (MR) broth, several drops of the indicator are added and the development of a red color is considered a positive result; yellow is considered negative while orange is inconclusive (Figure 71.1).

A second type of fermentation can occur in some organisms where acid products are quickly converted to the alcohols acetoin and 2,3-butanediol. The Voges-Proskauer (VP) test involves incubating an organism in VP broth (which is the same as MR broth) and then adding two reagents (Barritt's reagents A and B). Barritt's reagent A is α-napthol while reagent B is concentrated potassium hydroxide (KOH). These reagents first oxidize acetoin to diacetyl that then reacts further to produce a red color in the medium, which is considered a positive reaction (Figure 71.2). A brown or copper color is a negative reaction that can be easily mistaken for red if sufficient attention is not paid during the evaluation of the test.

The MR and VP tests are part of a common set of tests known as IMViC, an acronym representing the biochemical tests *i*ndole, *m*ethyl red, *V*oges-Proskauer, and *c*itrate. (The lower case "I" was added to the acronym long ago simply to make pronunciation easier.) This combination of tests is useful in distinguishing members of the family Enterobacteriaceae.

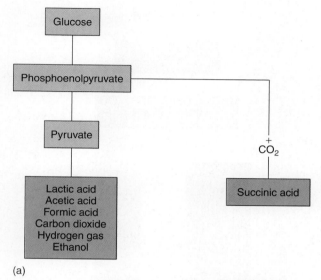

(a)

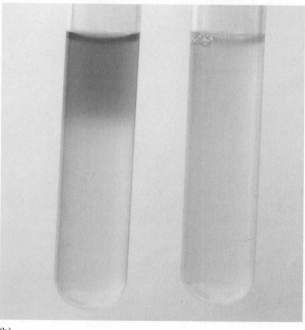

(b)

Figure 71.1 (a) Mixed acid fermentation. Fermentation of glucose during glycolysis (blue) leads to the production of pyruvate. Using pyruvate as a substrate, a variety of enzymatic reactions produce several acids as well as carbon dioxide, hydrogen gas, and ethanol. Phosphoenolpyruvate reacts with carbon dioxide to produce succinic acid. The combination of these acids is enough to drive the pH of the media below 5.0, which is indicated by a pink color after the addition of methyl red indicator. (b) The tube on the left (*Escherichia coli*) is positive while the tube on the right (*Enterobacter aerogenes*) is negative.

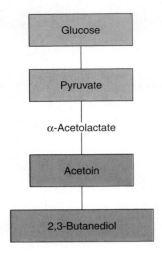

(a)

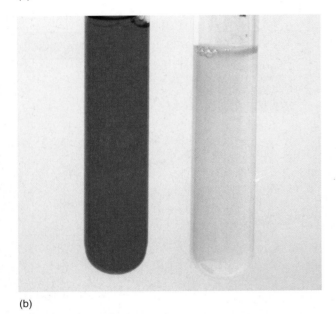

(b)

Figure 71.2 (a) Butanediol fermentation. Fermentation of glucose during glycolysis (blue) leads to the production of pyruvate. Pyruvate is converted to α-acetolactate and then to acetoin. Reduction of acetoin leads to the formation of 2,3-butanediol. In the Voges-Proskauer reaction, acetoin reacts with Barritt's reagents A and B to produce a red color. (b) The tube on the left (*Enterobacter aerogenes*) is positive while the tube on the right (*Escherichia coli*) is negative.

PRE-LAB QUESTION

1. *Define the following terms:*

  Ferment

  Carbohydrate

  Alcohol

PERIOD ONE

MATERIALS

Each student or group should obtain:
Three tubes of methyl red broth
Three tubes of Voges-Proskauer broth in screw-top tubes
Fresh broth cultures of:
  Escherichia coli
  Enterobacter aerogenes
Inoculating loop
Marking pen

PROCEDURE

1. Mark each tube with your name and lab time. Label one tube of each type of media with the name of each microorganism. Label the final tube of each type of media "control."
2. Use a loop to inoculate each tube with the appropriate organism. Do not inoculate the control tubes.
3. Incubate the tubes at 37°C for 2–5 days.

QUESTIONS—PERIOD ONE

1. What are the metabolic products being tested for in the MR test and the VP test?

2. How would you expect your control tubes to look?

PERIOD TWO

Retrieve your tubes from the incubator, and perform the following tests on them:

Methyl Red Test

1. Add 3–4 drops of methyl red indicator. An immediate red color is a positive reaction. Negative tubes may be reincubated to give a total incubation time of 5 days.

Voges-Proskauer Test

1. Add 30 drops of Barritt's reagent A (α-napthol). Tighten the top of the tube and shake vigorously to oxygenate the media.

2. Add 10 drops of Barritt's reagent B (KOH). Tighten the top of the tube and shake.

3. Place the tubes in a test tube rack, and allow them to incubate for up to 1 h. Observe the tubes for the development of a red color at 15-min intervals. Development of the color will be strongest at the top of the tube.

Possible results and interpretations for the MR and VP tests are summarized in Tables 71.1 and 71.2, respectively. Sketch the appearance of your tubes.

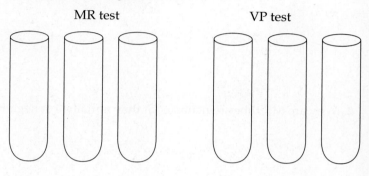

MR test   VP test

TABLE 71.1	Methyl red test results and interpretations
Interpretation (Symbol)	**Appearance**
Mixed acid fermentation (MR +)	Red
No mixed acid fermentation (MR −)	No color change*

*Continue incubation of negative or questionable tubes (orange in color) for up to 5 days.

TABLE 71.2	Voges-Proskauer test results and interpretations
Interpretation (Symbol)	**Appearance**
2,3-butanediol fermentation (VP +)	Red
No 2,3-butanediol fermentation (VP −)	Brown, copper, or no color change

EXERCISE 71 REVIEW QUESTIONS

1. How would you interpret a red color in either of your control tubes?

2. Why are MR tubes reincubated if they initially test negative?

REFERENCE

Wiley, J., Sherwood, L., and Woolverton, C. 2011. *Prescott, Harley & Klein's Microbiology, 8th ed.*, chap. 10. New York: McGraw-Hill.

Catalase Test

The catalase test detects the presence of the enzyme catalase, which catalyzes the breakdown of hydrogen peroxide to water and oxygen gas. The results of the catalase test are useful in the identification of many groups of bacteria and are particularly helpful in the differentiation of Staphylococcus *and* Streptococcus.

PRINCIPLES AND APPLICATIONS

When bacteria undergo aerobic respiration, oxygen is used as a terminal electron acceptor, where it is converted to water. At the same time, however, hydrogen peroxide (H_2O_2) is produced as a byproduct. Hydrogen peroxide is a highly reactive oxidizing agent that can damage proteins, DNA, and RNA within the cell. To prevent this damage, aerobic organisms produce the enzyme catalase, which enzymatically converts hydrogen peroxide to water and oxygen gas (Figure 72.1). Obligate anaerobes and aerotolerant bacteria lack catalase and are therefore unable to detoxify the hydrogen peroxide that accumulates as a result of metabolism in an aerobic environment. In the laboratory, determining whether or not catalase is produced can aid in the identification of bacteria. For example, staphylococci are easily differentiated from streptococci using the **catalase** test as the former produces catalase while the latter lacks the enzyme.

To test for catalase, a small amount of bacteria is transferred from a plate or slant to a glass slide and 1–2 drops of hydrogen peroxide is added to the bacteria. The production of bubbles indicates that catalase is degrading the hydrogen peroxide, producing oxygen gas, a positive reaction. A lack of bubbles means that catalase is lacking and is interpreted as a negative reaction.

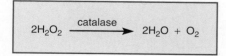

$$2H_2O_2 \xrightarrow{\text{catalase}} 2H_2O + O_2$$

Figure 72.1 The enzyme catalase catalyzes the breakdown of hydrogen peroxide to water and oxygen gas.

PRE-LAB QUESTION

1. *Define the following terms:*

Obligate anaerobe

Aerotolerant

Enzyme

MATERIALS

Each student or group should obtain:
Microscope slide
Hydrogen peroxide (3% solution)
Fresh solid cultures of:
    Staphylococcus epidermidis
    Enterococcus faecalis
Wooden applicator sticks

PROCEDURE

1. Use a wooden applicator to transfer a visible amount of one of the bacteria to the left-hand side of a glass slide. Dispose of the applicator in a hard-sided biohazard container.

2. Using a second applicator stick, transfer a visible quantity of bacteria to the right-hand side of the same slide. Dispose of the applicator properly.

3. Add 1 or 2 drops of hydrogen peroxide directly to the bacterial growth, and watch for the appearance of bubbles (Figure 72.2). If they do not appear immediately, use a magnifying glass or the scanning lens of your microscope to check more closely.

4. If no bubbles appear within 5 min, the bacteria can be considered negative for catalase activity.

5. Dispose of the slide in a hard-sided biohazard container or a container of disinfectant, as directed by your instructor.

Possible results for the catalase test are summarized in Table 72.1. Sketch the appearance of your catalase slide.

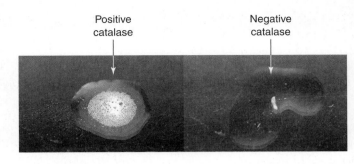

Figure 72.2 Interpretation of the catalase test. The production of bubbles of oxygen gas after the addition of hydrogen peroxide to a small amount of bacteria growth is considered a positive result in the catalase test. A lack of bubbles is considered a negative test result.

TABLE 72.1	Catalase test results and interpretations
Interpretation	Appearance
Catalase +	Bubbles
Catalase −	No bubbles

EXERCISE 72 REVIEW QUESTIONS

1. Hydrogen peroxide is used as an antiseptic to kill bacteria in cuts and abrasions. Many of these bacteria are catalase positive, yet they are still killed by the actions of hydrogen peroxide. Why does this occur?

2. An alternate method of performing the catalase test exists where hydrogen peroxide is dropped directly onto a colony growing on an agar plate. The test can be done if the bacterium is growing on nutrient agar but not blood agar. Why not?

REFERENCES

Cowan, M.K. 2012. *Microbiology: A Systems Approach, 3rd ed.*, chap. 8. New York: McGraw-Hill.

Talaro, K.P., and Chess, B. 2012. *Foundations in Microbiology, 8th ed.*, chap. 8. New York: McGraw-Hill.

Wiley, J., Sherwood, L., and Woolverton, C. 2011. *Prescott, Harley & Klein's Microbiology, 8th ed.*, chap. 16. New York: McGraw-Hill.

NOTES

Oxidase Test

The oxidase test is a means of determining whether a bacterium possesses cytochrome c oxidase, an enzyme found in the electron transport chain of some bacteria. The oxidase test is commonly used to differentiate oxidase-positive Pseudomonas from oxidase-negative Enterobacteriaceae.

PRINCIPLES AND APPLICATIONS

In **aerobic respiration,** the electron transport chain links the production of adenosine triphospate (ATP) to the oxidation of electron carriers such as nicotinamide adenine dinucleotide (NADH) and reduced flavin adenine dinucleotide (FADH$_2$). This process ends with the transfer of electrons to oxygen, forming water. In some bacteria, the enzyme responsible for transferring electrons to oxygen (the terminal electron acceptor) is cytochrome c oxidase. Other bacteria may still use oxygen as the final electron acceptor but use different terminal oxidases. The oxidase test distinguishes between these two types of bacteria by using an artificial electron acceptor (N,N,N',N'-tetramethyl-p-phenylenediamine) that changes from yellow to purple when electrons are transferred from reduced cytochrome c oxidase to the artificial acceptor (Figure 73.1). This reagent is unstable and can oxidize if left exposed to air for prolonged periods. Because of this, a commercial system is used in which the reagent is stored in a glass ampoule that is broken just prior to use.

Although the oxidase status of an organism is useful in many instances, it is particularly helpful in that it differentiates the mostly oxidase-negative *Pseudomonas* from the Enterobacteriaceae, all of which are oxidase positive.

PRE-LAB QUESTION

1. *Define the following terms:*

   Oxidize

   Reduce

MATERIALS

Each student or group should obtain:

Two sterile swabs

Fresh solid cultures of:

    Escherichia coli

    Pseudomonas aeruginosa

Ampoule of oxidase reagent,
    N,N,N',N'-tetramethyl-p-phenylenediamine

PROCEDURE

1. Holding an ampoule of oxidase reagent between your thumb and forefinger and being sure that it points away from you, squeeze until you feel the glass break.
2. Tap the ampoule on the benchtop several times.
3. Using a sterile swab, gather a small amount of growth from one of the two cultures. Squeeze the ampoule gently so as to allow several drops of oxidase reagent to fall on the tip of the swab (where the bacteria growth is located).

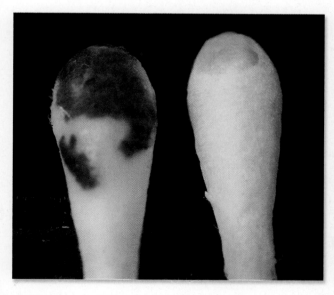

Figure 73.1 Interpretation of oxidase reaction. The left-hand swab shows a purple (positive) reaction due to the production of oxidase. The right-hand swab displays the yellow color typical of a negative reaction.

4. Observe the swab for up to 30 sec. A deep purple color developing within this time indicates a positive reaction. A yellow color, or any purple after 30 sec, is a negative reaction. Record your results.

5. Dispose of the swab in a hard-sided biohazard container.

6. Repeat this process using another swab and the other culture.

7. Possible results for the oxidase test are summarized in Table 73.1. Sketch the appearance of your oxidase tests.

TABLE 73.1	Oxidase test results and interpretations
Interpretation	Appearance
Oxidase +	Purple swab
Oxidase −	Yellow swab

EXERCISE 73 REVIEW QUESTIONS

1. When performing an oxidase test on an unknown organism, what controls should be included and what reactions would you expect for each control?

2. If an oxidase test turns purple after 30 sec, the result should not be used. Is such a test an example of a false negative or a false positive?

REFERENCES

Cowan, M.K. 2012. *Microbiology: A Systems Approach, 3rd ed.*, chap. 8. New York: McGraw-Hill.

Reddy, C.A., Beveridge, T.J., Breznak, J.A., Marzluf, G.A., Schmidt, T.M., and Snyder, L.R. (eds.). 2007. *Methods for General and Molecular Microbiology, 3rd ed.*, chap. 15. Washington D.C.: ASM Press.

Talaro, K.P., and Chess, B. 2012. *Foundations in Microbiology, 8th ed.*, chap. 8. New York: McGraw-Hill.

Wiley, J., Sherwood, L., and Woolverton, C. 2011. *Prescott, Harley & Klein's Microbiology, 8th ed.*, chap. 35. New York: McGraw-Hill.

Nitrate Reduction Test

The nitrate reduction test is used to determine if an organism performs nitrate respiration. The test allows for the differentiation of organisms based on whether they partially reduce nitrate to nitrite or to some other nitrogenous product, or whether they completely reduce nitrate, forming nitrogen gas.

PRINCIPLES AND APPLICATIONS

Anaerobic respiration involves the transfer of electrons to a final electron acceptor other than oxygen. Many bacteria are able to transfer electrons to (i.e., reduce) nitrate in a process called **nitrate respiration.** While some bacteria, such as *Escherichia coli*, partially reduce nitrate (NO_3^-) to nitrite (NO_2^-) through the use of the enzyme nitrate reductase (Figure 74.1), others reduce nitrate to another nitrogen-containing compound such as ammonia or hydroxylamine. Still other bacteria are capable of completely reducing nitrate, forming gaseous nitrogen (N_2) in a process called **denitrification.** Because organisms vary both in whether or not they can reduce nitrate and in the end products they produce, the nitrate reduction test is a useful aid in identifying bacteria.

Nitrate broth contains beef extract, peptone, and potassium nitrate. A small inverted test tube called a **Durham tube** is placed in the media prior to autoclaving. This tube will collect a portion of any gas produced by the organism. The media has no pH indicators, and all the other reagents are added after the incubation.

Following the incubation, the Durham tube is checked for gas. If the organism being studied is a fermenter, the gas could be from carbohydrate fermentation and no conclusions can be made as to nitrate reduction. If, however, the organism is known to be a nonfermenter and gas is present in the Durham tube, it must be nitrogen gas produced by denitrification of nitrate. If this is the case, then the organism completely reduces nitrate to molecular nitrogen and the test is over.

If no gas is present in the Durham tube, then reagents are added to the tube to determine if any reduction of nitrate has occurred. First, nitrate reagent A (sulfanilic acid) and nitrate reagent B (naphthylamine) are added to test for the reduction of nitrate to nitrite. If nitrite is present in the tube, a red color will develop, indicating the partial reduction of nitrate to nitrite and the presence of nitrate reductase. If a red color is not apparent after the addition of nitrate reagents A and B, two possibilities exist: either nitrate was not reduced at all, and is still present in the media, or nitrate was reduced to an end product other than nitrite. To differentiate between these two possibilities, nitrate reagent C (powdered zinc) is added to the media, which is mixed slightly and allowed to incubate

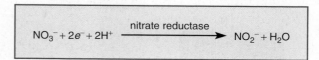

Figure 74.1 The enzyme nitrate reductase catalyzes the reduction of nitrate to nitrite.

for up to 10 min. Zinc catalyzes the reduction of nitrate to nitrite and will convert any nitrate in the tube to nitrite, which will in turn produce a red color because of the presence of nitrate reagents A and B. If the media remains clear after the addition of zinc, then nitrate was reduced to ammonia, nitric oxide, or a similar nonnitrite compound. The process of nitrate broth evaluation is illustrated in Figure 74.2.

PRE-LAB QUESTIONS

1. *Define the following terms:*

Reduction

Anaerobic respiration

2. *Please provide the chemical formulas for nitrate, nitrite, ammonia, and nitrogen gas.*

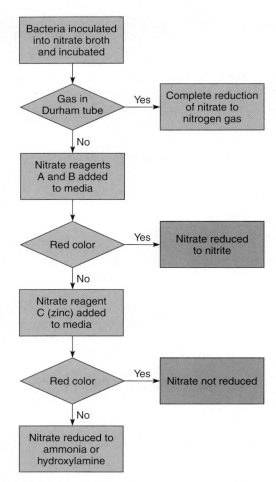

Figure 74.2 Flowchart for analysis of nitrate broth. After incubation, a gas bubble trapped in the Durham tube indicates that denitrification took place, i.e., the organism completely reduces nitrate to nitrogen gas *unless* the bacteria in question is known to be a fermenter, in which case no conclusions can be drawn about the source of the gas. If, after nitrate reagents A and B are added to the media a red color appears, then the organism reduced nitrate to nitrite. If the media remains clear after addition of nitrate reagents A and B, then zinc is added to catalyze the reduction of nitrate in the tube. A red color at this point indicates that nitrate was not reduced by the organism; if the media remains clear, then nitrate was reduced to a nonnitrite end product.

PERIOD ONE

MATERIALS

Each student or group should obtain:
Four tubes of nitrate broth
Fresh broth cultures of:
  Escherichia coli
  Pseudomonas aeruginosa
  Bacillus sphaericus

Nitrate reagents A, B, and C (zinc powder)
Inoculating loop
Laboratory marker

PROCEDURE

1. Mark each tube with your name and lab time. Label one tube with the name of each microorganism. Label the final tube "control."
2. Use a loop to inoculate each tube with the appropriate organism. Do not inoculate the control tube.
3. Incubate the tubes at 37°C for 48 h.

QUESTIONS—PERIOD ONE

1. How do you recognize each of the following:
   a. Denitrification of nitrate?

   b. Nitrate reduction to a compound other than nitrite?

   c. Nitrate reduction to nitrite?

2. How would you expect your control tube to look?

PERIOD TWO
Retrieve your tubes from the incubator, and examine them using the following procedure as well as Figure 74.2 as a guide. Treat the control tube exactly as you do the other tubes.

Nitrate Reduction Test

1. Look for the production of gas in the Durham tube.
2. Add 10 drops of nitrate reagents A and B to each tube. Mix gently, and allow 2 min for any color change to become apparent.

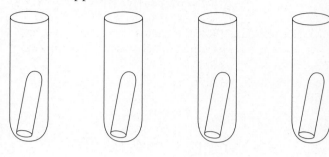

3. If the tube remains clear, add a small quantity of zinc, mixing gently to disperse the zinc throughout the media. Wait an additional 10 min for any color change.

Table 74.1 summarizes the possible reactions for the nitrate broth. Record your results in the space provided. Also include a sketch that conveys the appearance of each of your tubes.

TABLE 74.1	Nitrate broth results and interpretations
Interpretation	**Appearance**
Denitrification to produce nitrogen gas (+)	Bubble in Durham tube, bacteria is a known nonfermenter
Source of gas is unknown. Nitrate reagents A and B (and perhaps C) must be added	Bubble in Durham tube, bacteria may be a fermenter
Nitrate reduction to nitrite (+)	Red color after addition of nitrate reagents A and B
Add nitrate reagent C (zinc powder) to continue test	No color after addition of nitrate reagents A and B
Nitrate reduction to nongaseous, nonnitrite products (+)	No color after addition of nitrate reagent C
No nitrate reduction (−)	Red color after addition of nitrate reagent C

EXERCISE 74 REVIEW QUESTIONS

1. Explain the reactions you would see in your control tube at each step of the evaluation process.

2. How would you interpret a red color in your control tube after the addition of nitrate reagents A and B?

REFERENCE

Reddy, C.A., Beveridge, T.J., Breznak, J.A., Marzluf, G.A., Schmidt, T.M., and Snyder, L.R. (eds.). 2007. *Methods for General and Molecular Microbiology, 3rd ed.*, chap. 15. Washington D.C.: ASM Press.

Coagulase Test

The coagulase test detects whether or not a particular bacterium produces the enzyme coagulase. The presence of coagulase is commonly used to differentiate Staphylococcus aureus *from other species of* Staphylococcus.

PRINCIPLES AND APPLICATIONS

Coagulase is an enzyme that interacts with fibrinogen in blood plasma to cause **coagulation.** In the body, this thickening of the plasma around bacterial cells provides a great deal of protection, as these cells are then shielded from phagocytosis and other actions of the immune systems. The production of coagulase is one of the virulence factors used by *Staphylococcus aureus* to escape the body's defenses. Because 97% of *S. aureus* species are coagulase positive while other species of *Staphylococcus* are coagulase negative, the ability to detect the presence of coagulase is a powerful laboratory tool in the identification of *S. aureus*, easily differentiating it from other Gram-positive cocci.

The test itself is simple. A small tube of rabbit plasma is inoculated with the test organism and incubated. The tube is checked hourly for the next several hours and then again 24 h after the initial inoculation. Thickening of the plasma at any time is considered a positive reaction.

PRE-LAB QUESTION

1. *Define the following terms:*

    Fibrin

    Coagulation

PERIOD ONE

MATERIALS

Each student or group should obtain:
Three tubes of sterile rabbit plasma
Fresh slant cultures of:
  Staphylococcus aureus
  Staphylococcus epidermidis
Inoculating loop
Marking pen

PROCEDURE

1. Mark each tube with your name and lab time. Label one tube with the name of each microorganism. Label the third tube "control."
2. Use a loop to heavily inoculate each tube with the appropriate organism. Do not inoculate the control tube.
3. Incubate the tubes at 37°C. Check the tubes for thickening of the plasma or the formation of long fibrin threads hourly for the next 4 h. Any thickening is considered a positive reaction.
4. Tubes that have remained liquid can be incubated overnight but must be checked within 24 h of their initial inoculation as coagulated tubes may revert to a liquid within 24 h.

QUESTIONS—PERIOD ONE

1. How do you recognize each of the following:
    a. A positive coagulase result?

    b. A negative coagulase result?

2. How would you expect your control tube to look?

PERIOD TWO

Retrieve your tubes from the incubator, and examine them using Figure 75.1 as a guide. A solid clot or a loose clot suspended in the plasma is considered coagulase positive. Table 75.1 summarizes the possible reactions for the coagulase test. Sketch the appearance of your tubes and indicate for each whether it is coagulase negative or positive.

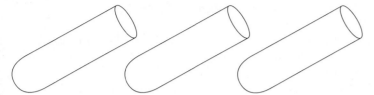

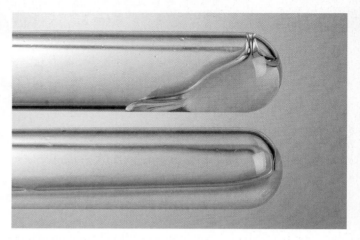

Figure 75.1   Coagulase test. Upper tube is positive, lower tube is negative.

TABLE 75.1	Coagulase test results and interpretations
Interpretation	Appearance
Coagulase +	Plasma is solid
Coagulase +	Plasma is partially liquid with a suspended clot
Coagulase −	Plasma is liquid

EXERCISE 75 REVIEW QUESTION

1. How would you explain a clot forming in your control tube?

REFERENCES

Cowan, M.K. 2012. *Microbiology: A Systems Approach, 3rd ed.,* chap. 18. New York: McGraw-Hill.

Reddy, C.A., Beveridge, T.J., Breznak, J.A., Marzluf, G.A., Schmidt, T.M., and Snyder, L.R. (eds.). 2007. *Methods for General and Molecular Microbiology, 3rd ed.,* chap. 15. Washington D.C.: ASM Press.

Talaro, K.P., and Chess, B. 2012. *Foundations in Microbiology, 8th ed.,* chap. 18. New York: McGraw-Hill.

Citrate Test

Citrate agar is a defined medium that includes citrate as its sole carbon source and ammonium phosphate as the sole nitrogen source. Citrate media is commonly used to differentiate Gram-negative bacteria.

PRINCIPLES AND APPLICATIONS

Simmons citrate agar is an example of a **utilization medium:** a defined media in which one or more essential nutrients is limited. Simmons citrate media, for example, contains citrate as its only source of carbon and ammonium phosphate as its only source of nitrogen. Bromthymol blue is a pH indicator included in the media; it is green at pH 6.9 and blue at pH 7.6. Organisms able to utilize citrate as their sole carbon source and ammonium phosphate as their sole source of nitrogen will produce ammonia as a metabolic byproduct. Ammonia raises the pH of the media and causes the indicator to turn from green to blue.

PRE-LAB QUESTIONS

1. What is IMViC?

2. What is a defined medium?

PERIOD ONE

MATERIALS

Each student or group should obtain:
Three tubes of Simmons citrate agar
Fresh broth cultures of:
  Escherichia coli
  Enterobacter aerogenes
Inoculating needle
Marking pen

PROCEDURE

1. Mark each tube with your name and lab time. Label one tube with the name of each microorganism. Label the final tube "control."
2. Use a needle to make a single streak up the slant of the media to inoculate each tube with the appropriate organism. Do not inoculate the control tube.
3. Incubate the tubes at 37°C for 48 h.

QUESTIONS—PERIOD ONE

1. How do you recognize a positive result on Simmons citrate agar?

2. How would you expect your control tube to look?

PERIOD TWO

Retrieve your tubes from the incubator, and examine them for growth and color change. Any color change from green to blue is indicative of citrate utilization and a positive reaction, as is growth on the slant (Figure 76.1). Tubes displaying neither growth nor color change may be reincubated up to a total of 7 days. Possible results for citrate utilization are summarized in Table 76.1. Sketch the appearance of your tubes.

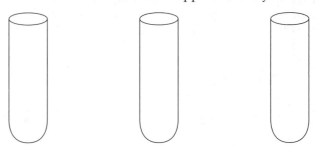

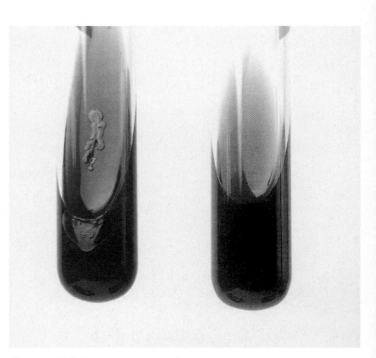

Figure 76.1 Citrate utilization. Tube on the left is positive (note both growth and color change) while the tube on the right is negative.

TABLE 76.1	Citrate test results and interpretations
Interpretation (Symbol)	**Appearance**
Citrate +	Growth, no color change (media is green)
Citrate +	Growth, with color change (media is blue)
Citrate −	No growth, no color change (media is green)

EXERCISE 76 REVIEW QUESTIONS

1. How would you interpret a blue color throughout your control tube?

2. Could a complex (undefined) media be used for a citrate test?

3. How would you interpret a tube that had growth but that was still green (hint: what would happen if you continued to incubate the tube)?

REFERENCE

Reddy, C.A., Beveridge, T.J., Breznak, J.A., Marzluf, G.A., Schmidt, T.M., and Snyder, L.R. (eds.). 2007. *Methods for General and Molecular Microbiology, 3rd ed.*, chap. 15. Washington D.C.: ASM Press.

Malonate Test

The malonate test differentiates organisms that can use malonate as their sole source of carbon from those that cannot. Bacteria that utilize malonate grow in the media, producing alkaline end products. Bromthymol blue is included as a pH indicator and will turn from green to blue under alkaline conditions, which is interpreted as a positive reaction. The test is used to differentiate many members of the Enterobacteriaceae.

PRINCIPLES AND APPLICATIONS

Malonate media is an example of a **utilization medium;** one which limits an essential nutrient. Organisms that cannot use the limited, provided nutrient will not grow. In most cases, this type of media has a defined formula that includes only one form of an essential nutrient. Malonate media, by contrast, is an undefined media, containing small amounts of both yeast extract and glucose, meaning several different carbon sources are available. The large amount of malonate in the media competitively inhibits a key enzyme in the Krebs cycle that catalyzes the conversion of succinate to fumarate. In most organisms, malonate binds to the enzyme succinate dehydrogenase, preventing the succinate to fumarate

reaction and thereby halting the Krebs cycle and eventually killing the cell. Some organisms, however, possess an alternative metabolic pathway that allows them to utilize malonate as their sole source of carbon (Figure 77.1). Malonate utilization produces alkaline end products, driving the pH of the media higher and causing the bromthymol blue pH indicator to change from green to blue.

Yeast extract and glucose in the media promote the growth of a wide array of organisms, many of which may produce alkaline end products, but these are negated by a phosphate buffering system that resists small changes in the pH of the media. Organisms that can ferment glucose may produce acid end products, which can be seen as a slight yellowing of the media. A change from green to blue is the only result that indicates the ability to utilize malonate (Figure 77.2).

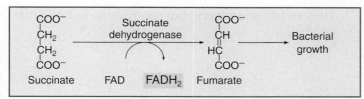

(a)

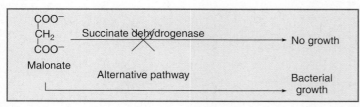

(b)

Figure 77.1 Competitive inhibition by malonate. (a) As part of the Krebs cycle, the conversion of succinate to fumarate is catalyzed by the enzyme succinate dehydrogenase. Fumarate then continues through the Krebs cycle, resulting in bacterial growth. (b) Malonate competes with succinate for the active site on the succinate dehydrogenase enzyme, blocking the metabolic step catalyzed by the enzyme and preventing the conversion of succinate to fumarate. This inhibition will result in the death of the cell *unless* the bacteria being tested can utilize an alternative pathway to metabolize malonate, resulting in continued bacterial growth.

Figure 77.2 Interpretation of malonate broth. A blue color throughout the media indicates that malonate can be utilized as the sole source of carbon, a positive reaction. A green color indicates that malonate could not be used as the sole source of carbon and is interpreted as a negative reaction.

PRE-LAB QUESTIONS

1. *What is the difference between a defined and an undefined medium?*

2. *What is a buffer?*

PERIOD ONE

MATERIALS

Each student or group should obtain:
Three tubes of malonate broth
Fresh broth cultures of:
  Escherichia coli
  Enterobacter aerogenes
Inoculating loop
Marking pen

PROCEDURE

1. Mark each tube with your name and lab time. Label one tube with the name of each microorganism, and label the final tube "control."
2. Use a loop to inoculate each tube with the appropriate organism. Do not inoculate the control tube.
3. Incubate the tubes at 37°C for 48 h.

QUESTIONS—PERIOD ONE

1. How do you recognize a positive result in the malonate broth?

2. How would you expect your control tube to look after incubation?

PERIOD TWO
Retrieve your tubes from the incubator and examine them for any color change. A color change from green to blue is indicative of malonate utilization and a positive reaction (Figure 77.2). Possible results for malonate utilization are

summarized in Table 77.1. Sketch the appearance of your tubes.

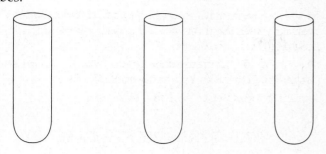

TABLE 77.1	Malonate test results and interpretations
Interpretation (Symbol)	Appearance
Malonate +	Media is blue
Malonate −	Media is green or yellow-green

EXERCISE 77 REVIEW QUESTIONS

1. How would you interpret a blue color in your control tube?

2. Malonate broth normally contains 3 g malonate/L. What do you think would be the result if a broth was made that contained only 3 mg malonate/L? Why?

REFERENCE

Reddy, C.A., Beveridge, T.J., Breznak, J.A., Marzluf, G.A., Schmidt, T.M., and Snyder, L.R. (eds.). 2007. *Methods for General and Molecular Microbiology*, 3rd ed., chap. 15. Washington D.C.: ASM Press.

NOTES

Decarboxylation Test

The decarboxylation test is used to determine whether or not an organism possesses the enzyme that catalyzes the removal of an amino acid's carboxyl group (−COOH). The base decarboxylation media is supplemented with a single amino acid as a substrate along with a pH indicator, bromcresol purple. Decarboxylation of the amino acid in the media produces alkaline end products, resulting in the media turning purple, which is interpreted as a positive reaction. The test is used in the differentiation of many Gram-negative enteric bacteria.

PRINCIPLES AND APPLICATIONS

Amino acids, the monomers from which proteins are made, all have the same basic structure. At one end of the molecule is the **amino group** (−NH$_2$) while the other end is referred to as the **carboxyl group** (−COOH). What makes the twenty or so amino acids different from one another is the "R" group, with the chemical nature of this group conferring very specific properties upon the amino acid. When an amino acid is degraded by removing the carboxyl end of the molecule (decarboxylation), an amine is produced (Figure 78.1), which raises the pH of the media, a reaction that is easily detectable. Carbon dioxide is also produced during the course of the reaction, but it is far more difficult to test for.

Each amino acid has a specific **decarboxylase** and produces a specific product. In the microbiology lab, the three most useful decarboxylation reactions are those involving lysine, ornithine, and arginine. Lysine is decarboxylated to produce cadaverine (Figure 78.2), and ornithine is decarboxylated to produce putrescine (Figure 78.3). Arginine goes through a multistep process in which it is first converted to agmatine and then to putrescine (Figure 78.4). In all the cases, however, alkaline end products are produced.

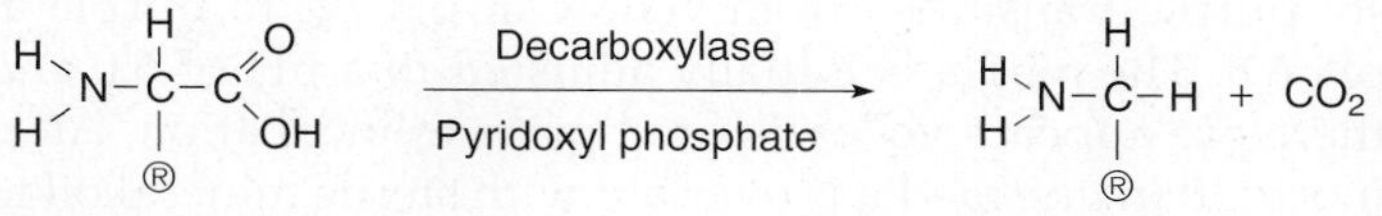

Figure 78.1 Amino acid decarboxylation. Decarboxylase enzymes remove the carboxyl group from an amino acid, producing an amine and carbon dioxide. Regardless of the specific amino acid being decarboxylated, pyridoxyl phosphate is a cofactor in the reaction.

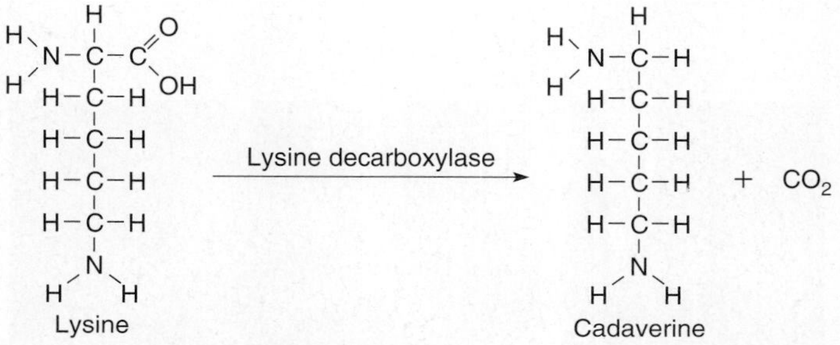

Figure 78.2 Lysine decarboxylation. Lysine decarboxylase removes the carboxyl group from lysine, producing the amine cadaverine and carbon dioxide.

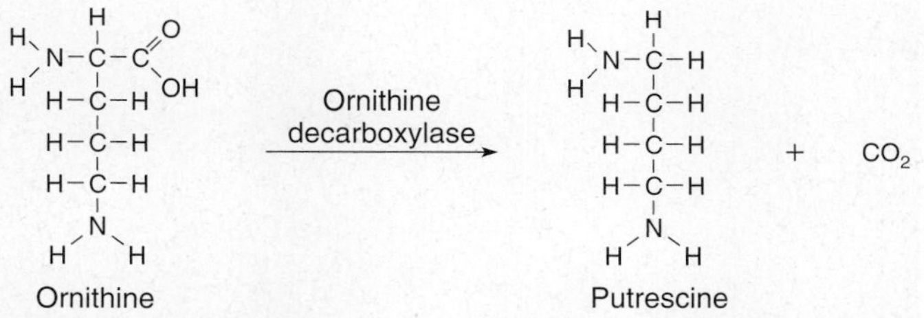

Figure 78.3 Ornithine decarboxylation. Ornithine decarboxylase removes the carboxyl group from ornithine, producing the amine putrescine and carbon dioxide.

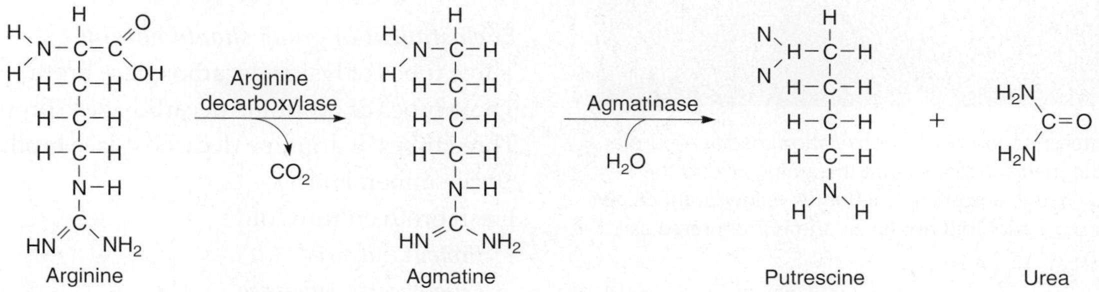

Figure 78.4 Arginine decarboxylation. Arginine decarboxylation is a multistep process that eventually produces the amine putrescine and urea. The urea may in turn be degraded into carbon dioxide and ammonia.

Decarboxylase media is used to determine an organism's ability to decarboxylate specific amino acids. The media contains beef extract and peptones to supply the nutrients needed to support bacterial growth. Also included in the media are pyridoxal, a cofactor needed by amino acid decarboxylase enzymes; glucose, a fermentable carbohydrate; and bromcresol purple, a pH indicator. Bromcresol purple transitions from yellow at pH 5.2 to purple at pH 6.8. The media is initially adjusted to a pH of 6.0 and therefore appears yellowish-red prior to incubation. After inoculation, the media is overlaid with sterile mineral oil to prevent the entry of oxygen and incubated for up to a week. Initially, fermentation of glucose, if it occurs, will produce acid end products, driving the pH downward and causing the media to take on a yellow appearance. If the decarboxylase reaction occurs, amine products will be produced, elevating the pH and turning the media purple (Figure 78.5).

Figure 78.5 Interpretation of decarboxylation media. A purple color throughout the media indicates that the amino acid in the media was decarboxylated, a positive reaction. A yellow/orange color indicates that decarboxylation did not occur and is interpreted as a negative reaction.

PRE-LAB QUESTIONS

1. *Draw the amino acid lysine. Circle and label the carboxyl and amino groups?*

2. *What is a cofactor?*

PERIOD ONE

MATERIALS

Each student or group should obtain:
Three tubes of lysine decarboxylase broth
Three tubes of ornithine decarboxylase broth
Three tubes of arginine decarboxylase broth
Sterile mineral oil
Fresh broth cultures of:
 Proteus vulgaris
 Enterobacter aerogenes
Inoculating loop
Marking pen

PROCEDURE

1. Mark each tube with your name and lab time. For each set of decarboxylase tubes, label one tube with the name of each microorganism, and label the final tube "control."
2. Use a loop to inoculate each tube with the appropriate organism. Do not inoculate the control tube.
3. Overlay each tube with 3–4 mm of sterile mineral oil.
4. Incubate the tubes at 37°C for 48 h.

QUESTIONS—PERIOD ONE

1. How do you recognize a positive result in decarboxylase broth?

2. How would you expect your control tubes to look after the incubation?

negative reaction may be incubated up to a total of 1 week. Possible results for decarboxylase broths are summarized in Table 78.1 Sketch the appearance of your tubes.

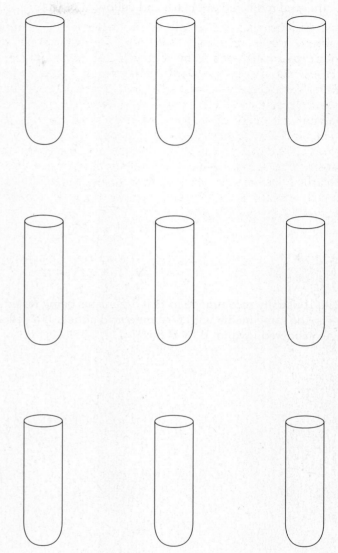

TABLE 78.1	Decarboxylation test results and interpretations
Interpretation (Symbol)	**Appearance**
Decarboxylation +	Media is purple
Decarboxylation −	Media is yellow
Decarboxylation −	Media is red-yellow (no change)

PERIOD TWO

Retrieve your tubes from the incubator, and examine them for any color change. Purple is the only color that should be interpreted as a positive reaction. Tubes displaying a

EXERCISE 78 REVIEW QUESTIONS

1. What color would a positive reaction produce if neutral red was used as a pH indicator instead of bromcresol purple (neutral red is red at pH 6.8 and yellow at pH 8.0)?

2. It is usually recommended that organisms being tested in decarboxylase media also be inoculated into base media, that is, the same media with no amino acid added. If a yellow color is seen in the base media, then the decarboxylation test is considered invalid. Why is this?

REFERENCE

Reddy, C.A., Beveridge, T.J., Breznak, J.A., Marzluf, G.A., Schmidt, T.M., and Snyder, L.R. (eds.). 2007. *Methods for General and Molecular Microbiology*, 3rd ed., chap. 15. Washington D.C.: ASM Press.

Phenylalanine Deaminase Test

The phenylalanine deaminase test determines whether or not an organism possesses the enzyme phenylalanine deaminase, which catalyzes the removal of the amine group from the amino acid phenylalanine. The addition of a ferric chloride-containing reagent to the media after incubation imparts a deep green color to the media if deamination has taken place. The medium is commonly used to differentiate members of the family Enterobacteriaceae.

PRINCIPLES AND APPLICATIONS

The degradation of amino acids, the monomers of which make up proteins, is a highly regulated event, with specific enzymes being used to remove the carboxyl and/or amino end of the molecule. Detection of the decarboxylation reaction was discussed in Exercise 78 while recognition of **deamination** is covered here. Because amino acid degradation is an enzyme-catalyzed event that produces specific end products, the ability of a bacterial species to deaminate a particular amino acid can be used as a clue to its identification. The deamination of phenylalanine is commonly used to differentiate organisms in the laboratory. In this reaction, the deamination of phenylalanine results in the production of phenylpyruvic acid along with ammonia (Figure 79.1).

Phenylalanine agar contains yeast extract for growth and phenylalanine as a substrate for the enzyme phenylalanine deaminase. The media is prepared as a slant and heavily inoculated. After incubation, a reagent containing ferric chloride is added to the slant. Ferric chloride reacts with phenylpyruvic acid to produce a green color that is interpreted as a positive reaction. A yellow color (from the ferric chloride reagent) is a negative reaction (Figure 79.2). The phenylalanine deaminase test is commonly used to differentiate the genera *Proteus, Providencia,* and *Morganella,* all of which test positive, from other members of the family Enterobacteriaceae, all of which test negative.

PRE-LAB QUESTION

1. *Draw the amino acid phenylalanine. Circle the chemical group removed by the enzyme phenylalanine deaminase.*

PERIOD ONE

MATERIALS

Each student or group should obtain:
Three phenylalanine deaminase agar slants
Phenylalanine deaminase test reagent
Fresh slant cultures of:
  Proteus vulgaris
  Escherichia coli
Inoculating needle
Marking pen

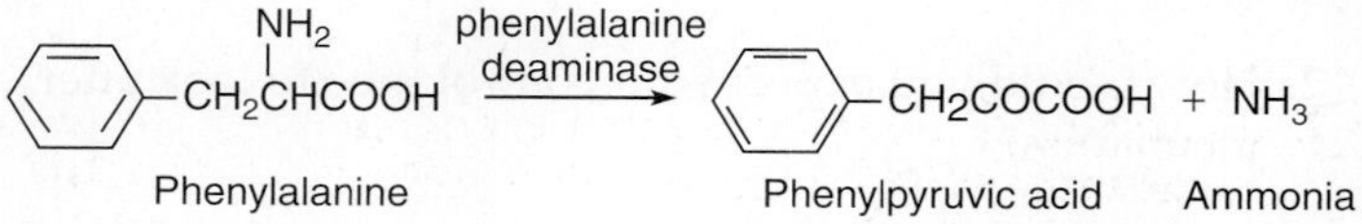

Figure 79.1 Deamination of phenylalanine by phenylalanine deaminase results in the production of phenylpyruvic acid and ammonia.

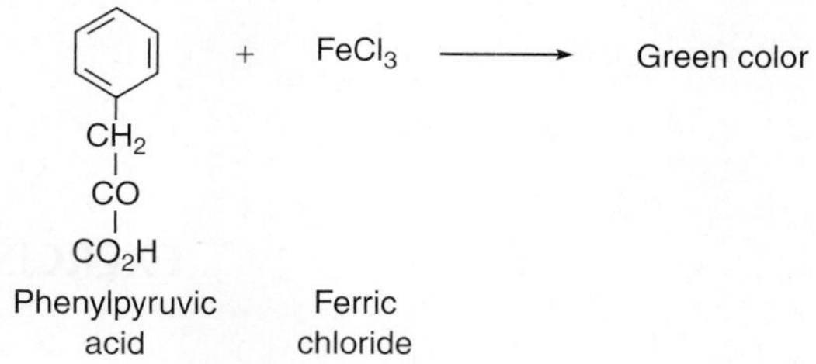

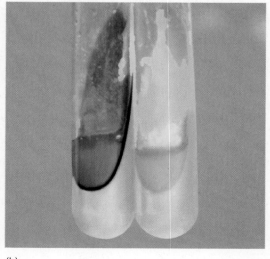

(a)

(b)

Figure 79.2 (a) Ferric chloride reacts with phenylpyruvic acid to produce a green color. (b) The tube on the left exhibits a positive reaction while the tube on the right is negative.

PROCEDURE

1. Mark each tube with your name and lab time. Label one tube with the name of each microorganism, and label the final tube "control."
2. Use a needle to inoculate each tube with the appropriate organism by streaking the surface of the slant twice. Do not inoculate the control tube.
3. Incubate the tubes at 37°C for 24–48 h.

QUESTIONS—PERIOD ONE

1. How do you recognize a positive result in phenylalanine agar?

2. How would you expect your control tube to look after incubation?

PERIOD TWO

Retrieve your tubes from the incubator. Add 4–5 drops of ferric chloride indicator to the surface of the slant. Gently roll the tube in your hands to loosen the growth. The appearance of a green color within 1–5 min is a positive reaction. Possible results for phenylalanine deaminase media are summarized in Table 79.1. Sketch the appearance of your tubes.

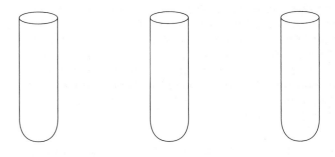

TABLE 79.1	Phenylalanine deaminase test results and interpretations
Interpretation (Symbol)	**Appearance**
Phenylalanine deaminase +	Media is green
Phenylalanine deaminase −	Media is yellow or no change

EXERCISE 79 REVIEW QUESTION

1. Along with phenylpyruvic acid, ammonia is produced as a result of phenylalanine deamination. Could you think of a way in which the production of ammonia could be used as an indication of a positive reaction?

REFERENCE

Reddy, C.A., Beveridge, T.J., Breznak, J.A., Marzluf, G.A., Schmidt, T.M., and Snyder, L.R. (eds.). 2007. *Methods for General and Molecular Microbiology, 3rd ed.,* chap. 15. Washington D.C.: ASM Press.

Bile Esculin Test

Bile esculin agar is a medium with both selective and differential properties that is used to distinguish enterococci and *Streptococcus bovis* from other streptococci. Oxgall (bile) accounts for the selective properties of the media, inhibiting the growth of Gram-negative bacteria and most Gram-positive bacteria. The media's differential aspects are due to the iclusion of esculin, which very few organisms can hydrolyze in the presence of bile. Hydrolysis of esculin leads to the formation of a deep brown to black color in the media, which is indicative of a positive reaction.

PRINCIPLES AND APPLICATIONS

Glycosides are molecules in which a sugar is bound by a glycosidic bond to some other chemical group. Esculin is an example of a glycoside in which the sugar glucose is bound to esculetin. Under acidic conditions, many bacteria are able to **hydrolyze** esculin to produce glucose and esculetin (Figure 80.1a), but few bacteria other than group D streptococci and enterococci can do this in the presence of bile. Ferric citrate (a component of the media) reacts with esculetin to produce a dark brown to black color indicative of a positive reaction (Figure 80.1b). Blackening of the media begins in the slant and proceeds downward into the butt of the tube. If more than one-half of the media is darkened within

48 h, the test is considered positive while a tube that is less than one-half darkened is considered a negative reaction (Figure 80.2). Medically important organisms giving positive reactions on the media include *Enterococcus faecium* and *E. faecalis* as well as group D streptococci such as *Streptococcus bovis*, all of which are opportunistic pathogens.

Figure 80.2 The tube on the left exhibits positive bile esculin hydrolysis while the tube on the right is negative.

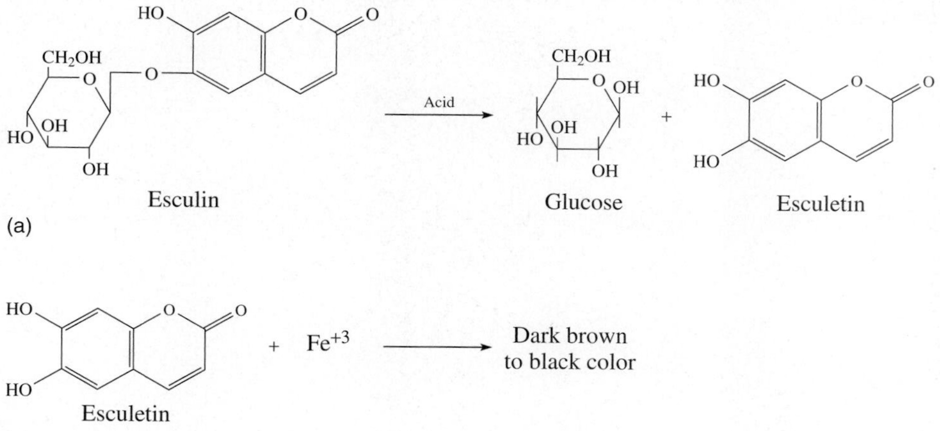

Esculin

(a)

Acid →

Glucose

Esculetin

Esculetin + Fe^{+3} ⟶ Dark brown to black color

(b)

Figure 80.1 (a) Esculin hydrolysis results in the production of esculetin and free glucose. Although many organisms can hydrolyze esculin under acidic conditions, only group D streptococci and enterococci can do this when bile is present in the medium. (b) Esculetin reacts with ferric citrate in the media to produce a dark brown to black color, which is interpreted as a positive reaction.

PRE-LAB QUESTIONS

1. *How does an opportunistic pathogen differ from a primary, or true, pathogen?*

2. *Mark each of the statements below with a D (for differential), S (for selective), or S/D (for both).*

a. _____ Oxgall (bile) inhibits most Gram-positive organisms.

b. _____ *Streptococcus bovis* grows on bile esculin agar.

c. _____ *Streptococcus bovis* turns bile esculin agar black.

d. _____ An unknown bacterial species grows on bile esculin agar and turns one-half of the media black in 48 h.

3. *Differentiate between group D enterococci, group D nonenterococci, and streptococci (reviewing the introduction to Exercise 34 would be helpful).*

PERIOD ONE

MATERIALS

Each student or group should obtain:
Three bile esculin agar slants
Fresh broth cultures of:
  Proteus vulgaris
  Enterococcus faecalis
Inoculating loop
Marking pen

PROCEDURE

1. Mark each tube with your name and lab time. Label one tube with the name of each microorganism, and label the final tube "control."
2. Use a loop to inoculate the surface of each tube with the appropriate organism by streaking the surface of the slant. Do not inoculate the control tube.
3. Incubate the tubes at 37°C for 48 h.

QUESTIONS—PERIOD ONE

1. How do you recognize a positive result in bile esculin agar?

2. How would you expect your control tube to look after incubation?

PERIOD TWO
Retrieve your tubes from the incubator. Evaluate them for growth and blackening of the slant. If more than one-half of the tube is darkened, the reaction should be considered positive. Possible results for bile esculin agar are summarized in Table 80.1. Sketch the appearance of your tubes.

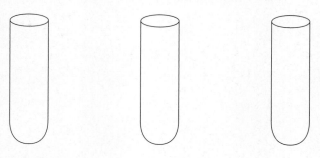

TABLE 80.1	Bile esculin test results and interpretations
Interpretation (Symbol)	Appearance
Organism is a group D *Streptococcus* or *Enterococcus* (bile esculin +)	Media more than half blackened
Organism is a not a group D *Streptococcus* or *Enterococcus* (bile esculin −)	Media is less than half blackened

EXERCISE 80 REVIEW QUESTIONS

1. How would leaving out each of the following ingredients affect bile esculin agar?

   a. Esculin

   b. Ferric ammonium citrate

2. Why is it necessary to carry out a Gram stain in addition to a bile esculin test when differentiating streptococci?

REFERENCE

Reddy, C.A., Beveridge, T.J., Breznak, J.A., Marzluf, G.A., Schmidt, T.M., and Snyder, L.R. (eds.). 2007. *Methods for General and Molecular Microbiology, 3rd ed.*, chap. 11. Washington D.C.: ASM Press.

Starch Hydrolysis

Starch agar is a differential medium used to distinguish bacteria that can hydrolyze starch from those that cannot. After incubating an organism on starch agar, the plate is flooded with iodine, which stains only those areas of the plate in which starch has not been hydrolyzed. A clear halo surrounding growth on a plate indicates that starch has been hydrolyzed.

PRINCIPLES AND APPLICATIONS

Starch is a macromolecule consisting of many glucose molecules bonded together into long chains. Two types of chains exist, amylose and amylopectin. In amylose, the glucose molecules are bonded between the first and fourth carbon atoms, forming 1,4-α-glycosidic linkages and resulting in a long, straight chain. Amylopectin occurs when a bond exists between the first and sixth carbon atoms of glucose, forming 1,6-α-glycosidic linkages and resulting in a branched arrangement. Together these two types of chains result in an enormous molecule made up of thousands of individual glucose units (Figure 81.1).

Organisms that produce the enzyme **amylase** are capable of hydrolyzing starch by breaking the glycosidic bonds between the individual glucose molecules, eventually producing many individual glucose molecules from a single molecule of starch. Amylase is usually excreted from the cell and is used to break down starch in the environment. The smaller glucose molecules can then be imported into the cell to be used as a metabolic substrate.

In the laboratory, starch hydrolysis is detected by streaking an organism onto a plate containing starch agar. After incubation, the plate is flooded with iodine, which stains starch, but not glucose, a dark blue. If the organism has hydrolyzed the starch in the media, a clear halo will form around the bacterial growth, representing the area where starch has been hydrolyzed. If no starch hydrolysis has occurred, the media immediately surrounding the bacteria will be stained uniformly blue, with no halo present (Figure 81.2).

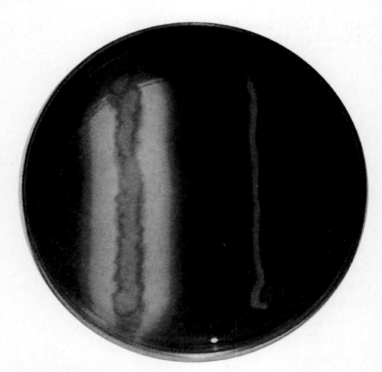

Figure 81.2 Starch hydrolysis. The clear zone surrounding the streak on the left is indicative of starch hydrolysis by amylase. The absence of a clear halo surrounding the streak on the right represents a negative reaction.

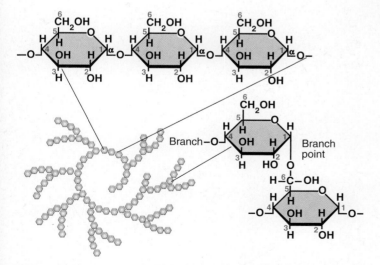

Figure 81.1 Structure of starch. Starch is composed of two types of glucose polymers, amylase and amylopectin. Amylose consists of glucose molecules bonded in a 1,4 pattern while the side chains of amylopectin are made up of glucose bonded in a 1,6 pattern.

PRE-LAB QUESTIONS

1. *What does the term hydrolyze mean? In what way are the starch hydrolysis test and the casease test similar?*

2. *In what way is starch agar a differential medium?*

PERIOD ONE

MATERIALS

Each student should obtain:
One starch agar plate
Fresh broth cultures of:
   Bacillus subtilis
   Escherichia coli
Inoculating loop
Marking pen

PROCEDURE

1. Mark the bottom of your Petri dish with your name and lab time. Label one half of the dish with the name of each organism.
2. Use a loop to inoculate the surface of the agar with a single streak of the appropriate organism.
3. Incubate the plate at 37°C for 48 h.

QUESTION—PERIOD ONE

1. How do you recognize a positive result on starch agar?

PERIOD TWO
1. Retrieve your plate from the incubator. Evaluate the plate for growth, noting the location and spread of growth for each organism.
2. Flood the plate with enough iodine (from your Gram-staining kit) to completely cover the media.
3. Over the next minute, the majority of the plate should darken; look for the appearance of a halo around either of your bacterial isolates. Production of a halo is interpreted as a positive reaction. The color will eventually disappear so be sure to read your results within 2 min. Possible results for starch agar are summarized in Table 81.1. Sketch the appearance of your plate.

TABLE 81.1	Starch agar test results and interpretations	
Interpretation (Symbol)	Appearance
Amylase is present (starch +)	Clear halo around growth
Amylase is absent (starch −)	No clear halo around growth

EXERCISE 81 REVIEW QUESTIONS

1. Why is it necessary to ensure that bacteria have grown adequately on the media before applying iodine to the plate?

2. Occasionally, when iodine is added to a starch agar plate, the bacterial growth washes loose from the media, becoming completely detached from the surface of the plate. If this were to happen, how would it affect your results?

REFERENCE

Reddy, C.A., Beveridge, T.J., Breznak, J.A., Marzluf, G.A., Schmidt, T.M., and Snyder, L.R. (eds.). 2007. *Methods for General and Molecular Microbiology, 3rd ed.*, chap. 15. Washington D.C.: ASM Press.

NOTES

ONPG Test

The ONPG (o-nitrophenyl-β-D-galactopyranoside) test is used to differentiate bacteria that slowly ferment lactose from those that are incapable of fermenting lactose. When incubated with ONPG, slow lactose fermenters will generate a yellow color as a result of the production of o-nitrophenol while tubes containing lactose nonfermenters will remain clear. The test is useful for the differentiation of many Gram-negative bacteria.

PRINCIPLES AND APPLICATIONS

Fermentation of lactose is a two-step process in which lactose is first transported across the cell membrane and then digested once inside the cell. Transport of the sugar is catalyzed by the enzyme **β-galactosidase permease** while another enzyme, **β-galactosidase,** is responsible for **hydrolyzing** lactose to produce glucose and galactose. Bacteria that are lactose fermenters possess both enzymes, allowing lactose to be imported and digested quickly. Some bacteria however possess the hydrolytic enzyme but not the permease, meaning that lactose could be digested but has difficulty entering the cell. For reasons that are somewhat unclear (probably as a result of a mutation that creates a crude permease), lactose fermentation eventually takes place in these bacteria. Because fermentation takes so long to occur, these organisms are referred to as **slow** (or **late**) **lactose fermenters.**

In the laboratory, slow lactose fermentation is detected by providing the cell with *o*-nitrophenyl-β-D-galactopyranoside (ONPG). ONPG can move across the cell membrane unaided and is then cleaved by β-galactosidase to produce *o*-nitrophenol, which has a yellow color. In most laboratories, a commercial disk or tablet containing ONPG is combined with a suspension of the bacteria to be tested. After a short incubation, a yellow color in the tube is considered positive for slow lactose fermentation while a clear color is negative (Figure 82.1). β-Galactosidase is an inducible enzyme that is only produced by a cell when lactose is present in the environment. Therefore, to ensure that the enzyme is produced by the cell, bacteria being tested are always grown on a lactose-containing media prior to the test.

PRE-LAB QUESTIONS

1. *What does the term hydrolyze mean?*

2. *How do you recognize a positive result in an ONPG test?*

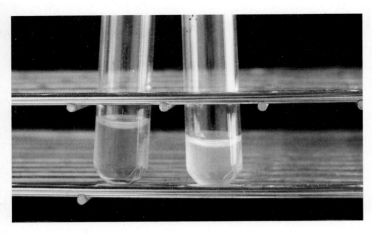

Figure 82.1 Interpretation of ONPG test. A yellow color throughout the test solution is interpreted as ONPG positive while a clear solution represents a negative result.

3. *How does the hydrolysis of starch differ in location from the hydrolysis of lactose (Exercise 81 may provide a clue)? Why do you think starch and lactose are digested in different ways?*

MATERIALS

This procedure utilizes ONPG WEE-TABS from Key Scientific Products Inc. Each WEE-TABS contains enough ONPG for one test. Your instructor will tell you if another procedure is to be used.

Each student should obtain:

Three WEE-TABS tubes[1]

Sterile distilled water

Fresh Kligler's iron agar cultures of:

> *Proteus mirablis*
>
> *Escherichia coli*

Inoculating loop

Marking pen

PROCEDURE

1. Mark each tube with your name and lab time. Label one tube with the name of each organism, and label the third tube "control."
2. Add sterile water to each tube until it reaches the bottom of the label (approximately 0.5 ml); allow the WEE-TABS to dissolve.
3. Use a loop to heavily inoculate each tube with the proper organism. Do not inoculate the third tube.
4. Incubate the tubes for at 37°C for 2 h.
5. A yellow color in the media is considered a positive reaction. Possible results for the ONPG test are summarized in Table 82.1. Sketch the appearance of your tubes.

TABLE 82.1	ONPG test results and interpretations
Interpretation (Symbol)	**Appearance**
β-galactosidase present (ONPG +)	Yellow color
β-galactosidase absent (ONPG −)	No color change

EXERCISE 82 REVIEW QUESTIONS

1. What is an inducible enzyme?

2. Why is it necessary to grow the bacteria on Kligler's iron agar prior to ONPG testing? Could another media be used?

3. How does the success of this test relate to the lac operon?

1. Key Scientific Products Inc. K1490 ONPG WEE-TABS. Round Rock, Texas.

REFERENCE

Forbes, B.A., Sahm, D.F., and Weissfeld, A. 2007. *Bailey and Scott's Diagnostic Microbiology, 12th ed.*, p. 237. St Louis: Mosby-Elsevier.

Urease Test

The enzyme urease catalyzes the breakdown of urea to carbon dioxide and ammonia. Urea broth contains urea as a substrate as well as the pH indicator phenol red. The production of ammonia from the hydrolysis of urea raises the pH, turning the media pink. The urease test is commonly used as an aid in the differentiation of members of the Enterobacteriaceae.

PRINCIPLES AND APPLICATIONS

When certain amino acids are **hydrolyzed, urea** is produced as a byproduct. Many bacteria, especially Gram-negative enterics, are able to further degrade urea to produce ammonia and carbon dioxide (Figure 83.1). Urea broth contains urea, a strong potassium phosphate buffer, and the pH indicator phenol red, which is peach colored at pH 6.6 and pink at pH 8.0. The media is initially adjusted to pH 6.8 and so appears peach colored. In most instances, as urea is broken down, ammonia raises the pH of the broth, but the buffer system within the media ultimately prevents alkalinization of the media. Bacteria in the genera *Proteus, Morganella,* and *Providencia* however produce ammonia quickly enough that the buffer system is overwhelmed, turning the media pink as the pH rises above 8.0. Because of their ability to produce copious amounts of ammonia so quickly, these bacteria are referred to as **rapid urease-positive organisms.** After inoculation, urea broths are incubated for only 24 h before being evaluated. A pink color is considered positive for rapid urease hydrolysis while a peach color is interpreted as a negative reaction (Figure 83.2). This test is commonly used to distinguish *Proteus,* a common cause of urinary tract infections, from other enteric bacteria.

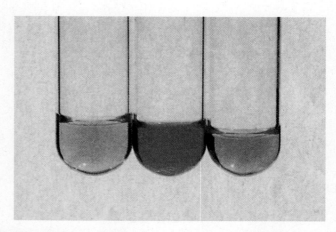

Figure 83.2 Urease test results. The middle tube displays a positive reaction. On the left is an uninoculated tube, and on the right is a negative result.

PRE-LAB QUESTIONS

1. *What purpose does a buffer serve, in general, and in urea broth specifically?*

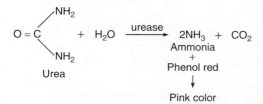

Figure 83.1 Urea hydrolysis. The hydrolysis of urea is catalyzed by the enzyme urease to produce carbon dioxide along with ammonia, which raises the pH of the media, allowing the reaction to be easily detected through the inclusion of the pH indicator phenol red.

2. *How would the media color change if neutral red was used as the pH indicator instead of phenol red. Neutral red appears pink at pH 6.8 and yellow at pH 8.0.*

2. How would you expect your control to look after incubation?

PERIOD ONE

MATERIALS

Each student should obtain:
Three tubes of urea broth
Fresh slant cultures of:
    Proteus vulgaris
    Escherichia coli
Inoculating needle
Marking pen

PROCEDURE

1. Mark each tube with your name and lab time. Label each tube with the name of an organism, and label the final tube "control."
2. Use a needle to heavily inoculate each tube with the appropriate organism. Do not inoculate the control.
3. Incubate the tubes at 37°C for 24 h.

QUESTIONS—PERIOD ONE

1. How do you recognize a positive result in urea broth?

PERIOD TWO

Retrieve your tubes from the incubator. Evaluate them for any color change. A pink color should be considered a positive reaction while a peach or yellow color is negative. Possible results for urea broth are summarized in Table 83.1. Sketch the appearance of your tubes.

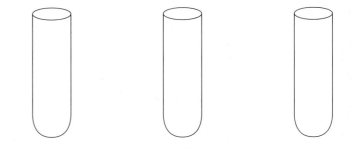

TABLE 83.1	Urea broth test results and interpretations	
Interpretation (Symbol)	**Appearance**
Strong urease producer (rapid urea hydrolysis +)	Pink
No urease hydrolysis (rapid urea hydrolysis −)	Peach or yellow

EXERCISE 83 REVIEW QUESTIONS

1. How would the test be affected if the buffer was left out of the media?

2. Urea broth is meant to be read after 24 h incubation. What could happen if the media was allowed to incubate for a longer period of time, such as 96 h?

REFERENCE

Forbes, B.A., Sahm, D.F., and Weissfeld, A. 2007. *Bailey and Scott's Diagnostic Microbiology, 12th ed.,* p. 110. St. Louis: Mosby-Elsevier.

NOTES

Casease Test

The casease test employs skim milk agar as a differential medium to distinguish bacteria that can hydrolyze casein (the protein that gives milk its white color) from those that cannot. After incubation on skim milk agar, organisms able to hydrolyze casein will be surrounded by a clear halo where the ordinarily opaque protein has been digested. Organisms unable to hydrolyze casein will have no such halo. Skim milk agar is commonly used for the enumeration of bacteria in liquid milk, dried milk, ice cream, and whey.

PRINCIPLES AND APPLICATIONS

Proteins are composed of long strings of **amino acids** held together by **peptide** bonds. Enzymatic hydrolysis of a protein results in the liberation of many individual amino acids (Figure 84.1). When the milk protein casein, which is responsible for the white color of milk, is **hydrolyzed,** it loses both its color and opacity, becoming clear. Some bacteria produce the proteolytic enzyme casease, which is exported from the cell and used to break extracellular casein into smaller peptides and amino acids that can be more easily transported across the cell membrane.

The casease test is performed by inoculating bacteria onto a plate of skim milk agar and incubating for 24–48 h; bacteria that produce casease will hydrolyze casein in the surrounding media, creating a clear zone around the bacteria. Such a clear halo indicates the production of casease in the organism while the lack of a halo is interpreted as a lack of casease (Figure 84.2). Because casease is not produced in every cell, it can be used as an aid in identification of many bacterial species.

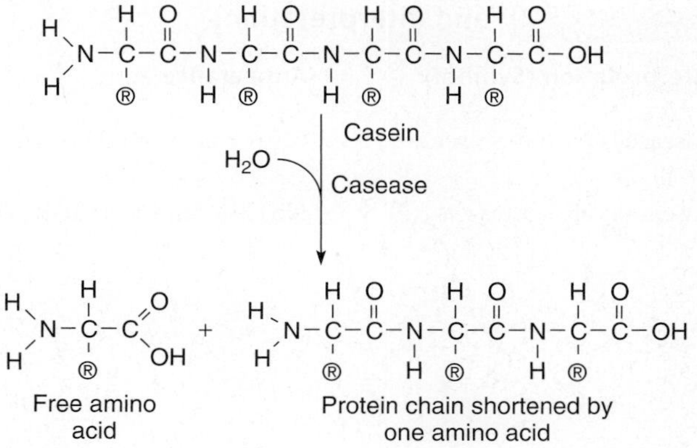

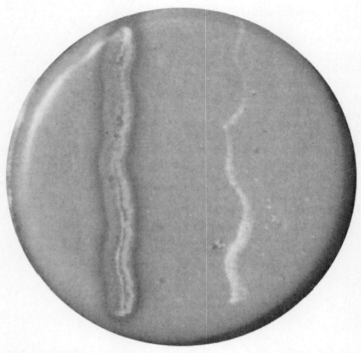

Figure 84.2 Casein hydrolysis. The clear zone surrounding the streak on the left is indicative of casein hydrolysis. The absence of a clear halo surrounding the streak on the right represents a negative reaction.

PRE-LAB QUESTIONS

1. *Define the following terms:*

   Proteolytic

   Hydrolysis

Figure 84.1 Casein hydrolysis. The enzyme casease hydrolyzes the milk protein casein, breaking it down to its component amino acids.

Protein

Amino acid

2. *In what way are the casease test and the starch hydrolysis test similar?*

PERIOD ONE

MATERIALS

Each student should obtain:
One skim milk agar plate
Fresh broth cultures of:
  Bacillus subtilis
  Escherichia coli
Inoculating loop
Marking pen

PROCEDURE

1. Mark the bottom of your Petri dish with your name and lab time. Label one half of the dish with the name of each organism to be tested.
2. Use a loop to inoculate the surface of the agar with a single streak of the appropriate organism.
3. Incubate the tubes at 37°C for 48 h.

QUESTION—PERIOD ONE

1. How do you recognize a positive result in a casease test?

PERIOD TWO
Retrieve your plate from the incubator. Evaluate the plate for growth of each organism and any clearing of the medium. The presence of a clear zone around an organism is a positive result for casease activity. Possible results for casease activity are summarized in Table 84.1. Sketch the appearance of your plate.

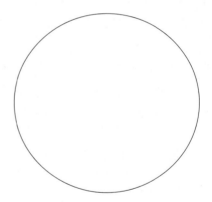

TABLE 84.1	Casease test results and interpretations	
Interpretation (Symbol)	**Appearance**
Casease is present (casease +)	Clear halo around growth
Casease is absent (casease −)	No clear halo around growth

EXERCISE 84 REVIEW QUESTIONS

1. Why is it necessary to ensure that bacteria have grown adequately on the media prior to checking for caseage activity?

2. A biochemical test called the *Biuret reaction* is commonly used to identify the presence of protein in a solution. Biuret reagent stains peptide bonds violet while a lack of peptide bonds results in a blue color. If you stained your caseage plate from this exercise with Biuret reagent, what would you expect to see?

REFERENCE

Reddy, C.A., Beveridge, T.J., Breznak, J.A., Marzluf, G.A., Schmidt, T.M., and Snyder, L.R. (eds.). 2007. *Methods for General and Molecular Microbiology, 3rd ed.*, chap. 15. Washington D.C.: ASM Press.

NOTES

Gelatinase Test

The gelatinase test is used to differentiate those organisms that can enzymatically liquefy gelatin from those that cannot. The media contains gelatin, peptone, and beef extract, with the gelatin serving both as the solidifying agent and as a substrate for the enzyme gelatinase. After an incubation of up to 7 days, the tubes are refrigerated, and those that fail to solidify are considered gelatinase positive; those that solidify upon cooling are gelatinase negative.

PRINCIPLES AND APPLICATIONS

The gelatin familiar to most of us is a protein derived from collagen, a component of connective tissue. One of the reasons gelatin is not used to solidify growth media is that some microorganisms contain **gelatinase,** a proteolytic enzyme that will liquefy gelatin to its component amino acids. Because not all bacteria possess gelatinase, the ability to liquefy gelatin can be used as an aid in identification.

Nutrient gelatin is used to test for the presence of gelatinase. This media contains beef extract and peptone along with 12% gelatin as both a solidifying agent and a substrate for gelatinase. The media is stabbed and incubated for as long as 7 days to provide gelatinase-positive organisms time to grow and secrete gelatinase, which is often produced very slowly. During the incubation, especially at temperatures above 25°C, the gelatin may simply melt, which must be distinguished from enzymatic degradation. This can be accomplished by placing the tube in an ice bath for 10 min, which will lower the temperature of the media below 28°C, the point at which gelatin undergoes a transition from a liquid to a solid. After this brief cooling period, any liquefaction of the media can be attributed to the production of gelatinase by the organism, a positive result. A solid media, on the other hand, is considered a negative result (Figure 85.1). The gelatinase test is often used as an aid in differentiation of enteric bacilli.

PRE-LAB QUESTION

1. *In what way is nutrient gelatin a differential media?*

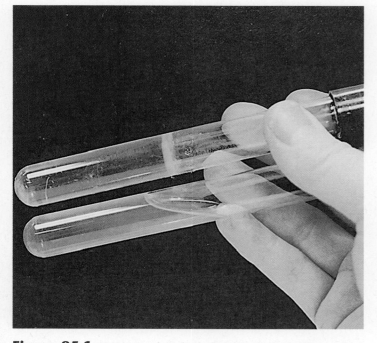

Figure 85.1 Nutrient gelatin hydrolysis. Media in the lower tube has liquefied as a result of the hydrolysis of gelatin by gelatinase. The upper tube displays a negative reaction.

PERIOD ONE

MATERIALS

Each student or group should obtain:
Three nutrient gelatin stabs
Fresh slant cultures of:
   Staphylococcus aureus
   Escherichia coli
Inoculating needle
Marking pen

PROCEDURE

1. Mark each tube with your name and lab time. Label one tube with the name of each microorganism, and label the final tube "control."
2. Use a needle to heavily inoculate each tube with a single stab to the bottom of the media. Do not inoculate the control tube.
3. Incubate the tubes at 37°C for 7 days.

QUESTIONS—PERIOD ONE

1. How do you recognize a positive result for gelatinase activity?

2. How would you expect your control tube to look after incubation? Would it have to be cooled prior to its evaluation?

PERIOD TWO

Retrieve your tubes from the incubator. Place them in an ice bath for 10 min or, alternatively, refrigerate for 20 min. Once the tubes have cooled, tilt them to determine if the media has liquefied. Any degree of liquefaction indicates the presence of gelatinase while a solid media indicates the lack of the enzyme. Possible results for the gelatinase test are summarized in Table 85.1. Sketch the appearance of your tubes.

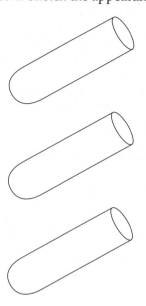

TABLE 85.1	Gelatinase test results and interpretations
Interpretation (Symbol)	**Appearance**
Gelatinase present (gelatinase +)	Media is liquid when cool (<25°C)
Gelatinase absent (gelatinase −)	Media is solid when cool (<25°C)

EXERCISE 85 REVIEW QUESTIONS

1. Another aspect used in evaluation of gelatinase activity is the pattern in which gelatinase liquefies the media, with parts of the media (the top, the area around the stab, etc.) becoming liquefied while other portions of the media remain solidified. How could the temperature at which the media is incubated have an effect on the pattern of gelatin liquefaction?

2. Why is agar typically used as a solidifying agent in microbiological media instead of gelatin? What could happen if gelatin was used?

REFERENCE

Reddy, C.A., Beveridge, T.J., Breznak, J.A., Marzluf, G.A., Schmidt, T.M., and Snyder, L.R. (eds.). 2007. *Methods for General and Molecular Microbiology, 3rd ed.*, chap. 15. Washington D.C.: ASM Press.

NOTES

DNase Test

DNase agar is a differential medium that distinguishes those bacteria that can enzymatically hydrolyze DNA from those that cannot. The medium contains DNA along with the dye methyl green, which can only bind to intact DNA. Plates of DNase agar are inoculated with a single streak and incubated for 24–48 h. After incubation, DNase-positive organisms will be surrounded by a clear halo in the green medium, a positive result. Organisms that are negative for DNase activity will have no halo.

PRINCIPLES AND APPLICATIONS

DNase agar is a differential medium that distinguishes those bacteria that enzymatically **hydrolyze** DNA from those that do not. DNA is a long molecule made up of tens of thousands of individual nucleotides bound end to end. Some bacteria produce DNases, enzymes that cleave the DNA molecule into short chains of 2–4 nucleotides in length. DNase agar contains DNA as a substrate for the DNase enzyme and the dye methyl green, which is able to bind to intact DNA molecules but cannot bind to the shorter oligonucleotides.

Plates of DNase agar are inoculated with a single heavy streak and incubated for 24–48 h. After incubation, organisms able to hydrolyze DNA (DNase +) will be surrounded by a clear halo, where methyl green is no longer able to bind to the hydrolyzed DNA. If an organism does not hydrolyze DNA (DNase −), the green color will be uniform throughout the media, running right to the edge of the bacterial growth (Figure 86.1). The DNase test is commonly used to help differentiate bacteria with similar Gram reactions from one another, such as *Staphylococcus aureus* from other staphylococci, *Serratia* from *Enterobacter,* and *Moraxella* from *Neisseria.*

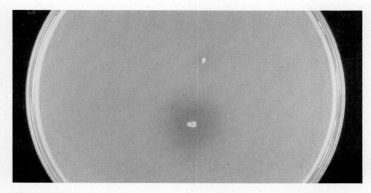

Figure 86.1 Interpretation of DNase media. A clear halo around the bacterial streak indicates the presence of DNase in the bacterium, a positive result. No halo is interpreted as a negative result.

PRE-LAB QUESTIONS

1. *Define the following terms:*

Hydrolysis

Oligonucleotide

2. *In what way are the casease test and the DNase test similar?*

PERIOD ONE

MATERIALS

Each student should obtain:
One DNase agar plate
Fresh broth cultures of:
   Staphylococcus aureus
   Staphylococcus epidermidis
Inoculating loop
Marking pen

PROCEDURE

1. Mark the bottom of your Petri dish with your name and lab time. Label one half of the dish with the name of each organism to be tested.
2. Use a loop to inoculate the surface of the agar with a single streak of the appropriate organism.
3. Incubate the tubes at 37°C for 48 h.

QUESTION—PERIOD ONE

1. How do you recognize a positive result in a DNase test?

PERIOD TWO
Retrieve your plate from the incubator. Evaluate the plate for growth of each organism and any clearing of the medium. The presence of a clear zone around an organism is a positive result for DNase activity while the lack of a clear zone is interpreted as a negative result. Possible results for DNase activity are summarized in Table 86.1. Sketch the appearance of your plate.

TABLE 86.1	DNase test results and interpretations
Interpretation (Symbol)	**Appearance**
DNase is present (DNase +)	Clear halo around growth
DNase is absent (DNase −)	No clear halo around growth (uniform green color throughout media)

EXERCISE 86 REVIEW QUESTION

1. Why is it necessary to ensure that bacteria have grown adequately on the media prior to checking for DNase activity?

REFERENCE

Reddy, C.A., Beveridge, T.J., Breznak, J.A., Marzluf, G.A., Schmidt, T.M., and Snyder, L.R. (eds.). 2007. *Methods for General and Molecular Microbiology, 3rd ed.*, chap. 15. Washington D.C.: ASM Press.

Lipase Test

The lipase test is used to distinguish bacteria that can enzymatically hydrolyze lipids from those that cannot. Spirit blue agar contains the lipid tributyrin along with spirit blue, a dye that is used to indicate lipolysis. Plates of spirit blue agar are inoculated with a single streak and incubated for 24–48 h. After incubation, lipase-positive organisms have a blue tint within the bacteria and are surrounded by a light halo where the medium has been depleted of fat. Lipase-negative organisms exhibit neither a halo nor any blue color within the growth of the bacteria.

PRINCIPLES AND APPLICATIONS

Fats and oils, more specifically known as **lipids,** serve as a ready source of energy for many cells. As is the case with proteins, starches, and other large biomolecules, the initial stages of lipid digestion take place outside the cell, where lipase enzymatically cleaves lipids to produce glycerol and fatty acids (Figure 87.1). Once inside the cell, glycerol eventually enters glycolysis while the breakdown products of fatty acids become substrates for the Krebs cycle. As lipase is not produced in every cell, it can be used as an aid in identification of many bacterial species.

Spirit blue agar contains peptone and yeast extract, which provide for the nutritional needs of most bacteria, while spirit blue serves as an indicator of lipolysis (i.e., hydrolysis of lipids). The media is supplemented with tributyrin, a simple fat, and polysorbate 80, an emulsifying agent that makes lipids more soluble in water.

Plates of spirit blue agar are inoculated with a single heavy streak and incubated for 24–48 h. After incubation, organisms able to lipolyze tributyrin will be surrounded by a clear halo, where oil droplets have been cleared from the media. The bacterial growth itself may in some cases be tinted blue as a result of a lowering of the pH of the media.

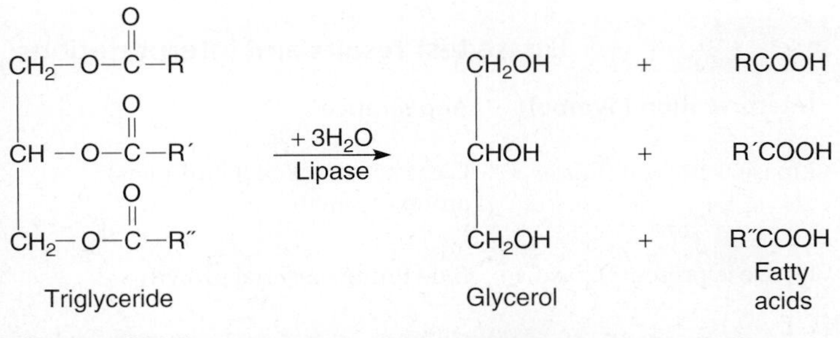

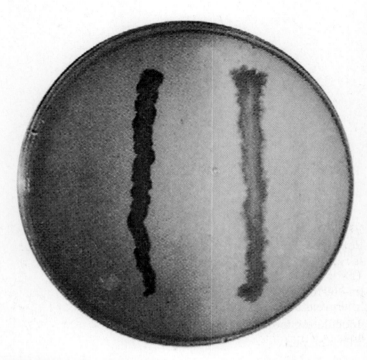

Figure 87.2 Lipid hydrolysis on spirit blue agar. Organism on the left is lipase positive, as indicated by the blue pigment. The organism on the right is lipase negative.

Either result is indicative of a positive reaction. Organisms that do not produce lipase will have neither a halo nor a blue tint within the bacterial growth (Figure 87.2).

PRE-LAB QUESTIONS

1. *Define the following terms:*

   Lipid

   Lipolysis

$$
\begin{array}{l}
\text{CH}_2\text{—O—}\overset{\displaystyle O}{\overset{\|}{\text{C}}}\text{—R} \\[4pt]
\text{CH—O—}\overset{\displaystyle O}{\overset{\|}{\text{C}}}\text{—R}' \\[4pt]
\text{CH}_2\text{—O—}\overset{\displaystyle O}{\overset{\|}{\text{C}}}\text{—R}''
\end{array}
\quad\xrightarrow[\text{Lipase}]{+\,3\text{H}_2\text{O}}\quad
\begin{array}{l}
\text{CH}_2\text{OH} \quad + \quad \text{RCOOH} \\[6pt]
\text{CHOH} \quad + \quad \text{R}'\text{COOH} \\[6pt]
\text{CH}_2\text{OH} \quad + \quad \text{R}''\text{COOH}
\end{array}
$$

Triglyceride   Glycerol   Fatty acids

Figure 87.1 Lipid hydrolysis (lipolysis). Lipase enzymatically cleaves each triglyceride molecule, resulting in the production of glycerol and three fatty acids.

2. *In what way are the starch hydrolysis test and the lipase test similar?*

2. What do you expect a negative result to look like?

PERIOD ONE

MATERIALS

Each student should obtain:
One spirit blue agar plate
Fresh broth cultures of:
  Staphylococcus aureus
  Proteus mirablis
Inoculating loop
Marking pen

PROCEDURE

1. Mark the bottom of your Petri dish with your name and lab time. Label one half of the dish with the name of each organism to be tested.
2. Use a loop to inoculate the surface of the agar with a single streak of the appropriate organism.
3. Incubate the tubes at 37°C for 48 h.

QUESTIONS—PERIOD ONE

1. How do you recognize a positive result in a lipase test?

PERIOD TWO

Retrieve your plate from the incubator. Evaluate the plate for growth of each organism. Carefully evaluate the plate for lipid hydrolysis as follows: (1) check for a lack of oil droplets immediately surrounding the bacterial growth on the plate, and (2) check for a blue tint to the bacterial growth. Either outcome is interpreted as a positive result. If no positive results are seen, the incubation may be continued for a total of 72 h. Possible results for lipase activity are summarized in Table 87.1. Sketch the appearance of your plate.

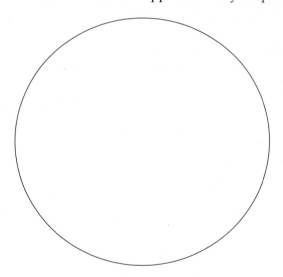

TABLE 87.1	Lipase test results and interpretations
Interpretation (Symbol)	**Appearance**
Lipase is present (lipase +)	Clear halo (lack of oil droplets) around growth
Lipase is present (lipase +)	Blue tint to bacterial growth
Lipase is absent (lipase −)	No clear halo (oil droplets are present) around growth

EXERCISE 87 REVIEW QUESTION

1. What causes the oil droplets to disappear from around the growth of a lipase-positive organism?

REFERENCE

Reddy, C.A., Beveridge, T.J., Breznak, J.A., Marzluf, G.A., Schmidt, T.M., and Snyder, L.R. (eds.). 2007. *Methods for General and Molecular Microbiology, 3rd ed.,* chap. 15. Washington D.C.: ASM Press.

NOTES

CAMP Test

The CAMP test is used to differentiate Streptococcus agalactiae *from other* Streptococcus sp.

PRINCIPLES AND APPLICATIONS

Streptococcus agalactiae, the most common representative of the group B streptococci, has been implicated in neonatal, skin, wound, and heart valve infections, and is a widespread nosocomial opportunist. In the laboratory, *S. agalactiae* may be presumptively identified through the use of the CAMP test. This test detects the production of a hemolytic protein called the **CAMP factor** (an acronym of the discoverers' last names) that interacts synergistically with the β-hemolysin produced by *Staphylococcus aureus* subspecies *aureus.* When the two organisms are streaked at right angles to one another on a blood agar plate, an arrowhead-shaped zone of hemolysis occurs where the two hemolysins interact. This is interpreted as a positive result (Figure 88.1).

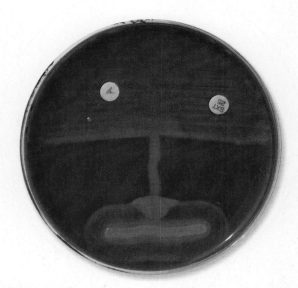

Figure 88.1 Positive CAMP reaction. The arrowhead-shaped zone of hemolysis is a result of the interaction between hemolysins produced by *Streptococcus agalactiae* (top) and *Staphylococcus aureus* (bottom). If the top organism was not *S. agalactiae,* a negative reaction would result in which hemolysis would be present around the growth of both bacteria but the arrowhead would be missing, as no synergistic interaction between the hemolysins would occur. The two discs in the photograph contain antibiotics that are also used in the classification of streptococci.

PRE-LAB QUESTION

1. *Define the following terms:*

   Hemolysin

   Synergism

   Nosocomial

PERIOD ONE

MATERIALS

Each student should obtain:
One blood agar plate
Fresh broth cultures of:
   Staphylococcus aureus subspecies *aureus*
   Unknown streptococcal species
Inoculating loop and needle
Laboratory marker

PROCEDURE

1. Streak your unknown bacteria heavily over approximately 40% of the plate and then down in a straight line to about 2 cm from the edge of the plate, as seen in Figure 88.2.
2. Use a needle to streak a single line of *S. aureus* subspecies *aureus* perpendicularly to the streak of your unknown organism. A gap of about 0.5 cm should separate the two streaks.
3. Incubate the plate at 37°C for 24 h.

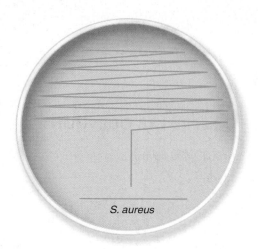

Figure 88.2 Streaking procedure for the CAMP test. Using a loop, the unknown organism is heavily streaked over approximately 40% of the media and then brought down in a straight line to within 2 cm of the edge of the plate. A needle is then used to streak a single thin line of *Staphylococcus aureus* subspecies *aureus* perpendicularly to the unknown, leaving a gap of about ½ cm between the two bacteria.

QUESTIONS—PERIOD ONE

1. How do you recognize a positive result in a CAMP test?

2. What does a positive result tell you about your unknown organism? What about a negative result?

PERIOD TWO

Retrieve your plate from the incubator. Look for the presence of an arrowhead-shaped region of hemolysis in the gap between the two organisms. This represents a positive reaction, and the organism can be presumptively identified as *S. agalactiae*. Sketch the appearance of your plate.

REFERENCE

Reddy, C.A., Beveridge, T.J., Breznak, J.A., Marzluf, G.A., Schmidt, T.M., and Snyder, L.R. (eds.). 2007. *Methods for General and Molecular Microbiology*, 3rd ed., chap. 15. Washington D.C.: ASM Press.

PYR Test

The PYR (L-pyrrolidonyl-β-naphthylamide) test identifies bacteria capable of enzymatically hydrolyzing L-pyrrolidonyl-β-naphthylamide to produce L-pyrrolidone and β-naphthylamine, the latter of which is easily detected through a color-producing secondary reaction. The test is used to presumptively identify group A streptococci and group D enterococci, both of which have a positive PYR reaction.

PRINCIPLES AND APPLICATIONS

The enzyme pyrrolidonyl peptidase catalyzes the **hydrolysis** of L-pyrrolidonyl-β-naphthylamide (PYR) to produce L-pyrrolidone and β-naphthylamine. This reaction is present in 98% of group A streptococci and 96% of group D enterococci while almost all other streptococcal species lack the enzyme to carry out this hydrolysis. Consequently, a positive reaction in the PYR test can serve as presumptive evidence that an unknown streptococcal species belongs to one of these two groups.

Several methods exist for performing the PYR test, but most labs use a rapid, commercially available method that can be completed in less than 5 min. In this technique, bacterial growth is applied to a piece of filter paper containing the PYR substrate and allowed to react for 2 min; a developing reagent (*p*-dimethyl-aminocinnamaldehyde) is then added to the paper. A deep pink color forming within 1 min after the addition of the reagent indicates a positive reaction while a yellow color (from the reagent) represents a negative reaction (Figure 89.1).

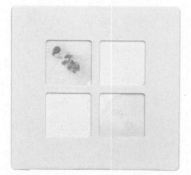

Figure 89.1 PYR DrySlide results. The upper left quadrant displays the pink color characteristic of a positive reaction while the lower right quadrant displays only the yellow color of the developing reagent, a negative reaction.

PRE-LAB QUESTIONS

1. *What does the term hydrolyze mean?*

2. *How do you recognize a positive result in a PYR test?*

MATERIALS

This procedure utilizes PYR DrySlides from Becton Dickinison Microbiology Systems. Your instructor will tell you if another procedure is to be used.

Each student should obtain:
One PYR DrySlide [1]
Sterile distilled water
Fresh slant cultures of:
    Enterococcus faecalis
    Streptococcus agalactiae
    Unknown streptococcal organism
Sterile wooden applicator sticks
Inoculating loop
Marking pen

PROCEDURE

1. Obtain a DrySlide and label it along the border, indicating which square corresponds to each of your three organisms.
2. For each organism, prepare the test as follows:
   • Apply a single loopful of sterile water to the filter paper in each square.
   • Using a wooden applicator stick, apply a heavy inoculum of the bacteria to be tested to the moistened area of the filter paper.
3. Allow the DrySlide to incubate for 2 min at room temperature.
4. Holding a dispenser of development solution upright and facing away from you, crush the glass ampoule inside between your thumb and forefinger. Apply 1 drop of the development solution to the inoculated area of the slide.
5. The appearance of a pink or fuchsia color within 1 min is considered a positive reaction.
6. Possible results for the PYR test are summarized in Table 89.1. Sketch the appearance of your PYR test.

TABLE 89.1	PYR test results and interpretations
Interpretation (Symbol)	**Appearance**
Pyrrolidonyl peptidase present (PYR +)	Pink color within 1 min
Pyrrolidonyl peptidase absent (PYR −)	No pink color within 1 min (yellow)

EXERCISE 89 REVIEW QUESTIONS

1. What other laboratory test would you like to know the result of before you conduct a PYR test?

2. What is the purpose of including known PYR-positive and PYR-negative organisms on the DrySlide?

1. Becton Dickinson Microbiology Systems. BBL DrySlide PYR Kit 231747. Sparks, Maryland. www.bd.com/ds/technicalCenter/inserts/8820381(0604).pdf.

REFERENCE

Forbes, B.A., Sahm, D.F., and Weissfeld, A. 2007. *Bailey and Scott's Diagnostic Microbiology, 12th ed.,* p. 111. St. Louis: Mosby-Elsevier.

API 20E System

The API 20E system is a multiple test system allowing the determination of 20 different biochemical tests nearly simultaneously. The tests chosen for inclusion are those most important in identifying Gram-negative bacteria, especially those found within the family Enterobacteriaceae.

PRINCIPLES AND APPLICATIONS

Differentiation of bacteria belonging to a single group, such as the family Enterobacteriaceae, can generally be accomplished by determining the reaction of an organism in a relatively small number of tests, often less than 20. Because the same tests are used time and again when these organisms are isolated from clinical specimens, commercial systems have been introduced that allow a common series of tests to be run very quickly, enhancing efficiency and speed dramatically. Several different multiple test systems exist, each containing the reagents needed for identification of a bacterial isolate within a specific group, that is, Gram-negative enterics, Gram-positive cocci, etc.

The **API (analytical profile index) 20E** system consists of a set of tests commonly used for identification of Gram-negative bacteria, such as *Escherichia, Salmonella,* and *Shigella.* All of the tests contained within the system have been discussed earlier in this book (Table 90.1). The test itself consists of a plastic strip with 20 separate compartments, each consisting of a depression or cupule, and a small tube containing dehydrated media specific for a single test. To inoculate each compartment, a saline suspension of a single isolated bacterial colony is made and added to each tube or cupule. The suspension flows through the cupule, filling the tube and allowing the saline to reconstitute the media. For those reactions requiring anaerobic conditions, the compartment is covered with a layer of sterile mineral oil.

After incubation for 18–24 h, the reaction for each test is recorded (Figure 90.1), which requires the addition of test reagents to some compartments. Finally, the test results are tabulated. In this process, positive tests are assigned a numerical value, and summation of these values results in every organism having a unique seven- or nine-digit profile number. The makers of the API 20E test provide several methods of identifying an organism based on the biochemical profile produced. Check with your lab instructor to see which resources are available to you. This exercise contains an identification table that can be used to identify your unknown organism if other resources are not available.

PRE-LAB QUESTIONS

1. *What piece of information about your unknown should you have prior to using the API 20E system?*

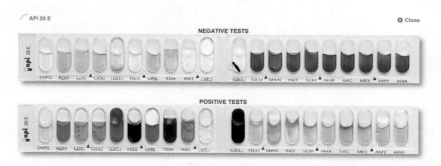

Figure 90.1 Negative and positive results of an API 20E test.

2. *How are aerobic reactions prepared differently than anaerobic reactions in the API 20E test?*

3. *Could this system be used to identify the organisms in a mixed culture?*

PERIOD ONE

MATERIALS

Each student should obtain:

Fresh slant or plate culture of your unknown organism

Test tube containing 5 ml of 0.85% sterile saline

API 20E test strip[1]

API incubation tray and cover

Sterile mineral oil

5-ml Pasteur pipette

Oxidase test system (Exercise 73)

10% ferric chloride

Barritt's reagent A (α-naphthol) and B (potassium hydroxide)

Kovac's reagent or James' reagent

Nitrate test reagents A (sulfanilic acid) and B (dimethyl-alpha-naphthylamine)

Zinc dust (nitrate reagent C)

Inoculating loop

Marking pen

PROCEDURE

This procedure is illustrated in Figure 90.2.

1. Prior to inoculating the API 20E test, you must have already established that your bacterium is Gram negative. Perform a Gram stain if this has not already been done.

2. Prepare an incubation tray for your test strip by adding approximately 5 ml of distilled water to the wells of the incubation tray (Figure 90.2, step 1).

3. Label the flap of the tray with your name and lab time as well as the identification number of your unknown.

4. Remove the API 20E test strip from its packaging, and place it in the incubation tray (Figure 90.2, step 2).

5. Prepare a saline suspension of your organism by transferring a single well-isolated colony to a tube of 0.85% saline solution. Pipette up and down several times to ensure that the bacteria are uniformly dispersed throughout the saline (Figure 90.2, step 3).

6. Immediately use the saline suspension to inoculate the test strip by filling the tube and or cupule of each test as follows: (Figure 90.2, step 4)

   • Slightly underfill the tubes that are *underlined*, that is, ADH, LDC, ODC, H₂S, and URE. Underfilling these tubes leaves room for the addition of oil to create an anaerobic environment.

   • Completely fill the cupule and tube of the *bracketed reactions*, that is, |CIT|, |VP|, and |GEL|. These reactions all require oxygen to proceed.

   • Fill the tubes, but not the cupules, for all the tests not indicated above.

   • To provide anaerobic conditions, add sterile mineral oil to the cupules of the ADH, LDC, ODC, H₂S, and URE compartments. (Figure 90.2, step 5)

7. Place the lid on the incubation tray and incubate the plate for 18–24 h at 37°C. If evaluation at 24 h is impossible, the trays should be removed from the incubator after 24 h and refrigerated until the next lab period.

QUESTION—PERIOD ONE

1. What type of media could help to ensure that a Gram-positive organism was not inadvertently used to inoculate an API 20E test?

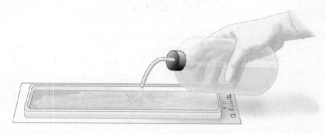

(1) After labeling the end tab of a tray with your name and unknown number, dispense approximately 5 ml of distilled water into bottom of tray.

(2) Place an API 20E test strip into the bottom of the moistened tray. Be sure to seal the pouch from which the test strip was removed to prevent contamination of remaining strips.

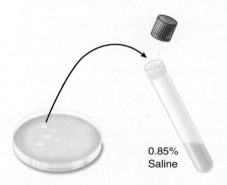

(3) Select one well-isolated colony to make a saline suspension of the unknown organism. Suspension should be well dispersed with a Vortex mixer.

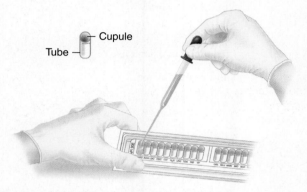

(4) Dispense saline suspension of organisms into cupules of all 20 compartments. Slightly *underfill* ADH, LDC, ODC, H₂S, and URE. *Completely fill* cupules of CIT, VP, and GEL.

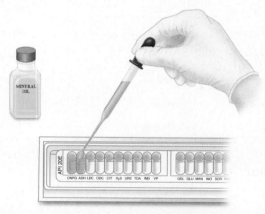

(5) To provide anaerobic conditions for chambers ADH, LDC, ODC, H₂S, and URE, completely fill cupules of these chambers with sterile mineral oil.

(6) After incubation and after adding test reagents to four compartments, record all the results and total numbers to arrive at 7-digit code. Consult the *Analytical Profile Index* to find the unknown.

Figure 90.2 Procedure for preparing and inoculating the API 20E test strip.

TABLE 90.1	Tests performed as part of the API 20E system			
API 20E Abbreviation	Biochemical substrate, reaction, or enzyme	Color negative reaction	Color postive reaction	See Exercise No.
ONPG	o-Nitrophenyl-β-D-galactopyranoside	Colorless	Yellow[1]	82
ADH	Arginine dihydrolase	Yellow	Red/orange[2]	78
LDC	Lysine decarboxylase	Yellow	Red/orange[2]	78
ODC	Ornithine decarboxylase	Yellow	Red/orange[2]	78
CIT	Sodium citrate utilization	Pale green/yellow	Blue green/blue[3]	76
H₂S	Sulfur reduction	Colorless/grayish	Black deposit/thin line	63
URE	Urease	Yellow	Red/orange[2]	83
TDA	Tryptophan deaminase	Yellow[4]	Reddish brown[4]	63
IND	Indole production	Colorless/pale green/ yellow[5]	Pink[5]	63
VP	Voges-Proskauer	Colorless[6]	Pink/red[6]	71
GEL	Gelatinase	No diffusion	Diffusion of black pigment	85
GLU	Glucose fermentation/oxidation[7]	Blue/blue green	Yellow	69 and 70[8]
MAN	Mannitol fermentation/oxidation[7]	Blue/blue green	Yellow	69 and 70[8]
INO	Inositol fermentation/oxidation[7]	Blue/blue green	Yellow	69 and 70[8]
SOR	Sorbitol fermentation/oxidation[7]	Blue/blue green	Yellow	69 and 70[8]
RHA	Rhamnose fermentation/oxidation[7]	Blue/blue green	Yellow	69 and 70[8]
SAC	Sucrose fermentation/oxidation[7]	Blue/blue green	Yellow	69 and 70[8]
MEL	Melibiose fermentation/oxidation[7]	Blue/blue green	Yellow	69 and 70[8]
AMY	Amygdalin fermentation/oxidation[7]	Blue/blue green	Yellow	69 and 70[8]
ARA	L-arabinose	Blue/blue green	Yellow	69 and 70[8]
OX	Cytochrome oxidase	Yellow	Purple	73
Nitrate reduction[9, 10]	• NO₂⁻ production • Reduction to N₂ gas	• Yellow[11] • Orange red[12]	• Red[11] • Yellow[12]	74
MOB[10]	Motility	Nonmotile	Motile	48 and 62
McC[10]	MacConkey agar	Growth −	Growth +	55
OF-F[10]	Oxidation-fermentation of glucose	Green[13]	Yellow[13]	68
OF-O[10]	Oxidation-fermentation of glucose	Green[14]	Yellow[14]	

[1]A very pale yellow should also be considered positive.
[2]An orange color after 36–48 h incubation must be considered negative.
[3]Reading made in the cupule (aerobic).
[4]Read color immediately after addition of TDA reagent (10% ferric chloride).
[5]Read the color immediately after addition of Kovac's or James' reagent.
[6]Read the color 10 min after addition of Barritt's reagent A and B. A slight pink color after 10 min should be considered negative.
[7]Fermentation begins in the lower portion of the tubes while oxidation begins in the cupule.
[8]The pH indicator used in the API strip for these tests is different from the one used for the same tests in this manual, leading to different colors for positive and negative results.
[9]Use the GLU tube after reading the results of glucose fermentation.
[10]Tests completed separately from the API 20E test strip, only if needed.
[11]Read the color 2–5 min after addition of nitrate reagents A and B.
[12]Read the color 5 min after the addition of zinc.
[13]Color for fermentation results is read under mineral oil.
[14]Color for oxidation result is read exposed to air.

PERIOD TWO

1. Retrieve your test strip from the incubator.

2. Prior to adding any reagents, evaluate the test by referring to Table 90.1. If fewer than three tests are positive, reincubate the strip for an additional 24 h without adding any reagents. At the end of the additional incubation, continue with the next step.

3. If three or more reactions are positive, record the results of all of the reactions that do not require the addition of reagents (i.e., all of the reactions except TDA, IND, and VP) in the Results section (as seen in Figure 90.2, step 6).

4. Add one drop of 10% ferric chloride to the TDA tube. A positive reaction is reddish/brown and will occur immediately in tryptophan deaminase-positive organisms. If no color change occurs, the test is negative.

5. Add one drop each of Barritt's A and B reagent to the VP tube. A pink or red color within 10 min is a positive reaction. A pale pink color, either immediately or after 10 min, should be considered a negative reaction for the Voges-Proskauer test.

6. Add one drop Kovac's or James' reagent to the IND tube. The appearance of a red color within 2 min is a positive reaction for indole production. Ignore any brownish-red color that appears after several minutes.

7. After recording all of your test results, be sure to dispose of the API 20E in a biohazard container.

RESULTS

Record your results as follows:

1. Indicate a positive result with a (+) and a negative result with a (−) for each of the 20 tests making up the API 20E. If your organism was oxidase positive, record a (+) in last column on the right side of the table (marked OXI). Tests are grouped in sets of three with assigned values of 1, 2, or 4. A positive reaction is given that value assigned to it, a negative reaction's value is always 0. For example, ONPG, ADH, (arginine dehydrolase), and LDC (lysine decarboxylase) are grouped together and are assigned 1, 2, and 4 points, respectively. For each group of three tests, add the individual numbers together to arrive at a number between 0 and 7. Continuing with the same example, if positive test results were seen for ONPG and LDC, then the values 1 and 4 would be added together to arrive at 5. This number would then be placed in the box below this set of tests. The series of seven numbers, read from left to right, is the seven-digit profile of your unknown organism (Figure 90.3). Place the seven-digit profile of your bacterium here:

_____

2. Use the manufacturer's provided API index if it is available in your laboratory. If not, use Table 90.2 to identify your organism.

3. If identification is not possible using the seven-digit profile, a nine-digit profile can be constructed by performing the following tests:

   - Nitrate reduction and the production of nitrogen gas (Exercise 74)
   - Motility (Exercise 48)
   - Growth on MacConkey agar (Exercise 55)
   - Fermentation of glucose (Exercise 68)
   - Oxidation of glucose (Exercise 68)

Discuss this option with your instructor if you are having difficulty identifying your organism.

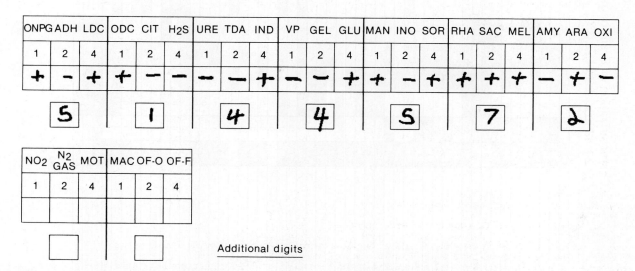

Figure 90.3 Example of API 20E profile construction. Each test result is recorded as positive or negative. Negative test results have a point value of zero while positive results are worth one, two, or four points. The oxidase test is recorded as the twenty-first test in the series. The positive results in each group of three tests are added together to arrive at a single digit. The seven single digits, when read left to right, make up the seven-digit profile of the organism. If identification is impossible based on the seven-digit code, an additional set of tests can be done to produce a nine-digit code. Here, the seven-digit code generated (5144572) identified the unknown organism as *E. coli*. No additional tests were needed.

TABLE 90.2 Characterization of Gram-negative rods—the API 20E identification system (percent of positive reactions after 18–24/48 hrs. at 36°C ± 2°C)

API 20E (V4.1)	ONPG	ADH	LDC	ODC	CIT	H2S	URE	TDA	IND	VP	GEL	GLU	MAN	INO	SOR	RHA	SAC	MEL	AMY	ARA	OX	NO2	N2	MOB	McC	OF/O	OFF
Buttiauxella agrestis	100	0	0	85	25	0	0	0	0	0	0	100	100	0	1	99	0	92	99	100	0	100	0	100	100	100	100
Cedecea davisae	99	89	0	99	75	0	0	0	0	89	0	100	100	10	0	0	100	0	100	1	0	99	0	87	100	100	100
Cedecea lapagei	99	99	0	0	75	0	0	0	0	90	0	100	99	0	0	100	1	1	100	99	0	99	0	87	100	100	100
Citrobacter braakii	50	45	0	99	75	81	1	0	4	0	0	100	100	1	100	99	1	91	99	99	0	100	0	95	100	100	100
Citrobacter freundii	90	24	0	0	75	75	1	0	0	0	0	100	99	25	99	99	99	82	40	99	0	98	0	95	100	100	100
Citrobacter koseri/amalonaticus	99	75	0	100	97	0	1	0	99	0	0	100	100	25	99	99	1	99	98	99	0	99	0	95	100	100	100
Citrobacter koseri/farmeri	99	2	0	100	25	0	1	0	99	0	0	100	100	1	95	100	99	80	99	99	0	100	0	95	100	100	100
Citrobacter youngae	100	50	0	1	80	80	0	0	1	0	0	100	100	1	95	100	1	90	25	100	0	85	0	95	100	100	100
Edwardsiella hoshinae	0	0	100	99	50	94	0	0	99	0	0	100	100	0	0	1	100	0	0	1	0	100	0	100	100	100	100
Edwardsiella tarda	0	0	100	99	1	75	0	0	99	0	0	100	0	0	0	1	0	0	0	0	0	100	0	98	100	100	100
Enterobacter aerogenes	99	0	99	98	82	0	1	0	0	85	0	99	99	99	99	99	99	99	99	99	0	100	0	97	100	100	100
Enterobacter amnigenus 1	99	25	0	99	40	0	0	0	0	75	0	100	100	1	1	100	1	99	99	99	0	100	0	92	100	100	100
Enterobacter amnigenus 2	99	80	0	99	80	0	0	0	0	75	0	100	100	0	100	100	99	99	100	99	0	100	0	100	100	100	100
Enterobacter asburiae	100	25	0	99	80	0	0	0	0	10	0	99	100	25	100	0	99	0	100	100	0	99	0	95	100	100	100
Enterobacter cancerogenus	98	75	0	99	99	0	1	0	0	89	0	99	100	1	1	100	1	90	100	100	0	100	0	99	100	100	100
Enterobacter cloacae	99	82	1	92	90	0	1	0	0	85	0	99	99	12	90	85	96	90	99	99	0	100	0	90	100	100	100
Enterobacter gergoviae	99	0	32	100	75	0	99	0	0	90	0	100	99	23	1	100	99	100	99	100	0	100	0	92	100	100	100
Enterobacter intermedius	99	0	0	99	1	0	0	0	0	2	0	100	97	88	88	99	40	100	99	99	0	100	0	96	100	100	100
Enterobacter sakazakii	100	96	0	91	94	0	0	0	25	91	10	100	100	75	1	99	99	99	99	99	0	100	0	95	100	100	100
Escherichia coli 1	90	1	74	70	0	0	3	0	89	0	0	99	98	1	91	82	36	75	3	99	0	98	0	95	100	100	100
Escherichia coli 2	26	1	45	20	0	0	1	0	50	0	0	99	90	1	42	30	3	3	1	70	0	98	0	5	100	100	100
Escherichia fergusonii	96	1	99	100	1	0	1	0	99	0	0	100	99	1	0	87	0	1	99	99	0	100	0	93	100	100	100
Escherichia hermannii	100	30	0	100	1	0	0	0	99	0	0	100	100	0	1	95	25	0	95	99	0	100	0	99	100	100	100
Escherichia vulneris	98	0	50	0	0	0	0	0	0	0	0	99	100	0	1	95	7	95	50	99	0	100	0	60	100	100	100
Ewingella americana	75	0	0	0	75	0	0	0	0	95	1	99	99	0	0	1	0	1	50	1	0	100	0	85	100	100	100
Hafnia alvei 1	50	0	99	98	50	0	10	0	0	50	0	99	99	0	0	99	0	0	25	99	0	100	0	85	100	100	100
Hafnia alvei 2	99	0	99	99	1	0	1	0	0	10	0	99	99	0	1	1	1	0	1	99	0	100	0	0	100	100	100
Klebsiella oxytoca	99	1	80	0	89	0	78	0	99	80	0	100	100	99	100	99	99	100	100	100	0	100	0	0	100	100	100
Klebsiella pneumoniae ssp ozaenae	94	18	25	1	18	0	1	0	1	9	1	99	96	57	66	58	20	80	97	85	0	92	0	0	100	100	100
Klebsiella pneumoniae ssp pneumoniae	99	1	73	0	86	0	75	0	1	62	0	100	99	99	99	99	99	99	99	99	0	100	0	0	100	100	100
Klebsiella pneumoniae ssp rhinoscleromatis	1	0	0	0	0	0	0	0	0	90	0	99	100	90	90	75	75	1	99	10	0	100	0	0	100	100	100
Kluyvera spp	95	0	25	99	60	0	0	0	80	0	0	100	99	0	25	93	89	99	99	99	0	95	0	94	100	100	100
Leclercia adecarboxylata	99	0	0	0	1	0	1	0	99	1	1	100	99	2	0	100	66	99	99	100	0	90	0	0	100	100	100
Moellerella wisconsensis	97	0	0	98	40	0	0	0	15	1	0	99	1	0	0	0	100	99	0	0	0	88	0	0	100	100	100
Morganella morganii	85	1	0	98	13	0	99	93	99	0	0	99	0	1	0	0	1	0	0	0	0	85	0	95	100	100	100
Pantoea spp 1	99	1	0	0	99	0	1	0	1	9	1	99	99	1	26	1	98	26	59	61	0	85	0	85	100	100	100
Pantoea spp 2	99	1	0	0	99	0	1	0	53	62	4	100	99	36	82	90	98	81	99	99	0	85	0	85	100	100	100
Pantoea spp 3	99	1	0	0	21	0	1	0	1	86	15	100	99	34	82	97	93	23	65	97	0	85	0	85	100	100	100
Pantoea spp 4	86	1	0	0	29	0	1	0	59	1	0	99	100	10	32	99	72	89	99	99	0	93	0	95	100	100	100
Proteus mirabilis	1	0	0	99	50	75	99	98	1	1	82	98	3	0	0	0	15	0	0	0	0	95	0	94	100	100	100
Proteus penneri	1	0	0	0	1	20	99	99	0	99	50	98	0	0	0	0	100	0	0	0	0	99	0	85	100	100	100
Proteus vulgaris group	1	0	0	0	12	83	99	99	92	0	74	99	1	1	0	1	89	0	66	1	0	99	0	94	100	100	100
Providencia alcalifaciens/rustigianii	0	0	0	0	80	0	0	99	99	0	0	100	1	1	0	0	0	0	0	0	0	100	0	96	100	100	100
Providencia rettgeri	1	1	0	0	74	0	99	99	90	0	0	98	82	78	1	50	1	0	40	1	0	98	0	94	100	100	100
Providencia stuartii	1	0	0	0	85	0	30	98	95	0	0	98	3	80	0	0	15	0	0	0	0	100	0	85	100	100	100
Rahnella aquatilis	100	0	0	0	50	0	0	0	0	99	0	100	100	0	98	99	100	97	100	98	0	99	0	6	100	100	100
Raoultella ornithinolytica	100	0	99	99	99	0	85	0	100	65	0	100	100	99	100	99	100	100	100	100	0	100	0	0	100	100	100
Raoultella terrigena	98	75	99	6	52	0	0	0	0	75	0	100	100	99	99	99	1	78	100	99	0	98	0	0	100	100	100
Salmonella choleraesuis ssp arizonae	98	75	97	98	75	99	0	0	1	0	0	100	100	0	98	99	1	99	100	99	0	100	0	99	100	100	100
Salmonella choleraesuis ssp choleraesuis	0	15	99	99	6	64	0	0	0	0	0	100	100	0	98	99	0	20	1	99	0	100	0	95	100	100	100
Salmonella ser. Gallinarum	0	1	100	1	0	25	0	0	0	0	0	100	100	0	0	1	0	0	0	100	0	100	0	0	100	100	100

API 20 E V4.1	ONPG	ADH	LDC	ODC	CIT	H2S	URE	TDA	IND	VP	GEL	GLU	MAN	INO	SOR	RHA	SAC	MEL	AMY	ARA	OX	NO2	N2	MOB	McC	OF/O	OF/F
Salmonella ser.Paratyphi A	0	5	0	99	0	1	0	0	0	0	0	100	99	0	99	98	0	96	0	99	0	100	0	95	100	100	100
Salmonella ser.Pullorum	0	1	75	100	0	85	0	0	0	0	0	100	100	0	0	100	0	99	0	75	0	100	0	0	100	100	100
Salmonella typhi	0	1	82	0	0	8	0	0	0	0	0	100	99	40	99	0	1	99	0	0	0	100	0	97	100	100	100
Salmonella spp	1	56	82	93	65	83	0	0	1	0	1	100	100	40	99	86	1	90	1	99	1	100	0	95	100	100	100
Serratia ficaria	99	0	0	0	100	0	0	0	0	40	90	100	100	50	99	74	99	99	100	99	0	92	0	100	100	100	100
Serratia fonticola	99	0	73	99	75	0	0	0	0	0	0	100	100	97	99	99	30	99	99	99	0	99	0	91	100	100	100
Serratia liquefaciens	95	1	78	98	80	0	0	0	0	59	65	100	99	80	98	2	99	72	97	97	0	100	0	95	100	100	100
Serratia marcescens	94	0	95	95	96	0	0	0	1	70	87	100	99	85	98	1	99	68	97	25	0	95	0	97	100	100	100
Serratia odorifera 1	95	0	95	99	95	0	2	0	50	50	99	100	99	99	99	99	99	99	99	99	0	99	0	100	100	100	100
Serratia odorifera 2	95	0	96	1	95	0	25	0	99	50	99	100	99	99	99	1	99	99	99	95	0	99	0	100	100	100	100
Serratia plymuthica	99	0	0	0	65	0	0	0	0	65	50	100	90	70	70	1	99	85	98	98	0	99	0	50	100	100	100
Serratia rubidaea	99	0	30	0	92	0	1	0	0	71	82	99	99	75	1	3	99	95	99	98	0	100	0	85	100	100	100
Shigella spp	1	0	0	1	0	0	0	0	29	0	0	99	63	0	7	7	1	20	0	50	0	100	0	0	100	100	100
Shigella sonnei	96	0	0	93	0	0	0	0	0	0	0	99	99	0	1	75	1	0	1	99	0	100	0	0	99	99	99
Yersinia enterocolitica	80	0	0	90	0	0	98	0	50	5	0	99	99	25	98	1	99	4	0	75	0	98	2	0	99	99	99
Yersinia frederiksenii/intermedia	99	0	0	75	1	0	99	0	99	1	0	100	99	25	99	1	99	1	75	99	0	98	5	2	99	99	99
Yersinia kristensenii	80	0	0	80	0	0	99	0	97	1	0	100	99	10	99	99	0	1	99	99	0	98	5	5	100	100	100
Yersinia pestis	68	0	0	0	0	0	0	0	0	0	0	99	99	70	70	0	0	0	30	99	0	47	0	0	99	100	100
Yersinia pseudotuberculosis	98	0	0	0	1	0	99	0	0	1	0	99	97	0	0	75	0	50	30	99	0	95	0	0	99	99	99
Aeromonas hydrophila gr. 1	98	90	25	1	25	0	0	0	85	25	90	99	99	1	3	5	97	0	75	50	100	97	0	95	100	99	99
Aeromonas hydrophila gr. 2	99	97	80	1	80	0	0	0	85	80	97	97	99	1	9	1	80	0	75	75	100	97	0	99	99	99	99
Aeromonas salmonicida ssp salmonicida	1	60	1	0	0	0	0	0	0	0	75	50	54	0	9	1	0	1	1	5	100	98	0	1	99	99	99
Grimontia hollisae	1	0	75	0	1	0	0	0	94	0	75	10	0	0	0	0	1	0	1	0	100	100	0	25	99	99	99
Photobacterium damselae	1	99	75	1	1	0	98	0	0	10	1	50	0	0	0	0	0	0	0	0	100	100	0	0	99	99	99
Plesiomonas shigelloides	95	99	100	100	0	0	1	0	100	0	0	99	0	99	0	0	0	25	0	0	100	99	0	95	99	99	99
Vibrio alginolyticus	0	0	98	75	60	0	1	0	100	75	75	99	100	0	0	0	100	10	99	1	100	47	0	100	94	94	94
Vibrio cholerae	98	1	94	97	60	0	0	0	99	58	92	98	98	1	1	0	94	5	10	1	100	96	0	100	94	94	94
Vibrio fluvialis	95	99	0	0	1	0	0	0	80	1	75	75	80	0	1	0	75	0	5	75	100	100	0	100	99	99	99
Vibrio mimicus	99	99	99	99	1	0	0	0	80	1	75	75	99	1	1	0	1	0	36	0	100	100	0	100	96	99	99
Vibrio parahaemolyticus	0	0	100	99	50	0	1	0	100	1	75	100	99	0	0	1	1	0	12	50	100	95	0	100	95	99	99
Vibrio vulnificus	99	0	91	90	25	0	0	0	99	1	99	99	75	0	0	0	1	0	90	0	99	63	0	100	98	99	99
Pasteurella aerogenes	4	0	0	80	0	0	99	0	0	0	0	99	0	0	0	0	99	0	0	75	75	54	0	0	99	100	100
Pasteurella multocida 1	7	0	0	25	0	0	0	0	99	0	0	29	1	97	1	0	75	0	0	75	99	100	0	0	100	100	100
Pasteurella multocida 2	60	0	1	0	0	0	0	0	80	0	0	44	99	12	99	0	99	0	0	0	89	90	0	0	96	99	99
Pasteurella pneumotropica/ Mannheimia haemolytica	0	0	0	0	0	0	25	0	15	0	0	35	12	12	12	1	35	0	1	1	80	99	0	0	100	100	100
Acinetobacter baumannii/calcoaceticus	0	0	0	0	51	0	0	0	0	0	0	0	0	0	0	0	0	0	0	0	0	3	0	0	2	23	23
Bordetella/Alcaligenes/Moraxella spp *	50	0	0	0	52	0	14	1	0	0	0	0	0	0	0	0	0	0	0	0	95	62	1	75	75	23	33
Burkholderia cepacia	0	0	25	16	78	0	0	0	0	0	1	60	1	99	1	0	13	0	7	20	90	40	0	99	88	98	0
Chromobacterium violaceum	0	99	0	0	75	0	0	0	14	43	99	99	0	0	0	0	10	0	0	0	90	75	0	99	99	99	0
Chryseobacterium indologenes	5	0	0	0	12	0	0	0	75	0	80	0	0	0	0	0	10	0	0	0	99	75	0	0	57	90	6
Chryseobacterium meningosepticum	77	0	0	0	20	0	1	0	85	0	90	0	0	0	0	0	0	0	0	0	99	6	0	0	48	93	10
Eikenella corrodens	0	0	75	99	0	0	0	0	0	0	0	0	0	0	0	0	0	0	0	0	100	95	0	1	1	6	6
Myroides (Chryseobacterium indologenes)	0	0	0	0	50	0	75	0	1	0	100	0	0	0	0	0	0	0	0	0	99	0	0	0	84	49	49
Ochrobactrum anthropi	15	0	0	0	30	0	75	1	0	0	0	1	0	0	0	0	0	0	0	10	90	42	60	99	99	2	2
Pseudomonas aeruginosa	0	89	0	0	92	0	25	0	0	0	75	50	0	0	0	0	0	0	0	25	97	12	56	97	100	98	0
Pseudomonas fluorescens/putida	0	75	0	0	75	0	25	0	27	10	75	25	0	0	0	0	1	0	20	20	99	26	0	100	96	93	0
Pseudomonas luteola	86	75	0	0	94	0	0	0	0	25	13	84	0	0	0	1	15	15	1	85	0	30	0	100	91	94	0
Pseudomonas oryzihabitans	1	1	0	0	89	0	0	0	15	25	9	10	0	0	0	1	1	10	1	45	0	7	0	100	99	99	0
Non-fermenter spp	1	1	0	0	37	1	1	0	0	0	9	1	0	0	0	1	1	0	1	1	93	48	35	1	85	49	49
Shewanella putrefaciens group	0	0	80	80	75	75	1	0	0	0	75	1	0	0	0	0	0	0	0	2	93	96	9	99	96	49	9
Stenotrophomonas maltophilia	70	0	75	1	75	1	0	0	0	0	90	1	0	0	0	0	0	0	0	0	1	26	1	100	91	0	0

Courtesy of bioMérieux, Inc.

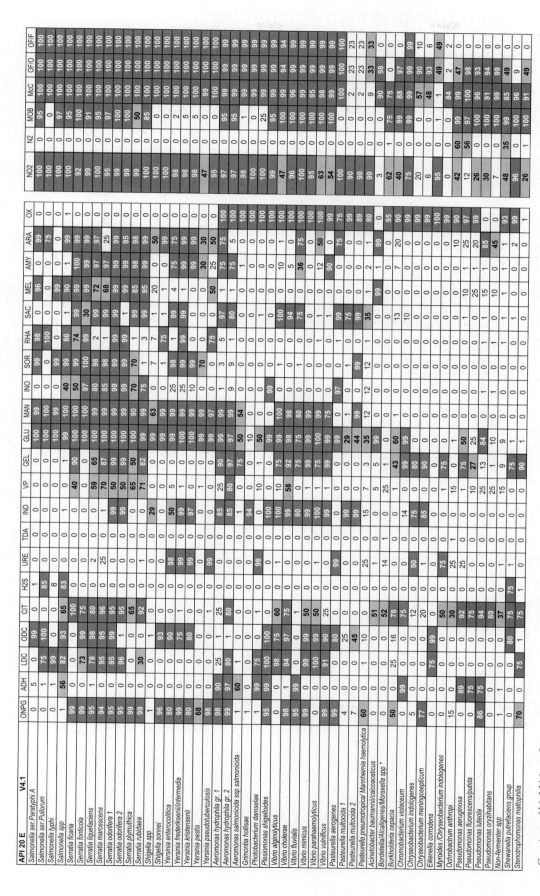

EXERCISE 90 REVIEW QUESTIONS

1. Why is the API 20E only useful for identifying Gram-negative rods and not other bacterial species as well?

2. The API 20E test strip has been designed to identify enteric bacteria in a clinical setting. With this is mind, design your own test strip to differentiate among the species of *Staphylococcus, Streptococcus,* and *Micrococcus,* using the flowcharts presented in Exercise 39 to help you design your test strip.

3. After inoculating all the tests in the strips, a streak-plate is often done using the same saline suspension used to inoculate the tests strip to ensure that a pure culture was used to inoculate the strip. What is the importance of doing this?

1. bioMérieux. *API (Clinical Microbiology Product).*
   http://www.biomerieux-usa.com/servlet/srt/bio/usa/dynPage?open=USA_PRD_LST&doc=USA_PRD_LST_G_PRD_USA_5&lang=en.

Enterotube II System

The Enterotube II system is a multiple test system allowing the determination of 15 different biochemical tests nearly simultaneously. The tests chosen for inclusion are those most important in identifying Gram-negative bacteria, especially those found within the family Enterobacteriaceae.

PRINCIPLES AND APPLICATIONS

Differentiation of bacteria belonging to a single group, such as the family Enterobacteriaceae, can generally be accomplished by determining the reaction of an organism in a relatively small number of tests, often less than 20. Because the same tests are used time and again when these organisms are isolated from clinical specimens, commercial systems have been introduced that allow a common series of tests to be run very quickly, enhancing efficiency and speed dramatically. Several different multiple test systems exist, each containing the reagents needed for identification of a bacterial isolate within a specific group, that is, Gram-negative enterics, Gram-positive cocci, etc.

The **Enterotube II system** consists of a set of tests commonly used for identification of Gram-negative, oxidase-negative bacteria, and all of the tests contained within the system have been discussed earlier in this book (Table 91.1). The test itself consists of a plastic tube with 12 separate compartments, each containing a different agar-based medium. Compartments that require anaerobic conditions have layers of paraffin wax over the media while those requiring aerobic conditions have small holes, allowing oxygen to enter. Extending through the entire length of the tube is an inoculating wire that is used to pick up a sample of bacteria. The wire is then withdrawn, pulling the sample through the tube and inoculating the media in each compartment.

After incubation, the reactions in all of the compartments are noted (Figure 91.1); an indole test is performed, and a Voges-Proskauer test may also be performed. Based on the number of positive reactions, a five-digit numerical code is generated. Identification of the bacterial species is achieved by consulting the **Enterotube II Interpretation Guide,** which lists the numerical codes assigned to each member of the Enterobacteriaceae.

TABLE 91.1	Tests performed as part of the Enterotube II system	
Enterotube II compartment no.	Biochemical test	See Exercise No.
1	Glucose fermentation	69 and 70[1]
2	Lysine decarboxylation	78
3	Ornithine decarboxylation	78
4	H$_2$S production	63
4 (after addition of Kovac's reagent)	Indole production	63
5	Adonitol fermentation	69
6	Lactose fermentation	69
7	Arabinose fermentation	69
8	Sorbitol fermentation	69
9	Voges-Proskauer	71
10	Dulcitol fermentation	69 and 70[1]
10	Phenylalanine deaminase	79
11	Urea hydrolysis	83
12	Citrate utilization	76

[1]The pH indicator used for these tests is different from the one used for the same tests in this manual, leading to different colors for positive and negative results.

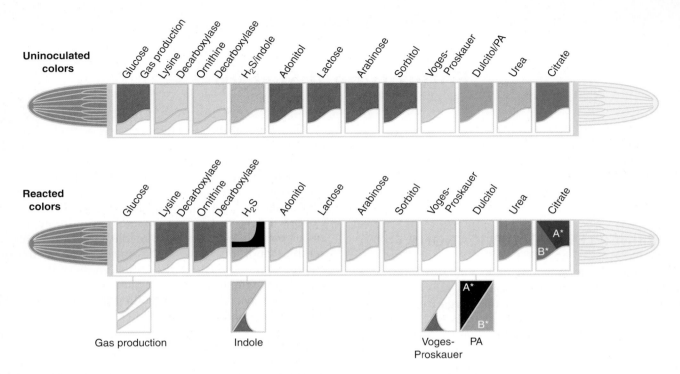

Figure 91.1 Enterotube II results. Tube on top is uninoculated. Tube on bottom displays all positive reactions.

PRE-LAB QUESTIONS

1. *What single piece of information about your unknown should you have prior to using the Enterotube II system?*

2. *How are aerobic and anaerobic reactions prepared differently in the Enterotube test?*

3. *Could a single Enterotube be used to identify multiple organisms present in a mixed sample? Explain.*

PERIOD ONE

MATERIALS

Each student should obtain:

Culture plate of your unknown organism
One Enterotube II[1]
Barritt's reagent A (α-naphthol) and B (potassium hydroxide)
Kovac's reagent
Syringes with needles or disposable pipettes
Marking pen

(1) Remove organisms from a well-isolated colony. Avoid touching the agar with the wire. To prevent damaging Enterotube II media, do not heat-sterilize the inoculating wire.

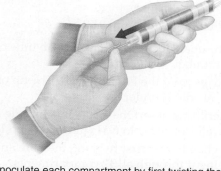

(2) Inoculate each compartment by first twisting the wire and then withdrawing it all the way out through the 12 compartments, using a turning movement.

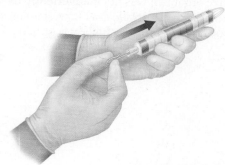

(3) Reinsert the wire (without sterilizing), using a turning motion through all 12 compartments until the notch on the wire is aligned with the opening of the tube.

(4) Break the wire at the notch by bending. The portion of the wire remaining in the tube maintains anaerobic conditions essential for true fermentation.

(5) Punch holes with broken-off part of wire through the thin plastic covering over depressions on sides of the last eight compartments (adonitol through citrate). Replace caps and incubate at 35°C for 18–24 h.

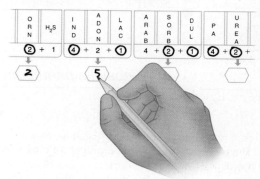

(6) After interpreting and recording positive results on the sides of the tube, perform the indole test by injecting 1 or 2 drops of Kovac's reagent into the H₂S/indole compartment.

(7) Perform the Voges-Proskauer test, if needed for confirmation, by injecting the reagents into the H₂S/indole compartment.
    After circling the numbers of the positive tests on the Laboratory Report, total up the numbers of each bracketed series to determine the five-digit code number. Refer to the *Enterotube II Interpretation Guide* for identification of the unknown by using the code number.

Figure 91.2 Procedure for preparing and inoculating the Enterotube II.

PROCEDURE

This procedure is illustrated in Figure 91.2.

1. Label the Enterotube with your name and lab time or an identifying number.

2. Unscrew both caps from the Enterotube II. The tip of the inoculating needle is under the white cap.

3. Without heat sterilizing, insert the inoculating needle into the center of a well-isolated colony (Figure 91.2, step 1).

4. Inoculate each chamber by first twisting the wire and then withdrawing it through all 12 compartments. Rotate the wire as you withdraw it through the tube (Figure 91.2, step 2).

5. Again, without sterilizing, reinsert the wire and, while turning, force it through all 12 compartments until the notch on the wire is aligned with the opening of the tube. About 1.5 in. of the wire should be extending from the tube, and the tip of the wire should be visible in the citrate compartment (Figure 91.2, step 3).

6. Break the wire at the notch by bending as shown in Figure 91.2, step 4. The portion of the wire remaining in the tube maintains anaerobic atmosphere for those reactions that require one.

7. Without touching the end of the broken wire, use the tip to punch holes through the air inlets of the last eight compartments (i.e., adonitol, lactose, arabinose, sorbitol, Voges-Proskauer, dulcitol/PA, urea, and citrate). The air inlets are located on the side of the tube opposite the

Enterotube label. The holes will allow oxygen to enter these compartments (Figure 91.2, step 5).

8. Replace the caps at both ends and incubate the Enterotubes for 18–24 h at 35°C. If many tubes are being incubated at once, allow some space between them for air to circulate. If evaluation at 24 h is impossible, the tubes should be removed from the incubator after 24 h and refrigerated until the next lab period.

9. Discard the broken end of the inoculating wire in a hard-sided biohazard container.

QUESTION—PERIOD ONE

1. What type of media could help to ensure that a Gram-positive organism was not inadvertently used to inoculate an Enterotube II test? Does a Gram stain need to be done if the bacteria used for inoculation is grown on this media?

PERIOD TWO

1. Retrieve your Enterotube from the incubator.

2. Using Figure 91.1 and Table 91.2, evaluate each compartment as positive or negative. Record your results.

Record the results of each test in the following table with a plus(+) or minus (–)

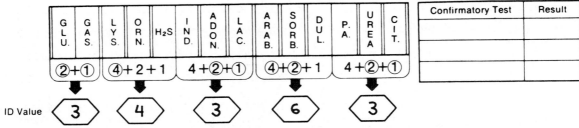

Figure 91.3 Example of Enterotube II ID value construction. Each test result is recorded as positive or negative. Negative test results have a point value of zero while positive results are worth one, two, or four points. The positive results in each group of tests are added together to arrive at a single digit. The five single digits, when read left to right, make up the five-digit ID value of the organism. If identification is impossible based on the five-digit code, a Voges-Proskauer reaction is performed as a confirmatory test. Here, the five-digit code that was generated (34363) represents the organism *Klebsiella pneumoniae*. The Voges-Proskauer test was not needed.

TABLE 91.2

Groups	Reactions	GLUCOSE	GAS PRODUCTION	LYSINE	ORNITHINE	H₂S	INDOLE	ADONITOL	LACTOSE	ARABINOSE	SORBITOL	VOGES-PROSKAUER	DULCITOL	PHENYLALANINE DEAMINASE	UREA	CITRATE
ESCHERICHIEAE	*Escherichia*	+ 100.0	+J 92.0	d 80.6	d 57.8	−K 4.0	+ 96.3	− 5.2	+J 91.6	+ 91.3	± 80.3	− 0.0	d 49.3	− 0.1	− 0.1	− 0.2
	Shigella	+ 100.0	−A 2.1	− 0.0	∓B 20.0	− 0.0	∓ 37.8	− 0.0	−B 0.3	± 67.8	∓ 29.1	− 0.0	d 5.4	− 0.0	− 0.0	− 0.0
EDWARDSIELLEAE	*Edwardsiella*	+ 100.0	+ 99.4	+ 100.0	+ 99.0	+ 99.6	+ 99.0	− 0.0	− 0.0	∓ 10.7	− 0.2	− 0.0	− 0.0	− 0.0	− 0.0	− 0.0
SALMONELLEAE	*Salmonella*	+ 100.0	+C 91.9	+H 94.6	+I 92.7	+E 91.6	− 1.1	− 0.0	− 0.8	+ 89.2	+ 94.1	− 0.0	dD 86.5	− 0.0	− 0.0	dF 80.1
	Arizona	+ 100.0	+ 99.7	+ 99.4	+ 100.0	+ 98.7	− 2.0	− 0.0	d 69.8	+ 99.1	+ 97.1	− 0.0	− 0.0	− 0.0	− 0.0	+ 96.8
CITROBACTER *freundii*		+ 100.0	+ 91.4	− 0.0	d 17.2	± 81.6	− 6.7	− 0.0	d 39.3	+ 100.0	+ 98.2	− 0.0	d 59.8	− 0.0	dw 89.4	+ 90.4
CITROBACTER *amalonaticus*		+ 100.0	+ 97.0	− 0.0	+ 97.0	− 0.0	+ 99.0	− 0.0	± 70.0	+ 99.0	+ 97.0	− 0.0	∓ 11.0	− 0.0	+ 81.0	+ 94.0
CITROBACTER *diversus*		+ 100.0	+ 97.3	− 0.0	+ 99.8	− 0.0	+ 100.0	+ 100.0	d 40.3	+ 98.0	+ 98.2	− 0.0	± 52.2	− 0.0	dw 85.8	+ 99.7
PROTEEAE — PROTEUS *vulgaris*		+ 100.0	±G 86.0	− 0.0	− 0.0	+ 95.0	+ 91.4	− 0.0	− 0.0	− 0.0	− 0.0	− 0.0	− 0.0	+ 100.0	+ 95.0	d 10.5
PROTEUS *mirabilis*		+ 100.0	+G 96.0	− 0.0	+ 99.0	+ 94.5	− 3.2	− 0.0	− 2.0	− 0.0	− 0.0	∓ 16.0	− 0.0	+ 99.6	± 89.3	± 58.7
MORGANELLA *morganii*		+ 100.0	±G 86.0	− 0.0	+ 97.0	− 0.0	+ 99.5	− 0.0	− 0.0	− 0.0	− 0.0	− 0.0	− 0.0	+ 95.0	+ 97.1	−L 0.0
PROVIDENCIA *alcalifaciens*		+ 100.0	dG 85.2	− 0.0	− 1.2	− 0.0	+ 99.4	+ 94.3	− 0.3	− 0.7	− 0.6	− 0.0	− 0.0	+ 97.4	− 0.0	+ 97.9
PROVIDENCIA *stuartii*		+ 100.0	− 0.0	− 0.0	− 0.0	− 0.0	+ 98.6	∓ 12.4	− 3.6	− 4.0	− 3.4	− 0.0	− 0.0	+ 94.5	∓ 20.0	+ 93.7
PROVIDENCIA *rettgeri*		+ 100.0	∓G 12.2	− 0.0	− 0.0	− 0.0	+ 95.9	+ 99.0	d 10.0	− 0.0	+ 1.0	− 0.0	− 0.0	+ 98.0	+ 100.0	+ 96.0
KLEBSIELLEAE — ENTEROBACTER *cloacae*		+ 100.0	+ 99.3	− 0.0	+ 93.7	− 0.0	− 0.0	∓ 28.0	± 94.0	+ 99.4	+ 100.0	+ 100.0	d 15.2	− 0.0	± 74.6	+ 98.9
ENTEROBACTER *sakazakii*		+ 100.0	+ 97.0	− 0.0	+ 97.0	− 0.0	∓ 16.0	− 0.0	+ 100.0	+ 100.0	− 0.0	+ 97.0	+ 6.0	− 0.0	− 0.0	+ 94.0
ENTEROBACTER *gergoviae*		+ 100.0	+ 93.0	± 64.0	+ 100.0	− 0.0	− 0.0	− 0.0	∓ 42.0	+ 100.0	− 0.0	+ 100.0	− 0.0	− 0.0	+ 100.0	+ 96.0
ENTEROBACTER *aerogenes*		+ 100.0	+ 95.9	+ 97.5	+ 95.9	− 0.0	− 0.8	+ 97.5	+ 92.5	+ 100.0	+ 98.3	+ 100.0	− 4.1	− 0.0	− 0.0	+ 92.6
ENTEROBACTER *agglomerans*		+ 100.0	∓ 24.1	− 0.0	− 0.0	− 0.0	∓ 19.7	− 7.5	d 52.9	+ 97.5	d 26.3	± 64.8	d 12.9	∓ 27.6	d 34.1	d 84.2
HAFNIA *alvei*		+ 100.0	+ 98.9	+ 99.6	+ 98.6	− 0.0	− 0.0	− 0.0	d 2.8	+ 99.3	− 0.0	± 65.0	− 2.4	− 0.0	d 3.0	d 5.6
SERRATIA *marcescens*		+ 100.0	±G 52.6	+ 99.6	+ 99.6	− 0.0	−w 0.1	∓ 56.0	− 1.3	− 0.0	+ 99.1	+ 98.7	− 0.0	− 0.0	dw 39.7	+ 97.6
SERRATIA *liquefaciens*		+ 100.0	d 72.5	± 64.2	+ 100.0	− 0.0	−w 1.8	− 8.3	d 15.6	+ 97.3	+ 97.3	∓ 49.5	− 0.0	− 0.9	dw 3.7	+ 93.6
SERRATIA *rubidaea*		+ 100.0	dG 35.0	+ 61.0	− 0.0	− 0.0	−w 2.0	± 88.0	+ 100.0	+ 100.0	− 8.0	+ 92.0	− 0.0	− 0.0	dw 4.0	± 88.0
KLEBSIELLA *pneumoniae*		+ 100.0	+ 96.0	+ 97.2	− 0.0	− 0.0	− 0.0	± 89.0	+ 98.7	+ 99.9	+ 99.4	+ 93.7	∓ 33.0	− 0.0	+ 95.4	+ 96.8
KLEBSIELLA *oxytoca*		+ 100.0	+ 96.0	+ 97.2	− 0.0	− 0.0	+ 100.0	± 89.0	∓ 98.7	+ 100.0	+ 98.0	+ 93.7	∓ 33.0	− 0.0	∓ 95.4	∓ 96.8
KLEBSIELLA *ozaenae*		+ 100.0	d 55.0	∓ 35.8	− 1.0	− 0.0	− 0.0	+ 91.8	± 26.2	+ 100.0	± 78.0	− 0.0	− 0.0	− 0.0	d 14.8	d 28.1
KLEBSIELLA *rhinoscleromatis*		+ 100.0	− 0.0	− 0.0	− 0.0	− 0.0	− 0.0	+ 98.0	d 6.0	+ 100.0	+ 98.0	− 0.0	− 0.0	− 0.0	− 0.0	− 0.0
YERSINIAE — YERSINIA *enterocolitica*		+ 100.0	− 0.0	− 0.0	+ 90.7	− 0.0	∓ 26.7	− 0.0	− 0.0	+ 98.7	+ 98.7	− 0.1	− 0.0	− 0.0	+ 90.7	− 0.0
YERSINIA *pseudotuberculosis*		+ 100.0	− 0.0	− 0.0	− 0.0	− 0.0	− 0.0	− 0.0	− 0.0	± 55.0	− 0.0	− 0.0	− 0.0	− 0.0	+ 100.0	− 0.0

E. *S. enteritidis* bioserotype Paratyphi A and some rare biotypes may be H₂S negative.

F. *S. typhi*, *S. enteritidis* bioserotype Paratyphi A and some rare biotypes are citrate-negative and *S. cholerae-suis* is usually delayed positive.

G. The amount of gas produced by *Serratia*, *Proteus* and *Providencia alcalifaciens* is slight; therefore, gas production may not be evident in the ENTEROTUBE II.

H. *S. enteritidis* bioserotype Paratyphi A is negative for lysine decarboxylase.

I. *S. typhi* and *S. gallinarum* are ornithine decarboxylase-negative.

J. The Alkalescens-Dispar (A-D) group is included as a biotype of *E. coli*. Members of the A-D group are generally anaerogenic, non-motile and do not ferment lactose.

K. An occasional strain may produce hydrogen sulfide.

L. An occasional strain may appear to utilize citrate.

Courtesy and © Becton, Dickinson, and Company

3. Add a small amount of Kovac's reagent to the H_2S/indole compartment by first puncturing the plastic membrane of the compartment and then using a needle and syringe or pipette to deliver one or two drops of reagent. A pink to red color within 10 sec is a positive reaction.

4. In some cases, the Enterotube II Interpretation Guide will instruct you to perform a Voges-Proskauer test. If this is the case for your organism, use a needle or pipette to inject 2 drops of Barritt's reagent B (KOH) and 3 drops of Barritt's reagent A (α-naphthol) into the

VP compartment. A positive result is indicated by formation of a red color within 20 min.

5. Once you have calculated the five-digit code for your unknown organism, as outlined in Figure 91.3, use the Enterotube II Interpretation Guide to identify your unknown. Alternatively, compare your results with the information given in Table 91.2. Record your Enterotube II results.

ONPG	ADH	LDC	ODC	CIT	H2S	URE	TDA	IND	VP	GEL	GLU	MAN	INO	SOR	RHA	SAC	MEL	AMY	ARA	OXI
1	2	4	1	2	4	1	2	4	1	2	4	1	2	4	1	2	4	1	2	4

NO2	N2 GAS	MOT	MAC	OF-O	OF-F
1	2	4	1	2	4

Additional Digits

EXERCISE 91 REVIEW QUESTIONS

1. Why is the Enterotube II only useful for identifying Gram-negative rods and not other bacterial species as well?

2. What reaction in the Enterotube II should always be positive if a member of the Enterobacteriaceae is being tested?

3. The API 20E test strip has been designed to identify enteric bacteria in a clinical setting. With this is mind, design your own test strip to differentiate among the species of *Staphylococcus, Streptococcus,* and *Micrococcus* presented in Exercise 39. Use the flowcharts presented in the exercise to help you design your test strip.

1. BD. *Enterotube II Interpretation Guide.* http://www.bd.com/ds/productCenter/243383.asp.

NOTES

Antibiotic Disk Sensitivity Tests

Paper disks impregnated with antibiotics can be used for the differentiation of many species of bacteria. Optochin, bacitracin, and SXT (sulfamethoxazole trimethoprim) are helpful in discriminating between streptococcal species while resistance to novobiocin is often used to presumptively identify Staphylococcus saprophyticus.

PRINCIPLES AND APPLICATIONS

Resistance to antibiotics is always a concern to the treating physician, who is attempting to rid the body of a troublesome bacterial infection, but antibiotic resistance also provides the microbiologist with a unique tool for the differentiation of closely related bacteria species. Resistance to optochin is commonly used to distinguish *Streptococcus pneumoniae* from other α-hemolytic bacteria while novobiocin is commonly used to differentiate *Staphylococcus saprophyticus* (resistant) from *S. aureus* and *S. epidermidis* (both sensitive). On occasion, even an unclear Gram reaction can be resolved by this method. The vast majority of Gram-positive organisms are susceptible to vancomycin while most Gram negatives are resistant. Conversely, very few Gram-positive organisms are susceptible to colistin or polymyxin, while Gram-negative species tend to be susceptible.

Several methods exist for determining antibiotic resistance but the simplest of these is often to streak the isolate onto a rich media such as blood agar so that a confluent lawn of bacteria is created and then to place a paper disk containing the antibiotic to be tested onto the media. After incubation, a zone of inhibition (of a certain size) indicates resistance to the antibiotic. Compared to plating bacteria on a media that already includes an antibiotic, this method has the advantage of allowing several drugs to be tested simultaneously and reduces the number of different types of media that must be kept on hand.

It is important to note that disks used for partial identification of a bacterial species are **differential disks,** not **susceptibility disks.** Differential disks have a lower concentration of antibiotic than the disks used in the Kirby-Bauer test for antimicrobial susceptibility (Exercise 20).

PRE-LAB QUESTIONS

1. *Define the following terms:*

Susceptible

Resistant

Confluent

Antibiotic

2. *Why is a rich media, such as blood agar, commonly used for susceptibility testing?*

PERIOD ONE

MATERIALS

Each student should obtain:
One plate of blood agar
Fresh broth cultures of:
   Staphylococcus aureus
   Staphylococcus saprophyticus
Novobiocin antibiotic disks
Forceps (not required if a disk dispenser is used)
Inoculating loop
Marking pen
Millimeter ruler

PROCEDURE

1. Mark the bottom of your Petri dish with your name and lab time. Label one half of the dish with the name of each organism.
2. Use a loop to inoculate the entire surface of the agar (on one-half of the plate) with the appropriate organism. Do the same to the other half of the plate using the other organism.
3. Use a disk dispenser or forceps to aseptically place a novobiocin disk on each half of the inoculated plate.
4. Incubate the plate at 37°C for 24 h.

QUESTIONS—PERIOD ONE

1. What is meant by the term **zone of inhibition?** What is the significance of the term in this exercise?

Figure 92.1 Antibiotic susceptibility is indicated by a zone of inhibition (an area of no growth) surrounding the antibiotic disk. To be considered positive, the zone must be larger than a specified cutoff size, which differs for each antibiotic tested.

2. Certain techniques call for you to measure the radius of the zone of inhibition while others ask for a measurement of the diameter. Describe how you would measure each of these.

PERIOD TWO

1. Retrieve your plate from the incubator.
2. Measure the diameter of the zone of inhibition in millimeters. Be sure to measure completely across the zone at its largest width, from the edge of bacterial growth on one side to the edge of bacterial growth on the other side.

3. Susceptibility to novobiocin is said to have occurred if the zone of inhibition surrounding the antibiotic disk is greater than 15 mm in diameter (Figure 92.1). For other antibiotics and organisms, the point at which susceptibility is indicated will be different.

4. Sketch the appearance of your plate.

EXERCISE 92 REVIEW QUESTIONS

1. How would you evaluate a plate on which the bacteria grew uniformly poorly across the surface of the agar?

2. Occasionally, when a plate is allowed to incubate too long, a zone of inhibition is seen, but within the zone is a very thin, uniform layer of bacterial growth. How would you interpret such a plate?

3. What could be the result of an overly heavy inoculation of your plate prior to applying the antibiotic disk?

REFERENCES

Cowan, M.K. 2012. *Microbiology: A Systems Approach, 3rd ed.,* chap. 18. New York: McGraw-Hill.

Forbes, B.A., Sahm, D.F., and Weissfeld, A. 2007. *Bailey and Scott's Diagnostic Microbiology, 12th ed.,* p. 111. St. Louis: Mosby-Elsevier.

Talaro, K.P., and Chess, B. 2012. *Foundations in Microbiology, 8th ed.,* chap. 18. New York: McGraw-Hill.

β-Lactamase Test

The β-lactamase test is used to identify those bacteria that produce the enzyme β-lactamase, which hydrolyzes the β-lactam ring in penicillin and cephalosporin, rendering these antibiotics ineffective. The test uses a cephalosporin (i.e., nitrocefin) that turns pink when hydrolyzed.

PRINCIPLES AND APPLICATIONS

Penicillins and cephalosporins are known as **β-lactamase** antibiotics because of the presence of a β-lactam ring at the center of their molecular structure. These antibiotics work by preventing cell wall synthesis, leaving affected bacterial cells very sensitive to osmotic changes and more prone to cell lysis. Organisms that produce the enzyme β-lactamase **hydrolyze** the β-lactam ring and inactivate the antibiotic. These organisms are resistant to penicillin and cephalosporin and represent the most commonly encountered example of antibiotic resistance. Many bacterial isolates, including *Neisseria gonorrhoeae* and *Enterococcus*, are routinely tested for resistance to β-lactam antibiotics.

The β-lactamase test is a means of determining whether or not a particular bacterial isolate is resistant to β-lactam antibiotics. The test is accomplished by providing bacteria with a substrate called **nitrocefin,** a cephalosporin that is converted from pale yellow to pink when it is hydrolyzed. The nitrocefin is provided on paper disks or filter paper that are then coated with the test organism. Development of a pink color is indicative of β-lactamase-catalyzed hydrolysis of the nitrocefin.

Several methods exist for performing the β-lactamase test, but most labs use a rapid, commercially available method that can be completed in less than an hour. In this technique, bacterial growth is applied to a piece of filter paper containing the nitrocefin substrate and allowed to react for up to 60 min. A deep pink color indicates a positive reaction while a yellow color represents a negative reaction (Figure 93.1).

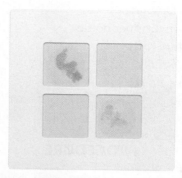

Figure 93.1 Nitrocefin DrySlide for the determination of β-lactamase production. Organisms that produce β-lactamase will result in a pink color (upper left) when allowed to react with nitrocefin, a type of cephalosporin. Nitrocefin remains yellow (lower right) in a negative reaction.

PRE-LAB QUESTIONS

1. *What does the term hydrolyze mean?*

2. *How do you recognize a positive result in a β-lactamase test?*

MATERIALS

This procedure utilizes Nitrocefin DrySlides from Becton Dickinson Microbiology Systems. Your instructor will tell you if another procedure is to be used.

Each student should obtain:

One Nitrocefin DrySlide[1]

Distilled water in dropper bottle

Fresh slant cultures of:

Enterococcus faecalis

Staphylococcus aureus

Inoculating loop

Marking pen

PROCEDURE

1. Obtain a DrySlide and label it along the border, indicating which square corresponds to each of your organisms.

2. For each organism, prepare the test as follows:
   - Apply a small drop of distilled water to the filter paper in each square to be used. Use only enough water to moisten the filter paper.
   - Using an inoculating loop, apply a heavy inoculum (several colonies) of the bacteria to be tested to the moistened area of the filter paper.

TABLE 93.1	β-Lactamase test results and interpretations
Interpretation (Symbol)	**Appearance**
β-lactamase present (+)	Pink color within 60 min
β-lactamase absent (−)	No pink color within 60 min (i.e., yellow)

3. Allow the DrySlide to incubate for up to 60 min at room temperature.

4. The appearance of a pink color at any time within 60 min is considered a positive reaction. A yellow color is considered a negative reaction.

5. Possible results for the β-lactamase test are summarized in Table 93.1. Sketch the appearance of your β-lactamase test.

EXERCISE 93 REVIEW QUESTION

1. If the β-lactamase test indicated that penicillin was not a viable treatment option, why would the Kirby-Bauer test be of interest to you?

1. BD. BBL DrySlide Nitrocefin. http://www.bd.com/ds/productCenter/231749.asp.

REFERENCE

Forbes, B.A., Sahm, D.F., and Weissfeld, A. 2007. *Bailey and Scott's Diagnostic Microbiology, 12th ed.,* p. 181. St. Louis: Mosby-Elsevier.

Viable Plate Count

The viable plate count is a method of estimating the number of cells in a liquid sample. A small volume of a bacterial sample is serially diluted to produce several samples with decreasing cell densities. Aliquots of these samples are then plated onto media, and the number of colonies arising after incubation is used to determine the approximate number of cells in the original sample. The test is referred to as a viable plate count because only living cells are counted.

PRINCIPLES AND APPLICATIONS

Having some idea of the microbial cell density of a liquid sample is oftentimes very important knowledge; the number of bacterial cells in milk, drinking water, even the ocean, directly affects human activities. Several methods exist to enumerate bacteria, with the viable plate count being among the simplest. In this method, a sample of liquid to be counted is serially diluted to produce several samples with decreasing cell densities. Aliquots of the dilutions are then plated onto media, and the number of colonies produced is counted after incubation. Once the number of colonies is known, it is a simple matter to mathematically calculate the number of cells in the original sample.

Each dilution is represented by a **dilution factor,** a term which corresponds to the amount of the original sample still present in the current sample. For example, if 1 ml of a sample is diluted into 9 ml of diluent (the fluid used for dilution), the dilution factor is 10^{-1} (1/10) as the sample is now 1/10th its original concentration. If 1 ml were diluted into 99 ml, the dilution factor would be 10^{-2} because the sample would be 1/100th as concentrated, and so on. The use of dilution factors, along with careful measurement allows the accurate determination of cell densities in any liquid.

When several dilutions are performed in sequence, a **serial dilution** is the end result, with each sample being less concentrated than the one before. The most important part of performing a serial dilution is being able to calculate exactly how dilute a sample is when compared to the original. This can easily be done by remembering two facts.

1. A dilution factor is always found by dividing the volume of the first sample by the volume of the first sample plus the volume of the diluent. For example, diluting 1 ml of a water sample into 99 ml of sterile water results in a 1/100, or 10^{-2}, dilution factor.

2. In a serial dilution, each successive dilution factor is simply the product of the current dilution factor multiplied by the previous dilution factor. If 1 ml of the 10^{-2} dilution above were diluted into 99 ml of sterile water, the dilution factor of the newest sample would be found as follows:

Dilution of sample	$\times$	previous dilution factor	$=$	dilution factor of new sample
1 ml/100 ml	$\times$	10^{-2}	$=$	dilution factor of new sample
10^{-2}	$\times$	10^{-2}	$=$	10^{-4}

Finally, the cell density of the original broth is a function of the volume of the sample plated onto the media, the dilution of the sample, and the number of colonies on the plate. The original cell density (OCD) can be calculated using the formula:

$$\text{OCD} = \frac{[\text{colonies on plate}]}{[\text{volume sample plated} \times \text{dilution factor}]}$$

For example, if 0.1 ml (100 µl) from a sample with a dilution factor of 10^{-4} is used to inoculate a plate and 122 colonies are counted after incubation, the OCD is:

$$\text{OCD} = \frac{[\text{colonies on plate}]}{[\text{volume sample plated} \times \text{dilution factor}]}$$

$$\text{OCD} = \frac{122 \text{ colonies}}{[0.1 \text{ ml} \times 10^{-4}]}$$

$$= \frac{1.22 \ 10^7 \ (12{,}200{,}000) \text{ colonies}}{\text{ml}}$$

One last point. Although we tend to think of a colony arising from a single cell, it is important to remember that for some organisms, single cells are rarely if ever seen. The term **colony forming unit (CFU)** allows us to sidestep the problem of exactly how many cells were involved in the formation of a single colony by grouping together pairs, tetrads, clusters, and chains under a common term. The fact remains that a colony arises from the smallest unit of colony formation, be it one cell, two, four, or more. For the sake of accuracy, therefore, the term CFU is preferable to cells and will be used from this point forward.

PRE-LAB QUESTION

1. *Define the following terms:*

Diluent

Aliquot

Serial dilution

PERIOD ONE

MATERIALS

Each group should obtain:
Twelve nutrient agar plates
Fresh broth cultures of either:
  Escherichia coli
  Staphylococcus aureus
Sterile 10 ml, 1 ml, and 0.1 ml pipettes
Six sterile dilution tubes
Beaker with ethanol and a bent glass rod (see Exercise 8)

Hand tally counter
Colony counter
Marking pen

PROCEDURE

This procedure is illustrated in Figure 94.1.

1. Mark the bottom of your Petri dish with your group number and lab time. Group the dishes into six sets labeled "A1, A2," "B1, B2," "C1, C2," "D1, D2," "E1, E2," and "F1, F2."

2. Label the six sterile dilution tubes "A," "B," "C," "D," "E," and "F."

3. To tube A, add 9.9 ml of sterile water. To tubes B–F, add 9.0 ml of sterile water.

4. Mix the broth culture until it seems uniformly turbid.

5. Transfer 0.1 ml from the culture to tube A. Pipette up and down several times to mix.

6. Transfer 1 ml from tube A to tube B. Pipette up and down several times to mix.

7. Transfer 1 ml from tube B to tube C. Pipette up and down several times to mix.

8. Transfer 1 ml from tube C to tube D. Pipette up and down several times to mix.

9. Transfer 1 ml from tube D to tube E. Pipette up and down several times to mix.

10. Transfer 1 ml from tube E to tube F. Pipette up and down several times to mix.

11. Transfer 0.1 ml from tube A to plate A1. Using the spread-plate technique (Exercise 8), spread the diluted sample across the entire surface of the plate. Prepare a replicate plate by transferring 0.1 ml to plate A2 and spreading in the same way.

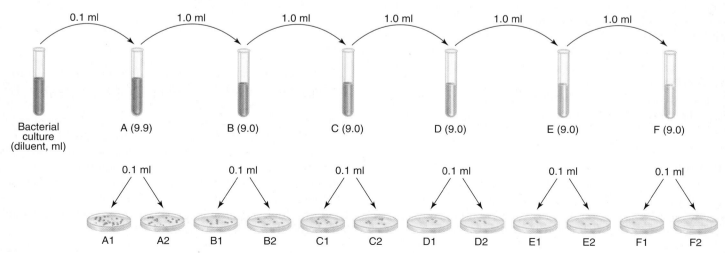

Figure 94.1 Procedure for serial dilution. The first diluent tube (A) contains 9.9 ml of sterile water, and tubes B–F contain 9.0 ml. From the bacterial culture, 0.1 ml is transferred to tube A and mixed. In turn, 1.0 ml is transferred from tube A to B, B to C, C to D, D to E, and E to F. Replicate plates of nutrient agar are inoculated with 0.1 ml from the corresponding dilution tube.

12. Repeat the above process for each dilution tube and its two plates.
13. Incubate the plate at 37°C for 24–48 h.

QUESTIONS—PERIOD ONE

1. What is the relationship between a cell, a CFU, and a colony.

0.001 ml

3. A 1-ml sample of milk is diluted 1:10 with sterile water (1 ml milk:9 ml water) and 0.1 ml is used to inoculate a plate of general purpose media. After a 24-h incubation, a total of 73 colonies are present on the plate.

   How many CFUs were present on the plate?

2. If a sample of water is required to have no more than 20,000 CFUs/ml, what is the maximum number of CFUs that are permissible in:

   0.1 ml

   How many CFUs/ml were in the diluted milk sample?

   0.01 ml

   How many CFUs are present in 1 ml of this milk?

PERIOD TWO

1. Retrieve your plate from the incubator.
2. Identify a set of plates with between 30 and 300 colonies. Plates outside of this range are statistically unreliable, with those containing fewer than 30 colonies being labeled as TFTC (too few to count) while those with greater than 300 colonies are labeled TMTC (too many to count). Examine your plates for a set that has between 30 and 300 colonies. Count the number of colonies appearing on each plate, and calculate the average. Record this number in the accompanying blank table.

3. Using the information given in the first part of the lab, calculate the density of the original sample. Table 94.1 contains the information used to calculate the dilution factor for each tube used in this exercise, with the required factor listed in the right-most column.
4. Complete the accompanying blank table.
5. What was the OCD of your culture? Does this agree with what others in the class determined?

	A1	A2	B1	B2	C1	C2	D1	D2	E1	E2	F1	F2
Colonies/plate												
Average												
Organism used												

TABLE 94.1   Calculation of dilution factors for tubes A–F

Sample	Volume of sample used (1)*	Volume of diluent (2)	Volume of diluent + sample (3)†	Dilution factor of previous sample (4)	Dilution factor of this sample (5)‡
Bacterial culture	0.1 ml	0 ml	0.1 ml	–	1
Tube A	0.1 ml	9.9 ml	10 ml	1	10^{-2}
Tube B	1.0 ml	9 ml	10 ml	10^{-2}	10^{-3}
Tube C	1.0 ml	9 ml	10 ml	10^{-3}	10^{-4}
Tube D	1.0 ml	9 ml	10 ml	10^{-4}	10^{-5}
Tube E	1.0 ml	9 ml	10 ml	10^{-5}	10^{-6}
Tube F	1.0 ml	9 ml	10 ml	10^{-6}	10^{-7}

*The numbers in parentheses represent the column numbers used in the calculations.
†Column three is the sum of columns one and two.
‡Column five is [column one × column four]/[column three].

EXERCISE 94 REVIEW QUESTIONS

1. You ask your lab partner to add 1 ml of a 10 ml water sample to 99 ml of diluent. Your partner inadvertently adds the entire 10 ml sample of water to the diluent bottle. What was the intended dilution factor, and what is the actual dilution factor?

2. You count 127 colonies on a plate. The plate was inoculated with 0.1 ml of a sample with a dilution factor of 10^{-6}? What is the OCD of the sample?

REFERENCES

Cowan, M.K. 2012. *Microbiology: A Systems Approach, 3rd ed.*, chap. 7. New York: McGraw-Hill.

Reddy, C.A., Beveridge, T.J., Breznak, J.A., Marzluf, G.A., Schmidt, T.M., and Snyder, L.R. (eds.). 2007. *Methods for General and Molecular Microbiology, 3rd ed.*, chap. 9. Washington D.C.: ASM Press.

Talaro, K.P., and Chess, B. 2012. *Foundations in Microbiology, 8th ed.*, chap. 7. New York: McGraw-Hill.

NOTES

Direct Cell Count

The direct cell count is a means of directly enumerating the cells present in a sample. A small amount of a liquid specimen is applied to a counting chamber that holds a fixed volume of fluid and then observed microscopically. The number of cells can be quickly determined, but no distinction can be made between living and dead cells.

PRINCIPLES AND APPLICATIONS

The ability to directly count the cells in a liquid sample seems, and often is, the simplest way of determining the cell density of a particular sample. The **Petroff-Hauser** counting chamber is a device that makes this possible. Similar in size and appearance to a microscope slide, the chamber has a small well in the center, 0.02 mm deep. An etched pattern on the bottom of the well divides it into 25 large squares, each of which is further divided into 16 smaller squares, all contained within a 1-mm square area of the well. When a coverslip is placed atop the well of the counting chamber and the chamber is allowed to fill with fluid, each small square encloses a volume of exactly $5.8 \ 10^{-8}$ ml. By counting the cells of a sample in a Petroff-Hauser counting chamber, the number of cells in a sample can be determined (Figure 95.1).

Even within the small volume of the counting chamber, the number of bacteria can be daunting, with many thousands of cells appearing in each small square. For improved accuracy, samples are diluted before being counting. The diluent is 0.1 N HCl, which both kills the cells and gives them a net positive charge, preventing them from clumping together. The formula used to calculate the **original cell density (OCD)** in a direct cell count is

$$OCD = \frac{[\text{cell count}]}{[\text{number of small squares} \times \text{dilution factor} \times 5.8 \ 10^{-8} \text{ ml}]}$$

The most statistically reliable results occur when at least 600 cells are counted and the sample has been diluted so that each square encloses between 5 and 15 cells. For example, if a sample was diluted 1:1000 (10^{-3}) and 645 cells were counted in a total of 48 small squares, both the total number of cells (>600) and the dilution of the sample (645/48 = 13.4 cells/square) would be optimized to provide a statistically reliable result. The OCD for this example would be:

$$OCD = \frac{645 \text{ cells}}{[48 \times 10^{-3} \times 5.8 \ 10^{-8} \text{ ml}]}$$

$$OCD = \frac{2.7 \ 10^{10} \text{ cells}}{\text{ml}}$$

The dilution factor seen in the equation above is a term that relates the amount of the original sample still present in the current sample. For example, if 1 ml of a sample is diluted into 9 ml of diluent, the dilution factor is 10^{-1} (1/10) as the sample is now 1/10th its original concentration. If 1 ml were diluted into 99 ml, the dilution factor would be 10^{-2} because the sample would be 1/100th as concentrated, and so on. The use of dilution factors, along with careful measurement, allows the accurate determination of cell densities in any liquid.

PRE-LAB QUESTION

1. *Define the following terms:*

Diluent

Dilution factor

MATERIALS

Each group should obtain:
Petroff-Hauser counting chamber with coverslip
1-ml pipettes with bulbs or pipette pumps
Disposable Pasteur pipettes
Hand tally counter
0.1 N HCl
Nonsterile test tube
Overnight culture of:
  Escherichia coli

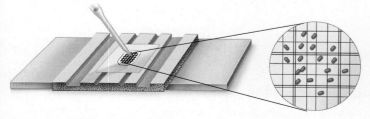

Figure 95.1 Bacterial cell-counting chamber. A small sample is placed on the grid under a coverslip. Individual cells, both living and dead, are counted.

PROCEDURE

1. Transfer 0.1 ml of the overnight culture to a nonsterile test tube.
2. Add 0.9 ml of 0.1 N HCl to the culture.
3. Place a coverslip on a Petroff-Hausser counting chamber, and place a single drop of the bacterial cell suspension at the edge of the coverslip; capillary action will draw the liquid into the counting chamber.
4. Place the counting chamber on the microscope, and, using the low power objective, focus on the counting grid. Reducing the light and closing the iris diaphragm will make the grid easier to see.
5. Increase the magnification until you are using the oil immersion lens to view the cells.
6. Count the number of cells in 16 small squares. Allow the other people in you group to do the same. Add together both the number of cells counted and the number of squares examined.

   Using the information, calculate the density of the original sample.

   OCD = _____ cells/ml

   What was the original cell density of your culture? Does this agree with what others in the class determined?

EXERCISE 95 REVIEW QUESTIONS

1. A 1-ml sample of lake water is diluted 1:100 with sterile water (1 ml water:99 ml diluent), and a sample of this dilution is counted using a Petroff-Hauser counter. Sixty-four small squares are counted yielding a total of 768 bacterial cells.

   What is the original cell density of the water? How many cells would be present in 1 ml of this water?

2. What is the primary difference in terms of information obtained between a direct cell count and viable cell count? Why is this important?

REFERENCES

Cowan, M.K. 2012. *Microbiology: A Systems Approach,* 3rd ed., chap. 7. New York: McGraw-Hill.

Reddy, C.A., Beveridge, T.J., Breznak, J.A., Marzluf, G.A., Schmidt, T.M., and Snyder, L.R. (eds.). 2007. *Methods for General and Molecular Microbiology,* 3rd ed., chap. 9. Washington D.C.: ASM Press.

Talaro, K.P., and Chess, B. 2012. *Foundations in Microbiology,* 8th ed., chap. 7. New York: McGraw-Hill.

Spectrophotometric Determination of Bacterial Growth: Use of the Spectrophotometer

The presence of turbidity in a liquid culture provides a simple, reliable indication of bacterial growth but does not provide the quantitative information seen in a direct cell count or a viable plate count. A spectrophotometer accurately measures the turbidity in a culture, thereby determining the level of growth of the cultured bacteria more precisely. As the number of cells within a culture increase, the ability of the culture to absorb light increases proportionally. The spectrophotometer compares the amount of light entering a culture with the amount of light exiting the culture; the difference between the two is a precise indication of the number of cells in the culture.

A spectrophotometer is a relatively simple instrument. A beam of light is focused using a set of lenses and a small entrance slit. A diffraction grating then splits the light into its component wavelengths, and a small exit slit allows a single wavelength of light to be selected for use. This monochromatic light (the incident light) is then passed through a sample tube, or cuvette, containing the culture to be tested and eventually strikes a phototube that measures the intensity of the transmitted light and compares it to the intensity of the incident light (Figure A.1). The light that passes through the sample tube is referred to as **transmitted (T)** light, which is measured on a linear scale from 0% to 100%, while the light that is lost to the sample is called the **absorbed (A)** light, which is measured on a logarithmic scale from 0 to 2. Together, the intensity of the absorbed light plus the intensity of the transmitted light equals the total intensity of the light prior to its passage through the sample tube. The equation relating absorbance and transmittance is:

$$A = 2 - \log T$$

If an analog spectrophotometer is used (i.e., one with a dial readout), a more precise reading is obtained if transmittance is read and then converted to absorbance mathematically. If a digital spectrophotometer is available, it is generally easier to read the absorbance directly from the machine. A near-proportional relationship exists between absorbance and growth, meaning that a higher absorbance value corresponds to a greater degree of growth within the culture.

When determining growth using a spectrophotometer, it is best to use a fresh culture for two reasons. First, a linear relationship between absorbance and growth only holds for fairly dilute cultures; past a certain point, the absorbance will no longer be linearly proportional to growth. Second, as a culture ages, it eventually enters the stationary and death phases of

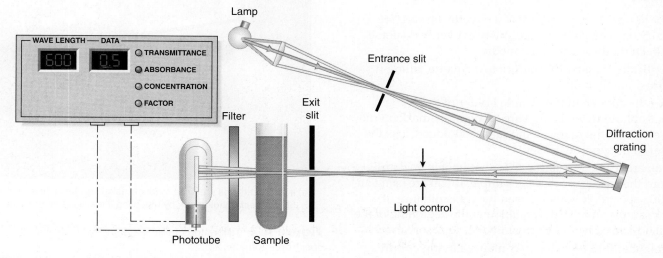

Figure A.1 Schematic of a spectrophotometer.

growth; at this point, viable cells are responsible for only a small percentage of the total absorbance. The procedure for measuring the absorbance of a liquid culture using a spectrophotometer such as the Bausch and Lomb Spectronic 20 digital is given here. Analog spectrophotometers as well as those made by other manufacturers are not significantly different.

Procedure

1. Turn on the spectrophotometer, and allow it to warm up for at least 15 min.

2. Proper handling of cuvettes during use of the spectrophotometer is essential to an accurate reading. Tips for handling cuvettes include:

   a. Rinse the cuvettes with distilled or deionized water prior to using them. Excess water should be drained or shaken from the cuvette, but they need not be dried.

   b. The light beam passes through the lower portion of the cuvette, so it is important to keep the bottom 1 in. or so of the cuvette scrupulously clean. Fingerprints, dirt, and scratches will contribute to the absorbance of your sample, interfering with a proper reading. Use only Kimwipes or lint-free tissues to clean cuvettes; like microscope lenses, cuvettes are made of optical glass and are easily scratched.

   c. When inserting the cuvette into the spectrophotometer, align the index mark on the cuvette with the index mark on the sample holder; be sure the cuvette is fully seated within the sample holder.

3. The spectrophotometer must be calibrated, or blanked, prior to use. Microbiological media are not generally clear, and the color in the media will absorb light. Calibrating the spectrophotometer ensures that any color in the media will be ignored and not counted as absorbance by the sample. The process involves placing a sterile tube of media into the machine and then setting the absorbance to zero. The process for calibrating the spectrophotometer is illustrated in Figure A.2.

4. Prior to transferring culture to a cuvette, be sure the culture has been mixed vigorously to evenly disperse the bacteria throughout the media.

5. Aseptically transfer the culture to a cuvette, filling it approximately halfway.

6. Place the cuvette in the sample holder of the spectrophotometer, being sure to align the mark on the cuvette with the mark on the sample holder. Close the cover to the sample holder.

7. If you are using a digital spectrophotometer, read and record the absorbance for the sample. If you are using an analog spectrophotometer, read the transmittance of your sample and use the equation in the beginning of the appendix to convert the transmittance to absorbance.

8. Remember that the cuvette contains a living culture. Follow your instructor's directors for decontamination of the cuvette when you are finished.

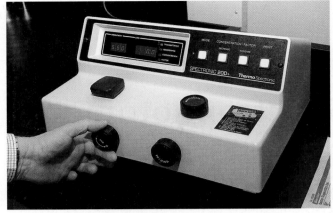

(a) The left front knob is used to turn on the machine; allow it to warm up for 15 min before proceeding. Set the filter to 600 nm and select the transmittance mode. Close the sample cover, and adjust the readout to zero percent transmittance by rotating the left front knob.

(b) Select absorbance mode. Place a cuvette containing sterile broth (the same type of broth used for your samples) into the sample holder, and close the cover. Adjust the absorbance to zero using the right front knob.

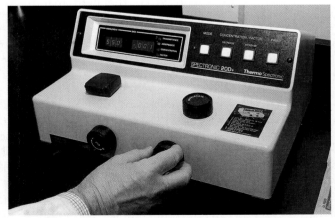

(c) If many samples are to be read, occasionally check the calibration of the spectrophotometer by repeating the calibration process.

Figure A.2 Calibration of the Spectronic 20 D digital spectrophotometer.

Use of Pipettes in the Laboratory

The study of microbiology routinely involves transferring precisely measured amounts of fluid between containers. Large pipettes are made of glass or plastic and resemble long drinking straws. They typically hold 1 to 10 ml (although larger sizes do exist) and can accurately dispense volumes as small as 1/100 ml (10 µl). Smaller micropipettes are made of plastic, generally hold no more than a single milliliter, and are capable of transferring volumes of less than a single microliter (1/1000 ml). Because each type of pipette has its own uses and associated techniques, they will be dealt with one at a time.

Part I: Serological and Mohr Pipettes

Large pipettes have been a staple of the microbiology laboratory for decades, allowing accurate transfer of fairly large (ml or more) volumes of liquids. In the past, pipettes were filled by sucking on them just as if they were drinking straws, the dangers inherent in this practice notwithstanding. Today, several types of pumps and bulbs are used to fill pipettes (Figure B.1), leading to a much safer lab experience. Pipettes always include certain information printed near their upper end (where the zero mark is located). In addition to the name of the manufacturer, this information always includes the maximum volume the pipette will hold, the smallest gradations on the barrel of the pipette, and whether the pipette is designed to deliver (TD) or to contain (TC) a specific volume (Figure B.2). Pipettes marked TD are called **serological pipettes** and are the most common type found in the microbiology laboratory. These pipettes should be filled to the appropriate mark for the volume to be delivered and then completely emptied, including blowing out any small amount remaining in the tip of the pipette. Mohr pipettes are designed to contain (TC) a specific volume and the volume dispensed must be mathematically determined. For example, if the pipette is filled to the 2.0 ml mark and liquid is dispensed until the pipette reads 4.2 ml, then 2.2 ml of liquid was dispensed. Fluid remaining in the tip of the pipette must not be dispensed as it is not accounted for in the pipette's calibration. In both types of pipettes, the volume of liquid should be measured at the base of the

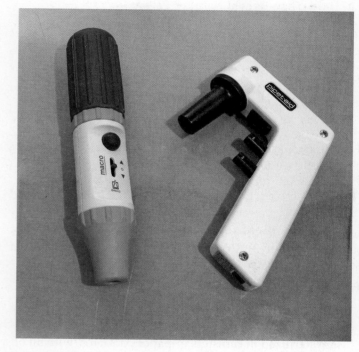

Figure B.1 Pipette pumps. Used to draw liquid into a pipette; manual (left) and electric (right) pumps have replaced the hazardous technique of "mouth pipetting."

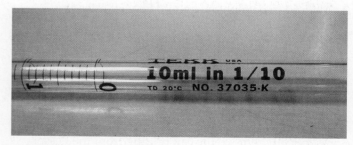

Figure B.2 Important information about a pipette is always located near the 0-ml mark. This information always includes a pipette's total volume, its smallest calibrated increments, and whether the pipette is calibrated to contain (TC) or to deliver (TD) its measured volume. The illustrated pipette has a total volume of 10 ml divided into 0.1-ml increments and is calibrated to deliver its measured volume.

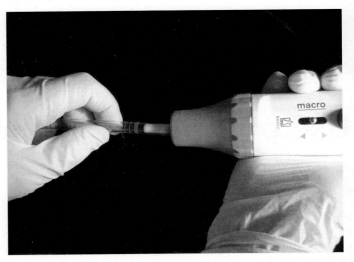

Figure B.4 Insert the pipette into the pump by holding it near the top and pressing it gently into the pipette pump. Do not use excess pressure when inserting the pipette as pipettes have been known to snap, causing injury.

Figure B.3 Reading the meniscus. The volume within a pipette is always read by aligning the base of the meniscus with the gradations on the outside of the pipette. The pipette on the left reads 3.3 ml while the pipette on the right reads 2.8 ml.

meniscus, the level of the fluid in the center of the pipettes lumen (Figure B.3). Procedures for filling and emptying a serological or Mohr pipette are described here.

Filling a Pipette

1. Retrieve a canister of the proper size pipettes and place it, on its side, on the edge of your lab bench. Pipettes are sterilized in canisters or paper packages, with all of the pipettes in a package being an identical size. Placing the canister horizontally lessens the chances of it falling over and reduces opportunities for contamination.

2. A liquid to be pipetted should be well mixed to disperse the organism evenly throughout the sample.

3. Remove a single pipette by grasping it near the top end. **Do not set the pipette down, and be sure not to touch the bottom end of the pipette.**

4. Gently insert the top end of the pipette into the pipette pump or bulb, friction will hold the pipette in place (Figure B.4).

5. Aseptically remove the cap or plug from the vessel containing your sample (test tube, flask, etc.) and hold it with the smallest finger of your pipetting hand. Orient the container at an angle to prevent contamination. Flame the lip of the vessel.

6. Just as you would with a loop or needle, move the culture container up and around the pipette. Withdraw the appropriate volume of liquid from the container, rotating the pipette and container so that the meniscus may be read with the pipette in the vertical position.

7. Remove the container from around the pipette, flame the opening, and replace the top. Set the container down.

8. Transfer the liquid to the receiving container (tube, plate, etc.) following the most appropriate of the following procedures.

Dispensing Liquid to a Tube

1. Remove the cap from the tube, holding it with the smallest finger of your pipetting hand. Be sure to flame the lip of the tube.

2. Hold the pipette steady while the tube is moved up and around the pipette. Be sure to hold the tube at an angle to prevent contamination.

3. Insert the pipette tip into the liquid (broth, diluent, etc.) in the tube and dispense the proper volume.

4. Withdraw the tube from the pipette, stopping briefly to touch the pipette tip to the inside of the tube, which will remove any excess broth from the tip of the pipette.

5. Flame the tube and replace its cap. Set the tube aside.

6. Carefully remove the pipette from the pipette pump or bulb. Dispose of the contaminated pipette according to your lab's protocols. Most commonly this involves placing the pipette, tip down, in a container of disinfectant. The pipettes will be autoclaved later.

Dispensing Liquid to a Plate

1. This procedure is most commonly used to inoculate a plate with a small amount of a liquid culture. Remove the top of the plate and hold it over the agar to protect against airborne contamination.

2. Without touching the agar with the pipette, dispense the correct volume onto the center of the agar surface.

After this, additional steps must be done quickly as the liquid will be absorbed by the agar in 20–30 sec.

3. Dispose of the contaminated pipette according to your lab's protocols. Most commonly this involves placing the pipette, tip down, in a container of disinfectant. The pipettes will be autoclaved later.

Part II: Micropipettes

Micropipettes are used to transfer very small volumes, as tiny as 0.5 µl, with precision. The micropipettes themselves are not calibrated but are filled using a pump (called a pipettor) that fills the pipette (called a tip) with a very precise volume of fluid. Pipettors are available in several sizes, calibrated to dispense volumes of 1 to 10 µl, 10 to 100 µl, and 100 to 1000 µl; each pipettor is designed for use with a specific tip (Figure B.5). Larger sizes are available but are not commonly used. The volume to be drawn into the pipette tip is adjusted by rotating the plunger on top of the pipettor until the correct volume, in µl, is displayed. The procedure for filling and emptying a micropipette is outlined here.

Filling a Micropipette

1. Determine the proper pipettor and tip for your needs based on the volume you wish to transfer.

2. A liquid to be pipetted should be well mixed to disperse the organisms evenly throughout the sample.

3. Adjust the pipettor by turning the plunger at the top of the instrument. For some pipettors, you may need to press and hold a small button to allow the plunger to turn. Do not attempt to adjust the pipettor beyond its designed range as you will damage it.

4. Hold the pipettor in the same hand in which you ordinarily hold a loop or needle. Open a rack of pipette tips (which are generally color coded to match the pipettor), and press the tip of the pipettor into a single sterile tip. Close the rack.

5. Aseptically remove the cap or plug from the vessel containing your sample (test tube, flask, etc.) and hold it with the smallest finger of your pipetting hand. Orient the container at an angle to prevent contamination. Flame the lip of the vessel.

6. Depress the plunger on top of the pipette until you feel resistance. This is the first "stop" on the plunger.

7. Insert the pipette tip into the container until it is just below the level of the liquid in the container.

8. Slowly release the plunger, drawing in exactly the volume of liquid indicated by the pipette's digital display. Remove the pipettor from the container, and set it aside, remembering to flame the lip and replace the top.

9. Transfer the liquid to the receiving container (tube, plate, etc.) following the most appropriate of the following procedures.

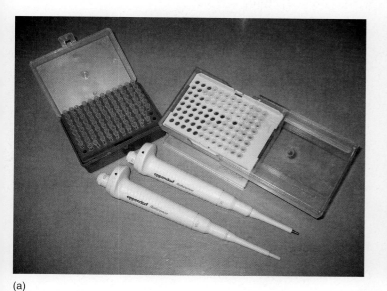

(a)

(b)

Figure B.5   (a) Micropipettors are found in different sizes, each of which uses a specific tip. (b) This pipettor has been adjusted to draw exactly 91.5 µl.

Dispensing Liquid to a Tube with a Micropipettor

1. Remove the cap from the tube, holding it with the small finger of your pipetting hand. Be sure to flame the lip of the tube.

2. Lower the pipettor into the tube, angling the pipette tip so that it touches the inside surface of the tube.

3. Slowly depress the plunger to the first stop, which should dispense all of the liquid in the pipettor, and then continue to the second stop, which blows out the tip.

4. Remove the pipettor from the tube, and then gently release the pressure on the plunger.

5. Flame the tube, and replace its cap. Set the tube aside.

6. Remove the tip from the pipettor by ejecting it into a biohazard container. To do this, place the tip of the pipettor directly over the container, and depress the plunger to the third stop, which will eject the tip.

Dispensing Liquid to a Plate with a Micropipettor

1. This procedure is most commonly used to inoculate a plate with a small amount of a liquid culture. Remove the top of the plate and hold it over the agar to protect against airborne contamination.

2. Gently touch the micropipette tip to the surface of the agar and depress the tip to the first stop to dispense the correct volume onto the center of the agar surface. Continue to the second stop to blow out the tip, ensuring that all liquid has been transferred. After this, additional steps must be done quickly as the liquid will be absorbed by the agar in 20–30 sec.

3. Remove the tip from the pipettor by ejecting it into a biohazard container. To do this, place the tip of the pipettor directly over the container, and depress the plunger to the third stop, which will eject the tip.

Preparation of Culture Media

A difficulty that arises when studying microorganisms in the laboratory is that they must be grown on artificial media. After all, a bacterium isolated from the human throat can hardly be grown on another throat for study; the best we can hope for is to design a media that replicates the nutrients and conditions within the throat to the greatest degree possible so that the organism will grow as it would in the body. To this end, thousands of different media formulations have been developed, each of which supplies a specific combination of nutrients and energy and creates an environment appropriate for growth.

All media can be classified as either **complex** or **defined.** A complex media contains one or more ingredients that are not precisely known, often an extract of animals, plants, or yeast. Most media commonly used in the microbiology laboratory fall into this category with trypticase soy agar and nutrient broth being examples of complex media used in this manual. Defined media have a precisely known chemical composition, with each ingredient weighed and added to the media during preparation. Minimal agar serves as an example of this type of media.

Nutritional Requirements of Bacteria

Since media is designed to satisfy the nutritional needs of bacterial cells, the first question that must be addressed concerns exactly what those needs are. In order to grow, a bacterium must have a source of carbon, energy, nitrogen, minerals, vitamins and growth factors, and water. Each of these components is addressed here.

Carbon Sources

Carbon forms the backbone of all organics molecules found in the bacterial cell, including proteins, carbohydrates, nucleic acids, and lipids. Bacteria that obtain carbon from organic compounds such as carbohydrates and proteins are referred to as **heterotrophs.** If a bacterium is able to use carbon dioxide as its sole source of carbon, it is know as an **autotroph.**

Energy Sources

Energy is required to assemble the raw materials found in the media into the biomolecules needed for continued cell growth. Bacteria may be classified into one of several groups based on the manner in which they derive energy. The prefix **chemo** indicates that energy is derived through the breakdown of chemical substrates while the prefix **photo** indicates that light is used to provide energy through photosynthesis. **Chemoorganotrophs** derive energy from the breakdown of organic molecules by fermentation or respiration. Most bacterial species, as well as all human beings for that matter, fall into this group. **Chemolithotrophs** rely on inorganic ions as an energy source, oxidizing inorganic substrates such as sulfur or iron to obtain energy. Chemolithotrophs such as the iron and sulfur bacteria are important in recycling inorganic nutrients in the environment. **Photoautotrophs** use photosynthetic pigments to convert sunlight into chemical energy through the process of photosynthesis. For these organisms, no energy source is found in the medium, but light must be supplied for growth to occur. Carbon dioxide is used as a carbon source. **Photoheterotrophs** also use light as a source of energy, but carbon is obtained from the breakdown of organic molecules such as glutamate. Table C.1 summarizes the grouping of bacteria by energy and carbon source.

Nitrogen

Nitrogen is essential for the synthesis of amino acids, nucleotides, and a few other cellular constituents. Depending on the bacteria, nitrogen sources may include inorganic nitrate (NO_3^-) and nitrite (NO_2^-) or organic ammonia (NH_3). A small number of bacteria are even capable of using atmospheric nitrogen (N_2) in a process called **nitrogen fixation.** Meat extracts and peptones (enzymatic digests of animal protein) are commonly used to supply nitrogen in microbiological media.

Minerals

Small quantities of several minerals are required to be part of any microbiological media. Bacterial metabolism utilizes minerals as cofactors for enzymatic reactions and includes them as part of the structure of cytochromes, bacteriochlorophyll, and vitamins. Minerals commonly required for bacterial metabolism include sodium, potassium, calcium, magnesium, manganese, iron, zinc, copper cobalt, and phosphorous. In the majority of media, the addition of meat or yeast extract provides the small amounts of minerals needed by most bacteria. If more than a catalytic amount is required, as with sodium chloride, for example, it may be added directly to the media.

TABLE C.1	Bacterial energy and carbon sources		
Bacterial group	**Energy source**	**Carbon source**	**Example**
Chemoorganotrophs	Organic molecules	Organic molecules	*E. coli* (and most other bacteria)
Chemolithotrophs	Inorganic molecules	Carbon dioxide	*Nitrosomonas*
Photoautotrophs	Light	Carbon dioxide	Green and purple sulfur bacteria
Photoheterotrophs	Light	Organic molecules	Purple nonsulfur bacteria

Vitamins and Growth Factors

Vitamins are complex organic molecules that serve as cofactors in many enzyme-catalyzed reactions. Because some bacteria, such as members of the genera *Streptococcus* and *Lactobacillus,* are unable to synthesize all the vitamins on their own, they are included in the media to ensure growth. The inclusion of meat or yeast extracts provides an adequate supply of most vitamins since only catalytic amounts are needed. **Growth factors** are complex organic compounds that are required for the growth of some fastidious organisms. Often, blood or serum is added to otherwise complete media to ensure the growth of certain bacteria; such a media is referred to as **enriched.**

Water

As bacterial cells consist of 70% or more water, it must obviously be present in the media. An aqueous environment is required for enzymatic reactions and to allow transport of materials within the cell. When preparing media, it is essential to use distilled or deionized water as tap water often contains ions such as calcium or magnesium that can interfere with bacterial growth.

Differential and Selective Media

Beyond providing nutrients for growth, media can be used to gather biological information or to pick a single organism from a complex mixture. A **selective** medium contains one or more ingredients that inhibits the growth of most bacteria, allowing only a single group to grow. Inhibitory agents used to create these selective conditions include dyes, salts, and antibiotics. For example, the dyes eosin and methylene blue are included in the formulation of EMB (*e*osin *m*ethylene *b*lue) agar to inhibit the growth of Gram-positive bacteria. Because the growth of Gram-negative bacteria is unaffected, the media serves to select only Gram-negative bacteria from a complex mixture. In a similar manner, the presence of 7.5% sodium chloride in mannitol salt agar selects for *Staphylococcus* but inhibits the growth of other bacteria that cannot tolerate so high a salt concentration.

A **differential** medium allows all species to grow, but the inclusion of a specific ingredient in the media results in some bacteria appearing different than others. When *Staphylococcus aureus* is grown on mannitol salt agar, it ferments the mannitol in the media to produce an acid, lowering the pH of the media around these colonies; a pH indicator in the media causes this portion of the media to turn from pink to yellow. Other *Staphylococcus* species do not ferment mannitol and do not change the pH of the media. Consequently, the media surrounding these colonies remains pink. It is also worth noting that mannitol salt agar, like many other types of media, has both differential and selective properties.

Preparation of Media

Long ago, laboratory workers made their own media from raw components, boiling plant materials or meat to prepare extracts. Today, most media is sold in powdered form, requiring only that the end user add water, heat, and perhaps adjust the final pH, making the task far easier. As a side benefit, commercially prepared media has a much better lot-to-lot consistency than media made from scratch. Even if you never make media during the course of your laboratory experience, knowing how it is made will give you the insight necessary to understand some of the biochemical reactions going on within the media and will certainly help you to troubleshoot unwanted results from time to time.

The first trait associated with a medium is its physical form, with media existing as liquids, solids, and semisolids. **Liquid media**—often referred to as **broths, milks,** or **infusions**—are, at their most basic, simply nutrients dissolved in water. Broths are used for growing large amounts of a specific bacterium but do not allow for any type of isolation. Streaking of bacteria and selection of isolated colonies is done on solid media. **Solid media** generally have the same formula as their liquid counterpart except that a solidifying agent is added so that the media is solid at temperatures used to incubate bacteria. The solidification agent used in almost all instances in the laboratory is **agar,** a polysaccharide isolated from the red algae *Gelidium*. Agar has several unique properties that make it ideal for use in microbiology. First, it melts at 100°C but does not solidify until about 42°C, allowing bacteria to be inoculated into melted agar at about 50°C without killing the cells. Once solidified, agar will not liquefy at normal incubation temperatures, allowing bacteria to be isolated from one another. Second, unlike solidifying agents like gelatin, very few bacteria can utilize agar as a nutrient. Any media containing more than 1% agar will be solid, with 1%–2% used in most instances. When a small quantity of agar, about 0.4%, is added to a liquid media, the result is a semisolid media, which has the consistency of a very thick liquid (think of jelly or hot fudge). Media of this consistency is used in motility studies as it is viscous enough to hold nonmotile bacteria in place but thin enough to allow motile bacteria freedom of movement.

Measurement and Mixing

Assuming that you are simply making a batch of media from a dehydrated powder, measurement is fairly easy. Follow the steps here.

- First determine the number of containers (pours, deeps, slants, broths, etc.) needed and multiply this by the volume required for each. The result is the total number of milliliters of media needed. Many different sized tubes are used in the laboratory, but the two sizes most commonly encountered are 16 or 20 mm in diameter by 15 cm in depth. Large tubes are used to prepare pours that will later be used to fill Petri dishes, while the smaller tubes are used for broths, deeps, and slants. Pours generally receive 12 ml of media, while deeps, slants, and broths receive 6, 4, and 5 ml, respectively. Broths that are to contain fermentation tubes receive 5–7 ml.

- The label on the media container will tell you how much powder is required for 1 L of media; you will have to calculate the amount of media based on what fraction of a liter you intend to make. This amount of media can be weighed out on a balance and added to a beaker of water (Figure C.1), along with a stirring bar.

- If the media does not contain agar, stirring alone will usually cause the media ingredients to go into solution; if the media does contain agar, it will need to be boiled to bring the agar into solution. In this case, mark the outside of the beaker with a lab marker to indicate the level of the water prior to heating.

- Heat the mixture on an electric hot plate, while stirring, until it comes to a boil (Figure C.2). Check the level of media against the mark you put on the beaker earlier, and add distilled water if some was lost to evaporation during the heating process. Keep the media at about 60°C to prevent solidification.

Adjusting the pH

Even though prepared media contain buffering agents to keep the pH in the desired range, it should always be checked, and adjusted if necessary, before being dispensed into smaller containers and autoclaved. A pH meter is ideal for this purpose, but pH paper is sufficient to the task in most cases.

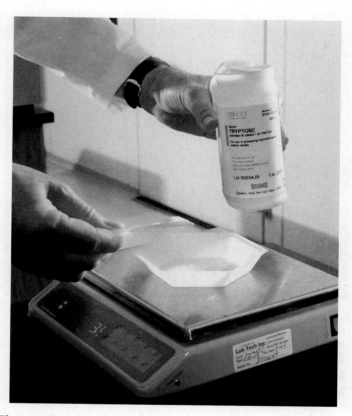

Figure C.1 A balance is used to weigh out enough dehydrated media for a single batch.

Figure C.2 Media that contains agar must be heated to boiling to dissolve the agar.

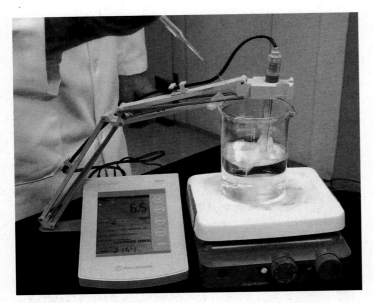

Figure C.3 When the media is entirely in solution, the pH may be checked with a pH meter or pH paper.

- If the pH is too low, use NaOH, which should be available as 1 N and 0.1 N solutions, to adjust the pH upward. Use the 1N solution for initial adjustments and the less concentrated solution for fine tuning of the pH. If the pH is too high, use 1 N and 0.1 N HCl to adjust the pH downward (Figure C.3). In both cases, a glass stirring rod or stir bar should be used to mix the media as the pH is adjusted.

Filling Test Tubes

After the pH of the media has been adjusted, it must be dispensed into test tubes.

- For small production runs of less than 100 tubes, an automatic pipettor may be used to fill tubes one at a time (Figure C.4), while larger batches may be dispensed more easily using an automatic dispenser (Figure C.5).
- If the media being dispensed contains agar, it should be stirred continually on a hot plate to keep the agar in solution. Keep the media warm, but do not allow it to boil.
- If the media is to be used for fermentation, insert a Durham tube, open end down, into each tube. Although the tubes may float initially, they will submerge during autoclaving.
- Prior to autoclaving, tubes must be capped. Slip-on caps made of polypropylene are most commonly used and these caps are manufactured with an air gap that allows steam to escape during autoclaving. If screw on caps are to be used, it is important to leave the caps loose so that steam can escape from the tubes during autoclaving (Figure C.6).

Figure C.4 Small batches of media may be dispensed using a pipettor.

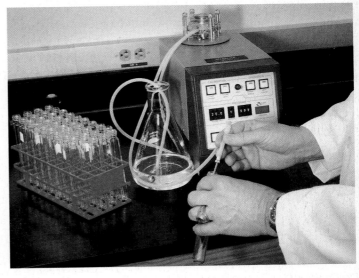

Figure C.5 Large batches of media are typically dispensed using an automatic dispenser.

Autoclaving

Once the media is prepared, it must be autoclaved immediately as bacteria in the water, on the inside walls of the tube, and in the dehydrated medium will soon begin to grow, destroying the medium. Tubes containing media are generally packed into baskets, which are in turn labeled with the type of media on the outside of the basket. The baskets are then loaded into the autoclave and sterilized (Figure C.7), paying attention to the following:

- Sterilization occurs at 121°C, which corresponds to a chamber pressure of 15 pounds per square inch (psi). All autoclaves have some means of checking the chamber pressure, and this should be done to make sure that pressure inside the chamber reaches 15 psi.
- Allow ample room for steam to circulate throughout the autoclave. If the autoclave is too tightly packed, sterilization may be incomplete.
- Adjust the time of sterilization to the size of the load. It takes time for the material in the autoclave to reach a temperature high enough to allow sterilization to begin. While a small load may take only 10–15 min, a full autoclave may require 30 min to reach operating temperature.

After Sterilization

After sterilization, media must be manipulated and stored according to its final purpose.

- If tubes are to be used to prepare slants, lay the baskets of tubes in an almost horizontal position immediately after removing them from the autoclave. The easiest way to do this is to use a piece of rubber tubing to support the capped ends of the tubes while they solidify (Figure C.8). The agar should be fully solidified in 30–60 min.
- Broths, deeps, and other similar media should be allowed to cool to room temperature after removal from the autoclave. Once cool, they should be stored in the refrigerator or cold room. Media tends to lose moisture at room temperature but will remain usable for months when stored at 4°C.

Figure C.7 Media is sterilized in at autoclave at 15 psi. Sterilization times usually run from 10–30 min.

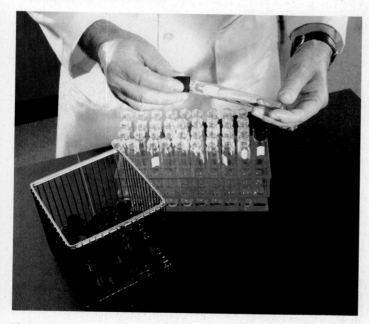

Figure C.6 When capping tubes of media, it is imperative that caps be loose enough that steam can escape during the autoclaving process.

Figure C.8 Slants are prepared by laying tubes almost horizontally and allowing them to solidify.

Media, Reagents, and Stain Formulas

Microbiological Media

Bile Esculin Agar

Pancreatic digest of gelatin	5.0 g
Beef extract	3.0 g
Oxgall	20.0 g
Ferric citrate	0.5 g
Esculin	1.0 g
Agar	14.0 g
Distilled or deionized water to	1.0 L

Adjust the pH to 6.8 ± 0.2 at 25°C

Brilliant Green Bile Broth

Peptone	10.0 g
Oxgall	20.0 g
Lactose	10.0 g
Brilliant green	13.3 mg
Distilled or deionized water to	1.0 L

Adjust the pH to 7.2 ± 0.2 at 25°C

Complete Agar (Nutrient Agar 1.5%)

Beef extract	3.0 g
Peptone	5.0 g
Sodium chloride	8.0 g
Agar	15.0 g
Distilled or deionized water to	1.0 L

Adjust the pH to 7.3 ± 0.2 at 25°C

Decarboxylase Broth Base (Moeller)

Peptone	5.0 g
Beef extract	5.0 g
Glucose	0.5 g
Bromcresol purple	10.0 mg
Cresol red	5.0 mg
Pyridoxal	5.0 mg
Distilled or deionized water to	1.0 L

Adjust the pH to 6.0 ± 0.2 at 25°C

Desoxycholate Agar

Peptone	10.0 g
Lactose	10.0 g
Sodium desoxycholate	1.0 g
Sodium chloride	5.0 g
Dipotassium phosphate	2.0 g
Ferric ammonium citrate	1.0 g
Sodium citrate	1.0 g
Agar	15.0 g
Neutral red	30.0 mg
Distilled or deionized water to	1.0 L

Adjust the pH to 7.1 ± 0.2 at 25°C

DNase Test Agar with Methyl Green

Pancreatic digest of casein	10.0 g
Proteose peptone no. 3	10.0 g
Deoxyribonucleic acid	2.0 g
Sodium chloride	5.0 g
Agar	15.0 g
Methyl green	50.0 mg
Distilled or deionized water to	1.0 L

Adjust the pH to 7.3 ± 0.2 at 25°C

EC Broth

Tryptose	20.0 g
Lactose	5.0 g
Bile salts no. 3	1.5 g
Dipotassium phosphate	4.0 g
Monopotassium phosphate	1.5 g
Sodium chloride	5.0 g
Distilled or deionized water to	1.0 L

Adjust the pH to 6.9 ± 0.2 at 25°C

Endo Agar

Dipotassium phosphate	3.5 g
Peptic digest of animal tissue	10.0 g
Agar	15.0 g
Lactose	10.0 g
Sodium sulfite	2.5 g
Basic fuchsin	0.5 g
Distilled or deionized water to	1.0 L

Adjust the pH to 7.5 ± 0.2 at 25°C

Eosin Methylene Blue Agar

Pancreatic digest of gelatin	10.0 g
Lactose	10.0 g
Dipotassium phosphate	1.0 g
Eosin Y	0.4 g
Lactose	10.0 g
Methylene blue	65.0 mg
Agar	15.0 g
Distilled or deionized water to	1.0 L

Adjust the pH to 7.1 ± 0.2 at 25°C

Fluid Thioglycollate Medium

Yeast extract	5.0 g
Pancreatic digest of casein	15.0 g
Dextrose	5.5 g
Sodium chloride	2.5 g
Sodium thioglycollate	0.5 g
L-Cystine	0.5 g
Agar	0.75 g
Resazurin	1.0 mg
Distilled or deionized water to	1.0 L

Adjust the pH to 7.1 ± 0.2 at 25°C

Halobacterium Agar

Casamino acids	5.0 g
Yeast extract	5.0 g
Sodium glutamate	1.0 g
Trisodium citrate	3.0 g
Magnesium sulfate heptahydrate	20.0 g
Ferrous chloride tetrahydrate	20.0 g
Manganese chloride tetrahydrate	0.36 g
Agar	20.0 g
Distilled or deionized water to	1.0 L

Adjust the pH to 7.1 ± 0.1 at 25°C

Hektoen Enteric Agar

Proteose peptone	12.0 g
Yeast extract	3.0 g
Bile salts no. 3	9.0 g
Lactose	12.0 g
Saccharose	12.0 g
Salicin	2.0 g
Sodium chloride	5.0 g
Sodium thiosulfate	5.0 g
Ferric ammonium citrate	1.5 g
Agar	13.5 g
Bromthymol blue	65.0 mg
Acid fuschin	100.0 mg
Distilled or deionized water to	1.0 L

Adjust the pH to 7.5 ± 0.2 at 25°C

Kligler's Iron Agar

Pancreatic digest of casein	10.0 g
Peptic digest of animal tissue	10.0 g
Lactose	10.0 g
Dextrose	1.0 g
Sodium chloride	5.0 g
Ferric ammonium citrate	0.5 g
Sodium thiosulfate	0.5 g
Agar	15.0 g
Crystal violet	25.0 mg
Distilled or deionized water to	1.0 L

Adjust the pH to 7.4 ± 0.2 at 25°C

Lauryl Tryptose Broth

Tryptose	20.0 g
Lactose	5.0 g
Dipotassium phosphate	2.75 g
Monopotassium phosphate	2.75 g
Sodium chloride	5.0 g
Sodium lauryl sulfate	0.1 g
Distilled or deionized water to	1.0 L
Adjust the pH to 6.8 ± 0.2 at 25°C	

LB (Luria-Bertani) Agar

Tryptone	10.0 g
Yeast extract	5.0 g
Sodium chloride	10.0 g
Agar	15.0 g
Distilled or deionized water to	1.0 L
Adjust the pH to 7.0 ± 0.2 at 25°C	

LB (Luria-Bertani) Broth

Tryptone	10.0 g
Yeast extract	5.0 g
Sodium chloride	5.0 g
Distilled or deionized water to	1.0 L
Adjust the pH to 7.5 ± 0.2 at 25°C	

Litmus Milk

Skim milk	100.0 g
Azolitmin	0.5 g
Sodium sulfite	0.5 g
Distilled or deionized water to	1.0 L
Adjust the pH to 6.5 ± 0.2 at 25°C	

Lysine Iron Agar

Peptone	5.0 g
Yeast extract	3.0 g
Dextrose	1.0 g
Lactose	10.0 g
L-Lysine HCl	10.0 g
Ferric ammonium citrate	0.5 g
Sodium thiosulfate	0.04 g
Bromcresol purple	0.02 g
Agar	15.0 g
Distilled or deionized water to	1.0 L
Adjust the pH to 6.7 ± 0.2 at 25°C	

MacConkey Agar

Peptone	17.0 g
Proteose peptone	3.0 g
Lactose	10.0 g
Bile salts no. 3	1.5 g
Sodium chloride	5.0 g
Agar	13.5 g
Neutral red	30.0 mg
Crystal violet	1.0 mg
Distilled or deionized water to	1.0 L
Adjust the pH to 7.1 ± 0.2 at 25°C	

m-Staphylococcus Broth

Pancreatic digest of casein	10.0 g
Yeast extract	2.5 g
Lactose	2.0 g
Mannitol	10.0 g
Dipotassium phosphate	5.0 g
Sodium chloride	75.0 g
Distilled or deionized water to	1.0 L
Adjust the pH to 7.0 ± 0.2 at 25°C	

Malonate Broth

Yeast extract	1.0 g
Ammonium sulfate	2.0 g
Dipotassium phosphate	0.6 g
Monopotassium phosphate	0.4 g
Sodium chloride	2.0 g
Sodium malonate	3.0 g
Gluocse	0.25 g
Bromthymol blue	25.0 mg
Distilled or deionized water to	1.0 L
Adjust the pH to 6.7 ± 0.2 at 25°C	

Mannitol Salt Agar

Proteose peptone no. 3	10.0 g
Beef extract	1.0 g
D-Mannitol	10.0 g
Sodium chloride	75.0 g
Agar	15.0 g
Phenol red	25.0 mg
Distilled or deionized water to	1.0 L
Adjust the pH to 7.4 ± 0.2 at 25°C	

Minimal Agar

Glucose	2.0 g
Ammonium sulfate	1.0 g
Dipotassium hydrogen phosphate	7.0 g
Magnesium sulfate	0.5 g
Agar	15.0 g
Distilled or deionized water to	1.0 L

Adjust the pH to 7.0 ± 0.2 at 25°C

Motility Media

Beef extract	3.0 g
Pancreatic digest of casein	10.0 g
Sodium chloride	5.0 g
Agar	4.0 g
Distilled or deionized water to	1.0 L

Adjust the pH to 7.3 ± 0.2 at 25°C

MR-VP Broth

Buffered peptone	7.0 g
Dipotassium phosphate	5.0 g
Glucose	5.0 g
Distilled or deionized water to	1.0 L

Adjust the pH to 6.9 ± 0.2 at 25°C

Mueller-Hinton II Agar

Beef extract	2.0 g
Acid hydrolysate of casein	17.5 g
Starch	1.5 g
Agar	17.0 g
Distilled or deionized water to	1.0 L

Adjust the pH to 7.3 ± 0.2 at 25°C

Nitrate Broth

Beef extract	3.0 g
Peptone	4.0 g
Proteose peptone no. 3	1.0 g
Potassium nitrate	1.0 g
Distilled or deionized water to	1.0 L

Adjust the pH to 7.0 ± 0.2 at 25°C

Nutrient Agar

Beef extract	3.0 g
Peptone	5.0 g
Agar	15.0 g
Distilled or deionized water to	1.0 L

Adjust the pH to 6.8 ± 0.2 at 25°C

Nutrient Broth

Beef extract	3.0 g
Peptone	5.0 g
Distilled or deionized water to	1.0 L

Adjust the pH to 6.8 ± 0.2 at 25°C

Nutrient Gelatin

Beef extract	3.0 g
Peptone	5.0 g
Gelatine	120.0 g
Distilled or deionized water to	1.0 L

Adjust the pH to 6.8 ± 0.2 at 25°C

OF Media

Pancreatic digest of casein	2.0 g
Sodium chloride	5.0 g
Dipotassium phosphate	0.3 g
Bromthymol blue	80.0 mg
Agar	2.0 g
Carbohydrate to be tested	10.0 g
Distilled or deionized water to	1.0 L

Adjust the pH to 6.8 ± 0.2 at 25°C

Phenol Red Broth

Pancreatic digest of casein	10.0 g
Sodium chloride	5.0 g
Carbohydrate to be tested	5.0 g
Phenol red	18.0 mg
Distilled or deionized water to	1.0 L

Adjust the pH to 7.3 ± 0.2 at 25°C

Phenylalanine Agar

D,L-Phenylalanine	2.0 g
Yeast extract	3.0 g
Sodium chloride	5.0 g
Dipotassium phosphate	1.0 g
Agar	12.0 g
Distilled or deionized water to	1.0 L

Adjust the pH to 7.3 ± 0.2 at 25°C

Purple Broth

Pancreatic digest of gelatin	10.0 g
Beef extract	1.0 g
Sodium chloride	5.0 g
Carbohydrate to be tested	10.0 g
Bromcresol purple	20.0 mg
Distilled or deionized water to	1.0 L

Adjust the pH to 6.8 ± 0.2 at 25°C

Sabouraud Dextrose Agar

Enzymatic digest of casein	10.0 g
Dextrose	40.0 g
Agar	15.0 g
Distilled or deionized water to	1.0 L

Adjust the pH to 5.6 ± 0.2 at 25°C

SIM Medium

Pancreatic digest of casein	20.0 g
Peptic digest of animal tissue	6.1 g
Ferrous ammonium sulfate	0.2 g
Sodium thiosulfate	0.2 g
Agar	3.5 g

Adjust the pH to 7.3 ± 0.2 at 25°C

Simmon's Citrate Agar

Ammonium dihydrogen phosphate	1.0 g
Dipotassium phosphate	1.0 g
Sodium chloride	5.0 g
Sodium citrate	2.0 g
Magnesium sulfate	0.2 g
Agar	15.0 g
Bromthymol blue	80.0 mg
Distilled or deionized water to	1.0 L

Adjust the pH to 6.9 ± 0.2 at 25°C

Skim Milk Agar

Skim milk powder	100.0 g
Agar	15.0 g
Distilled or deionized water to	1.0 L

Adjust the pH to 7.2 ± 0.2 at 25°C

Sodium Chloride Broth (Heart Infusion Broth Supplemented with 6.5% NaCl)

Beef heart, infusion from 500 g (commercially prepared and dehydrated)	10.0 g
Tryptose	10.0 g
Sodium chloride	65.0 g
Distilled or deionized water to	1.0 L

Adjust the pH to 7.4 ± 0.2 at 25°C

Spirit Blue Agar

Pancreatic digest of casein	10.0 g
Yeast extract	5.0 g
Agar	20.0 g
Spirit blue	0.15 g
Lipase reagent (tributyrin/polysorbate 80)	35.0 ml
Distilled or deionized water to	1.0 L

Adjust the pH to 6.8 ± 0.2 at 25°C

Starch Agar

Beef extract	3.0 g
Soluble starch	10.0 g
Agar	12.0 g
Distilled or deionized water to	1.0 L

Adjust the pH to 7.5 ± 0.2 at 25°C

Top Agar

Tubes containing 2 ml top agar should be prepared just prior to use from bottles of histidine/biotin stock solution and top agar base.

Histidine/Biotin Stock Solution

D-Biotin	30.9 mg
L-Histidine HCl	24.0 mg
Distilled or deionized water to	250 ml

Dissolve by boiling and sterilize by autoclaving for 20 min or filtering through a 0.22-μm membrane filter. Store the solution at 4°C.

Top Agar Base

Sodium chloride	5.0 gm
Agar	6.0 gm
Distilled or deionized water to	1.0 L

After autoclaving, store 100-ml aliquots in 250-ml flasks. Just prior to use, add 10 ml of the histidine/biotin stock solution to 100 ml of liquefied top agar base that has been cooled to 50°C. Mix thoroughly and aseptically distribute 2 ml of the mixture to sterile tubes. Hold the tubes in a 50°C incubator or water bath until they are used.

Triple Sugar Iron Agar

Beef extract	3.0 g
Yeast extract	3.0 g
Pancreatic digest of casein	15.0 g
Proteose peptone no. 3	5.0 g
Dextrose	1.0 g
Lactose	10.0 g
Sucrose	10.0 g
Ferrous sulfate	0.2 g
Sodium chloride	5.0 g
Sodium thiosulfate	0.3 g
Agar	13.0 g
Phenol red	24.0 mg
Distilled or deionized water to	1.0 L

Adjust the pH to 7.4 ± 0.2 at 25°C

Trypticase Soy Agar

Pancreatic digest of casein	15.0 g
Enzymatic digest of soybean meal	5.0 g
Sodium chloride	5.0 g
Agar	15.0 g
Distilled or deionized water to	1.0 L

Adjust the pH to 7.3 ± 0.2 at 25°C

Trypticase Soy Agar with 5% Sheep's Blood

Tryptone H	15.0 g
Soytone	5.0 g
Sodium chloride	5.0 g
Agar	15.0 g
Distilled or deionized water to	1.0 L

Adjust the pH to 7.4 ± 0.2 at 25°C

Add 5% sterile defibrinated blood to sterile agar that has been cooled to 45°C–50°C. Mix well prior to pouring the plates.

Trypticase Soy Broth

Pancreatic digest of casein	17.0 g
Papaic digest of soybean meal	3.0 g
Sodium chloride	5.0 g
Dipotassium phosphate	2.5 g
Dextrose	2.5 g
Distilled or deionized water to	1.0 L

Adjust the pH to 7.3 ± 0.2 at 25°C

Urea Broth

Yeast extract	0.1 g
Monosodium phosphate	9.1 g
Dipotassium phosphate	9.5 g
Urea	20.0 g
Phenol red	12.0 mg
Distilled or deionized water to	1.0 L

Adjust the pH to 6.8 ± 0.1 at 25°C

Veal Infusion Agar

Lean veal, infusion from 500 g (commercially prepared and dehydrated)	10.0 g
Proteose peptone no. 3	10.0 g
Sodium chloride	5.0 g
Agar	15.0 g
Distilled or deionized water to	1.0 L

Adjust the pH to 7.4 ± 0.2 at 25°C

Xylose Lysine Desoxycholate Agar

Xylose	3.75 g
L-Lysine	5.0 g
Lactose	7.5 g
Saccharose	7.5 g
Sodium chloride	5.0 g
Yeast extract	3.0 g
Phenol red	80.0 mg
Sodium desoxycholate	2.5 g
Sodium thiosulfate	6.8 g
Ferric ammonium citrate	0.8 g
Agar	15.0 g
Distilled or deionized water to	1.0 L

Adjust the pH to 7.4 ± 0.2 at 25°C

Stains and Other Reagents

Acid-Alcohol (Kinyoun Stain)

Ethanol (95%)	97 ml
Concentrated HCl	3 ml

Barritt's Reagent (Voges-Proskauer Test)

Reagent A: 6 g alpha-napthol in 100 ml ethyl alcohol

Reagent B: 16 g potassium hydroxide in 100 ml water

Brilliant Green (Kinyoun Acid-Fast Procedure)

Brilliant green dye	1.0 g
Sodium azide	10.0 mg
Distilled or deionized water to	100.0 ml

Carbol Fuschin (Kinyoun Acid-Fast Procedure)

Solution A: Dissolve 1.5 g basic fuschin in 5.0 ml of 95% ethanol. Add 20 ml of isopropanol.

Solution B: Dissolve 4.5 g phenol in 75.0 ml of distilled water.

Mix solutions A and B, and let stand for several days. Filter prior to use.

Carbol Fuschin (Ziel-Neelsen Procedure)

Solution A: Dissolve 0.3 g basic fuschin in 10.0 ml of 95% ethanol. Add 20 ml of isopropanol.

Solution B: Dissolve 5.0 g phenol in 95.0 ml of distilled water.

Mix solutions A and B and let stand for several days. Filter prior to use.

Crystal Violet Stain (Hucker Modification)

Solution A: Dissolve 2.0 g of crystal violet (85% dye content) in 20 ml of 95% ethyl alcohol.

Solution B: Dissolve 0.8 g ammonium oxalate in 80 ml distilled water.

Mix solutions A and B.

Ferric Chloride Reagent

Dissolve 10.0 g ferric chloride in approximately 90.0 ml distilled or deionized water. Add water to bring the final volume to 100 ml.

Gram's Iodine (Lugol's Formulation)

Dissolve 2.0 g of potassium iodide in 300 ml of distilled water and then add 1.0 g iodine crystals.

Kovac's Reagent (for Indole Test)

n-Amyl alcohol	75.0 ml
Concentrated hydrochloric acid	25.0 ml
p-Dimethyamine-benzaldehyde	5.0 g

Lactophenol Cotton Blue Stain

Phenol crystals	20.0 g
Lactic acid	20.0 ml
Glycerol	40.0 ml
Cotton blue	50.0 mg

Dissolve the phenol crystals in the other ingredients by heating the mixture gently using hot tap water.

Leifson Stain (Flagellar Staining)

Solution A: Dissolve 1.5 g sodium chloride in 100 ml distilled water.

Solution B: Dissolve 3.0 g tannic acid in 100 ml water.

Solution C: Dissolve 0.9 g pararosaniline acetate and 0.3 g pararosaniline hydrochloride in 95% ethanol.

Mix equal volumes of solution A and B, and then add 2 volumes of this mixture to one volume of solution C. May be refrigerated for up to 2 months.

Malachite Green Solution (Spore Stain)

Dissolve 5.0 g malachite green oxalate in 100 ml distilled water.

Methylene Blue (Loeffler's Formulation)

Solution A: Dissolve 0.3 g of methylene blue (90% dye content) in 30.0 ml 95% ethyl alcohol.

Solution B: Dissolve 0.01 g potassium hydroxide in 100 ml distilled water.

Mix solutions A and B.

Methylene Blue Solution (Simple and Acid-Fast Staining)

Methylene blue dye	10.0 mg
Distilled or deionized water to	225.0 ml

Methylene Blue Reductase Reagent

Dissolve 8.0 mg methylene blue dye in 200 ml of distilled water.

Napthol, Alpha

5% Alpha-napthol in 95% ethyl alcohol

Caution: Alpha-napthol is carcinogenic; avoid all contact with human tissue.

Nigrosin Solution

Nigrosin, water soluble	10.0 g
Distilled water	100 ml

Boil for 30 min. Add 0.5 ml formaldehyde as a preservative. Filter twice through two layers of filter paper and store under aseptic conditions.

Nitrate Test Reagents

Solution A: Dissolve 8 g sulfanilic acid in 1000 ml 5 N acetic acid (1 part glacial acetic acid to 2.5 parts water).

Solution B: Dissolve 5 g dimethyl-α-naphthylamine in 100 ml 5 N acetic acid. Do not mix solutions.

Caution: It is possible that dimethyl-α-naphthylamine is carcinogenic. Avoid all contact with the skin.

Safranin (Gram Staining)

Safranin O (2.5% solution in 95% ethyl alcohol)	10.0 ml
Distilled water	100.0 ml

Data Sheet for Unknown Identification

Student: _____

Lab Section: _____

Habitat: _____

Culture Number: _____

Morphological Characteristics
Cell shape:

Arrangement:

Size:

Endospore:

Motility:

Capsule:

Cultural Characteristics
Colony morphology:

Agar slant:

Nutrient broth:

Oxygen requirements:

Optimum temperature:

Physiological/Biochemical Characteristics
Fermentation tests:

Glucose:

Lactose:

Sucrose:

Mannitol:

Maltose:

Hydrolytic tests:

Gelatin:

Starch:

Casein:

Lipid:

DNase:

Other biochemical tests:

Indole:

Methyl red:

Voges-Proskauer:

Citrate utilization:

Nitrate reduction:

H_2S production:

Urease:

Catalase:

Oxidase:

Phenylalanase:

Litmus milk:

Glossary

A

abiotic Nonliving factors such as soil, water, temperature, and light that are studied when looking at an ecosystem.

ABO blood group system Developed by Karl Landsteiner in 1904; the identification of different blood groups based on differing isoantigen markers characteristic of each blood type.

abscess An inflamed, fibrous lesion enclosing a core of pus.

acellular vaccine A vaccine preparation that contains specific antigens such as the capsule or toxin from a pathogen and not the whole microbe. Acellular (without a cell).

acetylcholine A neurotransmitter that causes the contraction of muscles.

acid fast A term referring to the property of mycobacteria to retain carbol fuchsin even in the presence of acid-alcohol. The staining procedure is used to diagnose tuberculosis.

acidic A solution with a pH value below 7 on the pH scale.

acidophile Bacteria that grow at acid pH levels.

active immunity Immunity acquired through direct stimulation of the immune system by an antigen.

active site The specific region on an enzyme that binds substrate. The site for reaction catalysis.

acute Characterized by rapid onset and short duration.

adenine (A) One of the nitrogen bases found in DNA and RNA, with a purine form.

adenosine triphosphate (ATP) A nucleotide that is the primary source of energy for cells.

adhesion The process by which microbes gain a more stable foothold at the portal of entry; often involves a specific interaction between the molecules on the microbial surface and the receptors on the host cell.

adsorption A process of adhering one molecule onto the surface of another molecule.

aerobe A microorganism that lives and grows in the presence of free gaseous oxygen (O_2).

aerobic respiration Respiration in which the final electron acceptor in the electron transport chain is oxygen (O_2).

aerosols Suspensions of fine dust or moisture particles in the air that contain live pathogens.

aerotolerant The state of not utilizing oxygen but not being harmed by it.

aflatoxin From *Aspergillus flavus* toxin, a mycotoxin that typically poisons moldy animal feed and can cause liver cancer in humans and other animals.

agar A polysaccharide found in seaweed and commonly used to prepare solid culture media.

agarose gel electrophoresis A laboratory technique that separates fragments of DNA according to their size.

agglutination The aggregation by antibodies of suspended cells or similar-size particles (agglutinogens) into clumps that settle.

agranulocyte One form of leukocyte (white blood cells) having globular, nonlobed nuclei and lacking prominent cytoplasmic granules.

alcoholic fermentation An anaerobic degradation of pyruvic acid that results in alcohol production.

algae Photosynthetic, plantlike organisms that generally lack the complex structure of plants; they may be single-celled or multicellular and inhabit diverse habitats such as marine and freshwater environments, glaciers, and hot springs.

alkaliphile Bacteria that grow at alkaline pH values.

allograft Relatively compatible tissue exchange between nonidentical members of the same species. Also called *homograft*.

Ames test A method for detecting mutagenic and potentially carcinogenic agents based upon the genetic alteration of nutritionally defective bacteria.

amination The addition of an amine (—NH_2) group to a molecule.

amino acids The building blocks of protein. Amino acids exist in 20 naturally occurring forms that impart different characteristics to the various proteins they compose.

aminoglycoside A complex group of drugs derived from soil actinomycetes that impairs ribosome function and has antibiotic potential, e.g., streptomycin.

ammonification Phase of the nitrogen cycle in which ammonia is released from decomposing organic material.

amoeboid motion Movement marked by the formation and extension of transitory pseudopodia.

amphitrichous Having a single flagellum or a tuft of flagella at opposite poles of a microbial cell.

amylopectin A form of starch where a bond exists between the first and sixth carbon atoms of glucose (a 1,6-α-glycosidic linkage), resulting in a branched arrangement.

amylose A form of starch in which the glucose molecules are bonded between the first and fourth carbon atoms, forming a 1,4-α-glycosidic linkage and resulting in a long, straight chain.

anabolism The energy-consuming process of synthesizing cell molecules and structures.

anaerobe A microorganism that grows best, or exclusively, in the absence of oxygen.

anaerobe jar A sealed container used to grow bacteria under anaerobic conditions.

anaerobic respiration Respiration in which the final electron acceptor in the electron transport chain is an inorganic molecule containing sulfate, nitrate, nitrite, carbonate, and so on.

anamorph The asexual reproductive stage of a fungus.

anaphylaxis The unusual or exaggerated allergic reaction to antigen that leads to severe respiratory and cardiac complications.

anion A negatively charged ion.

anoxygenic Non-oxygen-producing.

anterior cruciate ligament One of the three major ligaments responsible for stability of the knee.

antibiotic A chemical substance from one microorganism that can inhibit or kill another microbe even in minute amounts.

antibody   A large protein molecule evoked in response to an antigen that interacts specifically with that antigen.

antigen   Any cell, particle, or chemical that induces a specific immune response by B cells or T cells and can stimulate resistance to an infection or a toxin. See *immunogen*.

antimicrobial   A special class of compounds capable of destroying or inhibiting microorganisms.

antimicrobial peptides   Short protein molecules found in epithelial cells; have the ability to kill bacteria.

antiparallel   Description of the orientation of the two strands of a DNA molecule, such that the two parallel strands run in different orientations, similar to two opposing lanes of traffic on a street.

antisepsis   Chemical treatments to kill or inhibit the growth of all vegetative microorganisms on body surfaces.

antiseptic   A growth-inhibiting agent used on tissues to prevent infection.

antiserum   Antibody-rich serum derived from the blood of animals (deliberately immunized against infectious or toxic antigen) or from people who have recovered from specific infections.

antitoxin   Globulin fraction of serum that neutralizes a specific toxin. Also refers to the specific antitoxin antibody itself.

apicomplexans   A group of protozoans that lack locomotion in the mature state.

appendages   Accessory structures that sprout from the surface of bacteria. They can be divided into two major groups: those that provide motility and those that enable adhesion.

ascospore   A spore formed within a saclike cell (ascus) of Ascomycota following nuclear fusion and meiosis.

ascus   Special fungal sac in which haploid spores are created.

asepsis   A condition free of viable pathogenic microorganisms.

aseptic technique   Methods of handling microbial cultures, patient specimens, and other sources of microbes in a way that prevents infection of the handler and others who may be exposed.

asymptomatic   An infection that produces no noticeable symptoms even though the microbe is active in the host tissue.

asymptomatic carrier   A person with an inapparent infection who shows no symptoms of being infected yet is able to pass the disease agent on to others.

atom   The smallest particle of an element to retain all the properties of that element.

AUG (start codon)   The codon that signals the point at which translation of a messenger RNA molecule is to begin.

autoantibody   An "anti-self" antibody having an affinity for tissue antigens of the subject in which it is formed.

autoclave   A sterilization chamber that allows the use of steam under pressure to sterilize materials. The most common temperature/pressure combination for an autoclave is 121°C and 15 psi.

autograft   Tissue or organ surgically transplanted to another site on the same subject.

autoimmune disease   The pathologic condition arising from the production of antibodies against autoantigens, e.g., rheumatoid arthritis. Also called *autoimmunity*.

autotroph   A microorganism that requires only inorganic nutrients and whose sole source of carbon is carbon dioxide.

auxotroph   A bacterium that is unable to synthesize an essential compound.

avirulent   A microbe that has lost its pathogenicity.

axial filament   A type of flagellum (called an endoflagellum) that lies in the periplasmic space of spirochetes and is responsible for locomotion. Also called *periplasmic flagellum*.

azole   Five-membered heterocyclic compounds, such as that found in histidine, which are used in antifungal therapy.

azolitmin   A litmus indicator that ranges from pink (below pH 4.5) to blue (above pH 8.3).

B

B lymphocyte (B cell)   A white blood cell that gives rise to plasma cells and antibodies.

back mutation   A genetic event that restores the sequence of a previously mutated gene.

bacillus   Bacterial cell shape that is cylindrical (longer than it is wide).

bacitracin   Antibiotic that targets the bacterial cell wall; component of over-the-counter topical antimicrobial ointments.

back-mutation   A mutation that counteracts an earlier mutation, resulting in the restoration of the original DNA sequence.

bacteremia   The presence of viable bacteria in circulating blood.

bacteremic   Bacteria present in the bloodstream.

Bacteria   When capitalized can refer to one of the three domains of living organisms proposed by Woese, containing all nonarchaea prokaryotes.

bacteria   (plural of bacterium) Category of prokaryotes with peptidoglycan in their cell walls and circular chromosome(s). This group of small cells is widely distributed in the earth's habitats.

bacterial chromosome   A circular body in bacteria that contains the primary genetic material. Also called *nucleoid*.

bactericide   An agent that kills bacteria.

bacteriophage   A virus that specifically infects bacteria.

bacteristatic   Any process or agent that inhibits bacterial growth.

bacterium   A tiny unicellular prokaryotic organism that usually reproduces by binary fission and usually has a peptidoglycan cell wall, has various shapes, and can be found in virtually any environment.

basic   A solution with a pH value above 7 on the pH scale.

basic dye   Any dye with a positively charged chromophore.

basidiospore   A sexual spore that arises from a basidium. Found in basidiomycota fungi.

basidium   A reproductive cell created when the swollen terminal cell of a hypha develops filaments (sterigmata) that form spores.

beta-galactosidase   An enzyme that catalyzes the hydrolysis of beta-galactosides such as lactose or ONPG.

beta-lactamase   An enzyme secreted by certain bacteria that cleaves the beta-lactam ring of penicillin and cephalosporin and thus provides for resistance against the antibiotic. See *penicillinase*.

bile   A bitter fluid secreted by the liver and stored in the gall bladder of most vertebrates. Bile aids in the digestion of lipids.

binary fission   The formation of two new cells of approximately equal size as the result of parent cell division.

binomial system   Scientific method of assigning names to organisms that employs two names to identify every organism—genus name plus species name.

biochemistry   The study of organic compounds produced by (or components of) living things. The four main categories of biochemicals are carbohydrates, lipids, proteins, and nucleic acids.

biofilm   A complex association that arises from a mixture of microorganisms growing together on the surface of a habitat.

biological vector   An animal that not only transports an infectious agent but plays a role in the life cycle of the pathogen, serving as a site in which it can multiply or complete its life cycle. It is usually an alternate host to the pathogen.

biosafety level (BSL)   A numeric designation that defines a set of minimum standards for a laboratory. Biosfafety levels range from 1 (least restrictive) to 4 (most restrictive).

biotechnology   The use of microbes or their products in the commercial or industrial realm.

biotic Living factors such as parasites, food substrates, or other living or once-living organisms that are studied when looking at an ecosystem.

blood cells Cellular components of the blood consisting of red blood cells, primarily responsible for the transport of oxygen and carbon dioxide, and white blood cells, primarily responsible for host defense and immune reactions.

boil A large red abscess that forms as a result of the infection of a hair follicle or sebaceous gland.

botulinum *Clostridium botulinum* toxin. Ingestion of this potent exotoxin leads to flaccid paralysis.

brackish Slightly salty.

broad spectrum Denotes drugs that have an effect on a wide variety of microorganisms.

brightfield microscope The most common type of laboratory microscope, in which specimens are viewed as dark objects against a bright field.

bromcresol purple A pH indicator that transitions from yellow at pH 5.2 to purple at pH 6.8.

bromthymol blue A pH indicator that transitions from green at pH 6.9 and blue at pH 7.6.

bronchoscopy A procedure in which a long thin tube is inserted through the mouth to examine the airway leading to the lungs.

brownian movement The passive, erratic, nondirectional motion exhibited by microscopic particles. The jostling comes from being randomly bumped by submicroscopic particles, usually water molecules, in which the visible particles are suspended.

brucellosis A zoonosis transmitted to humans from infected animals or animal products; causes a fluctuating pattern of severe fever in humans as well as muscle pain, weakness, headache, weight loss, and profuse sweating. Also called *undulant fever*.

bubo The swelling of one or more lymph nodes due to inflammation.

bubonic plague The form of plague in which bacterial growth is primarily restricted to the lymph and is characterized by the appearance of a swollen lymph node referred to as a bubo.

buffer A chemical compound that helps to prevent the change in the pH of a solution.

budding See *exocytosis*.

bullous Consisting of fluid-filled blisters.

C

cadaver A dead human body.

CAMP reaction A synergistic effect seen when the hemolysins from two different bacterial species combine, usually used to identify *Streptococcus agalactiae*.

capsid The protein covering of a virus's nucleic acid core. Capsids exhibit symmetry due to the regular arrangement of subunits called *capsomers*. See *icosahedron*.

capsomer A subunit of the virus capsid shaped as a triangle or disc.

capsule In bacteria, the loose, gel-like covering or slime made chiefly of polysaccharides. This layer is protective and can be associated with virulence.

carbohydrate A compound containing primarily carbon, hydrogen, and oxygen in a 1:2:1 ratio.

carbon cycle That pathway taken by carbon from its abiotic source to its use by producers to form organic compounds (biotic), followed by the breakdown of biotic compounds and their release to a nonliving reservoir in the environment (mostly carbon dioxide in the atmosphere).

carbon fixation Reactions in photosynthesis that incorporate inorganic carbon dioxide into organic compounds such as sugars. This occurs during the Calvin cycle and uses energy generated by the light reactions. This process is the source of all production on earth.

carbuncle A deep staphylococcal abscess joining several neighboring hair follicles.

carcinogen A cancer-causing substance.

cardinal temperatures The minimum, optimum, and maximum temperatures for a particular bacterial species.

carotenoid Yellow, orange, or red photosynthetic pigments.

carrier A person who harbors infections and inconspicuously spreads them to others. Also, a chemical agent that can accept an atom, chemical radical, or subatomic particle from one compound and pass it on to another.

casease An enzyme that hydrolyzes the milk protein casein.

caseous lesion Necrotic area of lung tubercle superficially resembling cheese. Typical of tuberculosis.

catabolism The chemical breakdown of complex compounds into simpler units to be used in cell metabolism.

catalase An enzyme that catalyzes the degradation of hydrogen peroxide to water and oxygen.

catalyst A substance that alters the rate of a reaction without being consumed or permanently changed by it. In cells, enzymes are catalysts.

catalytic site The niche in an enzyme where the substrate is converted to the product (also called an *active site*).

catheter A thin tube inserted into a body cavity to allow the draining of fluid.

cation A positively charged ion.

cell An individual membrane-bound living entity; the smallest unit capable of an independent existence.

cell wall In bacteria, a rigid structure made of peptidoglycan that lies just outside the cytoplasmic membrane; eukaryotes also have a cell wall but may be composed of a variety of materials.

cellulitis The spread of bacteria within necrotic tissue.

cellulose A long, fibrous polymer composed of glucose; one of the most common substances on earth.

Centers for Disease Control A government agency tasked with, among other things, tracking the spread of infectious disease.

cephalosporins A group of broad-spectrum antibiotics isolated from the fungus *Cephalosporium*.

cestode The common name for tapeworms that parasitize humans and domestic animals.

chemical bond A link formed between molecules when two or more atoms share, donate, or accept electrons.

chemoautotroph An organism that relies upon inorganic chemicals for its energy and carbon dioxide for its carbon. Also called a *chemolithotroph*.

chemoheterotroph Microorganisms that derive their nutritional needs from organic compounds.

chemotaxis The tendency of organisms to move in response to a chemical gradient (toward an attractant or to avoid adverse stimuli).

chemotherapy The use of chemical substances or drugs to treat or prevent disease.

chemotroph Organism that oxidizes compounds to feed on nutrients.

chitin A polysaccharide similar to cellulose in chemical structure. This polymer makes up the horny substance of the exoskeletons of arthropods and certain fungi.

chloramphenicol Antibiotic that inhibits protein synthesis by binding to the 50s subunit of the ribosome.

chlorophyll A group of mostly green pigments that are used by photosynthetic eukaryotic organisms and cyanobacteria to trap light energy to use in making chemical bonds.

chloroplast An organelle containing chlorophyll that is found in photosynthetic eukaryotes.

chromophore The color-bearing portion of a stain molecule.

chronic Any process or disease that persists over a long duration.

cilium (plural: *cilia*) Eukaryotic structure similar to flagella that propels a protozoan through the environment.

class In the levels of classification, the division of organisms that follows phylum.

cloning host An organism such as a bacterium or a yeast that receives and replicates a foreign piece of DNA inserted during a genetic engineering experiment.

coagulase A plasma-clotting enzyme secreted by *Staphylococcus aureus*. It contributes to virulence and is involved in forming a fibrin wall that surrounds staphylococcal lesions.

coccobacillus An elongated coccus; a short, thick, oval-shaped bacterial rod.

coccus A spherical-shaped bacterial cell.

codon A specific sequence of three nucleotides in mRNA (or the sense strand of DNA) that constitutes the genetic code for a particular amino acid.

coenzyme A complex organic molecule, several of which are derived from vitamins (e.g., nicotinamide and riboflavin). A coenzyme operates in conjunction with an enzyme. Coenzymes serve as transient carriers of specific atoms or functional groups during metabolic reactions.

cofactor An enzyme accessory. It can be organic, such as coenzymes, or inorganic, such as Fe^{+2}, Mn^{+2}, or Zn^{+2} ions.

coliform A collective term that includes normal enteric bacteria that are Gram negative and lactose fermenting.

colonize The establishment of a microbe in a new environment.

colony A macroscopic cluster of cells appearing on a solid medium, each arising from the multiplication of a single cell.

colony forming unit (cfu) Term for the group of cells that gives rise to a colony.

commensalism An unequal relationship in which one species derives benefit without harming the other.

common source epidemic An epidemic in which people are infected from a single source as opposed to infection that progresses person to person.

communicable infection Capable of being transmitted from one individual to another.

competent Cells that have been treated so that they are more likely to take up exogenous DNA through the process of transformation.

competitive inhibition Control process that relies on the ability of metabolic analogs to control microbial growth by successfully competing with a necessary enzyme to halt the growth of bacterial cells.

compounds Molecules that are a combination of two or more different elements.

concentration The expression of the amount of a solute dissolved in a certain amount of solvent. It may be defined by weight, volume, or percentage.

congenital Transmission of an infection from mother to fetus.

conidia Asexual fungal spores shed as free units from the tips of fertile hyphae.

conidiospore A type of asexual spore in fungi; not enclosed in a sac.

conjugation In bacteria, the contact between donor and recipient cells associated with the transfer of genetic material such as plasmids. Can involve special (sex) pili. Also a form of sexual recombination in ciliated protozoans.

conjunctiva The thin fluid-secreting tissue that covers the eye and lines the eyelid.

constitutive enzyme An enzyme present in bacterial cells in constant amounts, regardless of the presence of substrate. Enzymes of the central catabolic pathways are typical examples.

contagious Communicable; transmissible by direct contact with infected people and their fresh secretions or excretions.

contaminant An impurity; any undesirable material or organism.

contaminated culture A medium that once held a pure (single or mixed) culture but now contains unwanted microorganisms.

contrast The degree of difference between the lightest and darkest parts of an image.

convalescence Recovery; the period between the end of a disease and the complete restoration of health in a patient.

coryza Symptoms of a respiratory infection such as inflammation of and discharge from the mucous membranes of the upper respiratory tract, sinuses, and eyes.

counterstain The stain applied after the decolorization step in a differential stain to aid in distinguishing decolorized cells from those that have not decolorized.

covalent A type of chemical bond that involves the sharing of electrons between two atoms.

covalent bond A chemical bond formed by the sharing of electrons between two atoms.

Creutzfeldt-Jakob disease (CJD) A spongiform encephalopathy caused by infection with a prion. The disease is marked by dementia, impaired senses, and uncontrollable muscle contractions.

crista The infolded inner membrane of a mitochondrion that is the site of the respiratory chain and oxidative phosphorylation.

cryptosporidiosis A gastrointestinal disease caused by *Cryptosporidium parvum*, a protozoan.

CT (computed tomography) A medical imaging method used to generate a three dimensional image of the body.

culture The visible accumulation of microorganisms in or on a nutrient medium. Also, the propagation of microorganisms with various media.

curd The coagulated milk protein used in cheese making.

cutaneous Second level of skin, including the stratum corneum and occasionally the upper dermis.

cyanosis Blue discoloration of the skin or mucous membranes indicative of decreased oxygen concentration in blood.

cyst The resistant, dormant but infectious form of protozoans. Can be important in spread of infectious agents such as *Entamoeba histolytica* and *Giardia lamblia*.

cysteine A sulfur-containing amino acid.

cystine An amino acid, $HOOC—CH(NH_2)—CH_2—S—S—CH_2—CH(NH_2)COOH$. An oxidation product of two cysteine molecules in which the $–SH$ (sulfhydryl) groups form a disulfide union. Also called *dicysteine*.

cytochrome A group of heme-containing protein compounds whose chief role is in electron and/or hydrogen transport occurring in the last phase of aerobic respiration.

cytoplasm Dense fluid encased by the cell membrane; the site of many of the cell's biochemical and synthetic activities.

cytoplasmic membrane Lipid bilayer that encloses the cytoplasm of bacterial cells.

cytosine (C) One of the nitrogen bases found in DNA and RNA, with a pyrimidine form.

D

daptomycin A lipopetide antibiotic that disrupts the cytoplasmic membrane.

darkfield microscope A microscope in which specimens appear as light objects against a dark background.

deamination The removal of an amino group from a molecule, commonly an amino acid.

death phase End of the cell growth due to lack of nutrition, depletion of environment, and accumulation of wastes. Population of cells begins to die.

debridement Trimming away devitalized tissue and foreign matter from a wound.

decarboxylate The removal of a carboxyl group from a molecule, commonly an amino acid.

decomposer   A consumer that feeds on organic matter from the bodies of dead organisms. These microorganisms feed from all levels of the food pyramid and are responsible for recycling elements (also called *saprobes*).

decomposition   The breakdown of dead matter and wastes into simple compounds that can be directed back into the natural cycle of living things.

decontamination   The removal or neutralization of an infectious, poisonous, or injurious agent from a site.

definitive host   The organism in which a parasite develops into its adult or sexually mature stage. Also called the *final host*.

degerm   To physically remove surface oils, debris, and soil from skin to reduce the microbial load.

dehydration synthesis   A chemical process in which a polymer forms, as monomers are linked by the removal of water molecules.

denaturation   The loss of normal characteristics resulting from some molecular alteration. Usually in reference to the action of heat or chemicals on proteins whose function depends upon an unaltered tertiary structure.

denitrification   The end of the nitrogen cycle when nitrogen compounds are returned to the air.

deoxyribonucleic acid (DNA)   The nucleic acid often referred to as the "double helix." DNA carries the master plan for an organism's heredity.

deoxyribose   A 5-carbon sugar that is an important component of DNA.

dermatophytes   A group of fungi that cause infections of the skin and other integument components. They survive by metabolizing keratin.

desiccation   To dry thoroughly. To preserve by drying.

desquamate   To shed the cuticle in scales; to peel off the outer layer of a surface.

diatom   A type of algae found within the group Bacillarophyta.

dichotomous keys   Flowcharts that offer two choices or pathways at each level.

differential medium   A single substrate that discriminates between groups of microorganisms on the basis of differences in their appearance due to different chemical reactions.

differential stain   A technique that utilizes two dyes to distinguish between different microbial groups or cell parts by color reaction.

diffusion   The dispersal of molecules, ions, or microscopic particles propelled down a concentration gradient by spontaneous random motion to achieve a uniform distribution.

diluent   The liquid used to dilute a sample.

dilution factor   A number that relates the concentration of a diluted sample to the undiluted sample.

dimorphic   In mycology, the tendency of some pathogens to alter their growth form from mold to yeast in response to rising temperature.

dinoflagellate   An algae of the group Dinozoa.

diplobacilli   Rod-shaped cells found in pairs.

diplococci   Spherical or oval-shaped bacteria, typically found in pairs.

diplopia   Double vision.

direct or total cell count   1. Counting total numbers of individual cells being viewed with magnification. 2. Counting isolated colonies of organisms growing on a plate of media as a way to determine population size.

disaccharide   A sugar containing two monosaccharides, e.g., sucrose (made up of fructose and glucose).

disease   Any deviation from health, as when the effects of microbial infection damage or disrupt tissues and organs.

disinfection   The destruction of pathogenic nonsporulating microbes or their toxins, usually on inanimate surfaces.

division   In the levels of classification, an alternate term for phylum.

DNA   See *deoxyribonucleic acid*.

DNA polymerase   Enzyme responsible for the replication of DNA. Several versions of the enzyme exist, each completing a unique portion of the replication process.

DNA profile (fingerprint)   A pattern of restriction enzyme fragments that is unique for an individual organism.

DNA sequencing   Determining the exact order of nucleotides in a fragment of DNA. Most commonly done using the Sanger dideoxy sequencing method.

DNase   Enzyme that cleaves the DNA molecule into short chains of 2-4 nucleotides in length.

domoic acid   A toxin produced by some types of algae that can be concentrated in filter-feeding organisms such as clams or mussels, eventually causing disease in persons eating them.

doubling time   Time required for a complete fission cycle from parent cell to two new daughter cells. Also called *generation time*.

droplet nuclei   The dried residue of fine droplets produced by mucus and saliva sprayed while sneezing and coughing. Droplet nuclei are less than 5 mm in diameter (large enough to bear a single bacterium and small enough to remain airborne for a long time) and can be carried by air currents. Droplet nuclei are drawn deep into the air passages.

drug resistance   An adaptive response in which microorganisms begin to tolerate an amount of drug that would ordinarily be inhibitory.

Durham tube   A small test placed in a culture tube in an inverted position to catch a portion of any gas produced.

dyspnea   Difficulty in breathing.

E

E-test   An antibiotic sensitivity test that allows determination of the minimal inhibitory concentration of an antibiotic.

edema   The accumulation of excess fluid in cells, tissues, or serous cavities. Also called *swelling*.

electrolyte   Any compound that ionizes in solution and conducts current in an electrical field.

electron   A negatively charged subatomic particle that is distributed around the nucleus in an atom.

electrophoresis   The separation of molecules by size and charge through exposure to an electrical current.

electrostatic   Relating to the attraction of opposite charges and the repulsion of like charges. Electrical charge remains stationary as opposed to electrical flow or current.

element   A substance comprising only one kind of atom that cannot be degraded into two or more substances without losing its chemical characteristics.

ELISA   Abbreviation for *enzyme-linked immunosorbent assay*, a very sensitive serological test used to detect antibodies in diseases such as AIDS.

emerging disease   Newly identified diseases that are becoming more prominent.

encephalitis   An inflammation of the brain, usually caused by infection.

endemic disease   A native disease that prevails continuously in a geographic region.

endocytosis   The process whereby solid and liquid materials are taken into the cell through membrane invagination and engulfment into a vesicle.

endocarditis   Infection of the endocardium, the thin membrane lining the inside of the heart.

endoenzyme   An intracellular enzyme, as opposed to enzymes that are secreted.

endogenous   Originating or produced within an organism or one of its parts.

endospore   A small, dormant, resistant derivative of a bacterial cell that germinates under favorable growth conditions into a vegetative cell. The bacterial genera *Bacillus* and *Clostridium* are typical spore formers.

endosymbiosis   Relationship in which a microorganism resides within a host cell and provides a benefit to the host cell.

endotoxic shock   A massive drop in blood pressure caused by the release of endotoxin from Gram-negative bacteria multiplying in the bloodstream.

endotoxin   A bacterial toxin that is not ordinarily released (as is exotoxin). Endotoxin is composed of a phospholipid-polysaccharide complex that is an integral part of Gram-negative bacterial cell walls. Endotoxins can cause severe shock and fever.

energy of activation   The minimum energy input necessary for reactants to form products in a chemical reaction.

enriched medium   A nutrient medium supplemented with blood, serum, or some growth factor to promote the multiplication of fastidious microorganisms.

enteric   Pertaining to the intestine.

enterohemorrhagic *E. coli* **(EHEC)**   A group of *E. coli* species that induce bleeding in the intestines and also in other organs; *E. coli* 0157:H7 belongs to this group.

enteroinvasive   Predisposed to invade the intestinal tissues.

enteropathogenic   Pathogenic to the alimentary canal.

enterotoxigenic   Having the capacity to produce toxins that act on the intestinal tract.

enterotoxin   A bacterial toxin that specifically targets intestinal mucous membrane cells. Enterotoxigenic strains of *Escherichia coli* and *Staphylococcus aureus* are typical sources.

enveloped virus   A virus whose nucleocapsid is enclosed by a membrane derived in part from the host cell. It usually contains exposed glycoprotein spikes specific for the virus.

enzyme   A protein biocatalyst that facilitates metabolic reactions.

enzyme repression   The inhibition of an enzymatic reaction by an end product formed in the course of the pathway.

epidemic   A sudden and simultaneous outbreak or increase in the number of cases of disease in a community.

epidemiology   The study of the factors affecting the prevalence and spread of disease within a community.

epidermis   The outermost skin layer.

epitheca   The top half of the outer shell of a dinoflagellate.

epitope   The precise molecular group of an antigen that defines its specificity and triggers the immune response.

erysipelas   An acute, sharply defined inflammatory disease specifically caused by hemolytic *Streptococcus*. The eruption is limited to the skin but can be complicated by serious systemic symptoms.

erythema   Redness of the skin.

erythroblastosis fetalis   Hemolytic anemia of the newborn. The anemia comes from hemolysis of Rh-positive fetal erythrocytes by anti-Rh maternal antibodies. Erythroblasts are immature red blood cells prematurely released from the bone marrow.

erythrocytes (red blood cells)   Blood cells involved in the transport of oxygen and carbon dioxide.

erythrogenic toxin   An exotoxin produced by lysogenized group A strains of β-hemolytic streptococci that is responsible for the severe fever and rash of scarlet fever in the nonimmune individual. Also called a *pyrogenic toxin.*

eschar   A dark, sloughing scab that is the lesion of anthrax and certain rickettsioses.

essential nutrient   Any ingredient such as a certain amino acid, fatty acid, vitamin, or mineral that cannot be formed by an organism and must be supplied in the diet. A growth factor.

ester bond   A covalent bond formed by reacting carboxylic acid with an ROH group:

$$R-\overset{\overset{\textstyle O}{\|}}{C}-O-R'$$

Olive and corn oils, lard, and butter fat are examples of triacylglycerols—esters formed between glycerol and three fatty acids.

ethylene oxide   A potent, highly water-soluble gas invaluable for gaseous sterilization of heat-sensitive objects such as plastics, surgical and diagnostic appliances, and spices.

etiologic agent   The microbial cause of disease; the pathogen.

eubacteria   Term used for nonarchaea prokaryotes, means "true bacteria."

eukaryotic cell   A cell that differs from a prokaryotic cell chiefly by having a nuclear membrane (a well-defined nucleus), membrane-bounded subcellular organelles, and mitotic cell division.

eutrophication   The process whereby dissolved nutrients resulting from natural seasonal enrichment or industrial pollution of water cause overgrowth of algae and cyanobacteria to the detriment of fish and other large aquatic inhabitants.

evolution   Scientific principle that states that living things change gradually over time and these changes are expressed in structural and functional adaptations in each organism. Evolution presumes that those traits that favor survival are preserved and passed on to following generations, and those traits that do not favor survival are lost.

exanthem   An eruption or rash of the skin.

exfoliative toxin   A poisonous substance that causes superficial cells of an epithelium to detach and be shed, e.g., staphylococcal exfoliatin. Also called an *epidermolytic toxin.*

exocytosis   The process that releases enveloped viruses from the membrane of the host's cytoplasm.

exoenzyme   An extracellular enzyme chiefly for hydrolysis of nutrient macromolecules that are otherwise impervious to the cell membrane. It functions in saprobic decomposition of organic debris and can be a factor in invasiveness of pathogens.

exogenous   Originating outside the body.

exotoxin   A toxin (usually a protein) that is secreted and acts upon a specific cellular target, e.g., botulin, tetanospasmin, diphtheria toxin, and erythrogenic toxin.

exponential growth phase   The period of maximum growth rate in a growth curve. Cell population increases logarithmically.

extrapulmonary tuberculosis   A condition in which tuberculosis bacilli have spread to organs other than the lungs.

extremophiles   Organisms capable of living in harsh environments, such as extreme heat or cold.

F

facultative   Pertaining to the capacity of microbes to adapt or adjust to variations; not obligate, e.g., the presence of oxygen is not obligatory for a facultative anaerobe to grow. See *obligate.*

family   In the levels of classification, a midlevel division of organisms that groups more closely related organisms than previous levels. An order is divided into families.

fasciotomy   A surgical procedure where the soft tissue of an area is cut to relieve pressure, restoring circulation to distal tissue.

fastidious   Requiring special nutritional or environmental conditions for growth. Said of bacteria.

fecal coliforms   Any species of Gram-negative lactose-positive bacteria (primarily *Escherichia coli*) that live primarily in the intestinal tract and not the environment. Finding evidence of these bacteria in a water or food sample is substantial evidence of fecal contamination and potential for infection. See coliform.

feedback inhibition   Temporary end to enzyme action caused by an end product molecule binding to the regulatory site and preventing the enzyme's active site from binding to its substrate.

fermentation   The extraction of energy through anaerobic degradation of substrates into simpler, reduced metabolites. In large industrial processes, fermentation can mean any use of microbial metabolism to manufacture organic chemicals or other products.

fermentor   A large tank used in industrial microbiology to grow mass quantities of microbes that can synthesize desired products. These devices are equipped with means to stir, monitor, and harvest products such as drugs, enzymes, and proteins in very large quantities.

fibrin   A protein that polymerizes to form a blood clot after an injury.

filament   A helical structure, composed of proteins, that is part of bacterial flagella.

fimbria   A short, numerous-surface appendage on some bacteria that provides adhesion but not locomotion.

firmicutes   Taxonomic category of bacteria that have Gram-positive cell envelopes.

flaccid paralysis   An abnormal condition characterized by the weakening or loss of muscle tone.

flagellum   A structure that is used to propel the organism through a fluid environment.

fluid mosaic model   A conceptualization of the molecular architecture of cellular membranes as a bilipid layer containing proteins. Membrane proteins are embedded to some degree in this bilayer, where they float freely.

fluorescence   The property possessed by certain minerals and dyes to emit visible light when excited by, most commonly, ultraviolet radiation. A fluorescent dye combined with specific antibody provides a sensitive test for the presence of antigen.

fluoroquinolones   Synthetic antimicrobial drugs chemically related to quinine. They are broad spectrum and easily adsorbed from the intestine.

focal infection   Occurs when an infectious agent breaks loose from a localized infection and is carried by the circulation to other tissues.

folliculitis   An inflammatory reaction involving the formation of papules or pustules in clusters of hair follicles.

fomite   Virtually any inanimate object an infected individual has contact with that can serve as a vehicle for the spread of disease.

food poisoning   Symptoms in the intestines (which may include vomiting) induced by preformed exotoxin from bacteria.

formalin   A 37% aqueous solution of formaldehyde gas; a potent chemical fixative and microbicide.

fosfomycin trimethamine   Antibiotic that inhibits an enzyme necessary for cell wall synthesis.

frameshift mutation   An insertion or deletion mutation that changes the codon reading frame from the point of the mutation to the final codon. Almost always leads to a nonfunctional protein.

fructose   One of the carbohydrates commonly referred to as sugars. Fructose is commonly fruit sugar.

frustules   The silica cell wall of a diatom.

functional group   In chemistry, a particular molecular combination that reacts in predictable ways and confers particular properties on a compound, e.g., —COOH, —OH, and —CHO.

fungemia   The condition of fungi multiplying in the bloodstream.

fungi   Macroscopic and microscopic heterotrophic eukaryotic organisms that can be uni- or multicellular.

fungus   Heterotrophic unicellular or multicellular eukaryotic organism that may take the form of a larger macroscopic organism, as in the case of mushrooms, or a smaller microscopic organism, as in the case of yeasts and molds.

furuncle   A boil; a localized pyogenic infection arising from a hair follicle.

G

gamma globulin   The fraction of plasma proteins high in immunoglobulins (antibodies). Preparations from pooled human plasma containing normal antibodies make useful passive immunizing agents against pertussis, polio, measles, and several other diseases.

gas gangrene   Disease caused by a clostridial infection of soft tissue or wound. The name refers to the gas produced by the bacteria growing in the tissue. Unless treated early, it is fatal. Also called *myonecrosis*.

gastritis (gastroenteritis)   Pain and/or nausea, usually experienced after eating; result of inflammation of the lining of the stomach.

gel electrophoresis   A laboratory technique for separating DNA or protein fragments according to length by employing electricity to force the molecule through a gel-like matrix typically made of agarose. Smaller molecule fragments move more quickly through the gel, thereby moving farther than larger fragments during the same period of time.

gene   A site on a chromosome that provides information for a certain cell function. A specific segment of DNA that contains the necessary code to make a protein or RNA molecule.

generation time   Time required for a complete fission cycle from parent cell to two new daughter cells. Also called *doubling time.*

genetic engineering   A field involving deliberate alterations (recombinations) of the genomes of microbes, plants, and animals through special technological processes.

genetic fingerprinting   A technique that generates a set of DNA fragments by digesting a bacterial genome with restriction enzymes. The unique set of fragments can be used to identify a bacterial isolate.

genetics   The science of heredity.

genome   The complete set of chromosomes and genes in an organism.

genomics   The systematic study of an organism's genes and their functions.

genotype   The genetic makeup of an organism. The genotype is ultimately responsible for an organism's phenotype, or expressed characteristics.

genus   In the levels of classification, the second most specific level. A family is divided into several genera.

germ theory of disease   A theory first originating in the 1800s that proposed that microorganisms can be the cause of diseases. The concept is actually so well established in the present time that it is considered a fact.

germicide   An agent lethal to non-endospore-forming pathogens.

germination   Growth emerging from a period of dormancy.

giardiasis   Infection by the *Giardia* flagellate. Most common mode of transmission is contaminated food and water. Symptoms include diarrhea, abdominal pain, and flatulence.

gingivitis Inflammation of the gum tissue in contact with the roots of the teeth.

glomerulonephritis A type of kidney disease caused by inflammation of the internal kidney structures.

gluconeogenesis The formation of glucose (or glycogen) from noncarbohydrate sources such as protein or fat. Also called *glyconeogenesis*.

glucose One of the carbohydrates commonly referred to as sugars. Glucose is characterized by its 6-carbon structure.

glycerol A 3-carbon alcohol, with three OH groups that serve as binding sites.

glycocalyx A filamentous network of carbohydrate-rich molecules that coats cells.

glycogen A glucose polymer stored by cells.

glycolysis The energy-yielding breakdown (fermentation) of glucose to pyruvic or lactic acid. It is often called *anaerobic glycolysis* because no molecular oxygen is consumed in the degradation.

glycosides Molecules in which a sugar is bound by a glycosidic bond to some other chemical group.

glycosidic bond A bond that joins monosaccharides to form disaccharides and polymers.

Golgi apparatus An organelle of eukaryotes that participates in packaging and secretion of molecules.

gonococcus Common name for *Neisseria gonorrhoeae*, the agent of gonorrhea.

gracilicutes Taxonomic category of bacteria that have Gram-negative envelopes.

graft Live tissue taken from a donor and transplanted into a recipient to replace damaged or missing tissues such as skin, bone, blood vessels.

Gram stain A differential stain for bacteria useful in identification and taxonomy. Gram-positive organisms appear purple from crystal violet mordant retention, whereas Gram-negative organisms appear red after loss of crystal violet and absorbance of the safranin counterstain.

Gram negative A category of bacterial cells that describes bacteria with an outer membrane, a cytoplasmic membrane, and a thin cell wall.

Gram positive A category of bacterial cells that describes bacteria with a thick cell wall and no outer membrane.

granulomatous amebic encephalitis A necrotizing infection of the central nervous system caused by infection with one of several species of amoeba.

grana Discrete stacks of chlorophyll-containing thylakoids within chloroplasts.

growth curve A graphical representation of the change in population size over time.

This graph has four periods known as *lag phase*, *exponential* or *log phase*, *stationary phase*, and *death phase*.

growth factor An organic compound such as a vitamin or amino acid that must be provided in the diet to facilitate growth. An essential nutrient.

guanine (G) One of the nitrogen bases found in DNA and RNA, with a purine form.

H

habitat The environment to which an organism is adapted.

halogens A group of related chemicals with antimicrobial applications. The halogens most often used in disinfectants and antiseptics are chlorine and iodine.

halophile A microbe whose growth is either stimulated by salt or requires a high concentration of salt for growth.

halotolerant A microbe that can withstand high concentrations of salt without adverse effects.

Hansen's disease A chronic, progressive disease of the skin and nerves caused by infection by a mycobacterium that is a slow-growing, strict parasite. Hansen's disease is the preferred name for leprosy.

helminth A term that designates all parasitic worms.

hematopoiesis The process by which the various types of blood cells are formed, such as in the bone marrow.

hemolysin Any biological agent that is capable of destroying red blood cells and causing the release of hemoglobin. Many bacterial pathogens produce exotoxins that act as hemolysins.

hemolytic anemia Incompatible Rh factor between mother and fetus causing maternal antibodies to attack the fetus and trigger complement-mediated lysis in the fetus.

hemolytic uremic syndrome (HUS) Severe hemolytic anemia leading to kidney damage or failure; can accompany *E. coli* O157:H7 intestinal infection.

hemolyze When red blood cells burst and release hemoglobin pigment.

hepatitis Inflammation and necrosis of the liver, often the result of viral infection.

herd immunity A term describing vaccination of almost all people in a population. If nearly everyone is vaccinated, even the occasional unvaccinated person is safe as there is no one from whom the infectious agent can be acquired.

heredity Genetic inheritance.

heterotroph An organism that relies upon organic compounds for its carbon and energy needs.

hexose A 6-carbon sugar such as glucose and fructose.

host Organism in which smaller organisms or viruses live, feed, and reproduce.

host range The limitation imposed by the characteristics of the host cell on the type of virus that can successfully invade it.

hydrolysis A process in which water is used to break bonds in molecules. Usually occurs in conjunction with an enzyme.

hydrophilic The property of attracting water. Molecules that attract water to their surface are called *hydrophilic*.

hydrophobic The property of repelling water. Molecules that repel water are called *hydrophobic*.

hyperthermophile Microbes which grow best at temperatures between 80°C and 120°C.

hypertonic Having a greater osmotic pressure than a reference solution.

hyphae The tubular threads that make up filamentous fungi (molds). This web of branched and intertwining fibers is called a *mycelium*.

hypotheca The lower half of the outer shell of a dinoflagellate.

hypotonic Having a lower osmotic pressure than a reference solution.

I

icosahedron A regular geometric figure having 20 surfaces that meet to form 12 corners. Some virions have capsids that resemble icosahedral crystals.

immune complex reaction Type III hypersensitivity of the immune system. It is characterized by the reaction of soluble antigen with antibody, and the deposition of the resulting complexes in basement membranes of epithelial tissue.

immunity An acquired resistance to an infectious agent due to prior contact with that agent.

immunoassays Extremely sensitive tests that permit rapid and accurate measurement of trace antigen or antibody.

immunocompromised Having a less than adequate immune response.

immunodeficiency disease A form of immunopathology in which white blood cells are unable to mount a complete, effective immune response, which results in recurrent infections, e.g., AIDS and agammaglobulinemia.

immunoglobulin (Ig) The chemical class of proteins to which antibodies belong.

immunology The study of the system of body defenses that protect against infection.

immunotherapy   Preventing or treating infectious diseases by administering substances that produce artificial immunity. May be active or passive.

IMViC   Acronym for a series of laboratory tests (*i*ndole, *m*ethyl red, *V*oges-Proskauer, and *c*itrate) commonly used in the identification of Gram-negative organisms.

in vitro   Literally means "in glass," signifying a process or reaction occurring in an artificial environment, as in a test tube or culture medium.

in vivo   Literally means "in a living being," signifying a process or reaction occurring in a living thing.

incidence   In epidemiology, the number of new cases of a disease occurring during a period.

incineration   Destruction of microbes by subjecting them to extremes of dry heat. Microbes are reduced to ashes and gas by this process.

inclusion   A relatively inert body in the cytoplasm such as storage granules, glycogen, fat, or some other aggregated metabolic product.

incubate   To isolate a sample culture in a temperature-controlled environment to encourage growth.

incubation period   The period from the initial contact with an infectious agent to the appearance of the first symptoms.

index patient   The first person to present with an infection or disease; the initial case in an epidemic.

indicator bacteria   In water analysis, any easily cultured bacteria that may be found in the intestine and can be used as an index of fecal contamination. The category includes coliforms and enterococci. Discovery of these bacteria in a sample means that pathogens may also be present.

induced mutation   Any alteration in DNA that occurs as a consequence of exposure to chemical or physical mutagens.

inducible enzyme   An enzyme that increases in amount in direct proportion to the amount of substrate present.

inducible operon   An operon that under normal circumstances is not transcribed. The presence of a specific inducer molecule can cause transcription of the operon to begin.

induration   Area of hardened, reddened tissue associated with the tuberculin test.

infection   The entry, establishment, and multiplication of pathogenic organisms within a host.

infectious disease   The state of damage or toxicity in the body caused by an infectious agent.

inflammation   A natural, nonspecific response to tissue injury that protects the host from further damage. It stimulates immune reactivity and blocks the spread of an infectious agent.

inoculation   The implantation of microorganisms into or upon culture media.

inorganic chemicals   Molecules that lack the basic framework of the elements of carbon and hydrogen.

intoxication   Poisoning that results from the introduction of a toxin into body tissues through ingestion or injection.

iodophor   A combination of iodine and an organic carrier that is a moderate-level disinfectant and antiseptic.

ion   An unattached, charged particle.

ionic bond   A chemical bond in which electrons are transferred and not shared between atoms.

ionizing radiation   Radiant energy consisting of short-wave electromagnetic rays (X-ray) or high-speed electrons that cause dislodgment of electrons on target molecules and create ions.

irradiation   The application of radiant energy for diagnosis, therapy, disinfection, or sterilization.

isolate   The separation of individual cells from one another so that a mixed culture is obtained.

isoniazid   Older drug that targets the bacterial cell wall; used against *M. tuberculosis*.

isotonic   Two solutions having the same osmotic pressure such that, when separated by a semipermeable membrane, there is no net movement of solvent in either direction.

J

jaundice   The yellowish pigmentation of skin, mucous membranes, sclera, deeper tissues, and excretions due to abnormal deposition of bile pigments. Jaundice is associated with liver infection, as with hepatitis B virus and leptospirosis.

K

keratitis   Inflammation of the cornea.

kingdom   In the levels of classification, the second division from more general to more specific. Each domain is divided into kingdoms.

Koch's postulates   A procedure to establish the specific cause of disease. In all cases of infection: (1) the agent must be found; (2) inoculations of a pure culture must reproduce the same disease in animals; (3) the agent must again be present in the experimental animal; and (4) a pure culture must again be obtained.

Krebs cycle or tricarboxylic acid cycle (TCA)   The second pathway of the three pathways that complete the process of primary catabolism. Also called the *citric acid cycle.*

L

labile   In chemistry, molecules, or compounds that are chemically unstable in the presence of environmental changes.

lactose   One of the carbohydrates commonly referred to as sugars. Lactose is commonly found in milk.

lactose (*lac*) operon   Control system that manages the regulation of lactose metabolism. It is composed of three DNA segments, including a regulator, a control locus, and a structural locus.

lag phase   The early phase of population growth during which no signs of growth occur.

Lancefield group   A group within the Lancefield system of *Streptococcus* classification.

Legionnaire's disease   Infection by *Legionella* bacterium. Weakly Gram-negative rods that are able to survive in aquatic habitats. Some forms may be fatal.

lesion   A wound, injury, or some other pathologic change in tissues.

lipase   A fat-splitting enzyme, e.g., triacylglycerol lipase hydrolyzes the fatty acid chains from the glycerol backbone of triglycerides.

lipid   A term used to describe a variety of substances that are not soluble in polar solvents such as water but will dissolve in nonpolar solvents such as benzene and chloroform. Lipids include triglycerides, phospholipids, steroids, and waxes.

lipopolysaccharide   A molecular complex of lipid and carbohydrate found in the bacterial cell wall. The lipopolysaccharide (LPS) of Gram-negative bacteria is an endotoxin with generalized pathologic effects such as fever.

lithoautotroph   Bacteria that rely on inorganic minerals to supply their nutritional needs. Sometimes referred to as *chemoautotrophs.*

localized infection   Occurs when a microbe enters a specific tissue, infects it, and remains confined there.

locus (plural, loci)   A site on a chromosome occupied by a gene. Plural: *loci.*

log phase   Maximum rate of cell division during which growth is geometric in its rate of increase. Also called *exponential growth phase.*

loop dilution	A means of isolating a pure culture from a mixed culture by serially diluting a single loopful of bacteria through a series of tubes containing liquified agar.

lophotrichous	Describing bacteria having a tuft of flagella at one or both poles.

lyophilization	A method for preserving microorganisms (and other substances) by freezing and then drying them directly from the frozen state.

lyse	To burst.

lysis	The physical rupture or deterioration of a cell.

lysosome	A cytoplasmic organelle containing lysozyme and other hydrolytic enzymes.

lysozyme	An enzyme found in sweat, tears, and saliva that breaks down bacterial peptidoglycan.

M

macromolecules	Large, molecular compounds assembled from smaller subunits, most notably biochemicals.

macronutrient	A chemical substance required in large quantities (e.g., phosphate).

macrophage	A white blood cell derived from a monocyte that leaves the circulation and enters tissues. These cells are important in nonspecific phagocytosis and in regulating, stimulating, and cleaning up after immune responses.

macroscopic	Visible to the naked eye.

magnetic resonance imaging (MRI)	A medical imaging method used to generate a three dimensional image of the body.

malaise	A feeling of general discomfort or uneasiness.

maltose	One of the carbohydrates referred to as sugars. A fermentable sugar formed from starch.

Mantoux test	An intradermal screening test for tuberculin hypersensitivity. A red, firm patch of skin at the injection site greater than 10 mm in diameter after 48 h is a positive result that indicates current or prior exposure to the TB bacillus.

matrix	The dense ground substance between the cristae of a mitochondrion that serves as a site for metabolic reactions.

matter	All tangible materials that occupy space and have mass.

maximum temperature	The highest temperature at which an organism will grow.

McFarland standard	A series of tubes containing precise amounts of barium chloride. Each tube is progressively more turbid then the last, and the entire series is used to compare the extent of bacterial growth in a tube.

MDRTB	Multidrug-resistant tuberculosis.

mechanical vector	An animal that transports an infectious agent but is not infected by it, such as houseflies whose feet become contaminated with feces.

medium (plural, _media_)	A nutrient used to grow organisms outside of their natural habitats.

membrane	In a single cell, a thin double-layered sheet composed of lipids such as phospholipids and sterols and proteins.

membrane filtration method	A means of concentrating and enumerating the microbes present in a relatively clean sample of water.

Mendosicutes	Taxonomic category of bacteria that have unusual cell walls; archaea.

meninges	The tough trilayer membrane covering the brain and spinal cord. Consists of the dura mater, arachnoid mater, and pia mater.

meningitis	An inflammation of the membranes (meninges) that surround and protect the brain. It is often caused by bacteria such as _Neisseria meningitidis_ (the meningococcus) and _Haemophilus influenzae_.

merozoite	The motile, infective stage of an apicomplexan parasite that comes from a liver or red blood cell undergoing multiple fission.

mesophile	Microorganisms that grow at intermediate temperatures.

messenger RNA (mRNA)	A single-stranded transcript that is a copy of the DNA template and that corresponds to a gene.

metabolic analog	Enzyme that mimics the natural substrate of an enzyme and vies for its active site.

metabolism	A general term for the totality of chemical and physical processes occurring in a cell.

metabolites	Small organic molecules that are intermediates in the stepwise biosynthesis or breakdown of macromolecules.

metachromatic	Exhibiting a color other than that of the dye used to stain it.

metachromatic granules	A type of inclusion in storage compartments of some bacteria that stain a contrasting color when treated with colored dyes.

methanogens	Methane producers.

methyl red	A pH indicator that is red at pH 4.4, yellow at pH 6.2, and various shades of orange between the two endpoints.

MIC	Abbreviation for _m_inimum _i_nhibitory _c_oncentration. The lowest concentration of antibiotic needed to inhibit bacterial growth in a test system.

microaerophile	An aerobic bacterium that requires oxygen at a concentration less than that in the atmosphere.

microbe	See _microorganism._

microbial load	A reference to the number of microbes in a particular sample or at a specific location.

microbicides	Chemicals that kill microorganisms.

microbistatic	Preventing the growth of microbes.

microbiology	A specialized area of biology that deals with living things ordinarily too small to be seen without magnification, including bacteria, archaea, fungi, protozoa, and viruses.

micronutrient	A chemical substance required in small quantities (trace metals, for example).

microorganism	A living thing ordinarily too small to be seen without magnification; an organism of microscopic size.

microscopic	Invisible to the naked eye.

microscopy	Science that studies structure, magnification, lenses, and techniques related to use of a microscope.

minimum inhibitory concentration (MIC)	The smallest concentration of drug needed to visibly control microbial growth.

minimum temperature	The lowest temperature at which an organism will grow.

missense mutation	A mutation in which a change in the DNA sequence results in a different amino acid being incorporated into a protein, with varying results.

mitochondrion	A double-membrane organelle of eukaryotes that is the main site for aerobic respiration.

mitosis	Somatic cell division that preserves the somatic chromosome number.

mixed acid fermentation	An anaerobic degradation of pyruvic acid that results in more than one organic acid being produced (e.g., acetic acid, lactic acid, and succinic acid).

mixed culture	A container growing two or more different, known species of microbes.

mixed infection	Occurs when several different pathogens interact simultaneously to produce an infection. Also called a _synergistic infection._

molecule	A distinct chemical substance that results from the combination of two or more atoms.

mollusks	A group of shelled sea animals, including clams, oysters, and mussels.

monomer	A simple molecule that can be linked by chemical bonds to form larger molecules.

monosaccharide	A simple sugar such as glucose that is a basic building block for more complex carbohydrates.

monotrichous   Describing a microorganism that bears a single flagellum.

morbidity   A diseased condition.

Morbidity and Mortality Weekly Report   A publication of the CDC that tracks the incidence and prevalence of reportable disease.

morbidity rate   The number of persons afflicted with an illness under question or with illness in general, expressed as a numerator, with the denominator being some unit of population (as in $x/100{,}000$).

mordant   A chemical that fixes a dye in or on cells by forming an insoluble compound and thereby promoting retention of that dye, e.g., Gram's iodine in the Gram stain.

morphology   The study of organismic structure.

mortality rate   The number of persons who have died as the result of a particular cause or due to all causes, expressed as a numerator, with the denominator being some unit of population (as in $x/100{,}000$).

most probable number (MPN)   Test used to detect the concentration of contaminants in water and other fluids.

motility   Self-propulsion.

MRSA (methicillin-resistant *Staphylococcus aureus*)   A strain of *Staphylococcus aureus* resistant to the antibiotic methicillin.

mutagen   Any agent that induces genetic mutation, e.g., certain chemical substances, ultraviolet light, and radioactivity.

mutant strain   A subspecies of microorganism that has undergone a mutation, causing expression of a trait that differs from other members of that species.

mutation   A permanent inheritable alteration in the DNA sequence or content of a cell.

mutualism   Organisms living in an obligatory but mutually beneficial relationship.

myalgias   Muscle pain.

mycelium   The filamentous mass that makes up a mold. Composed of hyphae.

mycolic acid   A compound found in the cell walls of acid-fast bacteria.

mycology   The study of fungi.

mycosis   Any disease caused by a fungus.

myonecrosis   Death of muscle tissue.

N

narrow spectrum   Denotes drugs that are selective and limited in their effects, e.g., they inhibit either Gram-negative or Gram-positive bacteria but not both.

natural immunity   Any immunity that arises naturally in an organism via previous experience with the antigen.

necrosis   A pathologic process in which cells and tissues die and disintegrate.

negative stain   A staining technique that renders the background opaque or colored and leaves the object unstained so that it is outlined as a colorless area.

nematode   A common name for helminths called *roundworms*.

neonatal   Pertaining to the four weeks after birth.

nephritis   Inflammation of the kidney.

neutral red   A pH indicator which is red below pH 6.8 and yellow above pH 8.0.

neutralization   The process of combining an acid and a base until they reach a balanced proportion, with a pH value close to 7.

nomenclature   A set system for scientifically naming organisms, enzymes, anatomical structures, and so on.

noncommunicable   An infectious disease that does not arrive through transmission of an infectious agent from host to host.

noncompetitive inhibition   Form of enzyme inhibition that involves binding of a regulatory molecule to a site other than the active site.

nonsense codon   A triplet of mRNA bases that does not specify an amino acid but signals the end of a polypeptide chain.

nonsense mutation   A mutation that changes an amino-acid-producing codon into a stop codon, leading to premature termination of a protein.

nonseptate   Not divided by a septum.

normal flora   The native microbial forms that an individual harbors.

nosocomial infection   An infection not present upon admission to a hospital but incurred while being treated there.

nucleocapsid   In viruses, the close physical combination of the nucleic acid with its protective covering.

numerical aperture   In microscopy, the amount of light passing from the object and into the object in order to maximize optical clarity and resolution.

nutrient   Any chemical substance that must be provided to a cell for normal metabolism and growth. Macronutrients are required in large amounts, and micronutrients in small amounts.

nutrition   The acquisition of chemical substances by a cell or organism for use as an energy source or as building blocks of cellular structures.

O

objective lens   The lens closest to your eyes when looking through a microscope.

obligate   Without alternative; restricted to a particular characteristic, e.g., an obligate parasite survives and grows only in a host; an obligate aerobe must have oxygen to grow; and an obligate anaerobe is destroyed by oxygen.

opaque   A description of bacterial growth on a solid media which does not allow the passage of light.

opportunistic   In infection, ordinarily nonpathogenic or weakly pathogenic microbes that cause disease primarily in an immunologically compromised host.

optimum temperature   The temperature at which a species shows the most rapid growth rate.

order   In the levels of classification, the division of organisms that follows class. Increasing similarity may be noticed among organisms assigned to the same order.

organelle   A small component of eukaryotic cells that is bounded by a membrane and specialized in function.

origin of replication   The location in the DNA molecule at which replication begins.

osmophile   A microorganism that thrives in a medium having high osmotic pressure.

osmosis   The diffusion of water across a selectively permeable membrane in the direction of lower water concentration.

osteomyelitis   A focal infection of the internal structures of long bones, leading to pain and inflammation. Often caused by *Staphylococcus aureus*.

otitis media   Infection of the middle ear.

outer membrane   An additional membrane possessed by Gram-negative bacteria; a lipid bilayer containing specialized proteins and polysaccharides. It lies outside of the cell wall.

oxidation   In chemical reactions, the loss of electrons by one reactant.

oxidation-reduction   Redox reactions, in which paired sets of molecules participate in electron transfers.

oxidative phosphorylation   The synthesis of ATP using energy given off during the electron transport phase of respiration.

oxidizing agent   An atom or a compound that can receive electrons from another in a chemical reaction.

P

palisades   The characteristic arrangement of *Corynebacterium* cells resembling a row of fence posts and created by snapping.

pandemic   A disease afflicting an increased proportion of the population over a wide geographic area (often worldwide).

parasite   An organism that lives on or within another organism (the host), from which it obtains nutrients and enjoys protection. The parasite produces some degree of harm in the host.

parasitism   A relationship between two organisms in which the host is harmed in some way while the colonizer benefits.

parenteral   Administering a substance into a body compartment other than through the gastrointestinal tract, such as via intravenous, subcutaneous, intramuscular, or intramedullary injection.

passive carrier   Persons who mechanically transfer a pathogen without ever being infected by it, e.g., a health care worker who doesn't wash his/her hands adequately between patients.

pasteurization   Heat treatment of perishable fluids such as milk, fruit juices, or wine to destroy heat-sensitive vegetative cells, followed by rapid chilling to inhibit growth of survivors and germination of spores. The process prevents infection and spoilage.

pathogen   Any agent (usually a virus, bacterium, fungus, protozoan, or helminth) that causes disease.

pathogenicity   The capacity of microbes to cause disease.

pathologic   Capable of inducing physical damage on the host.

pathology   The structural and physiological effects of disease on the body.

pellicle   A membranous cover; a thin skin, film, or scum on a liquid surface; a thin film of salivary glycoproteins that forms over newly cleaned tooth enamel when exposed to saliva.

pelvic inflammatory disease (PID)   An infection of the uterus and fallopian tubes that has ascended from the lower reproductive tract. Caused by gonococci and chlamydias.

penetration (viral)   The step in viral multiplication in which virus enters the host cell.

penicillinase   An enzyme that hydrolyzes penicillin; found in penicillin-resistant strains of bacteria.

penicillins   A large group of naturally occurring and synthetic antibiotics produced by *Penicillium* mold and active against the cell wall of bacteria.

peptide   Molecule composed of short chains of amino acids, such as a dipeptide (two amino acids), a tripeptide (three), and a tetrapeptide (four).

peptide bond   The covalent union between two amino acids that forms between the amine group of one and the carboxyl group of the other. The basic bond of proteins.

peptidoglycan   A network of polysaccharide chains cross-linked by short peptides that forms the rigid part of bacterial cell walls. Gram-negative bacteria have a smaller amount of this rigid structure than do Gram-positive bacteria.

peptone   Protein derivative formed by partial hydrolysis of a protein with acid or proteolytic enzymes.

periplasmic space   The region between the cell wall and cell membrane of the cell envelopes of Gram-negative bacteria.

peritrichous   In bacterial morphology, having flagella distributed over the entire cell.

petechiae   Minute hemorrhagic spots in the skin that range from pinpoint- to pinhead-size.

pH   The symbol for the negative logarithm of the H ion concentration; p (power) to base 10 of $[H^+]$. A system for rating acidity and alkalinity.

phage   A bacteriophage; a virus that specifically parasitizes bacteria.

phagocyte   A class of white blood cells capable of infecting other cells and particles.

phenol red   A pH indicator that is red below pH 6.6 and yellow above pH 8.0.

phenotype   The observable characteristics of an organism produced by the interaction between its genetic potential (genotype) and the environment.

phosphorylation   Process in which inorganic phosphate is added to a compound.

photoactivation (light repair)   A mechanism for repairing DNA that have ultraviolet-light-induced mutations using an enzyme (photolyase) that is activated by visible light.

photoautotroph   An organism that utilizes light for its energy and carbon dioxide chiefly for its carbon needs.

photolysis   The splitting of water into hydrogen and oxygen during photosynthesis.

photophosphorylation   The process of electron transport during photosynthesis that results in the synthesis of ATP from ADP.

photosynthesis   A process occurring in plants, algae, and some bacteria that traps the sun's energy and converts it to ATP in the cell. This energy is used to fix CO_2 into organic compounds.

phototrophs   Microbes that use photosynthesis to feed.

phycobilin   Red or blue-green pigments that absorb light during photosynthesis.

phylum   In the levels of classification, the third level of classification from general to more specific. Each kingdom is divided into numerous phyla. Sometimes referred to as a *division.*

physiology   The study of the function of an organism.

phytoplankton   The collection of photosynthetic microorganisms (mainly algae and cyanobacteria) that float in the upper layers of aquatic habitats where sun penetrates. These microbes are the basis of aquatic food pyramids and, together with zooplankton, make up the plankton.

pili   Small, stiff filamentous appendages in Gram-negative bacteria that function in DNA exchange during bacterial conjugation.

pilus   A hollow appendage used to bring two bacterial cells together to transfer DNA.

pinocytosis   The engulfment, or endocytosis, of liquids by extensions of the cell membrane.

plague   Zoonotic disease caused by infection with *Yersinia pestis.* The pathogen is spread by flea vectors and harbored by various rodents.

plankton   Minute animals (zooplankton) or plants (phytoplankton) that float and drift in the limnetic zone of bodies of water.

plaque   In virus propagation methods, the clear zone of lysed cells in tissue culture or chick embryo membrane that corresponds to the area containing viruses. In dental application, the filamentous mass of microbes that adheres tenaciously to the tooth and predisposes to caries, calculus, or inflammation.

plasma   The carrier fluid element of blood.

plasmids   Extrachromosomal genetic units characterized by several features. A plasmid is a double-stranded DNA that is smaller than and replicates independently of the cell chromosome; it bears genes that are not essential for cell growth; it can bear genes that code for adaptive traits; and it is transmissible to other bacteria.

plasmolysis   Shrinkage or contraction of the protoplasm away from the wall of a bacterial cell, caused by loss of water through osmosis.

pleomorphism   Normal variability of cell shapes in a single species.

pleural effusion   The accumulation of excess fluid in the lungs.

pluripotential   Stem cells having the developmental plasticity to give rise to more than one type, e.g., undifferentiated blood cells in the bone marrow.

pneumococcus   Common name for *Streptococcus pneumoniae,* the major cause of bacterial pneumonia.

pneumonia   An inflammation of the lung leading to accumulation of fluid and respiratory compromise.

pneumonic plague The acute, frequently fatal form of pneumonia caused by *Yersinia pestis*.

point mutation A change that involves the loss, substitution, or addition of one or a few nucleotides.

polar Term to describe a molecule with an asymmetrical distribution of charges. Such a molecule has a negative pole and a positive pole.

poliomyelitis An acute enteroviral infection of the spinal cord that can cause neuromuscular paralysis.

polymer A macromolecule made up of a chain of repeating units, e.g., starch, protein, and DNA.

polymerase An enzyme that produces polymers through catalyzing bond formation between building blocks (polymerization).

polymerase chain reaction (PCR) A technique that amplifies segments of DNA for testing. Using denaturation, primers, and heat-resistant DNA polymerase, the number can be increased several millionfold.

polymicrobial Involving multiple distinct microorganisms.

polymyxin A mixture of antibiotic polypeptides from *Bacillus polymyxa* that are particularly effective against Gram-negative bacteria.

polypeptide A relatively large chain of amino acids linked by peptide bonds.

polysaccharide A carbohydrate that can be hydrolyzed into a number of monosaccharides, e.g., cellulose, starch, and glycogen.

population A group of organisms of the same species living simultaneously in the same habitat. A group of different populations living together constitutes the community level.

portal of entry Route of entry for an infectious agent; typically a cutaneous or membranous route.

portal of exit Route through which a pathogen departs from the host organism.

positive stain A method for coloring microbial specimens that involves a chemical that sticks to the specimen to give it color.

positive control A stain, test, or other procedure that is very similar to the actual experimental test, but which is known from previous experience to give a positive result.

potable Describing water that is relatively clear, odor-free, and safe to drink.

prevalence The total number of cases of a disease in a certain area and time period.

primary infection An initial infection in a previously healthy individual that is later complicated by an additional (secondary) infection.

primary stain The first stain used in a differential staining technique.

primary structure Initial protein organization described by type, number, and order of amino acids in the chain. The primary structure varies extensively from protein to protein.

prion A concocted word to denote *proteinaceous infectious agent*; a cytopathic protein associated with the slow-virus spongiform encephalopathies of humans and animals.

product(s) In a chemical reaction, the substance(s) that is (are) left after a reaction is completed.

proglottid The egg-generating segment of a tapeworm that contains both male and female organs.

promastigote A morphological variation of the trypanosome parasite responsible for leishmaniasis.

promoter A segment of DNA usually occurring upstream from a gene coding region and acting as a controlling element in the expression of that gene.

propagated transmission epidemic An epidemic where the infectious agent is passed from person to person.

prophage A lysogenized bacteriophage; a phage that is latently incorporated into the host chromosome instead of undergoing viral replication and lysis.

prophylactic Any device, method, or substance used to prevent disease.

protease Enzymes that act on proteins, breaking them down into component parts.

protease inhibitors Compounds that inhibit proteolytic enzymes. Commonly used to prevent the assembly of functioning viral particles.

proteolytic Protein digesting.

protein Predominant organic molecule in cells, formed by long chains of amino acids.

protoplast A bacterial cell whose cell wall is completely lacking and that is vulnerable to osmotic lysis.

protozoa A group of single-celled, eukaryotic organisms.

pseudohypha A chain of easily separated, spherical to sausage-shaped yeast cells partitioned by constrictions rather than by septa.

pseudopods Protozoan appendage responsible for motility. Also called *false feet*.

psychrophile A microorganism that thrives at low temperature (0°C–20°C), with a temperature optimum of 0°C–15°C.

pulmonary Occurring in the lungs. Examples include pulmonary anthrax and pulmonary nocardiosis.

purine Nitrogen bases that help form the genetic code on DNA and RNA. Adenine and guanine are the most important purines.

pure culture A container growing a single species of microbe whose identity is known.

pus The viscous, opaque, usually yellowish matter formed by an inflammatory infection. It consists of serum exudate, tissue debris, leukocytes, and microorganisms.

pyogenic Pertains to pus formers, especially the pyogenic cocci: pneumococci, streptococci, staphylococci, and neisseriae.

pyrimidine Nitrogen bases that help form the genetic code on DNA and RNA. Uracil, thymine, and cytosine are the most important pyrimidines.

pyrimidine dimer The union of two adjacent pyrimidines on the same DNA strand, brought about by exposure to ultraviolet light. It is a form of mutation.

pyrogen A substance that causes a rise in body temperature. It can come from pyrogenic microorganisms or from polymorphonuclear leukocytes (endogenous pyrogens).

Q

quaternary structure Most complex protein structure characterized by the formation of large, multiunit proteins by more than one of the polypeptides. This structure is typical of antibodies and some enzymes that act in cell synthesis.

quats A word that pertains to a family of surfactants called *quaternary ammonium compounds*. These detergents are only weakly microbicidal and are used as sanitizers and preservatives.

quinine A substance derived from cinchona trees that was used as an antimalarial treatment; has been replaced by synthetic derivatives.

quinolone A class of synthetic antimicrobic drugs with broad-spectrum effects.

R

radiation Electromagnetic waves or rays, such as those of light given off from an energy source.

raw milk Unpasteurized milk.

reactants Molecules entering or starting a chemical reaction.

real image An image formed at the focal plane of a convex lens. In the compound light microscope, it is the image created by the objective lens.

recombinant DNA technology A technology, also known as *genetic engineering,* that deliberately modifies the genetic structure of an organism to create novel products, microbes, animals, plants, and viruses.

redox Denoting an oxidation-reduction reaction.

red tides An algal bloom of dinoflagellates that imparts a red color to the water.

reducing agent An atom or a compound that can donate electrons in a chemical reaction.

reduction In chemistry, the gaining of electrons.

refraction In optics, the bending of light as it passes from one medium to another with a different index of refraction.

reportable disease Those diseases that must be reported to health authorities by law.

repressible operon An operon that under normal circumstances is transcribed. The buildup of the operon's amino acid product causes transcription of the operon to stop.

repressor The protein product of a repressor gene that combines with the operator and arrests the transcription and translation of structural genes.

resazurin An oxygen indicator used in fluid thioglycollate media. Resazurin turns pink in the presence of oxygen.

reservoir In disease communication, the natural host or habitat of a pathogen.

resident flora The deeper, more stable microbiota that inhabit the skin and exposed mucous membranes, as opposed to the superficial, variable, transient population.

resistance (R) factor Plasmids, typically shared among bacteria by conjugation, that provide resistance to the effects of antibiotics.

resolving power The capacity of a microscope lens system to accurately distinguish between two separate entities that lie close to each other. Also called *resolution.*

respiratory chain A series of enzymes that transfer electrons from one to another, resulting in the formation of ATP. It is also known as the *electron transport chain.* The chain is located in the cell membrane of bacteria and in the inner mitochondrial membrane of eukaryotes.

restriction endonuclease An enzyme present naturally in cells that cleaves specific locations on DNA. It is an important means of inactivating viral genomes, and it is also used to splice genes in genetic engineering.

retort A large airtight vessel used to steam sterilize canned food.

Rh factor An isoantigen that can trigger hemolytic disease in newborns due to incompatibility between maternal and infant blood factors.

ribonucleic acid (RNA) The nucleic acid responsible for carrying out the hereditary program transmitted by an organism's DNA.

ribose A 5-carbon monosaccharide found in RNA.

ribosomal RNA (rRNA) A single-stranded transcript that is a copy of part of the DNA template.

ribosome A bilobed macromolecular complex of ribonucleoprotein that coordinates the codons of mRNA with tRNA anticodons and, in so doing, constitutes the peptide assembly site.

RNA polymerase Enzyme process that translates the code of DNA to RNA.

S

saccharide Scientific term for sugar. Refers to a simple carbohydrate with a sweet taste.

sanitize To clean inanimate objects using soap and degerming agents so that they are safe and free of high levels of microorganisms.

saprobe A microbe that decomposes organic remains from dead organisms. Also known as a *saprophyte* or *saprotroph.*

sarcina A cubical packet of 8, 16, or more cells; the cellular arrangement of the genus *Sarcina* in the family Micrococcaceae.

satellitism A commensal interaction between two microbes in which one can grow in the vicinity of the other due to nutrients or protective factors released by that microbe.

saturation The complete occupation of the active site of a carrier protein or enzyme by the substrate.

scolex The anterior end of a tapeworm characterized by hooks and/or suckers for attachment to the host.

secondary infection An infection that compounds a preexisting one.

secondary structure Protein structure that occurs when the functional groups on the outer surface of the molecule interact by forming hydrogen bonds. These bonds cause the amino acid chain to either twist, forming a helix, or to pleat into an accordion pattern called a β-*pleated sheet.*

secretory antibody The immunoglobulin (IgA) that is found in secretions of mucous membranes and serves as a local immediate protection against infection.

selective media Nutrient media designed to favor the growth of certain microbes and to inhibit undesirable competitors.

selectively toxic Property of an antimicrobial agent to be highly toxic against its target microbe while being far less toxic to other cells, particularly those of the host organism.

self-limited Applies to an infection that runs its course without disease or residual effects.

semisolid media Nutrient media with a firmness midway between that of a broth (a liquid medium) and an ordinary solid medium; motility media.

semisynthetic Drugs that, after being naturally produced by bacteria, fungi, or other living sources, are chemically modified in the laboratory.

sentinel event An unanticipated event in a healthcare setting resulting in death or serious physical or psychological injury to a patient or patients, not related to the natural course of the patient's illness.

sepsis The state of putrefaction; the presence of pathogenic organisms or their toxins in tissue or blood.

septic shock Blood infection resulting in a pathological state of low blood pressure accompanied by a reduced amount of blood circulating to vital organs. Endotoxins of all Gram-negative bacteria can cause shock, but most clinical cases are due to Gram-negative enteric rods.

septicemia Systemic infection associated with microorganisms multiplying in circulating blood.

septum A partition or cellular cross wall, as in certain fungal hyphae.

sequela A morbid complication that follows a disease.

serology The branch of immunology that deals with in vitro diagnostic testing of serum.

serotyping The subdivision of a species or subspecies into an immunologic type, based upon antigenic characteristics.

severe acute respiratory syndrome (SARS) A severe respiratory disease caused by infection with a newly described coronavirus.

sex pilus A conjugative pilus.

sexually transmitted disease (STD) Infections resulting from pathogens that enter the body via sexual intercourse or intimate, direct contact.

shiga toxin Heat-labile exotoxin released by some *Shigella* species and by *E. coli* 0157:H7; responsible for worst symptoms of these infections.

sign Any abnormality uncovered upon physical diagnosis that indicates the presence of disease. A sign is an objective assessment of disease, as opposed to a symptom, which is the subjective assessment perceived by the patient.

simple stain Type of positive staining technique that uses a single dye to add color to cells so that they are easier to see. This technique tends to color all cells the same color.

solute A substance that is uniformly dispersed in a dissolving medium or solvent.

solution A mixture of one or more substances (solutes) that cannot be separated by filtration or ordinary settling.

solvent A dissolving medium.

somatic (O or cell wall antigen) One of the three major antigens commonly used to differentiate Gram-negative enteric bacteria.

source The person or item from which an infection is directly acquired. See *reservoir*.

species In the levels of classification, the most specific level of organization.

specificity Limited to a single, precise characteristic or action.

spontaneous mutation A mutation in the DNA of an organism with no known cause.

sporadic Description of a disease that exhibits new cases at irregular intervals in unpredictable geographic locales.

sporangiospore A form of asexual spore in fungi; enclosed in a sac.

sporangium A fungal cell in which asexual spores are formed by multiple cell cleavage.

spore A differentiated, specialized cell form that can be used for dissemination, for survival in times of adverse conditions, and/or for reproduction. Spores are usually unicellular and may develop into gametes or vegetative organisms.

sporicide A chemical agent capable of destroying bacterial endospores.

sporozoite One of many minute elongated bodies generated by multiple division of the oocyst. It is the infectious form of the malarial parasite that is harbored in the salivary gland of the mosquito and inoculated into the victim during feeding.

sporulation The process of spore formation.

spread-plate A means of isolating pure colonies from a mixed culture when the number of bacteria is a sample is very low.

sputum The matter found in the respiratory tract, including mucus, phlegm, and saliva.

stasis A state of rest or inactivity; applied to nongrowing microbial cultures. Also called *microbistasis*.

stationary growth phase Survival mode in which cells either stop growing or grow very slowly.

starch A carbohydrate polymer of glucose.

sterile Completely free of all lifeforms, including spores and viruses.

sterilization Any process that completely removes or destroys all viable microorganisms, including viruses, from an object or habitat. Material so treated is sterile.

strain In microbiology, a set of descendants cloned from a common ancestor that retain the original characteristics. Any deviation from the original is a different strain.

streak-plate A technique for isolating a pure culture from a mixed culture that involves streaking a small amount of the culture across the surface of a solid media.

streptolysin A hemolysin produced by streptococci.

strict or obligate anaerobe An organism that does not use oxygen gas in metabolism and cannot survive in oxygen's presence.

stroma The matrix of the chloroplast that is the site of the dark reactions.

subcellular vaccine A vaccine preparation that contains specific antigens such as the capsule or toxin from a pathogen and not the whole microbe.

subclinical A period of inapparent manifestations that occurs before symptoms and signs of disease appear.

subculture To make a second-generation culture from a well-established colony of organisms.

subcutaneous The deepest level of the skin structure.

substrate The specific molecule upon which an enzyme acts.

subunit vaccine A vaccine preparation that contains only antigenic fragments such as surface receptors from the microbe. Usually in reference to virus vaccines.

sucrose One of the carbohydrates commonly referred to as sugars. Common table or cane sugar.

sulfonamide Antimicrobial drugs that interfere with the essential metabolic process of bacteria and some fungi.

superficial mycosis A fungal infection located in hair, nails, and the epidermis of the skin.

superinfection An infection occurring during antimicrobial therapy that is caused by an overgrowth of drug-resistant microorganisms.

superoxide A toxic derivative of oxygen (O_2^-).

surfactant A surface-active agent that forms a water-soluble interface, e.g., detergents, wetting agents, dispersing agents, and surface tension depressants.

susceptibility testing A means of determining which antimicrobics are effective against a specific bacterial species.

symbiosis An intimate association between individuals from two species; used as a synonym for mutualism.

symptom The subjective evidence of infection and disease as perceived by the patient.

syndrome The collection of signs and symptoms that, taken together, paint a portrait of the disease.

synergism The coordinated or correlated action by two or more drugs or microbes that results in a heightened response or greater activity.

systemic Occurring throughout the body; said of infections that invade many compartments and organs via the circulation.

T

tachypenea An increase in the rate of respiration.

Taq polymerase DNA polymerase from the thermophilic bacterium *Thermus aquaticus* that enables high-temperature replication of DNA required for the polymerase chain reaction.

taxa Taxonomic categories.

taxonomy The formal system for organizing, classifying, and naming living things.

tertiary structure Protein structure that results from additional bonds forming between functional groups in a secondary structure, creating a three-dimensional mass.

tetanospasmin The neurotoxin of *Clostridium tetani*, the agent of tetanus. Its chief action is directed upon the inhibitory synapses of the anterior horn motor neurons.

tetracyclines A group of broad-spectrum antibiotics with a complex 4-ring structure.

tetrads Groups of four.

therapeutic index The ratio of the toxic dose to the effective therapeutic dose that is used to assess the safety and reliability of the drug.

thermal death point The lowest temperature that achieves sterilization in a given quantity of broth culture upon a 10-min exposure, e.g., 55°C for *Escherichia coli*, 60°C for *Mycobacterium tuberculosis*, and 120°C for spores.

thermal death time   The least time required to kill all cells of a culture at a specified temperature.

thermoduric   Resistant to the harmful effects of high temperature.

thermolabile   A substance whose physical structure or appearance changes when heated.

thermophile   A microorganism that thrives at a temperature of 50°C or higher.

thymine (T)   One of the nitrogen bases found in DNA but not in RNA. Thymine is in a pyrimidine form.

tincture   A medicinal substance dissolved in an alcoholic solvent.

titer   In immunochemistry, a measure of antibody level in a patient, determined by agglutination methods.

topical   On the skin.

toxin   A specific chemical product of microbes, plants, and some animals that is poisonous to other organisms.

toxinosis   Disease whose adverse effects are primarily due to the production and release of toxins.

toxoid   A toxin that has been rendered nontoxic but is still capable of eliciting the formation of protective antitoxin antibodies; used in vaccines.

trace elements   Micronutrients (e.g., zinc, nickel, and manganese) that occur in small amounts and are involved in enzyme function and maintenance of protein structure.

transfer RNA (tRNA)   A transcript of DNA that specializes in converting RNA into protein.

transformation   In microbial genetics, the transfer of genetic material contained in "naked" DNA fragments from a donor cell to a competent recipient cell.

transduction   The movement of DNA from one bacterial cell to another with the aid of a bacteriophage.

transient bacteria   Bacteria that may be present at a particular location on the body but do not normally reside there.

translation   Protein synthesis; the process of decoding the messenger RNA code into a polypeptide.

transcription   The process by which genetic information is copied from DNA to RNA.

translucent   Description of bacterial growth on a solid media which allows some diffuse light to pass through.

traveler's diarrhea   A type of gastroenteritis typically caused by infection with enterotoxigenic strains of *E. coli* that are ingested through contaminated food and water.

triglyceride   A lipid composed of a glycerol molecule bound to three fatty acids.

triplet   See *codon.*

trismus   An inability to open the mouth fully.

trophozoite   A vegetative protozoan (feeding form) as opposed to a resting (cyst) form.

true pathogen   A microbe capable of causing infection and disease in healthy persons with normal immune defenses.

trypomastigote   The infective morphological stage transmitted by the tsetse fly or the reduviid bug in African trypanosomiasis and Chagas disease.

tubercle   In tuberculosis, the granulomatous well-defined lung lesion that can serve as a focus for latent infection.

tuberculin   A glycerinated broth culture of *Mycobacterium tuberculosis* that is evaporated and filtered. Formerly used to treat tuberculosis, tuberculin is now used chiefly for diagnostic tests.

tuberculin reaction   A diagnostic test in which PPD, or purified protein derivative (of *M. tuberculosis*), is injected superficially under the skin and the area of reaction measured; also called the *Mantoux test.*

turbid   Cloudy appearance of nutrient solution in a test tube due to growth of microbe population.

tyndallization   Fractional (discontinuous, intermittent) sterilization designed to destroy spores indirectly. A preparation is exposed to flowing steam for an hour, and then the mineral is allowed to incubate to permit spore germination. The resultant vegetative cells are destroyed by repeated steaming and incubation.

U

ubiquitous   Present everywhere at the same time.

ultraviolet (UV) radiation   Radiation with an effective wavelength from 240 to 260 nm. UV radiation induces mutations readily but has very poor penetrating power.

undulant fever   See *brucellosis.*

universal donor   In blood grouping and transfusion, a group O individual whose erythrocytes bear neither agglutinogen A nor B.

universal precautions (UPs)   Centers for Disease Control and Prevention guidelines for health care workers regarding the prevention of disease transmission when handling patients and body substances.

uracil (U)   One of the nitrogen bases in RNA but not in DNA. Uracil is in a pyrimidine form.

urinary tract infection (UTI)   Invasion and infection of the urethra and bladder by bacterial residents, most often *E. coli.*

V

vaccination   Exposing a person to the antigenic components of a microbe without its pathogenic effects for the purpose of inducing a future protective response.

vaccine   Originally used in reference to inoculation with the cowpox or vaccinia virus to protect against smallpox. In general, the term now pertains to injection of whole microbes (killed or attenuated), toxoids, or parts of microbes as a prevention or cure for disease.

vancomycin   Antibiotic that targets the bacterial cell wall; used often in antibiotic-resistant infections.

variable region   The antigen binding fragment of an immunoglobulin molecule, consisting of a combination of heavy and light chains whose molecular conformation is specific for the antigen.

vector   An animal that transmits infectious agents from one host to another, usually a biting or piercing arthropod like the tick, mosquito, or fly. Infectious agents can be conveyed mechanically by simple contact or biologically whereby the parasite develops in the vector. A genetic element such as a plasmid or a bacteriophage used to introduce genetic material into a cloning host during recombinant DNA experiments.

vegetative   In describing microbial developmental stages, a metabolically active feeding and dividing form, as opposed to a dormant, seemingly inert, nondividing form, e.g., a bacterial cell versus its spore and a protozoan trophozoite versus its cyst.

vehicle   An inanimate material (solid object, liquid, or air) that serves as a transmission agent for pathogens.

vesicle   A blister characterized by a thin-skinned, elevated, superficial pocket filled with serum.

viable   Living.

viable nonculturable (VNC)   Describes microbes that cannot be cultivated in the laboratory but that maintain metabolic activity (i.e., are alive).

vibrio   A curved, rod-shaped bacterial cell.

virtual image   In optics, an image formed by diverging light rays; in the compound light microscope, the second, magnified visual impression formed by the ocular from the real image formed by the objective.

virulence   In infection, the relative capacity of a pathogen to invade and harm host cells.

virulence factors   A microbe's structures or capabilities that allow it to establish itself in a host and cause damage.

vitamins   Organic molecules that often function as coenzymes within the cell.

VRSA (vancomycin-resistant *Staphylococcus aureus*)   A very rare strain of *Staphylococcus aureus* that is resistant to the antimicrobic vancomycin.

W

wheal   A welt; a marked, slightly red, usually itchy area of the skin that changes in size and shape as it extends to adjacent areas. The reaction is triggered by cutaneous contact or intradermal injection of allergens in sensitive individuals.

whey   The residual fluid from milk coagulation that separates from the solidified curd.

whole blood   A liquid connective tissue consisting of blood cells suspended in plasma.

wild type   The natural, nonmutated form of a genetic trait.

X

XDRTB   Extensively drug-resistant tuberculosis (worse than multidrug-resistant tuberculosis).

xenograft   The transfer of a tissue or an organ from an animal of one species to a recipient of another species.

Y

yeast extract   Extract from killed and partially digested yeast; added to microbiological media as a source of nutrients.

Z

zoonosis   An infectious disease indigenous to animals that humans can acquire through direct or indirect contact with infected animals.

zooplankton   The collection of nonphotosynthetic microorganisms (protozoa, tiny animals) that float in the upper regions of aquatic habitat and together with phytoplankton comprise the plankton.

zygospore   A thick-walled sexual spore produced by the zygomycete fungi. It develops from the union of two hyphae, each bearing nuclei of opposite mating types.

Credits

Line Art, Text, and Tables

Exercise 2
Pages 9 & 18: The Collected Letters of Antoni van Leeuwenhoek, 1704-1707 Volume XV by can Leeuwenhoek, Antoni; Edited by Palm, L.C. Copyright by Swets & Zeitlinger, 1999.

Exercise 24
Figure 24.1: Plasmid pGlo from the Biotechnology Explorer Glo Bacteria Transformation Kit, p. 61. Reprinted with permission from Bio-Rad Laboratories, Inc.
Figure 24.2a: Arabinose Operon from the Biotechnology Explorer Glo Bacteria Transformation Kit, p. 61. Reprinted with permission from Bio-Rad Laboratories, Inc.
Figure 24.2b: Expression of Green Flourescent Protein from the Biotechnology Explorer Glo Bacteria Transformation Kit, p. 61. Reprinted with permission from Bio-Rad Laboratories, Inc.
Tables 24.1–24.10: From the Biotechnology Explorer Glo Bacteria Transformation Kit, p. 61. Reprinted with permission from Bio-Rad Laboratories, Inc.

Exercise 39
Pages 326–327: From *The New York Times*, © January 19, 1982 *The New York Times*. All rights reserved.

Exercise 90
Table 90.1: Reprinted with permission from bioMerieux, Inc.
Table 90.2: Reprinted with permission from bioMerieux, Inc.

Exercise 91
Table 91.2: Courtesy and © Becton, Dickinson and Company.

Photos

Chapter 1
Figure 1.1: © Royalty-Free/CORBIS; 1.2: © Barry Chess; 1.3A: © McGraw-Hill Companies, Inc./ Tim Fuller, photographer; 1.3B: © Royalty-Free/ CORBIS; 1.4: Photo courtesy AirClean® Systems.

Chapter 2
Page 9: © Kathy Park Talaro; 2.1: Courtesy of Leica Microsystems, Inc.; 2.3: © Stockbyte/Getty Images RF; 2.5A: © Michael Abbey/Visuals Unlimited; 2.5B: © George J. Wilder/Visuals Unlimited; 2.5C: © Invitrogen Corporation; 2.6: © The McGraw-Hill Companies, Inc./Mark Dierker, photographer; 2.8A: © A.M. Siegelman/ Visuals Unlimited; 2.8B: © Wim van Egmond/ Visuals Unlimited; 2.8C: © Stanley Flegler/Visuals Unlimited.

Chapter 3
Figure 3.1: © Oregon Department of Fish & Wildlife - Marine Resources Program/Brandon Ford; 3.2A: © Melba Photo Agency/PunchStock RF; 3.2B: © Yuuji Tsukii, Protist Information Server; 3.2C: © Steven P. Lynch; 3.4(top): © Melba Photo Agency/PunchStock RF; 3.5A: © Michele Bahr and D. J. Patterson, used under license to MBL (micro*scope); 3.5B: © Aurora Nedelcu; 3.6A-B: © Steven P. Lynch; 3.7A: © Luc Brient Université Rennes - France; 3.8-3.9A: © Chrysophytes, LLC/Peter A. Siver; 3.9B: © Dr. Peter Siver/Visuals Unlimited; 3.9C: © Chrysophytes, LLC/Peter A. Siver; 3.10: © Wim van Egmond/Visuals Unlimited; 3.11: © Elmer Frederick Fischer/CORBIS RF; p. 30: Photo by T. Archer, courtesy of NOAA, Great Lakes Environmental Research Laboratory.

Chapter 4
Figure 4.2: © David Phillips/Visuals Unlimited; 4.4A: © David Patterson/MBL/Biological Discovery in Woods Hole; 4.6A: © BioMEDIA ASSOCIATES; 4.7: © S. Aley, University of Texas at El Paso.

Chapter 5
Figure 5.1: © Barry Chess; 5.2B: © David Phillips/ Visuals Unlimited; 5.3A: © John D. Cunningham/ Visuals Unlimited; 5.3B: © Dr. Judy A Murphy, Microscopy & Imaging Consultant, Stockton, CA; 5.7: © imagebroker/Alamy RF; 5.8: © John D. Cunningham/Visuals Unlimited; 5.9: Image Courtesy of the Centers for Disease Control and Prevention; 5.10(1-16): © Harold J. Benson.

Chapter 6
Page 53: © Jim Kidd/Alamy.

Chapter 7
Figure 7.1-7.9: © Barry Chess.

Chapter 9
Figure 9.2A: CDC; 9.2B: © Dr. Gary Gaugler/Photo Researchers, Inc.; 9.2C: © A.M. Siegelman/Visuals Unlimited; 9.3A: © Dr. Jack M. Bostrack/Visuals Unlimited; 9.3B: © Lee W. Wilcox.

Chapter 10
Figure 10.1: Courtesy of Graham C. Walker; 10.2: © John D. Cunningham/Visuals Unlimited.

Chapter 11
Figure 11.1-11.3: CDC.

Chapter 12
Figure 12.2: © Barry Chess.

Chapter 13
Figure 13.1: © Dr. Jack M. Bostrack/Visuals Unlimited; 13.4: © Terese M. Barta, Ph.D.

Chapter 20
Figure 20.1: © Barry Chess; 20.2B: © Kathy Park Talaro; 20.2C: Image courtesy of BioTek Instruments, Inc., Winooski, VT, USA; 20.3: © AB BIODISK 2008, Reprinted with permission of AB BIODISK.

Chapter 21
Figure 21.1B: © Lee D. Simon/Photo Researchers, Inc.; 21.1C: © K.G. Murti/Visuals Unlimited.

Chapter 25
Figure 25.2A-C: © Barry Chess.

Chapter 26
Page 209: Photo by G.N. Miller/Rex USA, courtesy Everett Collection; 26.2: © Brad Mogen/Visuals Unlimited.

Chapter 27
Figure 27.1: © K.G. Murti/Visuals Unlimited; 27.4: © Barry Chess.

Chapter 28
Figure 28.1-28.3: © Barry Chess.

Chapter 29
Figure 29.1: © Barry Chess.

Chapter 30
Figure 30.2: © Barry Chess.

Chapter 32
Page 253: FEMA Photo/Andrea Booher; 32.1: © Barry Chess; 32.2A-32.3C: © The McGraw-Hill Companies, Inc./Auburn University Photographic Service; 32.4: © Barry Chess.

Chapter 33
Figure 33.1: Janice Haney Carr; 33.2-33.4: © Barry Chess.

Chapter 34
Figure 34.1: © Kathy Park Talaro; 34.2A: © L.M. Pope and D.R. Grote/Biological Photo Service; 34.2B: © Kathy Park Talaro; 34.3: © St. George University, School of Medicine/Zara Ross; 34.4-34.5B: © Harold J. Benson.

Chapter 35
Figure 35.2: © Barry Chess.

Chapter 36
Figure 36.1-36.2: © The McGraw-Hill Companies, Inc./Al Telser, photographer; 36.3: © Victor P. Eroschenko; 36.4-36.5: © The McGraw-Hill Companies, Inc./Al Telser, photographer.

Chapter 38
Figure 38.2C: © Hank Morgan/Science Source/ Photo Researchers, Inc.

Chapter 40
Figure 40.1: © The McGraw-Hill Companies, Inc./ Don Rubbelke, photographer; 40.3(all): © Barry Chess.

Chapter 41
Figure 41.1: © Biodisc/Visuals Unlimited.

Chapter 42
Figure 42.1B: © The McGraw-Hill Companies, Inc./ Auburn University Research Instrumentation Facility/Michael Miller, photographer; 42.4: © Harold J. Benson.

Chapter 43
Figure 43.1B: © The McGraw-Hill Companies, Inc./ Auburn University Research Instrumentation Facility/Michael Miller, photographer.

Chapter 44
Figure 44.1: © John D. Cunningham/Visuals Unlimited.

Chapter 45
Figure 45.1A-B: © The McGraw-Hill Companies, Inc./Auburn University Research Instrumentation Facility/Michael Miller, photographer.

Chapter 47
Figure 47.2: © The McGraw-Hill Companies, Inc./ Auburn University Research Instrumentation Facility/Michael Miller, photographer.

Chapter 49
Figure 49.1A-B: © Dr. E. C. S. Chan/Visuals Unlimited; 49.1C: © George J. Wilder/Visuals Unlimited.

Chapter 50
Figure 50.5: © Kathy Park Talaro.

Chapter 51
Figure 51.1B: © Kathy Park Talaro.

Chapter 52
Figure 52.1B: © Kathy Park Talaro.

Chapter 53
Figure 53.2: © Terese M. Barta, Ph.D.

Chapter 54
Figure 54.1: © Kathy Park Talaro.

Chapter 55
Figure 55.1: © Kathy Park Talaro.

Chapter 56
Figure 56.1: © Dr. E. C. S. Chan/Visuals Unlimited.

Chapter 57
Figure 57.1: Reprinted from *EPA Method 1604 (EPA-821-R-02-024)* courtesy of Dr. Kristen Brenner from the Microbial Exposure Research Branch, Microbiological and Chemical Exposure Assessment Research Division, National Exposure Research Laboratory, Office of Research and Development, U.S. Environmental Protection Agency.

Chapter 58
Figure 58.1: © Barry Chess.

Chapter 59
Figure 59.1A-C: CDC.

Chapter 60
Figure 60.1A-B: CDC.

Chapter 61
Figure 61.1A-C: © Fred E. Hossler/Visuals Unlimited.

Chapter 62
Figure 62.1: © Barry Chess.

Chapter 63
Figure 63.1-63.3(bottom): © The McGraw-Hill Companies, Inc./Auburn University Photographic Service.

Chapter 64
Figure 64.1A-E: © The McGraw-Hill Companies, Inc./Auburn University Photographic Service.

Chapter 65
Figure 65.1: © Kathy Park Talaro.

Chapter 66
Figure 66.1: CDC.

Chapter 67
Figure 67.1: © Harold J. Benson.

Chapter 68
Figure 68.1A-C: © The McGraw-Hill Companies, Inc./Auburn University Photographic Service.

Chapter 69
Figure 69.1: © Harold J. Benson.

Chapter 70
Figure 70.1: Courtesy and © Becton, Dickinson and Company.

Chapter 71
Figure 71.1B-71.2B: © The McGraw-Hill Companies, Inc./Auburn University Photographic Service.

Chapter 72
Figure 72.2A-B: Photo Compliments of Hardy Diagnostics.

Chapter 73
Figure 73.1: © The McGraw-Hill Companies, Inc./ Auburn University Photographic Service.

Chapter 75
Figure 75.1: Photo Compliments of Hardy Diagnostics.

Chapter 76
Figure 76.1: © The McGraw-Hill Companies, Inc./ Auburn University Photographic Service.

Chapter 77
Figure 77.2B: © Barry Chess.

Chapter 78
Figure 78.5: © Barry Chess.

Chapter 79
Figure 79.2B: © Harold J. Benson.

Chapter 80
Figure 80.2: © Harold J. Benson.

Chapter 81
Figure 81.2: © Harold J. Benson.

Chapter 82
Figure 82.1: © Dr. Martina Turk and Cene Gostincar.

Chapter 83
Figure 83.2: © Harold J. Benson.

Chapter 84
Figure 84.2: © Harold J. Benson.

Chapter 85
Figure 85.1: © Kathy Park Talaro.

Chapter 86
Figure 86.1: © Wellcome Photolibrary.

Chapter 87
Figure 87.2: © Harold J. Benson.

Chapter 88
Figure 88.1: © Harold J. Benson.

Chapter 89
Figure 89.1: Courtesy and © Becton, Dickinson and Company.

Chapter 90
Figure 90.1: Courtesy of bioMérieux, Inc.

Chapter 92
Figure 92.1: © Barry Chess.

Chapter 93
Figure 93.1: Courtesy and © Becton, Dickinson and Company.

Appendix A
Figure A.2A-C: © The McGraw-Hill Companies, Inc./Auburn University Photographic Service.

Appendix B
Figure B.1-B.5B: © Barry Chess.

Appendix C
Figure C.1-C.7: © The McGraw-Hill Companies, Inc./Auburn University Photographic Service; C.8: © Barry Chess.

Photo Atlas

2.4, 2.11, 2.26B, 2.32, 2.37, 2.42, 3.3, 3.6, 3.8, 3.14, 3.15A, 4.6, 4.13, 4.15, 8.4: © Barry Chess.
3.12: © Biodisc/Visuals Unlimited.
1.3, 1.6, 2.5, 2.12A-2.13B, 2.19, 3.1, 3.2, 3.4, 3.5, 3.7, 3.9-3.11, 3.13, 3.16-3.27, 4.1A-B, 4.3-4.5, 4.7-4.12B, 4.14A-B, 4.16-4.29, 5.1-5.22, 5.24-5.29B, 6.2A-6.4A, 6.4C-6.7B, 6.10-6.16, 7.1-7.14, 7.16-7.23C, 8.5: Centers for Disease Control and Prevention.
2.23, 2.31, 2.43, 2.44, 2.47-2.82: Courtesy and © Becton, Dickinson and Company.
2.26A, 2.29: Photo Compliments of Hardy Diagnostics.
2.45: Courtesy of bioMérieux, Inc.
6.29A, 6.32C: © Dan Ippolito.
2.10: Reprinted from *EPA Method 1604 (EPA-821-R-02-024)* courtesy of Dr. Kristen Brenner from the Microbial Exposure Research Branch, Microbiological and Chemical Exposure Assessment Research Division, National Exposure Research Laboratory, Office of Research and Development, U.S. Environmental Protection Agency.
2.9: © Dr. E. C. S. Chan/Visuals Unlimited.
1.1B: © Dr. Jack M. Bostrack/Visuals Unlimited.
2.36: © Dr. Martina Turk and Cene Gostincar.
6.29B: © Elmer Frederick Fischer/Corbis RF.
2.14A-C: © Fred E. Hossler/Visuals Unlimited.
5.23: © Glowimages/Getty Images RF.
2.15, 2.20, 2.22, 2.33-2.35, 2.38, 2.41: © Harold J. Benson.
2.1-2.3, 2.7, 2.8, 2.18, 2.39: © Kathy Park Talaro.
1.1A: © Lee W. Wilcox.
6.4B: © MedicalRF.com/Corbis.
6.1B, 6.9, 6.20A-B, 6.23A, 6.27, 6.31B-D: © Melba Photo Agency/PunchStock RF.
2.28: © Peggy Johnson.
6.31A: © PhotoLink/Getty Images RF.
7.15: © Science Photo Library RF/Getty Images.
6.1A, 6.1C, 6.8, 6.19A-B, 6.21, 6.23B, 6.25, 6.26, 6.28B: © Stephen Durr.
1.5, 4.2, 6.17, 6.18, 6.22, 6.30, 6.32A-B, 6.33: © Steven P. Lynch.
8.1B: © The McGraw-Hill Companies, Inc.
2.6: © Terese M. Barta, Ph.D.
6.28A, 8.1A, 8.2A-8.3B: © The McGraw-Hill Companies, Inc./Al Telser, photographer.
2.16A-2.17E, 2.21A-C, 2.24, 2.25, 2.27, 2.30: © The McGraw-Hill Companies, Inc./Auburn University Photographic Service.
1.2A-B, 1.4: © The McGraw-Hill Companies, Inc./ Auburn University Research Instrumentation Facility/Michael Miller, photographer.
3.15A, 6.24: © The McGraw-Hill Companies, Inc./ Don Rubbelke, photographer.
8.3C: © Victor P. Eroschenko.
2.40: © Wellcome Photolibrary.

Index

A Glossary of Terms starts on page 577. Color plates are found in the Photographic Atlas. The page locators for these plates are in **bold** (for example, *Clostridium butyricum*, **A-11**).